Advances in Experimental Medicine and Biology

Volume 918

Advances in Experimental Medicine and Biology presents multidisciplinary and dynamic findings in the broad fields of experimental medicine and biology. The wide variety in topics it presents offers readers multiple perspectives on a variety of disciplines including neuroscience, microbiology, immunology, biochemistry, biomedical engineering and cancer research. Advances in Experimental Medicine and Biology has been publishing exceptional works in the field for over 30 years and is indexed in Medline, Scopus, EMBASE, BIOSIS, Biological Abstracts, CSA, Biological Sciences and Living Resources (ASFA-1), and Biological Sciences. The series also provides scientists with up to date information on emerging topics and techniques.

More information about this series at http://www.springer.com/series/5584

Ruifu Yang • Andrey Anisimov
Editors

Yersinia pestis: Retrospective and Perspective

Springer

Editors
Ruifu Yang
Beijing Institute of Microbiology
and Epidemiology
Beijing, China

Andrey Anisimov
State Research Center for Applied
Microbiology
Moscow Region, Russia

ISSN 0065-2598 ISSN 2214-8019 (electronic)
Advances in Experimental Medicine and Biology
ISBN 978-94-024-0888-1 ISBN 978-94-024-0890-4 (eBook)
DOI 10.1007/978-94-024-0890-4

Library of Congress Control Number: 2016954202

Printed on acid-free paper

This Springer imprint is published by Springer Nature
The registered company is Springer Science+Business Media B.V. Dordrecht

Contents

1 **Plague: A Disease Which Changed the Path of Human Civilization** ... 1
Barbara Bramanti, Nils Chr. Stenseth, Lars Walløe, and Xu Lei

2 **Discovery of the Plague Pathogen: Lessons Learned** ... 27
Ruifu Yang and Thomas Butler

3 **Taxonomy of *Yersinia pestis*** ... 35
Zhizhen Qi, Yujun Cui, Qingwen Zhang, and Ruifu Yang

4 **Physiology of *Yersinia pestis*** ... 79
Robert R. Brubaker

5 **Ecology of *Yersinia pestis* and the Epidemiology of Plague** ... 101
Vladimir M. Dubyanskiy and Aidyn B. Yeszhanov

6 **Genome and Evolution of *Yersinia pestis*** ... 171
Yujun Cui and Yajun Song

7 **Pathology and Pathogenesis of *Yersinia pestis*** ... 193
Zongmin Du and Xiaoyi Wang

8 **Genetic Regulation of *Yersinia pestis*** ... 223
Yanping Han, Haihong Fang, Lei Liu, and Dongsheng Zhou

9 ***Yersinia pestis* in the Age of Big Data** ... 257
Ruifu Yang and Vladimir L. Motin

10 **Immunology of *Yersinia pestis* Infection** ... 273
Yujing Bi

11 **Plague: Clinics, Diagnosis and Treatment** ... 293
Vladimir V. Nikiforov, He Gao, Lei Zhou, and Andrey Anisimov

12 **Plague Vaccines: Status and Future** ... 313
Wei Sun

13 Bacteriophages of *Yersinia pestis* 361
Xiangna Zhao and Mikael Skurnik

14 Perspectives on *Yersinia pestis*: A Model for Studying Zoonotic Pathogens 377
Ruifu Yang, Yujun Cui, and Yujing Bi

Contributors

Andrey Anisimov State Research Center for Applied Microbiology, Obolensk, Moscow Region, Russia

Yujing Bi Beijing Institute of Microbiology and Epidemiology, Beijing, China

Barbara Bramanti Centre for Ecological and Evolutionary Synthesis (CEES), Department of Biosciences, University of Oslo, Oslo, Norway

Robert R. Brubaker Department of Microbiology, Immunology, and Molecular Genetics, Chandler Medical Center, The University of Kentucky, Lexington, KY, USA

Thomas Butler Department of Microbiology and Immunology, Ross University School of Medicine, Iselin, NJ, USA

Yujun Cui Beijing Institute of Microbiology and Epidemiology, Beijing, China

Zongmin Du Beijing Institute of Microbiology and Epidemiology, Beijing, China

Vladimir M. Dubyanskiy Stavropol Research Anti-Plague Institute, Stavropol, Russian Federation

Haihong Fang Beijing Institute of Microbiology and Epidemiology, Beijing, China

He Gao National Institute for Communicable Disease Control and Prevention, China CDC, Beijing, China

Yanping Han Beijing Institute of Microbiology and Epidemiology, Beijing, China

Xu Lei Centre for Ecological and Evolutionary Synthesis (CEES), Department of Biosciences, University of Oslo, Oslo, Norway

State Key Laboratory for Infectious Disease Prevention and Control, National Institute for Communicable Disease Control and Prevention, Chinese Center for Disease Control and Prevention, Changping, Beijing, People's Republic of China

Lei Liu Beijing Institute of Microbiology and Epidemiology, Beijing, China

Vladimir L. Motin Departments of Pathology and Microbiology & Immunology, University of Texas Medical Branch, Galveston, TX, USA

Vladimir V. Nikiforov Institute of Advanced Training, Federal Medical-Biological Agency of Russia, Moscow, Russia

Zhizhen Qi Qinghai Provincial Key Laboratory for Plague Control and Research, Qinghai Institute for Endemic Disease Prevention and Control, Xining, Qinghai Province, China

Mikael Skurnik Department of Bacteriology and Immunology, Medicum, Research Programs Unit, University of Helsinki, Helsinki, Finland

Yajun Song Beijing Institute of Microbiology and Epidemiology, Beijing, China

Nils Chr. Stenseth Centre for Ecological and Evolutionary Synthesis (CEES), Department of Biosciences, University of Oslo, Oslo, Norway

Wei Sun Department of Infectious Diseases and Pathology, College of Veterinary Medicine, University of Florida, Gainesville, FL, USA

Lars Walløe Department of Physiology, Institute of Basic Medical Sciences, University of Oslo, Oslo, Norway

Xiaoyi Wang Beijing Institute of Microbiology and Epidemiology, Beijing, China

Ruifu Yang Beijing Institute of Microbiology and Epidemiology, Beijing, China

Aidyn B. Yeszhanov M. Aikimbayev Kazakh Scientific Center of Quarantine and Zoonotic Diseases, Almaty, Kazakhstan

Qingwen Zhang Qinghai Provincial Key Laboratory for Plague Control and Research, Qinghai Institute for Endemic Disease Prevention and Control, Xining, Qinghai Province, China

Xiangna Zhao Institute of Disease Control and Prevention, Academy of Military Medical Sciences, Beijing, China

Dongsheng Zhou Beijing Institute of Microbiology and Epidemiology, Beijing, China

Lei Zhou Beijing Institute of Microbiology and Epidemiology, Beijing, China

Chapter 1
Plague: A Disease Which Changed the Path of Human Civilization

Barbara Bramanti, Nils Chr. Stenseth, Lars Walløe, and Xu Lei

Abstract Plague caused by *Yersinia pestis* is a zoonotic infection, i.e., it is maintained in wildlife by animal reservoirs and on occasion spills over into human populations, causing outbreaks of different entities. Large epidemics of plague, which have had significant demographic, social, and economic consequences, have been recorded in Western European historical documents since the sixth century. Plague has remained in Europe for over 1400 years, intermittently disappearing, yet it is not clear if there were reservoirs for *Y. pestis* in Western Europe or if the pathogen was rather reimported on different occasions from Asian reservoirs by human agency. The latter hypothesis thus far seems to be the most plausible one, as it is sustained by both ecological and climatological evidence, helping to interpret the phylogeny of this bacterium.

Keywords *Yersinia pestis* • Outbreak • Pandemics • Historic • Civilization

When traveling around in Europe today, we can still find monuments everywhere evoking the ancient pestilences, monuments such as churches, chapels, columns,

B. Bramanti (✉) • N.C. Stenseth
Centre for Ecological and Evolutionary Synthesis (CEES), Department of Biosciences, University of Oslo, Oslo, Norway
e-mail: barbara.bramanti@ibv.uio.no; n.c.stenseth@ibv.uio.no

L. Walløe
Department of Physiology, Institute of Basic Medical Sciences, University of Oslo, Oslo, Norway
e-mail: lars.walloe@medisin.uio.no

X. Lei
Centre for Ecological and Evolutionary Synthesis (CEES), Department of Biosciences, University of Oslo, Oslo, Norway

State Key Laboratory for Infectious Disease Prevention and Control, National Institute for Communicable Disease Control and Prevention, Chinese Center for Disease Control and Prevention, 102206 Changping, Beijing, People's Republic of China
e-mail: xulei@icdc.cn; lei.xu@ibv.uio.no

R. Yang, A. Anisimov (eds.), *Yersinia pestis: Retrospective and Perspective*, Advances in Experimental Medicine and Biology 918,
DOI 10.1007/978-94-024-0890-4_1

and altarpieces [1]. In addition, historical hospitals and lazarettos, and other buildings,[1] recall the past plagues, as do plenty of paintings, tableaus, pictures, sculptures, and wax reliefs. Plague is also depicted in certain famous novels, such as Boccaccio's *The Decameron*, Manzoni's *I Promessi Sposi*, Daniel Defoe's *A Journal of the Plague Year*, and Camus' *La Peste*, as well as in fables such as "Les Animaux Malades de la Peste" by La Fontaine. The terms "pest," "plague," and "pestilence" are also so commonly incorporated in expressions and proverbs in any European language[2] that, more or less consciously, the collective memory refers to past epidemics continuously. Even in drugstores we can find memories of the past plagues: from antiquity onward perfumes were often used as medicine and additionally in the Middle Ages to ward off the bubonic plague. This medicinal property was, for instance, attributed to the original Eau de Cologne that was launched in Cologne in 1709 by Giovanni Maria Farina (and is still commercialized to this

[1] Among the monuments recalling plague, we can mention Castel Sant'Angelo in Rome, which was originally the mausoleum of Hadrian and was converted into a castle at the beginning of the fourteenth century. Legend holds that the Archangel Michael appeared atop the mausoleum in 590 CE, sheathing his sword as a sign of the end of the plague in Rome. Other monuments of note include five plague churches in Venice, erected between the fifteenth and seventeenth centuries as vows for deliverance from some of the 70 epidemics of plague recorded in this port to the Levant. Of particular relevance is the School and Church of San Rocco (1485–1550), decorated with paintings by Tintoretto of Saint Roch's life. St. Roch, the Christian saint especially invoked against the plague, set out from Montpellier as a mendicant pilgrim for Rome during a poorly defined epidemic of plague in the second half of the fourteenth century. He could heal sick people in the public hospitals by prayer, the sign of the cross, and the touch of his hand. To St. Roch is devoted also the Rochuscapelle in Bingen on the Rhine, a chapel erected by the city in 1677 in thanksgiving for surviving the plague of the year before. The festival of St. Roch, which is still held there every year, was notably celebrated by Goethe (Sankt-Rochus-Fest zu Bingen, Hamburger Ausgabe, Band 10, Am 16. August 1814). Other monuments of importance include the Pestsäule of Vienna, erected by Emperor Leopold I in 1687 to celebrate the end of the epidemic, and again in Vienna, a city repeatedly ravaged by the plague, one can see the Karlskirche, built after the plague of 1713 and dedicated to the plague saint, St. Carlo Borromeo. In Florence the loggia Orsanmichele (originally housing a grain market) was transformed into a church to protect the picture of the Virgin who was claimed to have saved the city from the Black Death of 1348. Also in Florence, as a consequence of the plague which badly hit the city during the second half of the fourteenth century, the cultural period of the Rinascimento (Renaissance, from the Italian for "rebirth") took root and spread, eventually characterizing European history of the fifteenth–seventeenth centuries.

[2] Some examples are "to avoid or hate someone or something like the plague" (Italian idiomatic expression); "to stink like the plague" and "the choice between plague and cholera" (German idiomatic expressions); "A Sunday's child never dies of the plague" and "He who trusts a woman and leads an ass will never be free from plague" (French proverbs); "Nothing is ever well done in a hurry, except flying from the plague or from quarrels, and catching fleas" (Italian proverb); "Plague seize the hindmost" (Latin proverb); "Lies are the plague of speech" and "Forgetfulness is the plague of knowledge" (Arabic proverbs); "A stupid friend is a greater plague than a wise enemy" (Turkish proverb); "An honest wife is a treasure that lasts, a sad wife is worse than the plague" and "A new doctor is a plague on the country" (Sicilian proverbs); "It is better to murder during time of plague" (English proverb); and "If God should listen to rooks, he should send a plague to horses" (Bulgarian proverb).

day).[3] Although perfumes were not efficient disinfectants, their scent at least might have helped to reduce the unbearable odors linked to the pestilence. In vivid depictions by contemporary chroniclers, painters, and sculptors, victims were often portrayed as affected by strong coughing followed by hemoptysis (i.e., spitting of blood) and sometimes exhibiting suppurative lymphadenopathy (buboes) or necrotic extremities. Plague victims were depicted as suddenly falling down dead in the street, their bodies carted off by carriage and thrown in large plague pits containing a number of corpses deposited in layers "like lasagna."[4] The smell must have been truly horrific and not easily masked by the use of perfume.

Besides leaving a horrific and distressing impression, plague had a long-lasting impact on the history of Europe (in terms of social, political, and economical developments) as over centuries it repeatedly hit cities and villages, with devastating demographic consequences. The Black Death itself (1346–1353) has been estimated to have reduced the population in the affected areas of Europe by about one third, by the middle of the fourteenth century. A second plague epidemic occurred in 1358 in parts of Italy and Germany and spread to large parts of Europe (e.g., to Norway in 1359–1361), whereas a third epidemic spread in the mid-1370s (again including Norway). As a result, for areas of Europe where estimates are available, we know that by 1400, the population in these areas decreased to less than half of what it was before the Black Death [2, 3]. The sequence of plague epidemics continued in Western Europe until the mid-seventeenth century. In countries for which records are available, we know that there were between 20 and 30 epidemics of plague during the course of 300 years. The population of Europe was kept low during this whole period and only started to increase substantially again after the mid-seventeenth century.

1.1 Three Historical Plague Pandemics

"Plague," "pestilence," and "pest" are general terms often used in the past to describe catastrophic events. By reviewing historical records, historians have proposed that only three pandemics (Fig. 1.1) were effectively consistent with the infectious disease currently called plague, caused by the gram-negative coccobacillus *Yersinia pestis*. A caveat here is that, as we go further back in time, historic sources become more rare and more succinct, which limits our ability to tie a particular pathogen to an epidemic. We cannot therefore exclude the possibility that there have been earlier pandemics of plague, but can merely assert that there is currently no clear historic evidence for such pandemics.

The first plague pandemic, as described by Procopius, started in the middle of the sixth century in Egypt and spread rapidly to the whole Mediterranean Basin. From

[3] Farina, Jean-Marie (1825). *Précis sur les propriétés médicales de l'Eau de Cologne*. Warin-Thierry.

[4] di Coppo Stefani, Marchionne (1903). *Cronica Fiorentina*. Citta di Costella: Lapi. 88 pp.

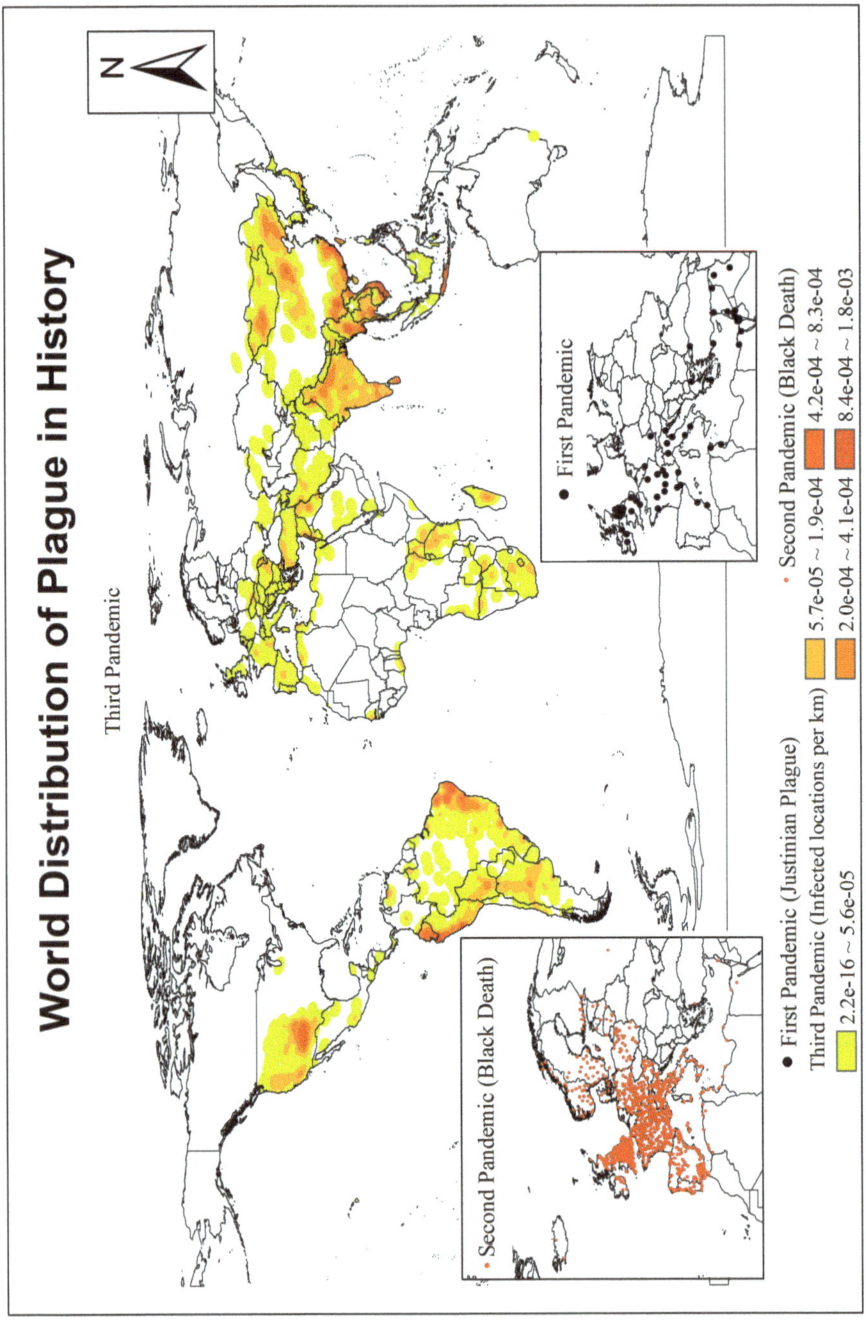
World Distribution of Plague in History
N
Third Pandemic
First Pandemic
Second Pandemic (Black Death)
First Pandemic (Justinian Plague)
Second Pandemic (Black Death)
Third Pandemic (Infected locations per km)
2.2e-16 ~ 5.6e-05
5.7e-05 ~ 1.9e-04
2.0e-04 ~ 4.1e-04
4.2e-04 ~ 8.3e-04
8.4e-04 ~ 1.8e-03

Fig. 1.1 Spatial distribution of the three plague pandemics. The distinction among the three plague pandemics and their classification is still under debate. We use here the historical definition of the three plague pandemics without taking into consideration the genetic differentiation of the distinct strains causing the outbreaks. *Black dots* indicate the known distribution of plague outbreaks in the period of the first pandemic (Source: Little and Rome [5] in main text), *red dots* those during the second pandemic (Source: Schmid et al. [79] in main text, Haesser 1875 [87], Ilmoni 1853 [88], Walløe 1995 [89]) and *colored shadows* those during the third pandemic. (The dataset used for the third pandemic was obtained from worldwide WHO records, and other published papers or databases. The compilation of this catalog was supported by the NSFC Project No.31420103913)

Pelusium on the Nile Delta, it reached Constantinople by ship in 541–542 CE: this first outbreak, which marks the onset of the first pandemic, is generally referred to as the "Justinian plague" due to the name of the Byzantine emperor at the time, Justinian. Although the source of this pandemic is not known, contemporaneous chronicles placed the origin of the plague in Axum, in the southeastern region of Africa, which was an important naval and trading connection to the East. India has also been proposed by historians as a possible source for this plague, as well as the Great Steppe region in Asia [4]. As for its spreading, few reports attest to the presence of plague in the northern part of Europe, but it is generally considered that plague quickly spread to the Rhine region and Ireland [5]. Repeated cycles of plague epidemics are known to have occurred in Europe following the Justinian outbreak and to have continued until 750 CE.

The second pandemic began with the devastating outbreak of the fourteenth century, now referred to as the Black Death (1346–1353). From Caffa on the Caspian Sea, plague spread to all of Europe and persisted there with a series of intermittent epidemics until the middle of the eighteenth century, after which it virtually disappeared from the continent [6]. The last official epidemic is considered to have occurred in Moscow in 1771, although other outbreaks were also recorded later and the pandemic may have persisted in parts of Europe[5] until the end of the nineteenth century [7]. The origin of the Black Death is not well known, but according to historical records, the disease may have traveled with Mongol armies to the Crimea where it was further transported via Genoese ships to the West. Inscriptions found on Nestorian graves near Lake Issyk-Kul in Kyrgyzstan (1338–1339), Central Asia, refer to an unknown epidemic, but there is no direct proof that it was due to plague.

Historians regard the third pandemic[6] to have started in 1772 in Yunnan Province, southwest China, infecting approximately 2.6 million people and causing 2.2 million deaths in China alone [8]. The pandemic's time of origin is still debated – there is evidence of the presence of plague in China at least as far back as 1644 and 1754, but from 1772 onward, continuous records of epidemics were reported. From Yunnan Province, plague spread to the southeast coast and reached Beihai in 1867. Plague was recorded in Taiwan Province in 1869 and Hainan Island in 1882. In February 1894, plague spread to Guangzhou Province (Canton) and resulted in the death of approximately 70,000 people. In the same year, the infection reached Hong Kong and from there spread globally via maritime shipping [9]. In northeast China, plague killed tens of thousands of people between the years of 1910 and 1920 [9].

[5] In this text, Europe is defined as a geographical entity: the northwestern peninsula of the larger landmass known as Eurasia, separated from Asia by the Ural Mountains, the Ural River, and the crest of the Caucasus Mountains.

[6] The distinction between the second and third pandemic is relevant for the western European countries where the last epidemics of the second pandemic occurred in the eighteenth century; these include the Baltic pandemic of 1709–1713, the devastating plague of Messina in 1743, and the plague of Moscow in 1770–1771. Smaller outbreaks were recorded at, e.g., Noja and Malta, ca. 1815. In Eastern Europe there were repeated epidemics throughout the eighteenth and nineteenth centuries.

Europe was not significantly infected during this pandemic,[7] whereas Asia, America, Australia, and Africa were impacted. In these years, plague reached at least two new continents where the disease had never been recorded before: Australia and South and North America.

Officially, the third pandemic is still ongoing, although human cases are mainly limited to those resulting from contact with rodents from wildlife plague reservoirs, followed by small localized outbreaks [10]. Nevertheless, the number of human cases is substantial: between 1954 and 1997, the World Health Organization (WHO) received reports of plague from 38 countries, with notification of 80,613 cases and 6587 deaths.[8] In some places like Madagascar, where plague was introduced in 1898, epidemics have struck nearly every year: in early 2015 the death toll was 79 and in early 2016, it was 63 out of at least 174 detected cases.[9] Plague continues to be closely monitored, as it is listed by the WHO in its report on the global surveillance of epidemic-prone infectious diseases,[10] along with eight other diseases which pose important public health threats. Its prevalence is controlled in endemic regions of 478 countries, regions which cover substantial areas. In China alone, where 24 human cases of plague were reported from 2009 to 2014, the endemic area covers approximately 1.5 million km^2.

1.2 A Brief History of the Study of Plague

From antiquity until the nineteenth century, medical knowledge based its explanation for infectious diseases on the miasmatic theory. This theory postulated that diseases and epidemics were caused by poisoned air (miasma) and were detectable by the accompanying foul smell. The miasma was thought to contain dangerous particles, particles originating from decomposing biological matter and dispersed by the winds under particular astrological and climatic conditions. This bad air then

---

[7] Several countries in Europe have reported outbreaks of plague (see also Fig. 1.1), the first being Portugal in 1899 with the epidemic of Oporto, which likely arrived from Alexandria (Echenberg, Myron (2007). *Plague ports: the global urban impact of bubonic plague, 1894–1901*. New York: New York University Press. 366 pp.). Generally, the outbreaks in Europe were rapidly brought under control, with a low number of confirmed cases and deaths.

[8] World Health Organization. *WHO Report on global surveillance of epidemic-prone infectious diseases – plague* [cited 2013 Apr 15]. http://www.who.int/csr/resources/publications/plague/CSR_ISR_2000_1/en/index5.html

According to another interpretation, the third pandemic, caused by *Y. pestis* biovar Orientalis, ended at the middle of the last century when the WHO received notification of less than 2000–3000 human cases (during the period 1954–1997). By this interpretation, we are currently in a period of endemic plague caused by representatives of different biovars, including Orientalis (for indication about the phylogeny of current strains, see [70–72]).

[9] http://www.reuters.com/article/2015/02/11/us-health-plague-madagascar-idUSKBN0LF1LI20150211; http://outbreaknewstoday.com/plague-death-toll-in-madagascar-reaches-63-58932/

[10] http://www.who.int/csr/resources/publications/surveillance/WHO_CDS_CSR_ISR_2000_1/en/

entered the human body through the skin's open pores and produced disease, plague included.

In order to combat the spread of plague, several measures were taken based on the medical knowledge of the day. To reduce the risk of dangerous particles entering the skin's pores to begin with, warm baths were discouraged and thermal and common baths were in general closed during plague outbreaks.[11] The belongings of victims as well as their clothing were also considered poisoned, and contact with the fur of pets was not recommended – in fact the historical records show that cats were killed during plague outbreaks. In the late Middle Ages, other measures were introduced in places such as Ragusa (and later Milan and Venice) to fight the spread of the infection. These measures included the implementation of quarantine and cordon sanitaire and the prohibition of human assemblages (e.g., [11]).

From the middle of the eighteenth century, the early part of the Enlightenment era, scientific and medical information about plague started to be collected in a more rational and empirical way, in contrast to previous data which was largely anecdotal. Evidence for this shift can be found in a large number of annotations related to plague by scholars in different fields: medicine (symptoms, disease progression, autoptic evidence[12]), epidemiology (ethnic groups and social classes involved, age, gender, and type of exposed worker), ecology (seasonality of the outbreaks, climatic conditions, susceptible animals and their ecological niches), hygiene (information about public and private hygiene, infections within the household, effectiveness of measures such as quarantine and cordon sanitaire), and geography (places of origin and spread of the plague). Experiments were even carried out in the nineteenth century, mostly to determine the susceptibility of different animals to plague [12], but also in some cases to test susceptibility in humans. The goal of these experiments was primarily to disclose the nature and the mechanisms of the human contagion and assess whether it was a contagious disease at all.[13] As a result, a large body of accurate knowledge on the disease and how it spread was accumulated, even before the dawn of the bacteriological era (see, e.g., de Merens).

At the end of the nineteenth century, thanks to Robert Koch and Louis Pasteur, the introduction of a new discipline, medical bacteriology, provided a challenge to the miasma theory, though consensus was not reached immediately. During the plague of Hong Kong in 1894, the etiological agent of the epidemic was discovered by Alexander Yersin: following the advice in his publication [13], plague was officially recognized as an infectious disease produced by the bacterium *Pasteurella pestis*. The bacterium's name was later changed to *Yersinia pestis* in the 1950s, in

[11] Erasmus in 1526 observed: "Twenty-five years ago, nothing was more fashionable in Brabant than public baths; today there are none, the new plague has taught us to avoid them." Cited by Ashenburg, K. (2007). *The dirt on clean: an unsanitized history*. New York: North Point Press.

[12] Records of autopsies can be found even earlier, see Cohn, Samuel K., Jr. (2010). *Cultures of plague: medical thinking at the end of the Renaissance*. Oxford: Oxford University Press. 342 pp.

[13] Contemporaneous descriptions might have been based on the miasma framework, but seem to have been very aware that plague was contagious through human interactions. This is according to records collected in Horrox, Rosemary (1994). *The Black Death*. Manchester: Manchester University Press. 364 pp.

part to honor Yersin's contributions, but also as microbiologists needed a new family name due to changes in the systematic division. The discovery of *Y. pestis*, supported by microscopical and experimental evidence, led to a drastic change in medical plague perspectives, as the medical community realized that a microorganism – and not polluted air – was the culprit of the common symptoms of the pathology.

From the clinical point of view, three different presentations of the plague were eventually recognized: bubonic, septicemic, and pneumonic. In bubonic plague, small numbers of *Y. pestis* penetrate a lymphatic vessel, interact with macrophages, and rapidly migrate to the local draining lymph node[14] where they multiply and give rise to enlarged purulent abscesses called "buboes" [14]. In the attempt to fight against the pathogen, the body reacts with a severe inflammatory response and symptoms such as fever, headache, confusion, or delirium. The period of incubation is generally 1–2 weeks and symptoms can persist some days. Without any antibiotic treatment, about 40–70 % of those affected by bubonic plague die because of cardiotoxins released from the buboes. Bubonic plague can also evolve into septicemic plague, which results in a high-level bacteremia and septicemia, characterized by hemorrhagic and necrotic spots on the skin and rapidly followed by septicemic shock and *exitus*. Septicemic plague will also develop as primary form if *Y. pestis* is injected directly into a blood vessel. Pneumonic plague can either develop when *Y. pestis* is directly inhaled, causing a fatal pneumonia, or it can emerge as a secondary form from either bubonic or septicemic plague; in any case, and as with septicemic plague, this severe form of the disease is always fatal and kills its victims within 2–3 days after the onset of the first symptoms.

Returning now to the outbreak in Hong Kong in 1894, Yersin made another important observation[15]: he noted that the streets were littered with dead black rats (*Rattus rattus*). A similar observation was also annotated by the Japanese physician Ogata Masanori on Formosa in 1897: he defined the outbreak as a "rat pest" and showed that rat fleas (*Xenopsylla cheopis*) carried the plague bacillus. Shortly after, in 1898, the French physician Paul-Louis Simond refined this concept: besides the unusually large number of dead rats on the streets of Karachi, he also reported the presence of lesions on the legs and feet of plague patients that could have been caused by the bite of a hematophagous insect, likely the flea. With a very simple experiment, using the infected fleas of an infected rat, he showed that the bacillus could be transmitted by the bites of the rat fleas to a healthy rat. However, Simond's conclusion that plague can be transmitted to humans by the bites of infected fleas was only accepted about 10 years later by medical authorities, when independent

[14] New observations support the theory that it is not the bacteria engulfed by local macrophages that cause infection. Rather, it is the few bacteria that are freely transported to the lymph nodes within minutes after a victim is bitten which are responsible. Gonzalez R.J., Lane M.C., Wagner N.J., Weening E.H., Miller V.L. (2015). Dissemination of a highly virulent pathogen: tracking the early events that define infection. *PLoS Pathogens*, 11(1):e1004587.

[15] Yersin also observed that the disease was equally carried by many other mammals, including water buffaloes.

Fig. 1.2 Rat guards for ships. With the ordinance of July 9, 1912 in the United States, the use of rat guards for plague control became widespread on commercial vessels. A rat guard is a sort of round metal "shield," placed over mooring lines to make it nearly impossible for rats to climb over and get onto or off the vessel when docked (Credit: Thomas Bredel). Black rats were very common on all commercial ships from far back in history (and up to 1940s). Black rats on ships were likely the main way the third pandemic spread from Hong Kong to India and later to the rest of the world. Possibly, black rats on vessels also were responsible for the rapid spread of plague during the second pandemic in the Mediterranean, along the coasts of Northern Europe as well as on rivers

experiments irrefutably confirmed his results: when a rat is dead and its blood temperature decreases, its infected fleas abandon the cadaver and infest the next host, a rat or another mammal, humans not excluded, and transmit the plague bacillus with their bites.

This advancement in scientific understanding was what public health authorities needed to effectively implement measures in the fight against plague. Worldwide, following the motto "avoid the rat-flea-human transmission chain," massive rat-proofing measures were instituted [15]. In maritime vessels (Fig. 1.2) and port facilities, rats were caught and killed, vessels were fumigated, and insecticides were used in any suspected area, while putative infected patients were isolated in quarantine stations. Hygiene was improved in areas close to wharves, as streets were cleansed and houses disinfected as soon as new cases were reported. The population was alerted and kept informed via newspaper advertisements, posters, and pamphlets. Throughout the outbreaks, some rudimentary types of vaccines were tentatively inoculated, although supplies soon ran out among the public in panic [16].

Thanks to all these measures and the introduction of antibiotics, plague was stopped in the middle of the twentieth century in several of the countries where it had been imported from Hong Kong or India, for instance, in Australia and in Europe. But plague did not disappear from the world.

1.3 Plague Is a Zoonosis: A Disease Transmitted Between Animals and Humans

Despite the measures introduced to combat it globally, plague has still persisted. In the United States, the WHO reported about 999 confirmed or probable human plague cases for the period between 1900 and 2010, mostly in the bubonic form. In recent decades, a range of 1–17 human cases per year has been reported in the United States, typically in the rural West (Fig. 1.1). The reason for the persistence of plague was disclosed in the first years of the twentieth century. Although plague quickly disappeared from the great harbors of California, three human cases were reported in the rural part of this state: some suspected that burrowing communities of wild rodents had picked up the plague bacillus by contact with infected rats and their fleas in the Californian ports. Two scientists, Wherry [17] and McCoy [18], provided evidence for this theory with the isolation of *Y. pestis* in California ground squirrels (*Spermophilus beecheyi*) from the same region. Today, the region west of the 100th meridian is regarded by the WHO as one of the largest world reservoirs of wild rodent plague. In these types of regions, also called plague foci, plague is enzootic, i.e., it persists in communities of partially resistant rodents (reservoir species) and in their fleas.

In wildlife, different modes of transmission impact the rate at which plague spreads in varying ways. For example, in the United States, ever since the first introduction of plague in 1894, transfer among animals has been estimated to occur at a rate of about 25 km (16 miles) per year [19]. This dissemination has been partially accelerated by human agency, but is mainly due to the life patterns of ground-burrowing wild rodents: as a rule, young rodents abandon their familial burrow and wander across country, sometimes over several miles to seek a new home [20]. As they move across an extensive range, they can join other communities and foster the dissemination of plague mediated by fleas transferred in the new burrows. Direct transmission is also not unusual, and it is believed that *Y. pestis* can be passed across mucous membranes and broken skin. The pneumonic form of plague may also be directly disseminated among animals in respiratory droplets (although this is not the case with rodents [21]). Additionally, *Y. pestis* can be orally transmitted via bites or by eating infected meals. Carnivores and omnivores may be infected by eating the tissue from contaminated animals and thereby contribute to maintaining plague in a region and occasionally transmit the pathology to humans. But since predators generally live isolated or in small communities, this mechanism cannot influence the modality and speed of dissemination of plague to a larger extent than the transmission among rodents does [22]. It has also been proposed that the dissemination of the plague bacterium can be due to its persistence in the soil of infected burrows (the so-called telluric hypothesis [23]), and a study seems to demonstrate that *Y. pestis* can remain viable and virulent after 40 weeks of incubation in sterilized humidified sand [24]. Survival in soil would be a model alternative to enzoonotics[16] for explaining the perseverance of plague in a region during inter-epizootic periods, but the matter is still controversial. Independent of the source of contagion in wildlife, by 1940 about 34 species of rodents and even more species of fleas were discovered to be infected by *Y. pestis* in the United States alone [20].

Elsewhere in the world, animal reservoirs for plague were already observed in Mongolia as early as 1895 and were later also discovered in other parts of Asia, Africa, and South America. In present-day China, 86 species of vertebrate host (including 14 major rodent hosts) and 63 species of arthropod vector (including the major flea vector) have so far been identified [25, 26].

Different local hosts, vectors, and *Y. pestis* strains give rise to different natural plague foci, which can be classified into 12 types and 19 subtypes [27, 28]. In Europe and its surrounding regions, known plague foci only exist in Iran, Georgia, Russia, Armenia, Azerbaijan [29], and North Africa[17], in territories placed on the desert belt or at the boundaries with deserts. In plague foci, epizootic periods may occasionally alternate with enzootic periods, inducing the death of a large number of plague hosts. Typically, three conditions are considered necessary for the development of an epizootic: more susceptible hosts, a large number of bacteria in the infected mammals (fatal high-level bacteremia), and a large number of infected fleas [30]. Several factors which regulate the interactions between the plague bacillus and its hosts and vectors can influence the alternation of enzootic/epizootic plague cycles [31]. However, the trophic cascade hypothesis, driven by a typical climatic pattern, seems to be the most appropriate model, especially in the plague sylvatic foci of the Asiatic desert belt [32–35]. A trophic cascade occurs when increased precipitation produces greater plant growth, which in turn results in an abundance of food production for epizootic rodents. This phenomenon leads to an increased animal population density and consequently to a greater number of fleas [34] and plague bacteria. If this period of plenty is followed by a severe drought, some individuals (the ones more susceptible or possibly debilitated by starvation) in the wild rodent populations develop septicemia and die of plague. The infected fleas on these individuals, after partaking of blood meals rich in *Y. pestis* bacteria, abandon the cold corpses of the victims and crowd onto the survivors, thereby contributing to the self-maintaining process of the epizootic. Depending on climatic conditions, the trophic cascade can simultaneously occur on a landscape scale in regions that have a similar biotope to one another yet are geographically separated (the Moran effect [36, 37]).

---

[16] The terms "enzootic" and "epizootic" are equivalent to nonhuman endemic and epidemic conditions, respectively.

[17] http://www.who.int/csr/resources/publications/plague/whocdscsredc992a.pdf

During an epizootic period, the chance of infective contact is higher, and transmission between animals and humans more readily occurs. In regions where plague is present in the sylvatic cycle, humans can acquire plague by direct contact with infected animals (by skinning or eating them or by accidental flea bites [10]). Once infected, the person in question can transport the disease back to his or her village or city. But plague can also be introduced into the urban cycle by wild rodents scavenging for food, as what happened in Los Angeles during the epidemic of 1924–1925, when both infected ground squirrels and rats were detected in the city [38]. With the exception of some communities of rats in Madagascar (which have become resistant to plague [39]), rats are generally very susceptible to the disease, which can lead to great die-offs such as the one observed by Yersin in Hong Kong, marking the onset of an epidemic among humans. Moreover, other commensal, peridomestic, and domestic animals are sensitive to plague and can directly – or indirectly with the mediation of their parasites – transmit the bacillus to humans. Domestic cats and dogs are known to carry plague, although dogs tend to have a higher resistance to the bacteria in comparison to cats and thus do not develop the high-level bacteremia, which is generally considered to be necessary for transmission.[18] In contrast, cats are quite susceptible and can transmit plague by ectoparasites or respiratory droplets. Camels, often used in caravans to transport humans and goods in Asia, the Near East, and North Africa along the desert belt, can become infected when their feed is contaminated by dead rodents or their excretions.[19] Camel-to-human contagion episodes have repeatedly been demonstrated, in particular related to the practice among nomads of eating dead animals regardless of the cause of death [40]. Finally, as early as the 1940s, observed outbreaks in rural settlements in Iran led researchers to suggest that plague epidemics can also be directly promoted by wild rodents without any mediation of commensal or peridomestic animals [41].

1.4 Plague Is a Vector-Borne Infectious Disease

Given the persistence of plague in enzootic animals in several continents and the recurrent epizootic episodes driven by climatic conditions, we should expect epidemics to strike with some regularity. Fortunately, since the 1920s, the majority of human plague cases reported by the WHO have been sporadic, and these cases have originated by humans coming into casual contact with wild rodents within enzootic foci [38]. Only in some places – recently primarily in Africa – has human plague infection more frequently developed into larger epidemics. Intriguingly, we still do

[18] A case of dog-to-human transmission of pneumonic plague was recently reported (Runfola J.K., House J., Miller L., et al. (2015). Outbreak of human pneumonic plague with dog-to-human and possible human-to-human transmission – Colorado, June–July 2014. *Morbidity and Mortality Weekly Report*, 64:429–434).

[19] Records from the former USSR detail 38 plague outbreaks among camels within the period 1907–1967. http://cns.miis.edu/opapers/pdfs/130904_soviet_antiplague_pdf

not understand exactly how the mechanisms underlying the spread and maintenance of human contagion operate once plague has entered the urban cycle.

In the urban cycle, one source of interpersonal transmission is via the inhalation of infected respiratory droplets of pneumonic plague (aerosol transmission). However, epidemics of pneumonic plague are rare and only spread rapidly in localized and contained outbreaks (1–117 cases [42–44]). The largest known outbreaks of primary pneumonic plague occurred in Manchuria in 1910–1911, 1917–1918, and 1920–1921, with 60,000, 16,000, and 9000 deaths, respectively [45]. Contagion via inhalation seems to require particular conditions to effectively spread, such as crowded and poorly ventilated spaces and a cold climate.

In addition to aerosol transmission, casual contact with infected (animal or human) tissue represents a real risk of infection for humans. Contagion by direct contact could cause a contained spread to occur among relatives and close communities of people who take care of their dead [46], but it is unknown to what extent this mechanism can maintain an epidemic, since death practices and rituals are usually abandoned after a short period of time.

So if aerosol transmission and contagion by direct contact alone are unlikely to maintain plague epidemics in human populations, what can? The experience of the last pandemic has shown that the great pathogenicity of *Y. pestis*, unique within the Enterobacteriaceae family, can be explained by an adaptation to two different systems and different temperatures: on the one hand, the blood, lymphoid, and reticuloendothelial systems of mammalian hosts (body temperature, typically 37 °C) and, on the other hand, the digestive tract of various flea species at ambient temperature [47, 48]. In other words, both mammalian hosts and ectoparasites can be the promoters of any pestilence, which leads us back to the role fleas play in the spread of plague.

In nature, there are about 80 species of fleas that can be infected by *Y. pestis* [49], and in the 1950s, mathematical models were first used to calculate the vector efficiency of different flea species [50, 51]. The results indicated that *X. cheopis* was the most efficient vector for the transmission of plague, likely due to the underlying mechanism of infection. Taking a closer look at the mechanism, it is thought that fleas in general become infected with *Y. pestis* after taking blood meals from mammals with high-level bacteremia or septicemia. *Y. pestis* bacteria are generally eliminated by fleas via the digestion process, but in some flea species, such as *X. cheopis*, another mechanism can be activated. In this alternative mechanism, *Y. pestis* can survive and multiply in the stomach (midgut) of an infected flea and form agglomerates of bacterial cells that can attach to the proventriculus, a valve-like chamber between the midgut and esophagus [52]. After an 8–16-day incubation period [53], this agglomerate of *Y. pestis* bacteria, protected by a biofilm, can efficiently block the proventriculus and hence block the passage of blood meal in the gut. As the digestive tract is blocked, the flea repeatedly attempts to feed but can only regurgitate into the bitten host a number of bacteria that are dislodged from the biofilm [30, 54].

Although considered the most efficient method of transmission, this method does have its own limitations [30]. First, a high bacterial load is required to cause the digestive tract of fleas to become blocked. Second, even if this condition is satisfied, only 50 % of the individual infected fleas can be effectively blocked. Third, only 45 % of the flea bites of an individual blocked flea are effective in transmitting the bacterium. Fourth, a prerequisite for efficient flea-borne transmission is the maintenance of the vectorial capacity of the infected flea in question [55], but the proventricular blockage quickly leads to dehydration of the flea and subsequently death by starvation [49]. For this reason, the infective period of a blocked flea is only a few days. In addition, the median number of *Y. pestis* cells transmitted by blocked *X. cheopis* is less than 100 [21]. Intriguingly, this is an efficient enough number to transmit disease, since less than ten to twenty bacteria were demonstrated experimentally to produce plague in mammal hosts [30]. This observation therefore paves the way for other hypotheses regarding possible mechanisms of infection: poor vectors should not be discarded as insignificant in the dissemination of plague [47].

Indeed, alternative flea-borne mechanisms for mammalian infection have been demonstrated to be efficient enough in maintaining plague in host populations. The ground squirrel flea, *Oropsylla montana*, the primary vector for plague transmission to humans in North America, transmits *Y. pestis* by so-called "early-phase" transmission (EPT [53, 56]), which does not require the 8–16 days of extrinsic incubation period necessary to develop an efficient proventricular blockage [53]. The mechanism of this early-phase transmission is not fully understood: it could be due to mechanical transmission of few *Y. pestis* bacilli surviving on the bloodstained mouthparts of fleas between consecutive feedings or, more likely, be driven by regurgitation of the infectious remnants from a previous blood meal [53]. This model was proposed for fleas that do not develop proventricular blockage, such as the so-called human flea,[20] *Pulex irritans*. Recent findings on an outbreak in Madagascar support the role of *P. irritans* in causing outbreaks of plague in the absence of infected *X. cheopis* [57]. The EPT model seems to support the idea that plague is a self-sustained vector-borne infectious disease in humans, an alternative mode of transmission first proposed by Blanc and Baltazard [41, 58] and often reproposed in the literature [e.g., 49, 59–61]. Blanc and Baltazard [41] also showed that human body lice (*Pediculus humanus*) from plague victims had *Y. pestis* in the gut and could infect shaven guinea pigs. Therefore, when considering plague epidemic transmission in humans, body lice, which can be transmitted by the exchange of clothing and can infect humans through feeding and feces, could also be a good vector candidate [62, 63]. What plague research shows us is that the methods and scale of transmission are varied and complex for this vector-borne disease.

[20] *P. irritans* is a cosmopolitan insect with a wide host spectrum, even though its common name suggests a primary affiliation with humans.

1.5 Discrepancies Between the Third Pandemic and the Two First Pandemics

When taking into account the evidence collected on plague since the end of the nineteenth century, doubt has been cast on the actual nature of the first two pandemics. Discrepancies observed in the historical records regarding the description of the symptoms can be attributed to the belief in the miasmatic theory (see a debate about this in [3]). But besides the symptoms, Cohn [7] recognized a series of additional discrepancies between the current third pandemic and the first and second ones. The most surprising anomaly regards the absence of rats: no historical report on the past pestilences mentions dead rats, although other animals are specifically mentioned as having died of plague (e.g., pigs in Boccaccio's *The Decameron*). Paintings and other artistic representations as well don't hint at the presence of urban rodents, although as described by Yersin [13] and other scholars, the phenomenon could hardly have been overseen. Further, black rats and *X. cheopis* were proposed to have been absent or very rare in some regions of Medieval Europe [61, 64], a hypothesis which has recently received support from archeological evidence [60]. Moreover, plague caused by black rats and blocked fleas like *X. cheopis* is efficient but slow, but we know that previous pestilences were fast movers [7]: sophisticated modeling methods have demonstrated that the Black Death spread by 1.5–6 km per day [65]. Yet another discrepancy is that the plague of the third pandemic has historically mostly been a plague of the harbors, whereas the two first ones penetrated rapidly beyond docklands. Finally, as observed by Cohn, outbreaks during the first and second pandemics could hit at any time of the year "before usually settling into a summer pattern for the southern Mediterranean and the Near East, that is during the hottest and driest points of the year, the least hospitable season for the most efficient flea vector of *Y. pestis* – *X. cheopis*" [7].

1.6 *Y. pestis* Was the Causative Agent of All Known Plague Pandemics

While Yersin [13] was strongly convinced that the bacterium he had isolated was also responsible for the first and second pandemics, his assertion was repeatedly challenged over the last decades due to the arguments cited above. The debate was recently settled by two molecular studies carried out on human skeletons from plague pits of the first [66] and the second pandemics [67]. The results were later independently confirmed by other studies [68, 69]. Following the theory that a septicemic event could have preceded the exitus and using the technique of ancient DNA analysis based on the retrieval of genetic material from the teeth, scientists were able to isolate *Y. pestis* DNA from putative victims buried in multiple or mass graves. By using the same genetic markers employed for the characterization of extant strains, they could also properly classify the ancient strains. In particular, the identification of single nucleotide polymorphisms (SNPs) allowed the reconstruction of phylogenetic trees [70–72] and the placement of the genealogical position of the ancient strains in relationship to the modern ones. This method unveiled essential information: first, it was confirmed that the strains of the first and second pandemics, strains that are now extinct, were ancestral to those responsible for the third pandemic (referred to as 1.ORI). This finding also confirmed the historical information of a recent introduction of *Y. pestis* 1.ORI strains in North America and Madagascar in newly established reservoirs. Second, related but different clones of *Y. pestis* were simultaneously circulating in Europe during the historical outbreaks [67, 69, 72]. Finally, it could be postulated that the origin of the bacteria responsible for the three pandemics was probably in Asia, although some scholars have proposed that the etiological agent of the third pandemic developed in Europe before it was introduced into China [69].

1.7 The Ecological and Climatic Dynamics of Plague

Since humans are an endpoint for the bacterium, it is not surprising that the historical strains of *Y. pestis* are now extinct. Instead, the genetic variability observed in extant strains [71, 72] mostly represents the evolution that occurred in wildlife reservoirs under the effect of different population dynamics over time, such as population expansions and contractions, isolation by distance phenomena, and episodes of bottlenecks or founder effects. As plague is primarily a zoonosis, its dynamics cannot be understood unless we understand the dynamics of the disease in its wildlife reservoir [10, 73].

Ecological studies of the plague system have shown that in order for the disease to spill over to humans, the abundance of the host wild rodent population needs to be above a certain threshold for a few years [74] and simultaneously spread out over a large geographic area [35]. The dynamic link between wildlife plague reservoirs and human populations has been studied by Samia et al. [32]. Their study demonstrated that, in addition to the wildlife threshold, another threshold exists: the plague bacterium spills over into the human population particularly when the ratio between fleas (the vector) and rodents (the host) is itself above a certain threshold. Biologically, this implies that the flea population on each individual rodent host becomes highly crowded, likely leading to higher flea mobility. Yet external drivers such as climate also play a role in the outbreak and dissemination of plague.

One of the most important external drivers in the outbreak of plague is climate, which affects both the location of plague foci and the intensity of the epidemic. For Central Asia it has been demonstrated that an increase of 1 °C will lead to a doubling of the prevalence of plague in its host [33] (see also [75, 76]). It has also been shown that climate variation primarily affects the behavior of the fleas (the vector) and their population dynamics [32]. In northern and southern China, the locations of natural plague foci are separately distributed, and the effects of climate factors on plague dynamics may show spatial heterogeneity between these two different regions. As regards the effect climate has on the intensity of an outbreak, Xu et al. [76] investigated the association of human plague intensity with proxy data on climate condition, using spatial and temporal human plague records in China from 1850 to 1964. The results demonstrated that the nonlinear responses of plague intensity to dry/wet conditions were different in northern and southern China [76]. Subsequently, by using a two-dimensional wavelet analysis, clustering groups were more precisely identified in the same database, and at least four well-defined independent foci were identified. In addition, environmental differences between plague territories – including but not limited to climate factors – were defined using ecological niche modeling [77]. A previous study has shown that the spatial temporal dynamics of plague are associated with a time-lagged global climate index [78]. The study employed cross-wavelet analysis to reveal that increasing rates of human plague oscillate in phase with the southern oscillation index[21] (SOI), but in antiphase with the sea surface temperature (SST) over periods of 2–4 years and approximately 8 years (6–10 years), respectively. This suggests that El Niño and the Southern Oscillation (ENSO)-driven climate variation might be an important factor for the outbreak of human plague in China [78].

Climate factors not only influence the outbreak of plague but also drive its spatial spread. The pattern of plague dissemination in China has been studied using two methods: nearest neighbor approach, a novel method which deals with both short- and long-distance transmissions, and trend surface analysis. These two methods were also used to investigate the spread of plague in Europe and the western United States [75]. The results indicated that the spread of plague has been slower on mainland China than it was in Europe during the Black Death and has been in the western United States since the 1900s. Factors such as floods and the presence of major roads, rivers, and coastlines accelerated the spread of plague and shaped transmission patterns [75, 79]. Additionally, the spread of climate-driven plague has been found to occur on a large spatial scale, on an intercontinental or global one. Research by Schmid et al. [79], synthesizing the ecologically dynamic effects of climate variation, indicates that plague in Europe has probably been repeatedly introduced from Central Asia in response to climatic effects occurring in Central Asia – not ones locally in Western Europe.

As already discussed, climate may impact plague due to its ability to initiate a trophic cascade, as climate affects the plant-rodent-flea-pathogen system in natural plague foci. For instance, surveillance data of two types of natural plague foci in Inner Mongolia has shown that plague epizootics in the Mongolian gerbil (*Meriones unguiculatus*) were driven by the effect of climate variation on the growth of local vegetation and consequently the density of the host population [80]. Furthermore, this phenomenon shows similarities with what has been observed in great gerbil

[21] The SOI is a standardized index calculated on sea level pressure differences between Tahiti and Australia and gives information about large-scale fluctuations in air pressure across the tropical Pacific. Abnormalities of the SST in the same part of the ocean can drive El Niño (and La Niña) episodes. El Niño (La Niña) and the Southern Oscillation (ENSO) affects weather conditions in many parts of the world by sufficient modifications of the atmosphere.

(*Rhombomys opimus*) plague foci in pre-Balkhash, Kazakhstan: the prevalence of plague in rodents increases with warmer springs and wetter summers [33].

1.8 The Historical Plagues Could Have Been Self-Sustained Vector-Borne Pandemics

It has long been assumed that previous pandemics were initiated by the arrival of black rats originating from Asian enzootic regions, allowing the establishment of new plague reservoirs in European wildlife. In particular, the Black Death period is often considered to have started as an isolated episode of plague introduced into the local fauna which then, with enzootic-epizootic cycles, led to the reinfection of human populations over the course of four centuries. Nevertheless, the locations of the foci responsible for the plague pandemics, as well as the nature of the mammalian reservoirs responsible for them, have never been established. To this day, there is speculation around these questions, as well as to what caused the disappearance of those putative foci from Western Europe. Plague has affected Europe and Constantinople until the nineteenth century, yet later plague episodes at the time of the third pandemic have been clearly attributed to introductions from outside, either directly from China or indirectly from Asia and Africa, and not to reactivation of local reservoirs. At present, there is no plague focus in Western Europe [81].

Both the topography and climate of Western Europe consistently differ from those of all other plague foci of the world, which makes it difficult to explain the presence of a local reservoir in the past. Perhaps because of this reason, it was often proposed that no wild rodents but the commensal or peridomestic rats themselves developed a reservoir for plague in Europe by acquiring immunity to the disease, as has been seen in present-day Malagasy populations of black rats and brown rats (*Rattus norvegicus*) [28]. The disappearance of the black rat from most European countries could explain the absence of current rat reservoirs, but we currently find high numbers of the brown rat in present-day Europe. Yet the brown rat was only introduced to European countries in the late eighteenth century, and there is no evidence that it was present in Western Europe during previous plague outbreaks, so its role remains uncertain. Additionally, while immunity could explain the absence of a mass die-off of rats in the historical records, it does not explain the lack of annual recurrence of outbreaks in Europe which are otherwise typical of endemic regions. For instance, in Madagascar, plague has resurfaced nearly every year[22] since 1980 in the same localities, despite plague surveillance and the use of insecticides and antibiotics. Even in the Malagasy port city of Mahajanga, where there were no environmental conditions for the establishment of a permanent reservoir, outbreaks of plague that were not preceded by an epizootic of rats were recorded annually for a decade between 1991 and 1999 [81]. In contrast, in Medieval Europe, the recur-

[22] http://www.who.int/csr/disease/plague/madagascar-outbreak/en/

rence of plague in the same localities typically showed epidemiological cycles of up to 3 years of outbreak followed by 11 years of disappearance [12].

Assuming that in Europe there once were well-established plague foci, Wagner et al. [69] proposed that *Y. pestis* strains from European foci reinfected wild rodents in China. The conclusion reached is one possible interpretation of the *Y. pestis* extant phylogeny, but there is another possible explanation [71, 72], which suggests that plague did not establish itself in permanent reservoirs in Europe, but rather was imported again and again from Asian reservoirs. This interpretation of the phylogeny of *Y. pestis* strains is now supported by new climatological data [79], refining the hypothesis of recurrent reintroductions to Europe from the outside by developing a new consistent model: on several occasions, geographically widespread climatic patterns in Asia were observed to precede the reintroduction of plague in Europe by about 15 years [79], namely, a pluvial period followed by a drought. This climatic pattern, which could have started the trophic cascade described in Sect. 1.3, has also been observed at the time of the Black Death, confirming that this model is credible. Fifteen years is not a long time for spillover from a reservoir to occur and for plague to spread over large distances in the Middle Ages. Yet, once the necessary time had elapsed for epizootics to develop, plague could have been picked up on the trading routes of the Silk Road by caravans and their camels and traveled with humans from one caravanserai to the next or from city to city. At the start of the Black Death, a similar process occurred, as plague spread from city to city, moving by human agency from Astrakhan near the Caspian Sea to the Crimea, Constantinople, Messina, and the Mediterranean harbors of Europe over a period of 4 years – a considerable amount of time [82]. In Asian human settlements, plague could have started new epidemics before being transported to Europe by other caravans or by maritime vessels, epidemics of which we have no historical record.

The proposed model of recurrent introductions of plague into Europe doesn't completely exclude the role of rats (perhaps often reintroduced into European harbors by maritime vessels [64]). Rather, it dismisses the notion that black rats and their fleas (*X. cheopis*) are *necessary* to explain the recurrence of pestilences in Europe. Moving away from previous assumptions about plague reservoirs in Europe, the resulting model proposes a different scenario for the historic pandemics: plague as a self-supporting system of infection, recirculating for a while in human European populations after its importation from enzootic places in Central Asia. Following maritime and fluvial routes or mainland roads, bubonic plague might have spread with humans, at their speed, possibly by means of hematophagous parasites. Since humans move in different directions, it is even theoretically possible under this model that strains of *Y. pestis* were reintroduced by humans in the Chinese reservoirs, as proposed by Wagner et al. [69].

Indeed, multiple historical records strongly suggest that traveled routes were responsible for the spread of plague. These records directly attribute the entrance of plague into a healthy settlement to soldiers, pilgrims, seamen, merchants, or travelers coming from regions with plague (the so-called Case 1). From the symptoms described in the reports of contemporaneous chroniclers, pneumonic plague is also often recognizable, which could have been responsible for some outbreaks or single cases of human-to-human transmission. The time between first symptoms and death is 2 days, which is rather short, but the time between being infected and dying is 2 days plus 3–4 days. Therefore, the possibility that infected people could carry and spread pneumonic plague over long distances seems plausible [42]. A single traveler infected by bubonic plague could have covered at least 40 km (25 miles) by land and an even greater distance by sea, before developing symptoms or dying. Additionally, the arrival via commercial routes of plague-infested belongings and clothing of people who had died of the disease has also often been linked to the further spread of plague. An example of this can be seen in the case of the tailor George Viccars, who first contracted bubonic plague in Eyam (in 1665) after having imported from London clothing from people claimed by the plague [83].

It has been suggested by Cohn [7] that the seasonality of plague in Europe was not consistent with the temperatures usually required to activate the mechanism of transmission mediated by *X. cheopis*. However, as previously mentioned, there are at least 80 known species of fleas that can transport plague besides *X. cheopis*. The early-phase transmission mechanism [53, 56], proposed specifically for these other flea species, could potentially explain the rapid spread of human plague across the European continent during the second pandemic. Additionally, there are several other potential vectors yet to be discovered, and the absence of seasonality in northern Europe suggests that other parasites, e.g., human body lice, could have been vectors for this region, while the inverse seasonality [75] observed in temperate regions (compared to desert belt ones) suggests that fleas were the likely vectors in southern Europe.

The model of frequent reintroductions, which negates the need for the presence of a local plague reservoir, can also easily explain the differences observed between the second and the third pandemic: no droughts were consistently connected with past outbreaks in Europe [7, 79] since drought wasn't necessary to introduce wild rodents into the urban cycle – simply put there was no local plague reservoir of wild rodents in Europe to start with. Finally, the model of (vector-borne) human-to-human transmission explains why plague is largely absent from the European continent today (with the exception of the regions already mentioned) – the absence of local rodent reservoirs diminishes the possibility of accidental contact, and the potential for the importation of plague from outside of Europe is severely limited by present-day regulations, improved communications, and improved environmental and personal hygienic conditions.

We have come a long way since the days when people believed that plague was caused by poisonous air. Yet, despite the large number of studies and the data that exists on *Y. pestis* and plague, not everything is decoded. Even more work is required to clarify what happened in the past pandemics, particularly during the first one where little historical information is available. In the case of the second pandemic, two recent aDNA studies [84, 85] have proposed that the phylogeny of *Y. pestis* strains from the eighteenth century demonstrates a continuity of plague in local Western European reservoirs from the time of the Black Death. However, in the discussion, the authors had to admit that this genetic continuity could not unambiguously be interpreted as proof for ancient reservoirs within Europe. This is especially questionable when considering the plague of 1720–1722 in Marseille [85], a city which was an active harbor and a place of recurrent epidemics. From historical records[23] it is known that plague was imported in 1720 from outside of Marseille, from regions closer to and economically connected with Asia [86]. In any case, regardless of which pandemic is under discussion, further genetic studies are required to refine the phylogenetic information on ancient strains, in order to support or refute certain theories, and other studies (e.g., ecological, climatological at different scales, historical, epidemiological, microbiological) are essential to correctly interpret the genetic information available.[24]

References

1. Avery H. Plague churches, monuments and memorials. Proc R Soc Med. 1966;59(2):110–6.
2. Hatcher J. The Black Death: an intimate history. London: Weidenfeld & Nicolson; 2010.
3. Walløe L. Medieval and modern bubonic plague: some clinical continuities. Med Hist Suppl. 2008;52(27):59–73.
4. Rosen W. Justinian's flea: plague, empire, and the birth of Europe. London: Penguin; 2007.
5. Little LK, Rome AAI. Plague and the end of antiquity: the pandemic of 541–750. Cambridge: Cambridge University Press in association with The American Academy in Rome; 2007.
6. Galvani AP, Slatkin M. Evaluating plague and smallpox as historical selective pressures for the CCR5-Δ32 HIV-resistance allele. Proc Natl Acad Sci. 2003;100(25):15276–9.
7. Cohn Jr SK. Epidemiology of the Black Death and successive waves of plague. Med Hist Suppl. 2008;52(27):74–100.
8. Liu Y. The atlas of plague and its environment in the People's Republic of China. Beijing: Science Press; 2000.
9. Lien-Teh W, Chun J, Pollitzer R, Wu C. Plague: a manual for medical and public health workers. Am J Public Health Nations Health. 1936;26(10):1049–50.
10. Stenseth NC, Atshabar BB, Begon M, Belmain SR, Bertherat E, Carniel E, Gage KL, Leirs H, Rahalison L. Plague: past, present, and future. PLoS Med. 2008;5(1):e3.
11. Cipolla CM. Cristofano e la peste: un caso di storia del sistema sanitario in Toscana nell'età di Galileo, vol. 166. Bologna: Il mulino; 1976.
12. Vasold M. Die Pest. Ende eines Mythos. Theiss Verlag, Stuttgart; 2003.
13. Yersin A. La peste bubonique à Hong Kong. Archives de médecine navale et coloniale. 1894;62:256–61.
14. Shannon JG, Bosio CF, Hinnebusch BJ. Dermal neutrophil, macrophage and dendritic cell responses to Yersinia pestis transmitted by fleas. PLoS Pathog. 2015;11(3):e1004734.
15. Barnett SA. Rat control in a plague outbreak in Malta. J Hyg. 1948;46(1):10–8.
16. Mafart B, Brisou P, Bertherat E. Epidémiologie et prise en charge des épidémies de peste en Méditerranée au cours de la seconde guerre mondiale. Bulletin de la Société de Pathologie Exotique (1990). 2004;97(4):306–10.

[23] http://infectiousdiseases.edwardworthlibrary.ie/plague/marseilles-case-study/

[24] This is the goal of an ERC Advanced Grant (MedPlag – The medieval plagues: ecology, transmission modalities and routes of the infections. Project No.324249), of which the authors are either members or partners (Barbara Bramanti is the PI). We would like to acknowledge the ERC for support and Samuel K. Cohn Jr., Andrey Anisimov, and Boris V. Schmid for relevant comments and suggestions. The authors are indebted to Sari C. Cunningham for her valuable improvement of the manuscript.

17. Wherry WB. Plague among the ground squirrels of California. J Infect Dis. 1908;5(5):485–506.
18. McCoy GW. Plague among ground squirrels in America. J Hyg. 1910;10(4):589–601.
19. Benedictow OJ. Yersinia pestis, the bacterium of plague, arose in East Asia. Did it spread westwards via the Silk Roads, the Chinese maritime expeditions of Zheng He or over the vast Eurasian populations of sylvatic (wild) rodents? J Asian Hist. 2013;47(1):1–31.
20. McNeill WH. Plagues and peoples, A peregrine book, vol. 2. Harmondsworth: Penguin; 1979. p. 99.
21. Lorange EA, Race BL, Sebbane F, Hinnebusch BJ. Poor vector competence of fleas and the evolution of hypervirulence in Yersinia pestis. J Infect Dis. 2005;191(11):1907–12.
22. Salkeld DJ, Stapp P. Seroprevalence rates and transmission of plague (Yersinia pestis) in mammalian carnivores. Vector Borne Zoonotic Dis. 2006;6(3):231–9.
23. Mollaret HH. Remarques sur la communication de MM. Brygoo et Dodin à propos de la peste tellurique et de la peste de fouissement, Données malgaches. Bull Soc Pathol Exot Filiales (Paris). 1965;58(2):140–54.
24. Ayyadurai S, Houhamdi L, Lepidi H, Nappez C, Raoult D, Drancourt M. Long-term persistence of virulent Yersinia pestis in soil. Microbiology. 2008;154(Pt 9):2865–71.
25. Gong Z, Yu X, Liu Q, Ye R, Lu L, Xu l, Zhang J, Li C, Bai X, Fang X. Ecological-geographic landscapes of natural plague foci in China VI. Biological characteristics of natural vectors of Yersinia pestis. Chin J Epidemiol. 2012;33(008):818–22.
26. Qin C, Xu L, Zhang R, Liu Q, Li G, Fang X. Ecological-geographic landscapes of natural plague foci in China V. Biological characteristics of major natural reservoirs of Yersinia pestis. Chin J Epidemiol. 2012;33(7):692–7.
27. Fang X, Xu L, Liu Q, Zhang R. Ecological-geographic landscapes of natural plague foci in China I. Eco-geographic landscapes of natural plague foci. Chin J Epidemiol. 2011;32(12):1232–6.
28. Chanteau S, Ratsifasoamanana L, Rasoamanana B, Rahalison L, Randriambelosoa J, Roux J, Rabeson D. Plague, a reemerging disease in Madagascar. Emerg Infect Dis. 1998;4(1):101–4.
29. Anisimov AP, Lindler LE, Pier GB. Intraspecific diversity of Yersinia pestis. Clin Microbiol Rev. 2004;17(2):434–64.
30. Hinnebusch BJ. The evolution of flea-borne transmission in Yersinia pestis. Curr Issues Mol Biol. 2005;7(2):197–212.
31. Gage KL, Kosoy MY. Natural history of plague: perspectives from more than a century of research. Annu Rev Entomol. 2005;50(50):505–28.
32. Samia NI, Kausrud KL, Heesterbeek H, Ageyev V, Begon M, Chan KS, Stenseth NC. Dynamics of the plague-wildlife-human system in Central Asia are controlled by two epidemiological thresholds. Proc Natl Acad Sci. 2011;108(35):14527–32.
33. Stenseth NC, Samia NI, Viljugrein H, Kausrud KL, Begon M, Davis S, Leirs H, Dubyanskiy VM, Esper J, Ageyev VS, et al. Plague dynamics are driven by climate variation. Proc Natl Acad Sci U S A. 2006;103(35):13110–5.
34. Reijniers J, Davis S, Begon M, Heesterbeek JA, Ageyev VS, Leirs H. A curve of thresholds governs plague epizootics in Central Asia. Ecol Lett. 2012;15(6):554–60.
35. Kausrud KL, Viljugrein H, Frigessi A, Begon M, Davis S, Leirs H, Dubyanskiy V, Stenseth NC. Climatically driven synchrony of gerbil populations allows large-scale plague outbreaks. Proc R Soc B Biol Sci. 2007;274(1621):1963–9.
36. Hudson PJ, Cattadori IM. The Moran effect: a cause of population synchrony. Trends Ecol Evol. 1999;14(1):1–2.
37. Koenig WD. Global patterns of environmental synchrony and the Moran effect. Ecography. 2002;25(25):283–8.
38. Caten JL, Kartman L. Human plague in the United States, 1900–1966. J Am Med Assoc. 1968;205(6):333–6.
39. Tollenaere C, Rahalison L, Ranjalahy M, Duplantier JM, Rahelinirina S, Telfer S, Brouat C. Susceptibility to Yersinia pestis experimental infection in wild Rattus rattus, reservoir of plague in Madagascar. Ecohealth. 2010;7(2):242–7.

40. Mustafa I. Bacterial diseases of dromedaries and bactrian camels. Rev Sci Tech Int Off Epizoot. 1987;6:391–405.
41. Blanc G, Baltazard M. Rôle des ectoparasites humains dans la transmission de la peste. Bull Acad Natl Med. 1942;126:446–8.
42. Gani R, Leach S. Epidemiologic determinants for modeling pneumonic plague outbreaks. Emerg Infect Dis. 2004;10(4):608–14.
43. Butler T. Plague gives surprises in the first decade of the 21st century in the United States and worldwide. Am J Trop Med Hyg. 2013;89(4):788–93.
44. Richard V, Riehm JM, Herindrainy P, Soanandrasana R, Ratsitoharina M, Rakotomanana F, Andrianalimanana S, Scholz HC, Rajerison M. Pneumonic plague outbreak, northern Madagascar, 2011. Emerg Infect Dis. 2015;21(1):8–15.
45. Teh WL. The second pneumonic plague epidemic in Manchuria, 1920–21: I. A general survey of the outbreak and its course. J of Hyg (Lond). 1923;21(3):262–88.
46. Persson B. Pestens gåta: Farsoter i det tidiga 1700-talets Skåne, vol. 5. Lund: Lund University; 2001.
47. Perry RD, Fetherston JD. Yersinia pestis – etiologic agent of plague. Clin Microbiol Rev. 1997;10(1):35–66.
48. Prentice MB, Rahalison L. Plague. Lancet. 2007;369(9568):1196–207.
49. Pollitzer R. Plague. Geneva: World Health Organization; 1954.
50. Pollitzer R. Plague studies. IX. Epidemiology. Bull World Health Organ. 1953;9(1):131–70.
51. Burroughs AL. Sylvatic plague studies: the vector efficiency of nine species of fleas compared with Xenopsylla cheopis. J Hyg (Lond). 1947;45(3):371–96.
52. Bacot AW, Martin CJ. LXVII. Observations on the mechanism of the transmission of plague by fleas. J Hyg (Lond). 1914;13(Suppl):423–39.
53. Eisen RJ, Dennis DT, Gage KL. The role of early-phase transmission in the spread of Yersinia pestis. J Med Entomol. 2015;52(6):1183–92.
54. Jarrett CO, Deak E, Isherwood KE, Oyston PC, Fischer ER, Whitney AR, Kobayashi SD, DeLeo FR, Hinnebusch BJ. Transmission of Yersinia pestis from an infectious biofilm in the flea vector. J Infect Dis. 2004;190(4):783–92.
55. Carniel E. Subtle genetic modifications transformed an enteropathogen into a flea-borne pathogen. Proc Natl Acad Sci U S A. 2014;111(52):18409–10.
56. Eisen RJ, Bearden SW, Wilder AP, Montenieri JA, Antolin MF, Gage KL. Early-phase transmission of Yersinia pestis by unblocked fleas as a mechanism explaining rapidly spreading plague epizootics. Proc Natl Acad Sci U S A. 2006;103(42):15380–5.
57. Ratovonjato J, Rajerison M, Rahelinirina S, Boyer S. Yersinia pestis in Pulex irritans fleas during plague outbreak, Madagascar. Emerg Infect Dis. 2014;20(8):1414–5.
58. Baltazard M, Bahmanyar M, Mostachfi P, Eftekhari M, Mofidi C. Recherches sur la peste en Inde. Bull World Health Organ. 1960;23(2–3):169–215.
59. Laudisoit A, Leirs H, Makundi RH, Van Dongen S, Davis S, Neerinckx S, Deckers J, Libois R. Plague and the human flea, Tanzania. Emerg Infect Dis. 2007;13(5):687–93.
60. Hufthammer AK, Walløe L. Rats cannot have been intermediate hosts for Yersinia pestis during medieval plague epidemics in northern Europe. J Archaeol Sci. 2013;40(4):1752–9.
61. Drancourt M, Houhamdi L, Raoult D. Yersinia pestis as a telluric, human ectoparasite-borne organism. Lancet Infect Dis. 2006;6(4):234–41.
62. Houhamdi L, Raoult D. Different genes govern Yersinia pestis pathogenicity in Caenorhabditis elegans and human lice. Microb Pathog. 2008;44(5):435–7.
63. Houhamdi L, Lepidi H, Drancourt M, Raoult D. Experimental model to evaluate the human body louse as a vector of plague. J Infect Dis. 2006;194(11):1589–96.
64. Davis DE. The scarcity of rats and the Black Death: an ecological history. J Interdiscip Hist. 1986;16(3):455–70.
65. Christakos G, Olea RA, Yu HL. Recent results on the spatiotemporal modelling and comparative analysis of Black Death and bubonic plague epidemics. Public Health. 2007;121(9):700–20.

66. Harbeck M, Seifert L, Hänsch S, Wagner DM, Birdsell D, Parise KL, Wiechmann I, Grupe G, Thomas A, Keim P, Zöller L, Bramanti B, Riehm JM, Scholz HC. Yersinia pestis DNA from skeletal remains from the 6th century AD reveals insights into Justinianic Plague. PLoS Pathog. 2013;9(5):e1003349.
67. Haensch S, Bianucci R, Signoli M, Rajerison M, Schultz M, Kacki S, Vermunt M, Weston DA, Hurst D, Achtman M, Carniel E, Bramanti B. Distinct clones of Yersinia pestis caused the Black Death. PLoS Pathog. 2010;6(10):e1001134.
68. Bos KI, Schuenemann VJ, Golding GB, Burbano HA, Waglechner N, Coombes BK, McPhee JB, DeWitte SN, Meyer M, Schmedes S, Wood J, Earn DJ, Herring DA, Bauer P, Poinar HN, Krause J. A draft genome of Yersinia pestis from victims of the Black Death. Nature. 2011;478(7370):506–10.
69. Wagner DM, Klunk J, Harbeck M, Devault A, Waglechner N, Sahl JW, Enk J, Birdsell DN, Kuch M, Lumibao C, Poinar D, Pearson T, Fourment M, Golding B, Riehm JM, Earn DJ, Dewitte S, Rouillard JM, Grupe G, Wiechmann I, Bliska JB, Keim PS, Scholz HC, Holmes EC, Poinar H. Yersinia pestis and the plague of Justinian 541–543 AD: a genomic analysis. Lancet Infect Dis. 2014;14(4):319–26.
70. Achtman M, Morelli G, Zhu P, Wirth T, Diehl I, Kusecek B, Vogler AJ, Wagner DM, Allender CJ, Easterday WR, Chenal-Francisque V, Worsham P, Thomson NR, Parkhill J, Lindler LE, Carniel E, Keim P. Microevolution and history of the plague bacillus, Yersinia pestis. Proc Natl Acad Sci U S A. 2004;101(51):17837–42.
71. Morelli G, Song Y, Mazzoni CJ, Eppinger M, Roumagnac P, Wagner DM, Feldkamp M, Kusecek B, Vogler AJ, Li Y, Cui Y, Thomson NR, Jombart T, Leblois R, Lichtner P, Rahalison L, Petersen JM, Balloux F, Keim P, Wirth T, Ravel J, Yang R, Carniel E, Achtman M. Yersinia pestis genome sequencing identifies patterns of global phylogenetic diversity. Nat Genet. 2010;42(12):1140–3.
72. Cui Y, Yu C, Yan Y, Li D, Li Y, Jombart T, Weinert LA, Wang Z, Guo Z, Xu L, Zhang Y, Zheng H, Qin N, Xiao X, Wu M, Wang X, Zhou D, Qi Z, Du Z, Wu H, Yang X, Cao H, Wang H, Wang J, Yao S, Rakin A, Li Y, Falush D, Balloux F, Achtman M, Song Y, Wang J, Yang R. Historical variations in mutation rate in an epidemic pathogen, Yersinia pestis. Proc Natl Acad Sci U S A. 2013;110(2):577–82.
73. Ben-Ari T, Neerinckx S, Gage KL, Kreppel K, Laudisoit A, Leirs H, Stenseth NC. Plague and climate: scales matter. PLoS Pathog. 2011;7(9):e1002160.
74. Davis S, Trapman P, Leirs H, Begon M, Heesterbeek JAP. The abundance threshold for plague as a critical percolation phenomenon. Nature. 2008;454(7204):634–7.
75. Xu L, Stige LC, Kausrud KL, Ben AT, Wang S, Fang X, Schmid BV, Liu Q, Stenseth NC, Zhang Z. Wet climate and transportation routes accelerate spread of human plague. Proc R Soc B Biol Sci. 2014;281(1780):20133159.
76. Xu L, Liu Q, Stige LC, Ben Ari T, Fang X, Chan KS, Wang S, Stenseth NC, Zhang Z. Nonlinear effect of climate on plague during the third pandemic in China. Proc Natl Acad Sci U S A. 2011;108(25):10214–9.
77. Ben-Ari T, Neerinckx S, Agier L, Cazelles B, Xu L, Zhang Z, Fang X, Wang S, Liu Q, Stenseth NC. Identification of Chinese plague foci from long-term epidemiological data. Proc Natl Acad Sci U S A. 2012;109(21):8196–201.
78. Zhang Z, Li Z, Tao Y, Chen M, Wen X, Xu L, Tian H, Stenseth NC. Relationship between increase rate of human plague in China and global climate index as revealed by cross-spectral and cross-wavelet analyses. Integr Zool. 2007;2(3):144–53.
79. Schmid BV, Büntgen U, Easterday WR, Ginzler C, Walløe L, Bramanti B, Stenseth NC. Climate-driven introduction of the Black Death and successive plague reintroductions into Europe. Proc Natl Acad Sci U S A. 2015;112(10):3020–5.
80. Xu L, Schmid BV, Liu J, Si X, Stenseth NC, Zhang Z. The trophic responses of two different rodent-vector-plague systems to climate change. Proc Biol Sci. 2015;282(1800):20141846.

81. Vogler AJ, Chan F, Nottingha R, Andersen G, Drees K, Beckstrom-Sternberg SM, Wagner DM, Chanteau S, Keim P. A decade of plague in Mahajanga, Madagascar: insights into the global maritime spread of pandemic plague. mBio. 2013;4(1):e00623–12.
82. Wheelis M. Biological warfare at the 1346 Siege of Caffa. Emerg Infect Dis. 2002;8(9):971–5.
83. Coleman MP. A plague epidemic in voluntary quarantine. Int J Epidemiol. 1986;15(3):379–85.
84. Seifert L, Wiechmann I, Harbeck M, Thomas A, Grupe G, Projahn M, Scholz HC, Riehm JM. Genotyping Yersinia pestis in historical plague: evidence for long-term persistence of *Y. pestis* in Europe from the 14th to the 17th century. PLoS One. 2016;11(1):e0145194.
85. Bos KI, Herbig A, Sahl J, Waglechner N, Fourment M, Forrest SA, Klunk J, Schuenemann VJ, Poinar D, Kuch M, Golding GB, Dutour O, Keim P, Wagner DM, Holmes EC, Krause J, Poinar HN. Eighteenth century *Yersinia pestis* genomes reveal the long-term persistence of an historical plague focus. eLife. 2016. doi:10.7554/eLife.12994.
86. Devaux CA. Small oversights that led to the Great Plague of Marseille (1720–1723): lessons from the past. Infect Genet Evol. 2013;14:169–85.
87. Haesser H. Lehrbuch der Geschichte der Medizin und der epidemischen Krankheiten. Jena: Hermann Duft Verlag; 1875 (Dritte Bearbeitung, Band 1, 2 und 3).
88. Ilmoni I. Bidrag til Nordens sjukdoms-historia [Contributions to the history of diseases in the Nordic countries]. Helsingfors [Helsinki]: J Simelii Arfvingar;1846, 1849, 1853 (1, 2 ock 3).
89. Walløe L. Plague and population: Norway 1350–1750. Avhandlinger (Norske videnskapsakademi), new series, No. 17. Oslo: University of Oslo, Department of Physiology; 1995.

Chapter 2
Discovery of the Plague Pathogen: Lessons Learned

Ruifu Yang and Thomas Butler

Abstract Plague resulted in three pandemics in history; however, its causative pathogen was isolated until the third pandemic in Hong Kong in 1894. At that time, two famous researchers, Dr. Alexandre Yersin and Dr. Shibasaburo Kitasato, went to HK, in order to identify the pathogen. The two great researchers had done a lot of work to isolate and identify the causative pathogen. However, Dr. Alexandre Yersin reported the real pathogen for plague, and we now acknowledge his work by nominating the pathogen's genus as *Yersinia*. In this chapter, we discussed the lessons learned from the two researchers' experience on isolation and identification of plague pathogen.

Keywords *Yersinia pestis* • Alexandre Yersin • Shibasaburo Kitasato • Lesson • Isolation

There are three plague pandemics recorded in human history, which have changed the path of human civilization. However, the causative agent of the plague was not identified as *Yersinia pestis* until the beginning of the third pandemic in Hong Kong (HK) in 1894. Although the solid culture method had been invented 11 years before isolation of the plague pathogen [1], only Dr. Alexandre Yersin and Dr. Shibasaburo Kitasato grasped the opportunity to isolate this pathogen in HK during the plague outbreak [2]. The outbreak of bubonic plague in HK occurred on May 5, 1894 [3]. Dr. James Alfred Lowson, a 28-year-old Scottish doctor, diagnosed the first case of bubonic plague in HK and also helped Dr. Kitasato in his efforts isolate the causative agent. Dr. Kitasato arrived in HK on June 12, 1894, where he was welcomed

R. Yang (✉)
Beijing Institute of Microbiology and Epidemiology,
No. Dongdajie, Fengtai, Beijing 100071, China
e-mail: ruifuyang@gmail.com

T. Butler
Department of Microbiology and Immunology, Ross University School of Medicine,
485 US Highway 1 South, Building B, 4th Floor, Iselin, NJ 08830, USA
e-mail: tbutler@rossmed.edu.dm; tbutler@rossu.edu

R. Yang, A. Anisimov (eds.), *Yersinia pestis: Retrospective and Perspective*,
Advances in Experimental Medicine and Biology 918,
DOI 10.1007/978-94-024-0890-4_2

by the local government and Dr. Lowson. Dr. Yersin arrived in HK 3 days later (on June 15), but was not treated as warmly as Dr. Kitasato. The two experts competed to identify the causative agent of the plague, with support from the Koch Institute and the Pasteur Institute, two famous institutes in microbiology, respectively [4]. This chapter will take readers back more than 120 years to better understand the two pioneers' work at that time and summarize the lessons we should learn from their story.

Research on bacterial pathogens greatly advanced in the nineteenth century after the invention of solid culture media by Dr. Robert Koch (1882–1901) in the period of 1876–1882 for bacterial isolation and by developing bacterial staining and disinfection methods [1]. *Bacillus anthracis*, the agent for anthrax, was successfully isolated in 1876, and *Mycobacterium tuberculosis* and *Vibrio cholerae* were also isolated in 1882 and 1884, respectively. These advances paved the way for isolating the plague pathogen. During the plague outbreak in HK, Dr. Yersin and Dr. Kitasato both claimed the initial identification of the responsible pathogen; however, contradictory descriptions were initially reported.

2.1 Dr. Yersin and His Work in HK

Dr. Yersin was born in 1863 in Switzerland. He studied medicine at Lausanne, Switzerland, from 1883 to 1884. Then, he went to Marburg, Germany, and Paris, France, between 1884 and 1886 to continue his studies. He also worked in Louis Pasteur's research laboratory with Emile Roux on the development of the anti-rabies serum. He obtained French nationality in 1888 to practice medicine in France. After receiving his Ph.D. in 1888, Dr. Yersin worked with Dr. Koch for 2 months in Germany. In 1889 he joined the newly created Pasteur Institute as part of Dr. Emile Roux's group, where he discovered the diphtheric toxin produced by a bacillus called *Corynebacterium diphtheria*. In 1890 he left the Pasteur Institute for French Indochina in Southeast Asia as a physician.

During the bubonic plague outbreak in 1894 in HK, the French government and the Pasteur Institute requested Dr. Yersin to go to HK to investigate the epidemic. Although Dr. Yersin was quite famous in France at that time, he did not receive a welcome party by the local government upon his arrival at June 15, 1894, possibly because Dr. Kitasato was a worldwide-known microbiologist and the local government already regarded him as the official investigator of the plague epidemic [4]. Dr. Yersin arrived in HK alone with only simple instruments, including a microscope and autoclave, and no official or personal support for his investigation. He hired local people to help him to build a small hut, just next to the hospital where Dr. Kitasato's laboratory was located [2, 4, 5]. At the beginning of his investigation, he could not obtain specimens for detecting potential pathogens. He had to bribe corpse removers and English sailors to "steal" samples for investigation. After repeated appeals to the HK government, he was finally allowed to collect corpse samples. He collected samples of swollen bubo for observation under a microscope and isolating the potential pathogen using rodents (rats, mice, and guinea pigs) and solid culture. Dr. Yersin demonstrated that short bacilli were present in both patient specimens and rodent organs, proving that a bacillus caused the bubonic plague in HK.

After this great discovery, Dr. Yersin continued his work on an antiplague serum preparation although its efficacy was disappointing in the field.

2.2 Dr. Kitasato and His Work in HK

Dr. Kitasato (January 29, 1853–June 13, 1931) was a Japanese physician and bacteriologist. He is remembered as the codiscoverer of the infectious agent of bubonic plague in HK in 1894, almost simultaneously with Dr. Yersin.

Dr. Kitasato was born in Okuni village, Higo Province (present-day Kumamoto prefecture), Kyushu. He was educated at Kumamoto Medical School and Tokyo Imperial University. He studied under Dr. Koch in Germany from 1885 to 1891. In 1889, he was the first person to grow the tetanus bacillus in pure culture and in 1890 cooperated with Emil von Behring in developing a serum therapy for tetanus using this pure culture. He also worked on antitoxins for diphtheria and anthrax. Dr. Kitasato and Behring demonstrated the value of antitoxins in preventing disease by producing passive immunity to tetanus in an animal that received graded injections of blood serum from another animal infected with the disease.

After returning to Japan in 1891, he founded the Institute for Study of Infectious Diseases with the assistance of Fukuzawa Yukichi. One of his early assistants was August von Wassermann. Notably, Dr. Kitasato demonstrated how dead cultures can be used in vaccination. He also studied the mode of infection in tuberculosis.

Dr. Kitasato and his assistants arrived in HK on June 12, 1894, and, as mentioned above, were warmly welcomed by local government and hospitals. The local officials even prepared a laboratory in a hospital for his investigation [2].

2.3 Difference in Results Between the Two Pioneers

Yersin isolated the plague bacilli on June 20, 1894, and described the pathogen as follows [2, 6]:

> The pulp of the bubo, in every case, was filled with a thick puree of short, thick bacilli with rounded ends, easily colored with aniline dyes but not by the method of Gram. The ends of the bacilli are more stained than the center. Often the bacilli appear to be covered by a capsule. One can recover a great amount from the buboes and lymph nodes of the diseased. The blood also contains them but not in such great numbers and only in very grave and deadly cases.
>
> The pulp of the bubo, when inoculated on agar, gives rise to white transparent colonies whose edges seem iridescent when examined under reflected light. The culture does even better on glycerin agar. The bacilli also grow in coagulated serum.

> In broth, the bacilli show a very characteristic aspect similar to erysipelas: clear liquid with flocked particles along the length and bottom of the tube....
>
> These cultures examined under the microscope show true chains of short bacilli, some appearing like a ball. On agar, if you examine the cultures with great care and high magnification, one can see bacilli among normal forms that sometimes are thin and sometimes fat chains of rods joined laterally. These swollen and abnormal forms become more and more numerous in old cultures and stain poorly.

Dr. Yersin also performed animal experiments, confirming that mice, rats, and guinea pigs died 3–5 days after inoculation with bacterial cultures of the suspected pathogen [2]. The animals' spleen and liver were swollen, and the pathogen could be re-isolated from the animals' blood, lymph nodes, spleens, and livers.

Dr. Kitasato published two papers in regard to the HK plague investigation. One was an editorial comment based on Dr. James A. Lawson's data [2, 7]:

> The organism, which is a bacterium resembling the bacilli found in the hemorrhagic septicemias, except that the ends are somewhat more rounded when stained lightly appears almost like an encapsulated diplococcus, but when more deeply stained it has the appearance of an ovoid bacillus, with a somewhat lighter center, especially when not accurately focused. When, however, it is focused more accurately it is still possible to make out the diplococcus form.

Two weeks later, Dr. Kitasato published his report on the plague pathogen in the *Lancet* [8]. He described his findings as follows:

> The different colonies are of a whitish-grey colour and by a reflected light have a bluish appearance; under the microscope they appear everywhere as if piled up with 'glass-wool', later as if having dense, large centres. If a cover-glass preparation is made from a cultivation on agar-agar, and having been stained, is observed under the microscope long threads of bacilli are seen, which might, by inspection, be mistaken for a coccus chain, but are recognized with certainty as 'threads of bacilli' under closer observations.

However, this report might be translated from Japanese or German, and the inaccuracy of translation might lead to the current misunderstanding [2]. Dr. Kitasato also found similar bacterial pathogens during the autopsy of 15 corpses who died from plague, and he performed similar animal experiments as Dr. Yersin [2].

In summary, Dr. Yersin and Dr. Kitasato described similar findings, although there are the following discrepancies:

First, the growth characteristics in both are different: Dr. Yersin described typical growth features as currently observed for *Y. pestis* including "clear liquid with flocked particles"; however, Dr. Kitasato reported it as "turbulent cultures."

Second, the Gram staining reactions are different: Dr. Yersin reported that the pathogen was not easily stained but that Gram staining could be used, while Dr. Kitasato thought Gram staining could not be used for identifying the plague pathogen.

Third, Dr. Yersin found morphological variation in the plague pathogen cultures, especially in old cultures, but Dr. Kitasato did not report any variation.

Fourth, the descriptions of the bacterial colonies and infected lymph nodes and other organs in animal experiments are also slightly different between the two pioneers' reports.

Although there are descriptive discrepancies between their works, the two scientists made obvious advances in the finding of the plague pathogen. Therefore, we should credit both pioneers as the co-finders of plague pathogen. However, researchers in the field have different ideas. In 1895, one of Dr. Kitasato's assistants in HK, Dr. Tanemichi Aoyama, published a paper in a Japanese journal in German, mentioning that the morphology of the possible pathogen found by Dr. Kitasato was different from that isolated by Dr. Yersin [2]. Dr. Kitasato isolated the bacterium from patient blood and it might therefore be a bacterium resulting from a secondary infection, such as *Streptococcus* ssp. Dr. Kitasato himself also wrote a paper admitting that the bacteria he found might be contaminated with other bacteria, likely *Staphylococcus* spp. or *Streptococcus* spp.

2.4 Lessons Learned

Recollection of the past is not only to settle disputes, such as who was the finder of the plague pathogen, but also to learn lessons and strengthen our scientific enquires.

Scientific behaviors and attitudes determine one's reputation. International colleagues in the field of science discredited Dr. Kitasato's contribution to the finding of the plague pathogen because he disputed the differing opinions from international colleagues to save face and his position in the field of infectious diseases. We acknowledge that Dr. Kitasato has done a lot of work in isolation and identification of the plague pathogen during the outbreak of plague in HK. In response to solid evidence that Dr. Kitasato's isolate was a type of *Streptococcus*, he still pointed out in a paper that his isolate was not a typical plague pathogen, and he thought it still played a critical role in development of the plague. His attitude of refusing to admit contamination is understandable at the time if we look into his background. As mentioned earlier, he made great contributions to the field of infectious diseases during his work in Dr. Koch's laboratory. He was the first foreign person to be endowed a professor position by the German government at the age of 40 years when he returned to Japan. He was also preparing to create his own research institute in Japan with great aspirations at that time. If he admitted that his isolate of plague pathogen was wrong because of contamination, his ambitions would come to grief. It is the self-protection of his prestige that made only international colleagues accredit the finder of plague pathogen to Dr. Yersin. If Dr. Kitasato admitted the contamination in his response to international colleagues' query and reported that he indeed found the plague pathogen but it was unfortunately contaminated by *Streptococcus*, then, he would at least be partially credited as one of the finders of plague pathogen. Only *Streptococcus* was left because possibly that the subculture operations only picked out the contaminated strains to be left to the field of science.

Monopoly of scientific research and personal interests in solving scientific issues need to be addressed. Dr. Kitasato was an official researcher invited by HK officials and assisted by a local hospital and professionals. However, in contrast, Dr. Yersin was an independent investigator researching the cause of the plague outbreak without official invitation or assistance. Dr. Yersin worked hard in isolating and identifying the causative agent. The final report from Dr. Yersin is not as long as that of Dr. Kitasato, but it stands the test of time. This indicates that attitude and spirit are critical for solving a scientific issue, but external conditions and supports are not indispensable. At that time, Dr. Kitasato was anxious to achieve quick success and achieve instant benefits; hence, he reported unreliable data with mistakes. It is regrettable that he denied himself the honor of being the co-finder of the plague pathogen. Dr. Wu Lien-teh (1879–1960), a plague investigator in China, evaluated their works in HK as follows: Dr. Kitasato should be attributed as the first person to describe the plague pathogen; however, Dr. Yersin was the first one to describe it correctly and accurately.

In different nations, there are different degrees of research monopoly in infectious diseases; therefore, how to integrate domestic and international intelligence to cope with an outbreak of severe infectious disease is an important topic. The outbreak of Ebola in West Africa at the end of 2013 awakened international support for these impoverished countries [9]. Similarly, during the German outbreak of food poisoning caused by *Escherichia coli* O104:H4 in 2011, an open-source genomics strategy was developed for coping with it using genome sequencing and Internet-based crowdsourcing techniques [10]. These two examples show different strategies to integrate international intelligence to solve a common threat.

The accuracy of scientific reporting is also of paramount importance. The only criterion for recognition of scientific achievement by international colleagues is its accuracy in the report. Although Dr. Kitasato's paper was longer and reported earlier than Dr. Yersin's, the self-contradictory data in Dr. Kitasato's reports reflected his non-stringent attitude to scientific data. Dr. Kitasato only isolated the potential pathogens from blood, and he did not use swollen lymph nodes for bacterial isolation, indicating that he did not think the feature of importance at that time. If he thoroughly investigated the patients and collected swollen buboes from patients to isolate the bacterium, he would have successfully obtained the correct pathogen.

However, the plague pathogen is easily confused with other common bacterial pathogens in the clinical laboratory [2]. Another reason for disputes between Dr. Kitasato and Dr. Yersin is that *Y. pestis* was not easy to be differentiated from *Streptococcus* at that time. Notably, the two early papers by Dr. Kitasato described a bacterium resembling *Streptococcus*. We currently have different means, including Gram staining, biochemical assays, and 16S rDNA sequencing, to differentiate these two pathogens.

In summary, history provides lessons and in the case of *Y. pestis*, Dr. Yersin is now recognized as the discoverer, and Dr. Kitasato missed out on this honor because of his overconfident attitude in response to colleagues' queries. Importantly, whether you are famous or not, you should respect colleagues' different opinions. We also need to be vigilant to ensure that all published data are accurate. Readers interested in Dr. Yersin's life after this discovery and the impact of the discovery on vaccine development and antimicrobial therapy should consult previously published reviews [5, 11, 12].

References

1. Kaufmann SH, Schaible UE. 100th anniversary of Robert Koch's nobel prize for the discovery of the tubercle bacillus. Trends Microbiol. 2005;13(10):469–75.
2. Bibel DJ, Chen TH. Diagnosis of plaque: an analysis of the Yersin-Kitasato controversy. Bacteriol Rev. 1976;40(3):633–51.
3. Editorial. The epidemic of plague in Hong Kong. BMJ. 1894;1(1746):1326.
4. Solomon T. Hong Kong, 1894: the role of James A Lowson in the controversial discovery of the plague bacillus. Lancet. 1997;350(9070):59–62.
5. Hawgood BJ. Alexandre Yersin (1863–1943): discoverer of the plague bacillus, explorer and agronomist. J Med Biogr. 2008;16(3):167–72.
6. Yersin A. La peste bubonique a Hong-Kong. Ann Inst Pasteur Paris. 1894;8:662–7.
7. Editorial. The plague at Hong-Kong. Lancet. 1894;2:325.
8. Kitasato S. The bacillus of bubonic plague. Lancet. 1894;2:428–30.
9. Zhang WY, Chen Y, Kamara A, Chen ZL, Chang GH, Wurie I, Kargbo D, Kargbo B, Liu C. Field labs in action for Ebola control in Sierra Leone. Infect Dis Trans Med. 2015;1(1):2–5.
10. Rohde H, Qin J, Cui Y, Li D, Loman NJ, Hentschke M, Chen W, Pu F, Peng Y, Li J, et al. Open-source genomic analysis of Shiga-toxin-producing *E. coli* O104:H4. N Engl J Med. 2011;365(8):718–24.
11. Butler T. Yersinia infections: centennial of the discovery of the plague bacillus. Clin Infect Dis. 1994;19(4):655–61; quiz 662–653.
12. Butler T. Plague history: Yersin's discovery of the causative bacterium in 1894 enabled, in the subsequent century, scientific progress in understanding the disease and the development of treatments and vaccines. Clin Microbiol Infect. 2014;20:202–9.

Chapter 3
Taxonomy of *Yersinia pestis*

Zhizhen Qi, Yujun Cui, Qingwen Zhang, and Ruifu Yang

Abstract This chapter summarized the taxonomy and typing works of *Yersinia pestis* since it's firstly identified in Hong Kong in 1894. Phenotyping methods that based on phenotypic characteristics, including biotyping, serotyping, antibiogram analysis, bacteriocin typing, phage typing, and plasmid typing, were firstly applied in classification of *Y. pestis* in subspecies level. And then, with the advancement of molecular biological technology, the methods based on outer membrane protein profiles, fatty acid composition, and bacterial mass fingerprinting were also used to identify the populations within *Y. pestis*. However, *Y. pestis* is a highly homogenous species; therefore, the above typing methods could only provide low resolution, e.g., only one serotype and one phage type were observed for the whole species. Since the 1990s, molecular typing based on DNA variations, including single-nucleotide polymorphism, gene gain/loss, variable-number tandem repeats, clustered regularly interspaced short palindromic repeat, etc., was introduced and improved the resolution and robust of typing result. Especially in recent years, genotyping-based whole-genome-wide variations were successfully employed in *Y. pestis*, which built the "gold standard" of typing scheme of the species and could distinguish the samples under the strain level. The taxonomy and typing works leaved us enormous polymorphism data; therefore, a comprehensive fingerprint database of *Y. pestis* was needed to collect and standardize these data, for facilitating future works on evolution, plague surveillance and control, anti-bioterrorism, and microbial forensic researches.

Keywords Taxonomy • Phenotyping • Genotyping • Molecular typing • Population diversity

Z. Qi • Q. Zhang
Qinghai Provincial Key Laboratory for Plague Control and Research, Qinghai Institute for Endemic Disease Prevention and Control, Xining, Qinghai Province 811602, China

Y. Cui • R. Yang (✉)
Beijing Institute of Microbiology and Epidemiology,
No. Dongdajie, Fengtai, Beijing 100071, China
e-mail: cuiyujun.new@gmail.com; ruifuyang@gmail.com

R. Yang, A. Anisimov (eds.), *Yersinia pestis: Retrospective and Perspective*,
Advances in Experimental Medicine and Biology 918,
DOI 10.1007/978-94-024-0890-4_3

In terms of taxonomy, the genus *Yersinia* belongs to the family *Enterobacteriaceae*, order *Enterobacteriales*, class *Gammaproteobacteria*, phylum *Proteobacteria*, and domain *Bacteria*. By the end of March of 2015, 19 species and two subspecies were described validly in this genus (http://www.bacterio.net/yersinia.html) (Table 3.1). The genus *Yersinia* was named by van Loghem (1944) after the Swiss/French bacteriologist Alexandre Yersin, who first isolated the causal organism of plague, *Yersinia pestis* (the type species of this genus), in 1894 (Lehmann and Neumann 1896) [1, 2]. In addition to *Y. pestis*, two other *Yersinia* species are associated with human infections. *Yersinia pseudotuberculosis* and *Yersinia enterocolitica* are enteropathogens that cause mild diarrhea. After its isolation and identification, the plague pathogen was known by many names, including *Bacterium pestis* (Lehmann and Neumann 1896), *Bacillus pestis* (Lehmann and Neumann 1896) Migula 1900, *Pasteurella pestis* (Lehmann and Neumann 1896; Bergey et al. 1923), and *Pestisella pestis* (Lehmann and Neumann 1896) Dorofeev 1947.

Y. pestis is thought to have evolved from *Y. pseudotuberculosis* (possibly serogroup O:1b) 2000–6000 years ago [27–29]. Through stringent DNA-DNA hybridization, Bercovier et al. confirmed that the DNA homology (83 %) between these two species is much higher than the species definition criterion of 70 %. Therefore, they recommended that *Y. pestis* should be reclassified as *Yersinia pseudotuberculosis* subsp. *pestis* (Lehmann and Neumann 1896) Bercovier et al. 1981. However, the clinical outcome of the diseases caused by these two species is obviously different, which will cause confusion in the clinical and public health fields if this recommendation were to be accepted [17]. Hence, the proposal was denied by the Judicial Commission of the International Committee on Systematic Bacteriology [30].

3.1 Phenotyping

Phenotyping methods are based on phenotypic characteristics and include biotyping, serotyping, antibiogram analysis, bacteriocin typing, phage typing, and plasmid typing. *Y. pestis* is a homogenous species with only one serotype and one phage type [29], and only a few antibiotic-resistant strains have been found [31, 32]. A biotyping system based on biochemical features is widely acknowledged in the field of plague research. Because of the nature of plague and its causative agent, it is difficult to exchange strains of this bacterium between plague researchers. Some other typing systems, such as ecotyping and subspecies classification systems, are only studied and used in certain countries [33–35]. Although these traditional methods have been found to be insufficiently discriminatory, have poor reproducibility, suffer from a lack of availability of specific reagents, and are affected by physiological factors, they have contributed substantially to the prevention and control of plague and our understanding of *Y. pestis*.

Table 3.1 Taxonomy of the genus *Yersinia*

Species or subspecies	Type strain	Etymology
Yersinia aldovae Bercovier et al. 1984 [3]	Strain Al 19955=ATCC 35236=CCUG 18770=CDC 669-83=CIP 103162=CNY 6005=IP 6005=JCM 5892	N.L. gen. fem. n. *aldovae*, of Aldova, named in honor of Eva Aldova, the Czechoslovakian microbiologist who first isolated the bacterium
Yersinia aleksiciae Sprague and Neubauer 2005 [4]	Strain Y159=WA758=DSM 14987=LMG 22254	N.L. gen. *aleksiciae*, of Aleksic, in honor of Professor Stojanca Aleksic, Hamburg, Germany, to whom we owe much of our knowledge of the epidemiology and microbiology of *Yersinia*
Yersinia bercovieri Wauters et al. 1988 [5]	Strain WAIP 208=ATCC 43970=CCUG 26329=CDC 2475-87=CIP 103323=CNY 7506	N.L. gen. masc. n. *bercovieri*, of Bercovier, named in honor of Hervé Bercovier, who first described biogroups 3A and 3B, and who has made outstanding contributions to the taxonomy and ecology of yersiniae
Yersinia enterocolitica (Schleifstein and Coleman 1939) Frederiksen 1964 [2]	Strain ATCC 9610=CCUG 11291=CCUG 12369=CIP 80.27=DSM 4780=JCM 7577=LMG 7899=NCTC 12982	Gr. n. *enteron*, intestine; Gr. n. *kolon*, colon; Gr. suff. *-tikos -ê -on*, suffix used with the sense of pertaining to; N.L. fem. adj. *enterocolitica*, pertaining to the intestine and colon. Synonym: *Bacterium enterocoliticum* Schleifstein and Coleman 1939
Yersinia enterocolitica subsp. *enterocolitica* (Schleifstein and Coleman 1939) Neubauer et al. 2000 [6]	Strain ATCC 9610=CCUG 11291=CCUG 12369=CIP 80.27=DSM 4780=JCM 7577=LMG 7899=NCTC 12982	Gr. n. *enteron*, intestine; Gr. n. *kolon*, colon; Gr. suff. *-tikos -ê -on*, suffix used with the sense of pertaining to; N.L. fem. adj. *enterocolitica*, pertaining to the intestine and colon
Yersinia enterocolitica subsp. *palearctica* Neubauer et al. 2000 [6]	Strain Y11=CIP 106945=DSM 13030	N.L. fem. adj. *palearctica* (from Gr. adj. *palaios*, ancient; and L. adj. *arcticus -a -um*, northern, arctic), palaearctic (the bacterium is resident in Northern Europe and northern parts of Asia major)
Yersinia entomophaga Hurst et al. 2011 [7]	Strain MH96=MH-1=ATCC BAA-1678=DSM 22339	Gr. part. adj. *entomon*, cut up, segmented (animal) (used to refer to an insect); N.L. fem. n. *phaga* (from Gr. n. *phagos*, voracious), eater; N.L. fem. n. *entomophaga*, insect eater
Yersinia frederiksenii Ursing et al. 1981 [8]	Strain 6175=ATCC 33641=CCUG 11293=CIP 80.29=NCTC 11470	N.L. gen. masc. n. *frederiksenii*, of *Frederiksen*; named after the Danish microbiologist Wilhelm Frederiksen, who made a substantial contribution to the study of the genus *Yersinia*
Yersinia intermedia Brenner et al. 1981 [9]	Strain 3953=Bottone 48=Chester 48=ATCC 29909=CCUG 11292=CIP 80.28=JCM 7579=NCTC 11469	L. fem. adj. *intermedia*, intermediate; here it implies that the biochemical reactions of this species lie midway between those of *Yersinia enterocolitica* and *Yersinia pseudotuberculosis*

(continued)

Table 3.1 (continued)

Species or subspecies	Type strain	Etymology
Yersinia kristensenii Bercovier et al. 1981 [10]	Strain 105 = ATCC 33638 = CCUG 8241 = CCUG 11294 = CIP 80.30 = JCM 7576 = NCTC 11471	N.L. gen. masc. n. *kristensenii*, of Kristensen, named after the Danish microbiologist Martin Kristensen, who first isolated this organism
Yersinia massiliensis Merhej et al. 2008 emend. Souza et al. 2011 [11, 12]	Strain 50640 = CCUG 53443 = CIP 109351	L. fem. adj. *massiliensis*, pertaining to *Massilia*, the Roman name of Marseille, France, where the type strain was isolated
Yersinia mollaretii Wauters et al. 1988 [5]	Strain WAIP 204 = ATCC 43969 = CCUG 26331 = CDC 2465-87 = CIP 103324 = CNY 7263	N.L. gen. masc. n. *mollaretii*, of Mollaret, named in honor of Henri H. Mollaret, head of the National *Yersinia* Center at the Pasteur Institute in Paris, for his years of study of the classification and epidemiology of the genus *Yersinia*
Yersinia nurmii Murros-Kontiainen et al. 2011 [13]	Strain APN3a-c = DSM 22296 = LMG 25213	N.L. gen. masc. n. *nurmii*, of Nurmi, in honor of Professor Esko Nurmi, a distinguished researcher in the field of food hygiene, who, with his colleagues, introduced, among other things, the concept of competitive exclusion
Yersinia pekkanenii Murros-Kontiainen et al. 2011 [14]	Strain ÅYV7.1KOH2 = DSM 22769 = LMG 25369	N.L. gen. masc. n. *pekkanenii*, of Pekkanen, in honor of Professor Timo Pekkanen, who was a veterinarian and former head of the Department of Food and Environmental Hygiene in the Faculty of Veterinary Medicine of the University of Helsinki. He made a substantial contribution to the development of scientific research of food hygiene in Finland
Yersinia pestis (Lehmann and Neumann 1896) van Loghem 1944 [1, 2]	Strain ATCC 19428 = CIP 80.26 = NCTC 5923	L. n. *pestis -is*, an infectious or contagious disease, plague; L. gen. n. *pestis*, of plague.
Yersinia philomiragia Jensen et al. 1969 [2, 15] This species is now called *Francisella philomiragia* (Jensen et al. 1969) Hollis et al. 1990	Strain ATCC 25015 = CCUG 4992 = CCUG 19700 = CIP 82.98 = DSM 7535	N.L. adj. *philus -a -um* (from Gr. adj. *philos -ê -on*), friend, loving; N.L. n. *miragium* (from L. v. *miro* or *miror*, to wonder or marvel at, to be astonished or amazed at a thing, to admire); mirage; N.L. n. *philomiragia*, a friend of mirages, loving mirages, because of the mirages observed in the area where the first isolations of this species were made. **Synonyms:** "*Bacillus pseudotuberculosis*" (*sic*) Pfeiffer 1889, "*Pasteurella pseudotuberculosis*" (Pfeiffer 1889) Topley and Wilson 1929, "*Bacterium pseudotuberculosis*" (Pfeiffer 1889) Migula 1900, "*Shigella pseudotuberculosis*" (Pfeiffer 1889) Haupt 1935

Yersinia pseudotuberculosis (Pfeiffer 1889) Smith and Thal 1965 [2, 16]	Strain ATCC 29833 = CCUG 5855 = CIP 55.85 = DSM 8992 = JCM 1676 = NCTC 10275	Gr. adj. *pseudês*, false; L. n. *tuberculum*, a small swelling, bump, protuberance, or tubercle; N.L. n. *tuberculosis -is*, tuberculosis; N.L. gen. n. *pseudotuberculosis*, of false tuberculosis
Yersinia pseudotuberculosis subsp. *pestis* (Lehmann and Neumann 1896) Bercovier et al. 1981 nom. rejic. [17–20]	Strain ATCC 19428 = CIP 80.26 = NCTC 5923	L. n. *pestis -is*, an infectious or contagious disease, plague; L. gen. n. *pestis*, of plague. Basonym: *Yersinia pestis* (Lehmann and Neumann 1896) van Loghem 1944 (Approved Lists 1980) Synonyms: *Bacterium pestis* Lehmann and Neumann 1896, *Bacillus pestis* (Lehmann and Neumann 1896) Migula 1900, *Pasteurella pestis* (Lehmann and Neumann 1896) Bergey et al. 1923, *Pestisella pestis* (Lehmann and Neumann 1896) Dorofeev 1947
Yersinia pseudotuberculosis subsp. *pseudotuberculosis* (Pfeiffer 1889) Bercovier et al. 1981 [21]	Strain ATCC 29833 = CCUG 5855 = CIP 55.85 = DSM 8992 = JCM 1676 = NCTC 10275	Gr. adj. *pseudês*, false; L. n. *tuberculum*, a small swelling, bump, protuberance, or tubercle; N.L. n. *tuberculosis -is*, tuberculosis; N.L. gen. n. *pseudotuberculosis*, of false tuberculosis
Yersinia rohdei Aleksic et al. 1987 [22]	Strain H271-36/78 = ATCC 43380 = CCUG 38833 = CDC 3022-85 = CIP 103163 = JCM 7376 = LMG 8454	N.L. gen. masc. n. *rohdei*, of Rohde; named in honor of Rolf Rohde, who founded the National Reference Center for *Salmonella* in Hamburg, Federal Republic of Germany, and who made many significant contributions to the diagnostic and serological identification of *Enterobacteriaceae*, especially *Salmonella*
Yersinia ruckeri Ewing et al. 1978 [23, 24]	Strain ATCC 29473 = CCUG 14190 = CDC 2396-61 = CIP 82.80 = HAMBI 1298 = JCM 2429 = JCM 15110 = NCTC 12986	N.L. gen. masc. n. *ruckeri*, of Rucker; named after R.R. Rucker, who studied redmouth disease and its etiological agents
Yersinia similis Sprague et al. 2008 [25]	Strain Y228 = CCUG 52882 = LMG 23763	L. fem. adj. *similis*, similar, resembling, as the strains are similar to those of *Yersinia pseudotuberculosis*
Yersinia wautersii Savin et al. 2014 [26]	Strain 12-219 N1 = CIP 110607 = DSM 27350	N.L. gen. masc. n. *wautersii*, named in honor of George Wauters, a Belgian yersiniologist who performed an in depth characterization of the genus *Yersinia*, including the biotyping scheme of *Y. enterocolitica*

3.1.1 Biotyping

Since its discovery in 1894, *Y. pestis* has been studied intensively to understand its diversity of phenotypic traits. Early investigators used glycerol fermentation to classify *Y. pestis* strains into glycerol-positive and glycerol-negative strains [33]. The negative strains were termed the oceanic type because they were usually isolated from rats in seaports, and the positive ones were termed the continental type because they were isolated from "wild" rodents, ground squirrels, and gerbils from natural plague foci [33].

The widely used biovar system at present is the one raised by Devignat et al. [36], in which glycerol fermentation and nitrate reduction are used to classify *Y. pestis* into three biovars, Orientalis, Antiqua, and Medievalis. Based on historical data and the bacteriological characteristics of the strains isolated from remnant foci of ancient plague, Devignat [36] hypothesized that biovars Antiqua, Medievalis, and Orientalis caused the first, second, and third pandemics, respectively. It is hardly possible to confirm the link between the first two pandemics and biovars Antiqua and Medievalis, other than to argue for hypotheses on the basis of historical data. Melibiose fermentation was shown by Mollaret and Mollaret [37] to distinguish biovar Orientalis and biovar Antiqua strains, neither of which ferments this sugar, from most biovar Medievalis strains, which do ferment it. Additional biochemical characteristics, including rhamnose, arabinose, melibiose, melezitose, maltose, mannose, and trehalose use, as well as pesticin–fibrinolytic–coagulase activities, nutritional requirements, susceptibility to pesticin, and virulence for mice and guinea pigs, have been used to characterize *Y. pestis* strains from different natural plague foci as more *Y. pestis* variants were found [33, 34]. For example, a strain from Inner Mongolia, China, is avirulent to humans and was isolated from Brandt's vole (*Microtus brandti*). It should be classified as Medievalis according to the three-biovar system, but Zhou et al. recently proposed it to be a new biovar, Microtus, based on its glycerol fermentation, nitrate reduction, and arabinose utilization phenotypes, as well as a genetic analysis [38]. As discussed in the section of subspecies classification, another new biovar, Intermedium, was proposed according to genetic diversity research [39]. Therefore, *Y. pestis* can be assigned into five biovars—Antiqua (glycerol positive, arabinose positive, and nitrate positive), Medievalis (glycerol positive, arabinose positive, and nitrate negative), Orientalis (glycerol negative, arabinose positive, and nitrate positive), Microtus (glycerol positive, arabinose negative, and nitrate negative), and Intermedium (its biochemical features still need to be determined) [38]. The first three biovars are postulated to be linked to the first, second, and third pandemics, respectively, of human plague, while the fourth is avirulent to humans, only naturally causing plague in *Microtus* spp. and its associated epizootics. As mentioned below, there are some natural plague foci in Russia where no human cases of plague have been found. New biovars can be envisioned to be found in the future; however, it should be noted that biovar characteristics are unstable and that one strain can undergo spontaneous phenotypic variation that would cause it to be classified as another biovar [33]. For example, strain Nicholisk

51, an isolate from Manchuria, was assigned as a member of biovar Orientalis by IS*100* genotyping and the presence of specific phage remnants but is glycerol positive and should be classified as biovar Antiqua [40]. The authors considered strain Nicholisk 51 to be an ancestor of biovar Orientalis or a variant of this biovar that had undergone phenotypic reversion back to the glycerol-positive phenotype [40].

3.1.2 Ecotyping

An ecotyping system, which used several biochemical features, including glycerin, rhamnose, maltose, melibiose, and arabinose fermentation, nitrate reduction, amino acids utilization, mutation rate from pigmentation $(Pgm)^+$ to Pgm^-, and water-soluble protein patterns on sodium dodecyl sulfate-polyacrylamide gel electrophoresis (SDS-PAGE), was developed to group Chinese isolates of *Y. pestis* into 18 ecotypes [34, 35, 41] (Table 3.2). Each of these ecotypes is located in a particular geographic region. Most of the plague foci with different primary reservoirs have unique ecotypes [42]. For the other plague foci, there is more than one ecotype in a single focus with a single primary reservoir, and each of the ecotypes corresponds to a unique set of natural landscapes and primary vector(s). The ecotyping system, which has been used as a framework for ecological and epidemiological analyses of plague in China, gives a preliminary explanation of the relationships between the ecotypes of *Y. pestis* and the natural environment, reservoirs, and vectors.

3.1.3 Subspecies Classification

The subspecies classification system was raised by Russian scientists, and because of the limited availability of documents in English, the following description about this classification system is taken from Anisimov's review [33]. To ensure the integrity of this review, some authors mentioned in the following three paragraphs were cited from Anisimov's review [33], and their original papers were not cited again in this chapter. For details, please read Anisimov's review [33].

Using numerical taxonomy, Martinevskii classified *Y. pestis* into three varieties: *mediaasiatica montana* (corresponding to biovar Antiqua), *mediaasiatica deserta* (corresponding to biovar Medievalis), and *oceanica* (corresponding to biovar Orientalis). He also concluded that strains isolated from common voles in natural foci of infection in the Transcaucasian highlands or from Mongolian pikas in the Altai Mountain and Transbaikalian regions were a different species, *Yersinia pestoides*, and included three varieties: *Yersinia pestoides parvocaucasica*, *Yersinia pestoides altaica*, and *Yersinia pestoides transbaicalica*. "Pestoides" is now reappearing in publications emanating from the USA as a strain designation and as part of the nomenclature used to classify strains imported from the Former Soviet Union (FSU), although this classification was rarely used in the FSU. This nonetheless

Table 3.2 Ecotypes of *Y. pestis* in China

Ecotypes	Nitrate reduction	Fermentation					Amino acid utilization	Mutation rate of Pgm^+ to Pgm^- in 10 generations (%)	PAGE pattern of water-soluble protein
		Glycerol	Arabinose	Melibiose	Rhamnose	Maltose			
Xilingol grassland type	–	+	–	+	+	+	Arg^- Leu^-	0	I
Microtus fuscus type	–	+	–	+	+	+	Arg^- Leu^-	0	I
Pamir Plateau type	+	+	+	– or +	–	+		>85	II
Qinghai–Tibet Plateau type	+	+	+	–	–	+	Phe^-	>85	II
Qilian Mountain type	+	+	+	–	–	–	Phe^-	>85	II
Gangdise Range type	+	+	+	–	–	–	Phe^-	>85	II
Kunlun Mountain type A	–	+	+	–	–	+	Ile^- Glu^-	100	II
Kunlun Mountain type B	–	+	+	–	–	+	Ile^- Glu^-	?	II
Ordos Plateau type	–	+	+	– or +	–	–	Trp^-	>75	II
Songliao Plain type A	+	+	+	–	–	+		>30	III
Loess Plateau type A	–	+	+	–	–	+	Leu^- Phe^-	<50	IV
Loess Plateau type B	–	+	+	–	–	+		>80	III
Western section of North Tian Shan type A	+	+	+	+	+	+	Trp^-	0	IV
Western section of North Tian Shan type B	+	+	+	+	+	+		0	IV
Eastern section of North Tian Shan type	+	+	+	+	+	+	Trp^-Thr^-	<30	V
West Yunnan transverse valley type	+	+	+	–	–	–	Phe^-	>75	VI
Residential area of Yunnan and Fujian type	+	–	+	–	–	+	Glu^-Phe^-	100	VI
Songliao Plain type B	+	+	+	–	–	+		>20	VII

Notes:
1. *Arg* arginine, *Leu* leucine, *Trp* tryptophan, *Glu* glutamic acid, *Thr* threonine, *Ile* isoleucine, *Phe* phenylalanine, *PAGE* polyacrylamide gel electrophoresis
2. In the past three years, strains isolated from a newly identified plague focus, the *Microtus fuscus* plague focus of the Qinghai–Tibet Plateau (focus M), were compared with other ecotypes of *Y. pestis* strains. These strains were assigned to a new (the 18th) ecotype (the *M. fuscus* type)

further indicates the amount of genetic and phenotypic diversity of plague isolates in relation to the sylvatic areas where they circulate as epizootic pathogens.

Timofeeva proposed a new classification of *Y. pestis* into subgroups, which was based on numerical taxonomy, and used subspecies as a taxon designator. This classification was formulated according to the International Code of Bacterial Taxonomy and used the nomenclature indicating the main species and subspecies (i.e., *Yersinia pestis* subsp. *pestis*). She further divided the main *Yersinia pestis* subspecies into two more groups, continental and oceanic. Additionally, with the help of numerical taxonomy, Peisakhis and Stepanov proposed a classification of *Y. pestis* strains that were isolated in the FSU into groups based on 25 phenotypic features, including rhamnose, melibiose, and arabinose fermentation, nitrate reduction, pesticin, fibrinolytic, and coagulase activities, susceptibility to pesticin 1, and virulence to guinea pigs. To standardize the classification system of *Y. pestis* isolates, the Conference of Experts of the Anti-Plague Establishments of the Soviet Union recommended classifying all of the variants of the plague pathogen that were isolated from the territory of the FSU and Mongolia into the "subspecies" *Y. pestis* subsp. *pestis* (sometimes referred to as the "main" subspecies), *Y. pestis* subsp. *altaica*, *Y. pestis* subsp. *caucasica*, *Y. pestis* subsp. *hissarica*, and *Y. pestis* subsp. *ulegeica* based on numerical analyses of 60 phenotypic features, including rhamnose, melibiose, arabinose, glycerol, and melezitose fermentation, nitrate reduction, urease utilization, pesticin 1 production, susceptibility to pesticin 1, fibrinolytic and coagulase activities, nutrient dependence (leucine, methionine, arginine, thiamine, cysteine, phenylalanine, threonine, and tyrosine), and virulence to guinea pigs. In 1998, Sludskii proposed one more intraspecific group, *Y. pestis* subsp. *talassica*. The last five subspecies are sometimes referred to as the "non-main" subspecies, and they have also been referred to as the "pestoides" group of *Y. pestis* isolates. All pestoides strains ferment rhamnose and are dependent on additional nutrients and exhibit elective virulence (they are less virulent in guinea pigs, but highly virulent in mice). These strains are described in the literature as causes of plague epizootics in voles and Mongolian pikas, occasional human plague cases without further human-to-human transfer, but they have never been associated with plague outbreaks and epidemics [33, 43].

By comparing the properties of strains belonging to the main subspecies, *pestis*, with strains of other subspecies using numerical taxonomy, many significant differences were found between the *pestis* and *caucasica* subspecies, whereas subspecies *altaica* and *hissarica* were found to be closely related. It was also found that, in general, among the five non-*pestis* subspecies (i.e., the pestoides subgroup), subspecies *caucasica* was classified as biovar Antiqua and subspecies *altaica*, *hissarica*, *ulegeica*, and *talassica* were classified as biovar Medievalis.

There are two biovar Microtus-related foci in China: one is in Inner Mongolia and the other is in Qinghai province. The strains isolated from the primary host *M. brandti* and *M. fuscus* are very closely related. From a typing study of *Y. pestis* isolates using a multiple locus variable-number tandem repeats (VNTRs) analysis (MLVA) and clustered regularly interspaced short palindromic repeats (CRISPRs), the strains from *M. brandti* and *M. fuscus* were proposed to be reclassified into subspecies *xilingolensis* and *qinghaiensis*, respectively [39]. The isolates from these two foci cluster with the *hissarica* and *altaica* subspecies. These four subspecies are suggested to be the most closely related to *Y. pestis* subsp. *pestis* [39].

From the MLVA analysis, a new biovar, Intermedium, was proposed to describe rhamnose-positive *Y. pestis* subsp. *pestis* strains that occasionally infect humans and that are isolated mostly from marmots in the northern Tian Shan Mountains in China [39]. Table 3.3 lists the details that differentiate these biovar or subspecies.

To avoid confusion from so many biovar or subspecies designations, Anisimov proposed a simple subspecies and biovar nomenclature [44]. In this system, *Y. pestis* is divided into two subspecies, *pestis* and *microtus*; then, the subspecies are further classified into biovars. *Y. pestis* subspecies *pestis* contains four biovars, including Antiqua, Medievalis, Orientalis, and Intermedium. *Y. pestis* subspecies *microtus* contains eight biovars, including Caucasica, Angola, Talassica, Qinghaiensis, Xilingolensis, Altaica, Hissarica, and Ulegeica. This simple classification helps us to understand the nature of *Y. pestis*. The strains from subspecies *pestis* can cause humans outbreaks and epidemics, while subspecies *microtus* isolates might be a reason of rare sporadic human cases. They can also be differentiated by biological features, their geographic distribution, and genetic biomarkers (Table 3.3).

3.1.4 Plasmid Typing

There are different kinds of plasmids in *Y. pestis* strains from different regions around the world. Most of the bacteria contain three plasmids, i.e., pPCP1 (also designated pYP, pPla, or pPst, 9.5 kb or approximately 6MDa), pMT1 (also designated pFra, pTox, or pYT, approximately 100 kb or 65 MDa), and pCD1 (also designated pYV, pCad, pLcr, or pVW, approximately 70 kb or 45MDa). The first two plasmids are unique to *Y. pestis*, and their acquisitions are thought to have played key roles in the evolution of the plague bacillus. All pathogenic yersiniae contain the virulence-associated pCD1 plasmid, which encodes a finely-tuned type III secretion machinery that exports anti-phagocytic factors [46]. pPCP1 encodes the plasminogen activator and the bacteriocin pesticin. pMT1 is responsible for the synthesis of the fraction 1 capsular antigen and phospholipase D. The plasminogen activator is involved in the dissemination of bacteria from the site of the initial fleabite, while phospholipase D (previously termed the murine toxin) plays a major role in bacterial survival in fleas [47]. The greatest size variations witnessed in the pMT1 and pCD1 are greater than that in pPCP1 [48]. Filippov et al. investigated the plasmid content of 242 *Y. pestis* strains from various natural plague foci of the FSU and other countries [49]. Of these strains, 172 (71 %) were shown to carry the three plasmids described previously, which are approximately 6, 45–50, and 60 MDa, respectively. Twenty strains (8 %) from different foci harbored additional cryptic plasmids, most often of approximately 20 MDa in size. Plasmid pPCP1 displayed a relatively constant molecular mass. In contrast, size variations of pCD1 (45–49 MDa) and, especially, pMT1 (60–190 MDa) were found. The molecular masses of

Table 3.3 Taxonomic characters of strains, which distinguish different *Y. pestis* subspecies, and compliance of subspecies with biovars[a]

Subspecies	Biovars	Fermentation of: Rhamnose	Melibiose	Arabinose	Glycerol	Melezitose	Nitrate reduction	Urease activity	Pesticin I production	Susceptibility to pesticin[i]	Fibrinolytic activity	Coagulase activity	Dependency upon nutrition: Leucine	Methionine	Arginine	Thiamine	Cysteine	Pheny-lalanine	Threonine	Tyrosine	Virulence for guinea pigs	Region of circulation	Hosts	Pgm mutation[j]
Microtus	angola[b]	+[c]	?[d]	+[e]	+[f]	?	–[g]	?	?	?	?	?	?	?	?	?	?	?	?	?	–	Angola	?	0
	caucasica[h]	+	+	+	+	–	+	–	–	+	–	–	±	+	+	+	±	+	+	?	–	Trans-Caucasian highland, Armenia, Azerbaijan and Georgia; Mountain Dagestan, Russia	*Microtus arvalis*	0
	ulegeica	+	+	+	+	?	–	–	+	+/–[g]	+	+	–	–	–	–	+	+	?	?	–	Gobi Desert, Northeast Mongolia	*Microtus gregalis, Alticola strelzovi, Ochotona pallasi pricei*	0
	altaica	+	+	–	+	?	–	–	+	+	+	+	+	–	+	–	+	+	?	?	–	Mountain Altai, Russia and Mongolia	*M. gregalis, A. strelzovi, O. pallasi pricei*	0
	hissarica	+	+	–	+	±	–	±	+	±	+	+	+	+	–	–	+	+	–	–	–	Hissarian ridge, Tadjikistan (Tajikistan) and Uzbekistan	*Microtus carruthersi*	0
	xilingolensis	+	+	–	+	?	–	–	+	+	?	?	–	–	–	?	–	–	?	?	–	Xilingol Grassland, Inner Mongolia, China	*Microtus brandti*	0
	qinghaiensis	+	+	–	+	–	–	–	+	?	?	?	–	–	–	?	–	–	?	?	–	Qinghai–Tibet Plateau, Qinghai and Sichuan, China	*Microtus fuscus*	0
	talassica	+	+	–	+	–	–	+	+	–	+	+	+	?	+	?	+	+	?	+	–	Talassian ridge, Kirghizia	*M. gregalis, Marmota caudata*	0

(continued)

Table 3.3 (continued)

	Intermedium	+	+	+	+	?	+	?	+	?	?	?	–	+	–	?	+	±	±	?	+	North Tian Shan, Xinjiang, China	*Marmota baibacina*, *Spermophilus undulatus*	0
	Antiqua	–	–	+	+	–	+	–	+	–	+	+	±	+	–	–	±	±	+	–	+	Central and northern Asia, central Africa	*Marmota* spp.	> 20
	Medievalis	–	±	+	+	–	–	–	+	–	+	+	±	+	–	–	±	±	+	–	+	Central Asia	*Citellus* (*Spermophilus*) spp., *Meriones* spp., *Rhombomys* spp.	> 20
pestis	Orientalis	–	–	+	–	–	+	±	+	–	+	+	±	+	–	–	±	±	+	–	+	Global distribution	*Rattus* spp., *Cynomys* spp., *Spermophilus* spp., *Cavia* spp., *Peromyscus* spp., etc.	> 20

[a]This table was compiled from Refs. [33, 42, 45] and from the table proposed by the Conference of Experts of the Anti-Plague Establishments of the Soviet Union (Saratov, 1985) for plague pathogens that were isolated from the territory of the FSU and Mongolia

[b]The characteristics of Angola are similar to that of the Medievalis-like Microtus biovar (glycerol positive, arabinose positive, and nitrate negative) [38]

[c]+, presence of a character; –, absence of a character; ±, presence of a character, but not in all strains

[d]No data

[e]No 122-bpdeletion was detected in the *araC* gene of biovar Angola. This deletion may account for the inactivation of the *ara* operon that is involved in arabinose utilization in strain 91001

[f]The intact *glpD* gene of biovar Angola. A 93-bp deletion may result in the glycerol-negative phenotype of strain CO92

[g]A point mutation (C->T) at nucleotide 1021 of the *napA* gene most likely results in the nitrate-negative phenotype of biovar Angola

[h]*caucasica* strains do not carry the pPst plasmid, which encodes pesticin

[i]Strains are susceptible to pesticin from strains of subspecies *pestis* and *altaica*, and they are resistant to pesticin from strains of its own subspecies

[j]The mutation rate from Pgm^+ to Pgm^- in ten generations (%)

these plasmids correlated with the host strain origin. Strains from Vietnam, Indonesia, Brazil, and Africa that lack the ability to ferment glycerol contain a 60-MDa pMT1. Strains from Gissar natural plague focus #34, in the western part of Tajikistan, and Talas focus #40 contain a 68-MDa pMT1. Strains from Mountain Altai natural plague focus #36of, southern Siberia, contain a 69-MDa pMT1. Microtus strains from mountainous areas of the FSU contain an 80-MDa pMT1, while strain 924 from China contains a 92-MDa pMT1and strain 679 from plague foci in the Volga-Ural region contains a 190-MDa pMT1. For plasmid pCD1, strain EV76 from Madagascar and glycerol-negative oceanic strains possess a 45-MDa pMT1, while the majority of glycerol-positive strains have a 47-MDa pMT1. Strains from Mid-Asia desert areas contain a 48-MDapCD1, and Microtus strains from mountainous areas of the FSU possess a 49-MDa pCD1 [48]. Strains isolated from plague foci around Qinghai Lake in China possess a different plasmid profile than those isolated from the Tanggula Mountains.

There are also other plasmids in *Y. pestis* from different regions [43, 50–53]. A 6-kb cryptic plasmid (pYC; 5919 bp) was found in *Y. pestis* isolates from regions of Yunnan province in China [50]. This plasmid is increasingly harbored by *Y. pestis* isolates recovered from a domestic rodent cycle in the southern regions of Yunnan province [50]. In another study, Dong et al. screened 1020 strains of *Y. pestis* isolates from 44 counties of Yunnan province in China and the China-Myanmar border [51], and they found that these strains carried nine kinds of plasmids with molecular weights of approximately 3.93, 6.05, 22.97, 35.65, 45.35, 64.82, 74.59, 111.36, and 129.55 MDa. The plasmid profiles of the strains could be divided into ten types (plasmidovars, as deemed by Russian scientists) according to their plasmid content. Types I to V were common and clustered according to their geographic distribution [51]. Xu et al. analyzed 109 *Y. pestis* strains that were isolated from four natural plague foci in Xinjiang, China, and they divided these strains into five plasmidovars (I to V) [54]. They found a good correlation between plasmidovars and the geographic distributions of the isolates: plasmidovars I to V are found in a gray marmot (*Marmota baibacina*)-long-tailed ground squirrel (*Spermophilus undulatus*) plague focus of the northern Tian Shan Mountains; plasmidovars I and V are found in a *M. baibacina* plague focus of the southern Tian Shan Mountains; and plasmidovar I is only in a long-tailed marmot (*Marmota caudata*) plague focus of the Pamir Plateau and a Himalayan marmot(*Marmota himalayana*) plague focus of the Kunlun Mountains.

Song et al. [53] reported a 21,742-bp plasmid, pCRY, from *M. brandti*. A pair of primers targeting the *repA* gene in pCRY was used to screen 257 strains of *Y. pestis* from different regions and different hosts or vectors for the presence of this plasmid, and the results demonstrated that this plasmid is only presented in 11 strains, three from a *M. brandti* plague focus of the Xilingol Grassland, six from a Mongolian gerbil (*Meriones unguiculatus*) plague focus of the Inner Mongolian Plateau, one from a Daurian ground squirrel (*Spermophilus dauricus*) plague focus of the Songliao Plain, and one from a *M. himalayana* plague focus of the Gangdise Mountains. Plasmid pMT1 has been reported to be rearranged in different *Y. pestis* strains. Song et al. [53] and Golubov et al. [43] reported a 106,642-bp pMT1 and a

137,036-bp plasmid, pG8786, from *M. brandti* and the common vole (*Microtus arvalis*), respectively. They share a common 4642-bp region that is not present in other reported pMT1 plasmids. Another 32,617-bp is unique to isolates from voles in the high mountainous Caucasus region in Georgia [43]. Studies indicated that these special forms of pMT1 may represent an ancestral form of pMT1 [43].

Many studies have examined plasmid variations in different strains of *Y. pestis* from different regions in the FSU [33]. Variations in plasmid contents and sizes can be used to classify different strains of *Y. pestis* into different plasmidovars, which are associated to a high, but not exclusive, degree, with strain source and phenotype, thereby documenting the potential use of this method for epidemiologic investigations and for contributing to the determination of the pathogenic potential of a given isolate. However, plasmid variations in American strains of *Y. pestis* are not as obvious as those of isolates from the FSU, Mongolia, and China [33, 55]. In Brazil, a profile of four plasmids with molecular masses of 6.4 MDa (pPCP1), 14.9 MDa (cryptic), 44 MDa (pCD1), and 60 MDa (pMT1) was found in all 26 strains that were isolated from Paraiba State. DNA cleavage with EcoRI further demonstrated the uniform plasmid content of these *Y. pestis* isolates [33, 55]. An analysis of the plasmid contents of 250 Brazilian *Y. pestis* strains stored in culture collections for different periods confirmed that the majority of the strains contained a homogeneous pattern composed of the three classic *Y. pestis* plasmids: pPCP1, pCD1, and pMT1 [33, 55]. An analysis of the plasmid patterns of 53 *Y. pestis* isolates from three plague foci from the state of Ceará, Brazil, showed that 39 strains had the three classical plasmids, while seven isolates had an additional plasmid that was larger than 90 kb. The other seven lacked all or some of the plasmids [56]. Chu et al. from the US Centers for Disease Control and Prevention reported that geographic isolates of *Y. pestis* might be differentiated by their plasmid profiles and that isolates from the western USA harbored an additional plasmid, estimated to be approximately 19 kb in size, a dimer of the 9.5-kb plasmid [52]. This plasmid was found in isolates from Arizona, California, Colorado, New Mexico, and Texas [52].

Reports of the discovery of plasmid-mediated, high-level resistance to multiple antibiotics indicate the high potential of *Y. pestis* to acquire new genetic materials [57, 58], which makes clinical treatment and prophylaxis more difficult.

3.2 Genotyping

Genotyping is based on the analysis of nucleic acid fragments from target microorganisms for subtle variations to differentiate closely related bacterial strains. This can be achieved by fingerprinting chromosomal or plasmid DNA by restriction fragment length polymorphism (RFLP), RFLP-DNA probe hybridization patterns, polymerase chain reaction (PCR)-based amplification profiles, and DNA fragment sequencing comparisons. All these methods cannot provide a complete picture of chromosomal or plasmid polymorphisms, which are provided by genomic typing techniques, as mentioned in the following sections of this review. Different methods have been employed to differentiate *Y. pestis* strains from different regions around the world. It is difficult to say which method is better because most of the methods have not been compared in parallel. However, according to the data reported, the stabilities of pulsed-field gel electrophoresis (PFGE) profiles [59] and fluorescent amplified length polymorphisms (FAFLP) profiles [60] of subcultures have been questioned. Ribotyping did not classify isolates into their respective biovars [59, 61–63]. Specific insertion sequences (IS), including IS*100* [64], IS*285* [65], and IS*1541* [66, 67], were used as markers in RFLP analyses [27, 68] and in a PCR-based technique [40]. The last approach revealed low resolution and produced identical patterns of IS*100* distribution in Antiqua and Medievalis isolates, which might be caused by biased sampling in the experiment [40, 63]. Although a variable-number tandem repeat technique [63, 69, 70] was reported to have a greater discrimination capacity than did ribotyping, isolates from different areas were found to harbor identical types. An analysis of the fragments of five housekeeping genes in 36 *Y. pestis* isolates from various locations did not identify any sequence diversity [27], but when 19 of these 36 *Y. pestis* isolates were analyzed using the multiple spacer typing (MST) method, 12 MST profiles were found [63]. Although randomly amplified polymorphic DNA (RAPD) has been reported for *Y. pestis* typing, it was demonstrated to be less discriminative and it exhibited low typing ability [41, 71, 72]. Therefore, to obviate the typing bias caused by the less discriminative power of a particular method, it is recommended that polyphasic methods, which employ more than one of the aforementioned genetic techniques, should be used to differentiate *Y. pestis* strains.

3.2.1 Pulsed-Field Gel Electrophoresis

PFGE, which separates DNA fragments after digestion of the chromosome with restriction endonucleases that cleave infrequently [73], can facilitate a broad view of the whole genome of an organism. It is highly effective in molecular epidemiological studies of bacterial isolates, and it is superior to other methods in discriminating among isolates of *Escherichia coli*, *Staphylococcus aureus*, and many other species [74]. Additionally, it can be used to evaluate the clonal relatedness among bacterial isolates and to investigate outbreaks [75]. It has also been proven to be an effective method for qualitatively evaluating intraspecific genetic variation, as it enables the identification of individual isolates of a given species by comparing their macrorestriction patterns [76].

Using PFGE techniques, Lucier and Brubaker [62] found that the SpeI-digested DNA patterns of eight *Y. pestis* strains were closely related to their respective biovars. Similar results were obtained by Rakin and Heesemann [77] using the I-CeuI macrorestriction of nine *Y. pestis* strains and by Huang et al., who also used SpeI [78].

The choice of an appropriate macrorestriction enzyme is critical for PFGE analysis. Huang et al. compared the genomic restriction patterns of three different enzymes (NotI, SfiI, and SpeI) [78]. NotI digestion generated closely grouped bands of high molecular weight, as reported by Lucier and Brubaker [62], with many co-migrating fragments. In contrast, SfiI digestion produced many closely grouped lower-molecular-weight bands. However, SpeI produced a relatively wide range of DNA fragments that could easily be resolved by the PFGE separation conditions used by Huang et al. [78]. Therefore, SpeI digestion seems to be the best enzyme for PFGE analysis of *Y. pestis*.

To evaluate the extent of the variability of the PFGE pulsotype within one strain, Guiyoule et al. [59] randomly picked eight and four colonies from stock cultures of the Saigon 55–1239 and Kenya 169 strains, respectively, and their DNAs were subjected to PFGE. Five and three different SpeI restriction patterns, respectively, were observed for each of these two strains, indicating a high heterogeneity of the pulsotypes within a given strain. They also found similar phenomena using other restriction enzymes, such as XbaI or NotI, which reminds us of the limits of using PFGE to analyze *Y. pestis*. In our laboratory, we also found that the FAFLP profiles differed between three subcultures of the same strain [60].

3.2.2 Ribotyping

Ribotyping is a method that can identify and classify bacteria based upon differences in their rRNA genes [79–81]. It generates a highly reproducible and precise fingerprint that can be used to classify bacteria from the genus through and beyond the species level. DNA is extracted from a colony of bacteria and then restricted into discrete-sized fragments. Then, DNA is transferred to a membrane and probed with an rDNA probe to reveal the pattern of rRNA genes. Variations among bacteria in the position and intensity of the rRNA bands can be used for their classification and identification. The copy numbers of rRNA operons in *Y. pestis* differ. In the three sequenced strains of *Y. pestis*, the Orientalis strain CO92 has six copies [82], the Medievalis strain KIM has seven copies [83], and the Microtus strain 91001 has seven copies [53]. The arrangement of the three rRNA genes in CO92 and KIM is in the order of 16S-23S-5S, while that in 91001 differs, as an extra 5S rRNA gene follows the above arrangement. Yu et al. investigated the distribution of *Y. pestis* strains with different copy numbers of rRNA operons in Chinese isolates from different natural plague foci. The isolates from plague foci situated to the east of the *S. dauricus alaschanicus* plague focus of the Loess Plateau in Gansu and Ningxia provinces possess seven copies of rRNA operons, while those from plague foci to the west of the Gansu corridor have six copies, except isolates from the *M. fuscus* plague focus of the Qinghai–Tibet Plateau, which have seven copies. These results correlate excellently with the microevolutionary analysis of *Y. pestis* in China, as Zhou et al. demonstrated that *Y. pestis* had three microevolutionary routes in China. Branch one involved isolates from natural plague foci situated to the west of the Gansu corridor, and isolates from biovar Antiqua to biovar Orientalis had six copies of the rRNA operon. The second branch included strains from natural plague foci to the east of the Gansu corridor, and isolates from biovar Antiqua to biovar Orientalis had seven copies of the rRNA operon. The third branch contained isolates from *Microtus* plague foci that belong to the new biovar Microtus.

This technique has been automated by using a RiboPrint® instrument and a relevant database [84]. Databases for *Listeria* (80 pattern types), *Salmonella* (97 pattern types), *Escherichia* (65 pattern types), and *Staphylococcus* (252 pattern types) have been established, but do not include the EcoRI patterns of *Y. pestis*. Grif et al. [84] showed that the automated ribotyping system did not allow the reliable differentiation of *Y. pestis* from *Y. pseudotuberculosis*, but it did clearly demarcate *Y. pestis* from other species, such as *Y. enterocolitica*. The use of additional enzymes, such as PvuII, PstI, and AseI, allowed for correct identification and subtyping. In 1994, Guiyoule et al. [59] reported that the ribotypes of individual colonies within a given *Y. pestis* strain were stable and were not modified after five passages in vitro.

Different endonucleases, including EcoRI, EcoRV, HindIII, KpnI, and BamHI, were evaluated for their ability to yield the best DNA banding pattern [59], and the results showed that EcoRI and EcoRV were the most appropriate restriction enzymes for *Y. pestis*. DNAs from 70 strains from different regions around the world were analyzed by EcoRI and EcoRV digestion, followed by southern hybridization with a nonradioactive 2-acetylaminofluorene-labeled 16S-23S rRNA probe. The hybridization profiles of the 16S-23S rRNA probe with the genomic DNA digested with EcoRI or EcoRV were named the EcoRI and EcoRV ribopatterns and designated by the letters RI and RV, respectively, followed by a numeral. The combination of the EcoRI and EcoRV ribopatterns was renamed as the ribotype of a strain, and the ribotype was represented by capital letters A to P. Although 11 EcoRI (RI.1 to RI.11) and 11 EcoRV ribopatterns (RV.1 to RV.11) were found in the *Y. pestis* strains studied, the profiles obtained were relatively homogeneous, with most of them differing only by the presence or absence of one restriction fragment. The majority of the strains (74.3 % and 81.4 %) had two main EcoRI or EcoRV profiles, respectively, which were RI.1 (30 of 70) and RI.4 (22 of 70) and RV.2 (30 of 70) and RV.5 (27 of 70). However, some of the strains, including the Senegal Fa, Belgian Congo Li, and Kurdistan PKRV strains, had unique and specific patterns with both EcoRI and EcoRV. The results obtained with the two restriction enzymes were consistent: RI.1 was most often linked to RV.2, and RI.4 was most often linked to RV.5, which suggests some parallelism in the conservation (or the loss) of the EcoRI and EcoRV sites in the rRNA operons of *Y. pestis*.

When the EcoRI and EcoRV profiles were combined, 16 different ribotypes (A to P) were obtained. Ribotypes B (RI.1 plus RV.2) and O (RI.4 plus RV.5) were found in 65.7 % of the strains, and nine ribotypes were only found in one strain.

Good correlation was found between the ribotypes and the biovars of the strains studied (Table 3.4), and the relationship between the ribotypes and geographic origins of the strains, such as how the geographical distributions of ribotypes B and O, differed. B was found on five continents, while O was restricted to Central Africa

Table 3.4 Correlation between ribotypes and biovars

Ribotype	Biovar
F and G	Orientalis and Antiqua
O	Medievalis or Antiqua
A to E	Orientalis
H to N	Antiqua
P	Medievalis

and Central Asia, and the locations of ribotype O strains corresponded to the remaining plague foci of the first and second pandemics, while the distribution of ribotype B reflected the geographical spread of *Y. pestis* during the third pandemic [7, 18], which indicates that a ribotype B strain was responsible for the third pandemic. The results also showed that biovar Medievalis was much more homogeneous than biovars Orientalis and Antiqua.

In another study, Guiyoule et al. [61] used a representative sample of 59 *Y. pestis* strains isolated from 1939 to 1996 in 21 different geographical regions in Madagascar, and they found three new EcoRI patterns, RI.12, RI.13 and RI.15; one new EcoRV pattern, RV.12; and, hence, three new ribotypes, ribotypes Q, R, and T. The designations RI.14 and ribotype S were attributed to strains from a 1994 outbreak in India. An analysis of the epidemiological characteristics of the strains harboring the two new ribotypes demonstrated that all ribotype R strains came from different clinical samples of human patients, while ribotype Q came from rodents and human patients, essentially from bubo aspirates. More interestingly, a relationship was found between the geographical origins of the strains and their ribotypes: the two ribotype R strains were isolated in the Ambositra subprefecture, while the ribotype Q strains were isolated in the Ambohimahasoa subprefecture. These two neighboring regions are located on the high plateau, in the southern part of the Madagascar central "plague triangle."

The expansion of *Y. pestis* can be hypothesized based on the ribotyping analysis data: one main clone of *Y. pestis* of ribotype O (biovar Antiqua) spread from Central Asia to Central Africa and caused Justinian's plague, and later, a variant of the same clone (ribotype O), which lost the ability to reduce nitrate (biovar Medievalis), spread from Central Asia to Crimea and was responsible for the Black Death. It is probable that a *Y. pestis* ribotype B clone was responsible for the third pandemic and that it was derived from the first clone. This hypothesis is in complete agreement with Wu et al.'s and Devignat's proposition. Wu et al. stated that plague has been present since time immemorial in the Central Asiatic plateau, which is considered the original home of the infection [61]. Devignat [36] hypothesized that the Antiqua and Medievalis biovars might have originated from the above primary focus.

Four EcoRI types were found in *Y. pestis* isolates in China (professor Fan Zhenya, personal communication). Type D is only distributed in strains from the *M. brandti* plague focus of the Xilingol Grassland, and type C is distributed in isolates from the *M. unguiculatus* plague focus of the Inner Mongolian plateau. Types A and B are distributed in other foci.

3.2.3 IS-Based Typing

Insertion sequence (IS) elements have been loosely defined as small (<2.5 kb), phenotypically cryptic segments of DNA, with a simple genetic organization, which are capable of inserting at multiple sites in a target molecule [20]. ISs are involved in phenomena other than the acquisition of accessory functions.

McDonough and Hare identified IS*100* sequences in a specific subset of *Y. pseudotuberculosis* isolates that were also sensitive to the *Y. pestis*-produced bacteriocin, pesticin. In contrast, *Y. pseudotuberculosis* strains that did not contain IS*100* sequences were not sensitive to pesticin. They proposed that IS*100* serves as a molecular marker for identifying a subset of *Y. pseudotuberculosis* isolates that have a particularly close evolutionary and/or ecological relationship with *Y. pestis*.

Many ISs form an integral part of the chromosomes of most bacterial species, and they participate in chromosomal rearrangements and promote plasmid integration. In contrast, some specific IS elements at defined places in the chromosome are sufficiently stable to allow them to be used as markers in RFLP analyses for species typing and epidemiological studies [85].

Whole-genome sequencing of three strains of *Y. pestis*, CO92, KIM, and 91001 [53, 82, 83], revealed that they have abundant ISs in their genomes. They also have different copy numbers of IS elements [53]. For IS*100*, the IS copy number in relation to the strain is CO92 (44)>KIM (35)>91001 (30); for IS*1541*, it is CO92 (62)>KIM (49)>91001 (43); for IS*285*; it is 91001 (23)>CO92 (21)>KIM (19); and for IS*1661*, it is KIM (10)>CO92 (9)>91001 (8), which indicates that IS*100* played more critical roles than other IS elements in the microevolution of *Y. pestis*.

RFLP analyses provide information about the local genome environments of specific gene sequences on the basis of the probe used. IS*100* and IS*285* have been evaluated for typing *Y. pestis*. Huang et al. [78] used a RFLP analysis with IS*100* as a probe to analyze 37 strains of *Y. pestis*, and these strains were divided into 16 types, with 43 % belonging to IS*100* type 1. Typing with IS*285* as a probe was less specific, and it led to only four RFLP types, with 81 % belonging to type 1. Achtman et al. also probed RFLP patterns with IS*100* to analyze 36 *Y. pestis* isolates, and the resulting phylogenetic tree indicated that the three biovars formed distinct branches [27]. Just like ribotyping, a mixture of 16S and 23S rRNA genes as probes is more discriminative than any single probe [86]. Thus, it is beneficial to use a mixture of IS elements as probes to perform RFLP analyses that have the most discriminative power. Leclercq et al. developed a RFLP method based on the mixture of IS*100*, IS*285*, and IS*1541* elements for studying *Y. pestis* strains with worldwide diversity. The approach was designated as "3IS-RFLP," which could group *Y. pestis* isolates according to their geographical origin and therefore should be valuable in tracing the source of the plague outbreak [87].

Motin et al. developed a PCR-based genotyping system that detects the divergence of IS*100* locations within the *Y. pestis* genome, and they used it to characterize a large collection of isolates of different biovars from different geographical origins [40]. The dendrogram of *Y. pestis* isolates constructed by the unweighted pair group method with arithmetic mean (UPGMA) indicated that the Orientalis biovar formed a homogeneous group, while the Antiqua and Medievalis biovars clustered together. Biovar Antiqua isolates showed a variety of fingerprinting profiles, whereas Medievalis isolates clustered with the Antiqua isolates originating from Southeast Asia, which suggests that they have close phylogenetic relationships.

3.2.4 Multilocus Sequence Typing

Multilocus sequence typing (MLST) involves comparing the sequences of 5–7 housekeeping genes to epidemiologically type bacterial and parasitic strains. MLST typing has highlighted different mutation rates in various bacteria, and this makes the technique suitable for different purposes in each species. Its advantage is that sequence information is highly reproducible and easily deposited into growing banks of MLST sequences (e.g., see http://mlst.zoo.ox.ac.uk), which is aided by the rapid deposition and dissemination of information on the Internet. Furthermore, each laboratory will not need to obtain reference strains or conduct extensive validations before typing their own strains [88].

Achtman and colleagues compared the sequences of fragments of five housekeeping genes (*dmsA* (anaerobic dimethyl sulfoxide reductase chain A), *glnA* (glutamine synthetase), *thrA* (aspartokinase), *tmk* (thymidylate kinase), and *trpE* (anthranilate synthase component I)) and a gene involved in the synthesis of lipopolysaccharide (*manB* (phosphomannomutase)) in 36 strains representing the global diversity of *Y. pestis* and in 12–13 strains from each of the other related species, *Y. pseudotuberculosis* and *Y. enterocolitica*. They showed that all of the *Y. pestis* strains tested were identical and very similar to the *Y. pseudotuberculosis* strains. *Y. enterocolitica* strains were also highly conserved but were quite distinct from the former two species. This supports other evidence that *Y. pestis* and *Y. pseudotuberculosis* are closely related and should be reclassified as two subspecies. Clinical and historical considerations have prevented this so far. As the mutation rate is so low, Achtman and colleagues were able to estimate the point at which *Y. pestis* is likely to have diverged from *Y. pseudotuberculosis* as 1500–20,000 years ago, some time prior to the first plague pandemic in the sixth century, which is in agreement with Suntsov's ecology-based analysis that showed that the formation of the plague causative agent and natural foci took place in the Pleistocene, without the participation of humans [89].

3.2.5 Multiple Spacer Typing

Genome analysis of the closely related bacteria showed that intergenic spacers, which have been subjected to less evolutionary pressure than coding sequences, may be sufficiently variable to differentiate closely related microorganisms [90–95]. Drancourt et al. [63] developed an intergenic spacers sequencing method to determine the biovar-specific spacer pattern in *Y. pestis*. They named this method MST. After comparing the sequences of *Y. pestis* strains CO92 and KIM, primers flanking the intergenic sequences that were conserved in CO92 and KIM homologs were designed to screen 35 strains that are representative of the three *Y. pestis* biovars (11 Antiqua, 12 Medievalis, and 12 Orientalis isolates), which were isolated from 1947 to 1996 from various host species in 13 countries. This method was also used to analyze some ancient DNAs extracted from skeletons, which dated to the Justinian pandemic era, as determined by radiocarbon dating, at least one of which was recorded to be from a patient who died during the pandemic. The data showed that MST differentiates the three biovars of *Y. pestis*, and it supports the view that most *Y. pestis* isolates cluster according to their biovar, while biovar Antiqua isolates are more distantly related to each other than those of biovars Orientalis and Medievalis. They also found that one *Y. pestis* Antiqua isolate, number 611 form Japan, formed a fourth branch, which suggests that *Y. pestis* may comprise four different lineages, instead of the three that have been recognized thus far. A new lineage was confirmed recently as a new biovar, Microtus [38, 53, 96], indicating the need to exploit the diversities of *Y. pestis* by international collaborations on different collection of this deadly bacterium.

An analysis of ancient DNA from human remains yielded the surprising finding that biovar Orientalis, which now occurs worldwide, was involved in all three pandemics. The authors also detected *Y. pestis* in additional human remains from Black Death sites in Southern France and provided more evidence of Orientalis' role in the second pandemic [97, 98]. Combined with the historical record of the epidemiologic pattern of the Black Death, the authors argued that Black Death outbreaks in Northern Europe were not caused by the transmission of *Y. pestis* via blocked rat fleas, but rather by mechanical transmission of plague bacteria by another ectoparasite that used humans as their primary host, or another pathogen may have caused these outbreaks, as evidenced by the failure to detect *Y. pestis* DNA in the dental pulps of suspected plague victims in Copenhagen and Verdun that date from the eighteenth century. The absence of *Y. pestis*-specific DNA from multiple putative European plague burial sites [99], as well as other lines of evidence [100–103], supports this hypothesis. However, Raoult and Drancourt analyzed DNAs extracted from the teeth of a child and two adults from a fourteenth-century grave in France using suicide PCR. They concluded that the medieval Black Death was plague [97], and they confidently defended their results against other reports [99–103] that unless their work is proven to be wrong, or another agent is found, the above speculation that another agent was the cause of the Black Death is unsubstantiated [104].

3.2.6 Variable-Number Tandem Repeat

The polymorphism associated with tandem repeats has been instrumental in mammalian genetics for the construction of genetic maps, and it is still the basis of DNA fingerprinting in forensic applications [105, 106]. Tandem repeats are usually classified among satellites (spanning megabases of DNA, associated with heterochromatin), minisatellites (repeat units in the range of 6–100 bp, spanning hundreds of base pairs), and microsatellites (repeat units in the range 1–5 bp, spanning a few tens of nucleotides).

Recent studies showed that tandem repeat polymorphisms of mini- and microsatellites are likely to be a highly significant source of very informative markers for identifying pathogenic bacteria, even for recently emerged, highly monomorphic species [107–114]. Tandem repeats probably contributes to the pathogen's adaptation to its host, and they also account for bacterial phenotypic variations, depending on their location. For example, if a tandem repeat is located within the regulatory region of a gene, it can function as an on/off switch in gene expression at the transcriptional level [115]. Similarly, if it is located within a coding region and has a repeat unit length of three bases, it may cause antigenic variation for that bacterium, but if the repeat unit length is not a multiple of three, it can result in the production of truncated translation products.

VNTR sequences are common in the *Y. pestis* genome. They occur frequently in both intergene- and gene-coding regions and are distributed evenly throughout the genome and the two virulence plasmids, pCD1 and pMT1 [69]. Tandem-repeat typing may prove to be a powerful complement to the existing phylogenetic tools for studying *Y. pestis*. Adair et al. [70] identified a tetranucleotide repeat sequence, $(CAAA)_N$, in the genome of *Y. pestis*, and they demonstrated that this region has nine alleles and great diversity (calculated as 1 minus the sum of the squared allele frequencies) (diversity value, 0.82) within a set of 35 diverse *Y. pestis* strains.

MLVA was shown to be capable of distinguishing closely related strains and successfully classifying more distant relationships. Pourcel et al. [116] and Klevytska et al. [69] examined representative strains of *Y. pestis* using 25 and 42 VNTR loci, respectively, and vast differences in diversity were observed among these loci. Pourcel et al. [116] grouped 180 *Y. pestis* into 61 different genotypes, and the three biovars were distributed in the three main branches, which demonstrated the heterogeneity of biovar Medievalis. They proposed that seven selected markers could be used to quickly characterize a new strain. Compared with other genotyping methods, MLVA is easy to standardize for establishing databases. This is useful because *Y. pestis* is one of the most dangerous bioterrorism agents, which prevents the international exchange of bacterial strains and collaborative studies. This creates a barrier for researchers who are trying to better understand plague ecology and evolution around the world. Thus, it is a good idea to use a web-based comparison for identifying VNTR genotypes of *Y. pestis*.

Le Flèche et al. [115] created a database (http://minisatellites.u-psud.fr) of tandem repeats for pathogenic bacteria that is based on publicly available bacterial genomes, and they illustrated its application by characterizing minisatellites from two important human pathogens, *Y. pestis* and *Bacillus anthracis*. They found that *Y. pestis* contains 64 minisatellites with repeat units at least 9 bp long in which the unit is repeated at least seven times.

Then, Denoeud et al. [117] presented an Internet-based resource to help develop and perform tandem repeat-based bacterial strain typing. The tools are accessible through the above URL. There are four parts in the web page: The "Tandem Repeats Database" enables the identification of tandem repeats across entire genomes. The "Strain Comparison Page" identifies tandem repeats that differ between different genome sequences from the same species. The "Blast in the Tandem Repeats Database" facilitates the search for a known tandem repeat and the prediction of amplification product sizes. The "Bacterial Genotyping Page" is a service for strain identification at the subspecies level.

Limited by the availability of samples, the aforementioned MLVA genotyping studies mostly focused on biovar Orientalis *Y. pestis* strains. Li et al. [39] collected 383*Y. pestis* strains that were isolated from the majority of the known foci in China, the FSU, Mongolia, and a number of other foci around the world, which covered *Y. pestis* strains in all four biovars, and included rhamnose-positive strains. The 25 loci-based MLVA typing system developed by Pourcel et al. [116] was used to resolve a combined strain set that contained more than 500 strains, including previously published data for 180 strains, into 350 different genotypes [39]. The biovar Microtus strains were grouped together and were closely related with pestoides strains isolated from the FSU, and both these groups were located at the ancestral branches in the dendrogram based on the 25 MLVA loci, which agrees with estimations from previous studies [28, 45, 118]. Strains of the other three biovars were classified into two main branches, one of which contained Antiqua isolates from Africa and Northwest China, as well as the entire biovar Orientalis group, while the other contained the biovar Medievalis isolates and biovar Antiqua isolates from Mongolia, Northeast China, and Tibet. The two main branches were proven to correspond with branches 1 and 2 that were defined in subsequent research based on genome-wide single-nucleotide polymorphisms (SNPs) [119, 120]. The clustering results suggest a very close relationship between the strains in the corresponding foci, and they provide a database for source tracing *Y. pestis*.

The high mutation rates of VNTR loci increase the possibility of reversions and convergent mutations, which may lead to homoplasies and bias in phylogenetic reconstructions [121]. Additionally, although adding VNTR loci into an MLVA typing scheme would increase the discriminative power, using too many loci would largely increase the operation time and workload, which will not satisfy the need for rapid genotyping and source tracing when facing natural plague outbreaks or bioterrorism attacks. Therefore, Li et al. [122] further investigated the characteristics of VNTRs in five *Y. pestis* genomes, and they probed the relationship between the positions of VNTRs and their diversities in 97 representative *Y. pestis* strains. Then, according to the known SNP-based phylogeny of *Y. pestis*, Li et al. developed a simple hierarchical genotyping scheme based on optimized "14+12" VNTR loci, which provided high discriminative power and could classify the *Y. pestis* strains into subpopulations, as well as substantially reducing the operational time and cost. When a large number of strains need to be genotyped, a 14-VNTR loci analysis could be used to initially designate isolates into major populations, and then 1–5 additional loci could be used to classify them into subpopulations. Notably, if only a limited number of strains need to be genotyped, it would be more convenient to determine the polymorphism of all 26 VNTRs simultaneously before performing a hierarchical analysis. This "14+12" hierarchical scheme was used to determine the genotypes of 956 *Y. pestis* strains, and the generated population structure was consistent with a phylogeny based on a whole-genome SNP analysis [119, 120]. Therefore, this newly developed genotyping scheme would facilitate not only the global source tracing of future plague outbreaks but also microevolution studies of this important pathogen.

3.2.7 Clustered Regularly Interspaced Short Palindromic Repeat-Based Typing

The CRISPR/CRISPER-associated (Cas) system has become particularly well known since 2013 because of its application in multiplex genome engineering through RNA-guided, site-specific DNA cleavage [123–125]. CRISPR is a special type of repeat element that consists of noncontiguous direct repeat sequences (DR, 24–47 bp) separated by similarly sized unique sequences, which are widely distributed in bacterial and archaeal genomes [126, 127]. CRISPRs, Cas proteins, and leader sequences (the noncoding sequences that flank the CRISPRs on one side) constitute a prokaryotic immune system against bacteriophage attack via an RNA interference-like mechanism [128–132]. The DR sequences of CRISPRs are usually conserved within one species [133], while the unique parts, also called spacers/spacer arrays, seem to be one of the most rapidly diversified regions in prokaryotic genomes. Even in the same species, different spacer arrays often exist in different isolates [45, 134, 135], making them a powerful tool for genotyping and microevolutionary studies of prokaryotes. The first genotyping application that was based on CRISPR polymorphisms was the "spoligotyping" that is widely used in *Mycobacterium tuberculosis* [136]. Subsequently, CRISPR diversity-based genotyping and evolutionary studies were employed in many pathogen species, including *Streptococcus thermophilus*, *Salmonella enterica* serovar Virchow, *Salmonella enterica* subsp. *enterica*, *Campylobacter jejuni*, and *Y. pestis* [45, 129, 137–139].

In *Y. pestis*, three types of CRISPR loci were identified in the chromosome, which are designated as YPa, YPb, and YPc [134, 140]. Cui et al. determined the sequences of all three CRISPR loci from a collection of 125 *Y. pestis* strains that were isolated from 26 natural plague foci of China, the FSU, and Mongolia [45]. One hundred and thirty-one spacers were identified, and 77 (59 %) of them have homologous sequences in a prophage (YPO2096–2135 in the CO92 genome), whereas 22 spacers (17 %) are homologous to other nonviral regions in the *Y. pestis* chromosome. There are 83, 37, and 11 spacers in the YPa, YPb, and YPc loci, respectively, which suggests that the CRISPR loci have different degrees of activity and that the YPa locus is the most diverse. The spacer sizes range from 29 to 34 bp, and the majority are 32 bp (79 %). The longest CRISPRs in *Y. pestis* contain 14 spacers, while some contain only one spacer. Interestingly, in an Angola strain (which belongs to lineage 0.PE3), there is only one truncated DR, and the leader sequence was left in YPb and YPc, and no spacers were observed [45, 140]. The spacer was designated as a letter plus a number. The letter (a, b, or c) indicated the CRIRSPR locus that carried the spacer, and the number indicated a specific sequence that could be found from the predefined spacer dictionary [45]. Multiple spacers composed the spacer array in a CRISPR locus, which were represented as a letter plus a series number. Such as the spacer array "a-1-3-6-7," which means four spacers, a1, a3, a6, and a7, is arranged in order within the CRISPR locus YPa.

The diversity of CRISPRs provides fine discriminative power for genotyping *Y. pestis*. By combining spacer arrays of the three loci, 49 genotypes were identified in 125 *Y. pestis* strains. The first part of the spacer arrays (farside of the leader sequence) was conserved, with "a-1-2-3-4-5-6" in YPa, "b-1-2-3-4" in YPb, and "c-1-2-3" in YPc, and these conserved spacers were named species-specific spacers, while the spacers closer to the leader sequence often revealed geographical clustered characteristics, i.e., specific spacers were always found in isolates from the same natural plague focus. These spacers were subsequently called region-specific spacers (RSSs). Given the good correlation between spacer arrays and the geographical distribution of isolates, all of the studies' strains can be conveniently grouped into 12 clusters, designated by adding a prefix "C" to the name of a representative spacer (most of which are RSSs). Using this classification, most natural plague foci have one main cluster, except the one to the east of the Kunlun Mountains [45]. CRISPR-based typing provides a new solution for differentiating previously undistinguishable strains. For example, foci L and M are two geographically distinct natural plague foci in China. The phenotypic and biochemical features of the *Y. pestis* isolates from these two foci are almost identical, and they cannot be distinguished by conventional phenotyping and genotyping methods [96, 118]. There are approximately 120-bp differences (the length of the two spacers, as well as the DR sequences) in the YPc locus between the strains from these two foci, which are easily identified by a PCR–gel electrophoresis method [45].

The molecular typing method based on polymorphism of CRISPRs was also applied in determining the diversity of *Y. pestis* strains in the FSU and Mongolia. By analyzing three CRISPR loci on 59 rhamnose-positive *Y. pestis* strains, Platonov et al. identify more Mongolia isolates belonging to biovar qinghaiensis, which provide further evidence that ancient *Y. pestis* populations spread between China and Mongolia [44]. To comprehensively explore the diversity of *Y. pestis* strains in Mongolia, which harbors numerous highly active plague foci, Riehm et al. selected 100 strains from 37 areas and analyzed them by a combination of MLVA and CRISPR methods [141]. The study revealed that six main clusters, including three Microtus biovars, Ulegeica, Altaica, and Xilingolensis, were present in the samples and that they exhibited a distinct geographically specific distribution. The largest cluster comprises 78 strains, all belonging to the lineage 3. ANT, and it was the dominant *Y. pestis* population in Mongolia, but was rarely isolated elsewhere. The CRISPR analysis revealed that seven spacers, as well as six spacers that were present in all 14 different CRISPR genotypes, were newly observed. Riehm et al. presented a more reasonable nomenclature method to term each CRISPR genotype, which could reflect the diversity of all three CRISPR loci. First, all observed spacer arrays were given a number as a code, e.g., "a-1-2-3-4-5-6-7" was coded as "1" and "a-1-2-3" was coded as "2." Then, the CRISPR genotype was presented in a three-digit code, which represented the diversity of the YPa, YPb, and YPc loci, e.g., the code "1-1-1" represented that in the isolates, spacer array "a-1-2-3-4-5-6-7" was observed at YPa, "b-1-2-3-4-48" at YPb, and "c-1-2-3" at YPc. Notably, a lack of amplification for one locus was coded X, such as "1-X-1."

As increasing numbers of whole-genome sequences have been deciphered recently, the requirement to analyze the diversity of CRISPR loci in genome sequences has increased. Three online software programs at a CRISPR website (http://crispr.u-psud.fr/) could be used to perform such analyses. One of them is CRISPRFinder, which could identify the location and nucleotide sequence of known or new CRISPR loci. The other is CRISPRcompar, which could compare the variation in CRISPR loci of different isolates. The third is CRISPRtionary, which could be used to deposit and retrieve information pertaining to the spacers. All of these tools would substantially facilitate further genotyping or evolutionary analyses based on CRISPR polymorphisms [140, 142, 143].

3.2.8 Pseudogene Profiles

It is becoming increasingly clear that bacterial genomes constantly undergo structural changes due to horizontal gene acquisition, gene loss, intra-genome recombination, and mutational events [144, 145]. Investigations of genetic diversity based on these mechanisms will lead to a complete picture of the genomic diversity of *Y. pestis* during its focus expansion. In our laboratory, we used a pseudogene profile to investigate mutational events during the microevolution of *Y. pestis*, while a DNA rearrangement profile was used to analyze intra-genome recombination, and whole-genome microarray hybridization was used to assess horizontal gene acquisition and gene loss. These data provided us an opportunity to analyze the genetic diversity of *Y. pestis* in China during its focus expansion, and they revealed the relationship between genome structural changes, the mutation of key metabolic genes, and microevolution.

Tong et al. analyzed the pseudogene profile of *Y. pestis* based on three kinds of gene inactivation mechanisms: nonsense mutations, frameshift mutations, and disruption by IS elements. A total of 24 mutations in 24 genes, including 17 mutations that resulted from an IS element, two frameshift mutations, one in-frame deletion, one fragment deletion, two nonsense mutations, and one nonsynonymous mutation, were screened by PCR in 260 *Y. pestis* strains that were isolated from different foci in China, as well as in seven strains of *Y. pseudotuberculosis*. It was found that *Y. pestis* isolates from the same focus had an identical mutation profile, and the 260 *Y. pestis* isolates were divided into eight genotypes, while *Y. pseudotuberculosis* harbored wild-type alleles for all mutations. The isolates of the three known biovars were grouped into distinct branches in the phylogenetic tree, which supports the proposition that biovars Medievalis and Orientalis arose directly from biovar

Antiqua. The constructed phylogenetic tree suggests that isolates from focus B (the *M. baibacina–S. undulatus* plague focus of the Tian Shan Mountains) should be the oldest lineage of *Y. pestis* in China, except for isolates from foci L (the *M. brandti* plague focus of the Xilingol Grassland) and M (the *M. fuscus* plague focus of the Qinghai–Tibet Plateau), which might be a special lineage of *Y. pestis* with a unique origin.

3.2.9 Gene Loss and Acquisition Profiles

Bacteria can evolve via gene loss and acquisition to adapt to new niches. These events can be analyzed using whole-genome sequencing and subtractive hybridization, but they are too expensive and laborious to be applied to large numbers of bacterial strains. A whole-genome-based DNA microarray is an ideal technique for analyzing genomic fragment differences among large numbers of bacterial strains. Zhou et al. employed a whole-genome DNA microarray, comprising 4005 open reading frames (genes) from *Y. pestis* strains 91001 and CO92, to screen for differences among 43 strains of *Y. pestis* and seven strains of *Y. pseudotuberculosis* [96]. They found 29 genomic regions (termed different regions, DFRs) among these strains. Then, one or more genes were chosen from each DFR to represent the corresponding DFR, and PCR amplification of the selected genes was performed on 260 isolates of *Y. pestis*, which were obtained from different natural plague foci, to screen the distribution of DFRs in these strains. The DFR profiles of these isolates could be assigned into 14 genomovars. More recently, Dai et al. found a new DFR, DFR23, through suppression subtractive hybridization technology [146]. Li et al. used these 23 DFRs to genotype 909 *Y. pestis* that were isolated from 15 plague foci from 1943 to 2005, which presumably represented the most abundant diversity of *Y. pestis* strains in China. All strains were grouped into 32 genomovars. Most of the genomovars fell into three clusters, namely, A, B, and C, except for genomovars 01b and genomovars 04. All Orientalis strains (205 strains in three genomovars) were grouped in cluster A. All Medievalis strains (122 strains in eight genomovars) were grouped in cluster B, and Microtus strains (66 strains in three genomovars) were grouped in cluster C. This clearly illustrates the close relationships among strains of the same biovar in China. However, Antiqua strains (516 strains in 18 genomovars) were distributed in different branches of all three clusters, which revealed the considerable genomic diversities of Antiqua strains. Platonov et al. determined the genomovar of 275 *Y. pestis* strains from natural plague foci of the FSU and Mongolia based on the same 23 DFRs [147]. Interestingly, 64 novel genomovars were detected, which implied that the *Y. pestis* population in these foci occupied wider polymorphism in respect of gene gain/loss than the foci in China. The reason for the highly DFR diversity in the FSU isolates is worth to be further explored. The result also indicated that more *Y. pestis* isolates worldwide should be determined to obtain the comprehensive DFR profiles of the species.

The DFRs represent dynamic regions of the *Y. pestis* genome in natural populations, which suggests that they result from acquisition or deletion events during the adaptive evolution of *Y. pestis*. The parallel microevolution of the *Y. pestis* genome in natural populations was outlined by evidence of genome content flux through the loss/acquisition of plasmids and chromosomal segments. Biovar Orientalis evolved from biovar Antiqua by acquiring a genomic island, rather than arising from biovar Medievalis, which is in agreement with previous studies [34, 35]. Gene acquisition in this case promotes the dramatic evolutionary segregation within species, i.e., the emergence of biovar Orientalis strains. Opposite to the selective genome expansion by horizontal gene acquisition, genome reduction occurs through the loss of DFRs. Gene loss seems to be dominant in the continued microevolution after the rapid speciation of *Y. pestis*. The lost genes are always nonessential for bacterial survival (and, thus, unable to provide a selective benefit for bacterial growth or fitness in the host) in a specific host niche; i.e., if genes are rendered useless because of redundancy within the host niche, then loss bias occurs. The parallel loss of DFRs leads to the discrete segregation of progenitor and offspring strains. This genome reduction gradually enables the offspring strains to reside in a more host-specific niche (see below), without overlapping the niches of its progenitor strains.

The microevolution of genomovars is perfectly consistent with the expansion of plague foci. Each genomovar is confined to a specific geographic region, namely, a plague focus or part of a focus with a unique combination of natural environment, host reservoirs, and vectors. Most of the geographic regions with different primary reservoirs have unique genomovars. Sometimes, there is more than one genomovar in a single focus within a single primary reservoir, but each of the genomovars corresponds to a unique combination of natural environment and primary vector(s).

The long-term persistence of *Y. pestis* in nature is accompanied by its interactions with animal reservoirs and flea vectors. *Y. pestis* is pathogenic to its rodent reservoirs, while infections are limited to the entire population of the reservoir. Even during animal plague epidemics, the rodent population does not decline. There is a relative balance between *Y. pestis* and its reservoirs and vectors in natural foci. However, the specific natural environment in a defined geographic region (a plague focus or a part of the focus) will ultimately determine the food chain-based relationship among *Y. pestis* and its reservoirs and vectors. The unique set of hosts (reservoirs and vectors) in the specific natural environment and their interactions with *Y. pestis*, termed a host niche, determines the persistence and the type of *Y. pestis*.

It is logical to say that the origin of the ancestral *Y. pestis* strain is associated with only one kind of rodent but that it has the potential to infect other species of animals. Once the bacteria are exposed to new animals in new geographic regions by animal contacts or vector-borne routes, new host niches will come into being gradually. During this time, genetic variations, such as gene acquisition, gene loss, point mutations, and genome rearrangements, occur randomly in the *Y. pestis* genome. When *Y. pestis* temporarily colonizes a new host niche, the specific host niche directs natural selection, leading to the stabilization and vertical inheritance of beneficial genetic variations in the genome of the colonizing *Y. pestis* strain, which may be defined as directional microevolution of the genome. A specific host niche determines not only the long-term persistence of *Y. pestis* but also the genomovar of *Y. pestis* itself, i.e., the expansion of plague foci is part of the stepwise adaptation of *Y. pestis* to new host niches.

Each host niche in different natural plague foci provides a unique environment in which natural selection occurs, which thereby directs the parallel adaptation of *Y. pestis*. This led *Y. pestis* to evolve from a newly emerged single-host species to a multi-host pathogen. In other words, the parallel adaptation to various niches drives *Y. pestis* strains to diversify into different biovars or genomovars. Specific genomovars are limited to certain geographic regions (or host niches) through a course of within-species segregation, or the so-called pathogen specialization, which has the advantage of avoiding overlapping niches.

3.2.10 Genome-Wide SNP-Based Typing

SNPs have been increasingly used to genotype bacterial pathogens in recent years. Compared with other types of molecular targets, such as VNTRs, CRISPRs, and DFRs, SNPs reveal relatively low mutation rates and, hence, could provide more stable phylogenetic relationships across samples [121]. Many pathogens, such as *Y. pestis*, *S. enterica* serovar Typhi, *B. anthracis*, and *Burkholderia mallei*, are genetically monomorphic and harbor only a limited number of DNA variations. For example, in a globally representative collection of *S. enterica* Typhi strains, only 82 SNPs were identified within an 80-kb genomic fragment [148], and for *Y. pestis*, no SNPs were found in six genes with a total size of 2509 bp [27]. Therefore, to increase the discriminative power of an SNP-based typing method, longer genome regions need to be explored to find more SNPs, which serve as typing targets. Comparisons among whole-genome sequences would provide the most diverse information and the highest resolution power. Therefore, given decreasing sequencing costs, whole-genome-wide SNP analysis has become the new gold standard in genotyping and evolution analyses.

Prior to 2004, three complete genome sequences of *Y. pestis* strains CO92, KIM, and 91001 were released [53, 82, 83]. Achtman et al. compared 3250 orthologous coding sequences among these three genomes, and they found only 76 synonymous SNPs (sSNPs), which demonstrate the highly conservative nature of *Y. pestis* [28]. The ancestral nucleotide status of the sSNPs was defined as being identical to that of the *Y. pseudotuberculosis* genome (IP32953), while the statuses in the *Y. pestis* genomes were defined as accumulated mutations since the speciation of *Y. pestis* from *Y. pseudotuberculosis*. A three-branch phylogeny was deduced according to these sSNPs. One branch that contained strain 91001 (branch 0, branch length of 27 sSNPs) split off first, and then after thousands of years of evolution, the branches containing strain CO92 and KIM split into branch 1 (20 sSNPs) and branch 2 (15 sSNPs), respectively. As molecular markers to distinguish the three branches of *Y. pestis*, 44 sSNPs were selected to screen 105 *Y. pestis* strains by dHPLC. Eight phylogroups were identified, and to facilitate the transition between traditional designations based on biochemical phenotypes and phylogenetic relationships based on molecular markers, Achtman et al. designed a nomenclature system for *Y. pestis* that combined molecular relatedness and the traditional biovar designations. The populations were designated according to their branch plus an abbreviation for the biovar (abbreviations, ANT, Antiqua; PE, Pestoides (including Microtus isolates); MED, Medievalis; ORI, Orientalis; IN, intermediate strains between 1.ANT and 1.ORI) followed by a number of sequential populations, e.g., the group of bacteria related to Orientalis is referred to as 1.ORI to reflect the association of the Orientalis phenotype with branch 1, and almost all pestoides isolates were placed into groups 0.PE1, 0.PE2, 0.PE3, and 0.PE4 in branch 0.

Achtman et al. noted that biovar Antiqua strains were grouped into distinct groups that were designated as 1.ANT and 2.ANT, which were located on branches 1 and 2, respectively, although both strains can ferment glycerol and reduce nitrate. Similarly, the biovar Medievalis strain that can reduce nitrates is found among groups 2.MED, 2.ANT, 0.PE1, and 0.PE4, probably because of multiple, independent molecular events. Finally, the designation pestoides for organisms that can ferment melibiose and rhamnose combines a variety of diverse organisms from groups 0.PE1, 0.PE2, and 0.PE3. Therefore, they noted that the molecular populations are not fully compatible with the classical phenotypic categories that are used to designate biovars, and biovars are not necessarily monophyletic and should not be used for evolutionary or taxonomic purposes. Achtman et al. also noted that phylogroups were not necessarily a reliable indicator of phenotype, e.g., one 2.MED isolate (pestoides J) was unable to ferment glycerol, and 1.ORI includes one isolate (Nicholisk 51) that can ferment glycerol. Similarly, some 2.ANT isolates can reduce nitrate, whereas others cannot.

As more genome sequences were released, Morelli et al. [119] compared 17 *Y. pestis* genomes and identified 1364 SNPs from coding regions of their core genome (genome fragments that were present in all of the analyzed samples). By performing SNP discovery in up to a 185-kb genome region for other strain collections (99 from China, 90 from the USA, 82 from Madagascar, and 99 from other regions) via dHPLC, 141 SNPs were found. Sequenom MassARRAY SNP typing was performed on whole genomes that were amplified from the DNA of 288 global isolates. The states of 933 selected SNPs were used to construct a species phylogenetic tree by the minimal spanning method. The population structure was consistent with previous research, and it was composed of three main branches, except for a new population, 3.ANT, which was located at the end of branch 0 and may represent a fourth branch (branch 3, which was subsequently verified by whole-genome sequencing).

More new subpopulations were observed in Morelli et al.'s research, and interestingly, the subpopulations revealed a pattern of strong geographical clustering. Populations in branch 0 were mostly isolated from China (0.PE4, 0PE7, 0.ANT1, 0.ANT2, and 0.ANT3), except for 0.PE2, which was from the FSU. The 0.PE3 group contains only a single Angola strain with an unknown isolation location, but its name implied an African origin. Branch 1 contains three populations, 1.ANT, 1.IN, and 1.ORI, each of which had three subpopulations. Isolates in 1.ANT were restricted to east and central Africa. The next population on branch 1, 1.IN (termed because it was intermediate between biovars Orientalis and Antiqua in the phylogenetic tree), was from the Qing–Tibet Plateau in China. The youngest population in branch 1, 1.ORI, spread globally from Hong Kong, China, during the third plague pandemic. The subpopulation 1.ORI1 mostly appears in India and the USA now. 1.ORI2 experienced multiple, independent dissemination events that enabled it to reach Europe, South America, Africa, and Southeast Asia, and now it is mostly present in Southwest China and adjacent countries. 1.ORI3 spread to Madagascar and then Turkey, and it was still active in Madagascar in recent years. Branch 2 split into two populations, 2.ANT and 2.MED. All isolates in 2.ANT2 and 2.ANT3 were from China, whereas the two strains that belonged to 2.ANT1 were isolated in neighboring Nepal. Similarly, all 2.MED2 and 2.MED3 isolates were from China, whereas 2.MED1 strains were from Kurdistan in Central Asia. As strains isolated from China are scattered in multiple populations over the entire phylogenetic tree, Morelli et al. inferred that *Y. pestis* might have originated and evolved from China and then spread to other regions on multiple occasions [119].

Before the widespread use of next-generation sequencing technology in bacterial population genetic studies, most of the typing analyses were based on variations that arose by comparing a few of the strains for which whole-genome sequences were available, followed by screening for more isolates to determine the states of these variations in different loci. Such methods generate phylogenetic discovery bias [149], which can group divergent strains together, particularly if the two genomic sequences are from closely related strains [150]. Therefore, the real phylogenetic distances across samples would be ignored, thereby reducing the discriminative power in typing and evolution research. To better decipher the genetic diversity and historical evolution pattern of *Y. pestis* while avoiding the influence of phylogenetic discovery bias, Cui et al. [120] selected 118 *Y. pestis* strains that were isolated from China and Mongolia. They combined them with 15 published genome sequences of strains that were isolated globally to identify whole-genome-wide SNPs for each sample and then rebuilt the phylogenetic tree. This study differentiated all 133 strains into 132 genotypes based on 2298 high-quality SNPs. Only two strains that were isolated within 1 month of each other in the same local area were indistinguishable. Two novel branches, branches 3 and 4, which are as old as branches 1 and 2, were identified. The five-branch population structure was finally defined and used as the standard in other typing or evolutionary studies of *Y. pestis* [122, 151–153].

The sequenced isolates cover a wide geographic area of China, and they originated from a variety of hosts, providing a good chance to assess the extent of host specificity and geographic clustering of *Y. pestis* populations. Subpopulations were clearly not randomly distributed among hosts (Fig. 3.1a; nonparametricχ^2 test, 64.61; 999 permutations; $p=0.001$). Subpopulations were also clearly geographically clustered (Fig. 3.1b; multiple analysis of variance, Pillai–Bartlett trace = 1.75, $p<2.2\times10^{-16}$). Because these two factors are largely confounded statistically, it was impossible to evaluate the relative importance of geography and ecology in shaping the population structuring of *Y. pestis* in China.

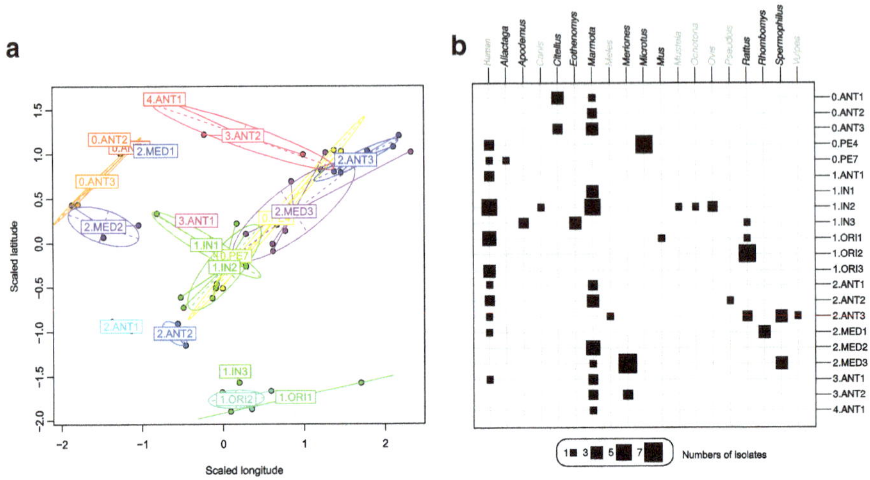

Fig. 3.1 Spatial and host distribution of the sampled *Y. pestis* isolates. (**a**). Spatial distribution of the sources of the isolates according to longitude and latitude that were scaled to a mean value of zero and standard deviation of 1.0 to account for differences between latitude and longitude. Ellipses encompass 95 % of the dispersion (inertia) of each group, and the geographic sources of singletons are indicated by points. (**b**). Numbers of isolates per bacterial population and host genus. Non-rodent mammalian hosts are indicated in a gray font

Cui et al. found that the deepest branching subpopulation, 0.PE7, was isolated in the Qinghai–Tibet Plateau, China, and that this region encompasses the most diverse isolates; therefore, it may be the original source of *Y. pestis*. In the western part of China, the geographical sources of many isolates overlap with, or are near, areas that were historically associated with subroutes I and II of the Silk Road, which were used over a 2000-year period for trading with Central Asia and Europe [154]. Other isolates overlap with the Tea Horse Road (also referred to as the Southwest Silk Road) [155], which was the primary trade route for over 1300 years to South Asia, including India and Nepal (subroute III), and Southeast Asia, including Vietnam and Myanmar (subroute IV). The spread of plague from China to other areas probably occurred via these four routes, e.g., 2.MED2 was isolated along subroute I, which probably reflects the manner of its spread to Western Asia [156]. Similarly, 2.ANT1 may have spread to Nepal along subroute III, because 2.ANT1 was isolated near that route at the Nepalese border. Finally, the third pandemic that spread globally from Hong Kong in 1894 was caused by 1.ORI and followed epidemic outbreaks in Yunnan province [157]. The predecessor of 1.ORI (1.IN2) and one of the lineages within 1.ORI (1.ORI2) were both isolated in Yunnan province in the vicinity of the Tea Horse Road (subroute IV).

3.3 Outer Membrane Protein Profiles

The outer membrane proteins of 38 *Y. pestis* isolates from all known plague foci of northeast Brazil were analyzed by SDS-PAGE [158]. Approximately 20 bands were consistently found in all strains analyzed, and 11 were selected for comparative studies. Although qualitative differences among the electrophoretic profiles of the outer membrane proteins of the *Y. pestis* isolates were not observed, quantitative alterations were clearly noted for most of these proteins. No particular quantitative alteration of the outer membrane protein electrophoretic profile could be associated with the period of isolation and geographic origin of the isolates. A 64-kDa outer membrane protein was expressed at significantly higher levels in *Y. pestis* strains that were isolated from a recent plague outbreak. The possible use of outer membrane protein electrophoretic profiles of *Y. pestis* isolates as a tool for epidemiological studies and for the analysis of virulence determinants is discussed below.

In a recent study, three techniques for obtaining outer membrane-enriched fractions from *Y. pestis* were evaluated [159]. The techniques analyzed were differential solubilization of the cytoplasmic membrane with Sarkosyl or Triton X-100 and centrifugation in sucrose density gradients. An SDS-PAGE analysis of an outer membrane isolated by the different methods resulted in similar protein patterns. The measurement of NADH dehydrogenase and succinate dehydrogenase (which are inner membrane enzymes) indicated that the outer membrane preparations obtained by the three methods were pure enough for analytical studies. In addition, preliminary evidence of the potential applicability of outer membrane proteins for identifying *Y. pestis* geographic variants was presented.

3.4 Fatty Acid-Based Typing

More than 300 fatty acids and related compounds have been found in bacteria, and the wealth of information contained in these compounds can be estimated not only by considering the presence or absence of each acid but also by quantitatively analyzing the data. An automatic gas chromatographic system, the Sherlock Microbial Identification System (MIS), has been developed to analyze bacterial fatty acids that are 9–20 carbons in length. The peaks are automatically named and quantitated by the system. Branched-chain acids predominate in some Gram-positive bacteria, while short-chain hydroxy acids often characterize the lipopolysaccharides of Gram-negative bacteria. The fatty acid profiles can be used to identify and trace bacteria.

Song et al. [160] evaluated the possibility of using cellular fatty acid (CFA) information to type *Y. pestis*. They used 58 strains, representing different ecotypes, which were isolated in different regions in China. Major CFAs of the tested strains are 16:0, cyclo 17:0, 3OH-14:0, ω7c 16:1, and monounsaturated 18-carbon acids, which account for more than 70 % of the total CFAs. A dendrogram of the tested strains revealed high homogeneity among these *Y. pestis* isolates, which suggests that a CFA analysis cannot serve as a powerful tool in typing *Y. pestis* strains. All of the strains used were identified as *Y. pseudotuberculosis* by the Sherlock system, which was caused by CFA differences between the tested strains and Sherlock standard library strains. The Chinese strains of *Y. pestis* lack ω5c 16:1 and have an extra cyclo ω8c 19:1 compared with the database used by the Sherlock system, implying that CFA analyses are not useful for typing *Y. pestis.* Tan et al. used normalized Sherlock MIS and Sherlock standard libraries to analyze the fatty acid composition of different strains of *Y. pestis* and *Y. pseudotuberculosis*. They found that ratios of certain CFA components could serve as chemical markers to differentiate the two closely related bacteria [161]. Specifically, the ratios of 14:0/18:0 vs. 18:1 × 7c/18:0, 3-OH-14:0/18:0 vs. 18:1 × c/18:0, 16:1 × 7c/18:0 vs. 18:1 × 7c/18:0, 12:0/18:0 vs. 18:1 × 7c/18:0, and 12:0 ALDE/18:0 vs. 16:1 × 7c/18:0 fatty acids could be used to differentiate *Y. pestis* from *Y. pseudotuberculosis*.

3.5 Bacterial Mass Fingerprinting

Bacterial mass fingerprinting is a typing method that is based on MALDI-TOF MS. MALDI-TOF MS has been increasingly used in recent years to identify bacteria by determining their protein profiles [162]. The technology is rapid and sensitive, and it can distinguish bacteria at the genus, species, subspecies, or even strain levels [163]. Specimen preparation for MALDI-TOF MS is relatively simple and can be finished within minutes, and the workflow can be standardized for most bacterial species. Liu et al. evaluated the effects of different experimental conditions, including the number of bacterial cells used, and treatment procedures with different solutions, matrix species, and solvents, on mass spectra, and they showed that an optimized protocol could identify different bacterial species, including *Y. pestis*, *E. coli*, *Burkholderia cepacia*, *B. anthracis*, and *S. aureus* [164]. However, only the EV76 *Y. pestis* strain, an avirulent vaccine strain, was used to represent the *Y. pestis* species in this study.

Ayyadurai et al. reassessed the usefulness of MALDI-TOF protein profiling, in combination with a protein profile database, for identifying 39 different *Yersinia* strains representing 12 different *Yersinia* species [165]. Thirteen *Y. pestis* strains, including strains of the three major biovars, Antiqua, Medievalis and Orientalis, were included in the database. The results demonstrated that the 12 species of the *Yersinia* genus could be unambiguously identified by their unique protein profile, and the three biovars could be well distinguished by MALDI-TOF MS. Furthermore, Ayyadurai et al. collected two more *Y. pestis* isolates and 11 more *Y. enterocolitica* isolates and compared them against the database. The results showed that all of the strains could be differentiated to the species level, thereby proving that MALDI-TOF MS could reliably identify *Y. pestis* strains.

3.6 The Future of *Y. pestis* Typing

Molecular typing and microbial DNA fingerprinting were introduced successfully during the 1990s. While contributing to the efficiency of the medical microbiology laboratory and enhancing our insights into the processes of microbial population genetics and pathogenesis of infectious diseases, both technologies will continue to play indispensable roles in tracing the sources of bioterrorism-related agents and in aiding the control of bioagents, as well as their use by bioterrorists, which is the main task of microbial forensics, a newly emerging discipline concerning bioterrorism countermeasures.

As mentioned above, many methods can be used to type *Y. pestis*. Which methods we should employ is a difficult choice because more methods give us more choices, which in turn often makes us more hesitant in our decision making. How we choose a method depends upon the type and aim of the investigations. For normal epidemiological investigations, the principle for choosing a suitable method is that the selected method should be feasible if it can help you solve your specific problem. It should archive the typing data at low cost in terms of consumables, technical expertise, and required equipment. The resulting data should be easily compared between different laboratories. For bioterrorism-related, pathogen tracing investigations, the methods selected should be stable, sensitive, and standardized, and there should be a database for correlated comparisons among different laboratories. The method should be high throughput and easily automated. For the first situation, phenotypic, genetic, and chemical typing methods are the best choice, while for the second situation, genomic and chemical analysis techniques are preferred. Until now, we did not have a standardized, comprehensive database for these techniques, although forensic scientists have worked to achieve this aim. However, many unpredictable difficulties await us in the future, which include the increasing control of pathogen transfers between laboratories, the standardized methodology endorsed by scientists from different countries, database construction and analysis, information sharing and inter-discipline collaborations, etc. Furthermore, we need to overcome political barriers to work together toward our final goal because studies of bioterrorism-related pathogens are often influenced by political policies.

The polymorphism databases for *Y. pestis* are not only used for the above field but also for evolutionary and pathogenesis studies. As shown by our and other laboratories, both genetic and genomic polymorphism analysis methods can be applied in these fields. Polyphasic polymorphism studies of *Y. pestis* will also create a model for future studies of other medically important microorganisms. This will teach us how to organize an international collaboration to construct a database of deadly pathogens.

References

1. Van Loghem JJ. The classification of the plague-bacillus. Antonie Van Leeuwenhoek. 1944;10(1–2):15.
2. Skerman VBD, McGowan V, Sneath PH. Approved lists of bacterial names. Int J Syst Bacteriol. 1980;30:225–420.
3. Bercovier H, Steigerwalt AG, Guiyoule A, Huntley-Carter G, Brenner DJ. Yersinia aldovae (formerly Yersinia enterocolitica-like group X2): a new species of Enterobacteriaceae isolated from aquatic ecosystems. Int J Syst Bacteriol. 1984;34:166–72.
4. Sprague LD, Neubauer H. Yersinia aleksiciae sp. nov. Int J Syst Evol Microbiol. 2005;55(Pt 2):831–5.
5. Wauters G, Janssens M, Steigerwalt AG, Brenner DJ. Yersinia mollaretii sp. nov. and Yersinia bercovieri sp. nov., formerly called Yersinia enterocolitica biogroups 3A and 3B. Int J Syst Bacteriol. 1988;38:424–9.
6. Neubauer H, Aleksic S, Hensel A, Finke EJ, Meyer H. Yersinia enterocolitica 16S rRNA gene types belong to the same genospecies but form three homology groups. Int J Med Microbiol. 2000;290(1):61–4.
7. Hurst MR, Becher SA, Young SD, Nelson TL, Glare TR. Yersinia entomophaga sp. nov., isolated from the New Zealand grass grub Costelytra zealandica. Int J Syst Evol Microbiol. 2011;61(Pt 4):844–9.
8. Ursing J, Brenner DJ, Bercovier H, Fanning GR, Steigerwalt AG, Brault J, Mollaret HH. Yersinia frederiksenii: a new species of Enterobacteriaceae composed of rhamnose-positive strains (formerly called atypical Yersinia enterocolitica or Yersinia enterocolitica-like). Curr Microbiol. 1980;4:213–7.
9. Brenner DJ, Bercovier H, Ursing J, Alonso JM, Steigerwalt AG, Fanning GR, Carter GP, Mollaret HH. Yersinia intermedia: a new species of Enterobacteriaceae composed of rhamnose-positive, melibiose-positive, raffinose-positive strains (formerly called Yersinia enterocolitica or Yersinia enterocolitica-like). Curr Microbiol. 1980;4:207–12.
10. Bercovier H, Ursing J, Brenner DJ, Steigerwalt AG, Fanning GR, Carter GP, Mollaret HH. Yersinia kristensenii: a new species of Enterobacteriaceae composed of sucrose-negative strains (formerly called Yersinia enterocolitica or Yersinia enterocolitica-like). Curr Microbiol. 1980;4:219–24.
11. Merhej V, Adekambi T, Pagnier I, Raoult D, Drancourt M. Yersinia massiliensis sp. nov., isolated from fresh water. Int J Syst Evol Microbiol. 2008;58(Pt 4):779–84.
12. Souza RA, Falcao DP, Falcao JP. Emended description of Yersinia massiliensis. Int J Syst Evol Microbiol. 2011;61(Pt 5):1094–7.
13. Murros-Kontiainen A, Fredriksson-Ahomaa M, Korkeala H, Johansson P, Rahkila R, Bjorkroth J. Yersinia nurmii sp. nov. Int J Syst Evol Microbiol. 2011;61(Pt 10):2368–72.
14. Murros-Kontiainen A, Johansson P, Niskanen T, Fredriksson-Ahomaa M, Korkeala H, Bjorkroth J. Yersinia pekkanenii sp. nov. Int J Syst Evol Microbiol. 2011;61(Pt 10):2363–7.
15. Jensen WI, Owen CR, Jellison WL. Yersinia philomiragia sp. n., a new member of the Pasteurella group of bacteria, naturally pathogenic for the muskrat (Ondatra zibethica). J Bacteriol. 1969;100(3):1237–41.
16. Smith JE, Thal E. A taxonomic study of the genus Pasteurella using a numeral technique. Acta Patholo et Bacteriol Scand. 1965;64:213–23.
17. Williams JE. Proposal to reject the new combination Yersinia pseudotuberculosis subsp. pestis for violation of the first principle of the International Code of Nomenclature of Bacteria. Request for an opinion. Int J Syst Bacteriol. 1984;34:268–9.
18. JUDICIAL OPINION 60: Rejection of the name Yersinia pseudotuberculosis subsp. pestis (van Loghem) Bercovier et al. 1981 and conservation of the name Yersinia pestis (Lehmann and Neumann) van Loghem 1944 for the plague bacillus. Int J Syst Bacteriol 1985;35:540.
19. Bercovier H, Mollaret HH, Alonso JM, Brault J, Fanning GR, Steigerwalt AG, Brenner DJ. Intra- and interspecies relatedness of Yersinia pestis by DNA hybridization and its relationship to Yersinia pseudotuberculosis. Curr Microbiol. 1980;4:225–9.

20. Wayne LG. Actions of the Judicial Commission of the International Committee on Systematic Bacteriology on requests for opinions published in 1983 and 1984. Int J Syst Bacteriol. 1986;1986(36):357–8.
21. Garrity GM, Labeda DP, Oren A. Judicial commission of the international committee on systematics of prokaryotes. XIIth international (IUMS) congress of bacteriology and applied microbiology. Minutes of the meetings, 3, 4 and 6 august 2008, Istanbul, Turkey. Int J Syst Evol Microbiol. 2011;61:2775–80.
22. Aleksic S, Steigerwalt AG, Bockemuhl J, Huntley-Carter GP, Brenner DJ. Yersinia rohdei sp. nov. isolated from human and dog feces and surface water. Int J Syst Bacteriol. 1987;37:327–32.
23. Gelev I, Khvoinev A, Dimitrov K, Strashimirova N. Hemorrhagic septicemia in rainbow trout due to Yersinia ruckeri sp. nov. Veterinarno-meditsinski nauki. 1984;21(6):84–90.
24. Ewing WH, Ross AJ, Brenner DJ, Fanning GR. Yersinia ruckeri sp. nov., the redmouth (RM) bacterium. Int J Syst Evol Microbiol. 1978;28:37–44.
25. Sprague LD, Scholz HC, Amann S, Busse HJ, Neubauer H. Yersinia similis sp. nov. Int J Syst Evol Microbiol. 2008;58(Pt 4):952–8.
26. Savin C, Martin L, Bouchier C, Filali S, Chenau J, Zhou Z, Becher F, Fukushima H, Thomson NR, Scholz HC, et al. The Yersinia pseudotuberculosis complex: characterization and delineation of a new species, Yersinia wautersii. Int J Med Microbiol. 2014;304(3-4):452–63.
27. Achtman M, Zurth K, Morelli G, Torrea G, Guiyoule A, Carniel E. *Yersinia pestis*, the cause of plague, is a recently emerged clone of *Yersinia pseudotuberculosis*. Proc Natl Acad Sci U S A. 1999;96(24):14043–8.
28. Achtman M, Morelli G, Zhu P, Wirth T, Diehl I, Kusecek B, Vogler AJ, Wagner DM, Allender CJ, Easterday WR, et al. Microevolution and history of the plague bacillus, *Yersinia pestis*. Proc Natl Acad Sci U S A. 2004;101(51):17837–42.
29. Perry RD, Fetherston JD. *Yersinia pestis*–etiologic agent of plague. Clin Microbiol Rev. 1997;10(1):35–66.
30. Judicial-Commission: JUDICIAL OPINION 60: Rejection of the name Yersinia pseudotuberculosis subsp. pestis (van Loghem) Bercovier et al. 1981 and conservation of the name Yersinia pestis (Lehmann and Neumann) van Loghem 1944 for the plague bacillus. Int J Syst Bacteriol. 1985;35:540.
31. Anisimov AP, Dyatlov IA. A novel mechanism of antibiotic resistance in plague. J Med Microbiol. 1997;46(10):887–9.
32. Dennis DT, Hughes JM. Multidrug resistance in plague. N Engl J Med. 1997;337(10):702–4.
33. Anisimov AP, Lindler LE, Pier GB. Intraspecific diversity of *Yersinia pestis*. Clin Microbiol Rev. 2004;17(2):434–64.
34. Ji S, Zhang H, Liu Y, He Y, Huang J, Song Y, Zhu J, et al. A study on typing of *Yersinia pestis* in China and its ecologioepidmiologica significance (in Chinese). Chi J Epidemiol. 1990;11(suppl):60–6.
35. Ji S, He J, Teng Y, Zhan X, Lei C, Wang W. The discovery and research of plague natural foci in China (in Chinese). Chi J Epidemiol. 1990;11(Suppl):1–41.
36. Devignat R. Varieties of Pasteurella pestis; new hypothesis. Bull World Health Organ. 1951;4(2):247–63.
37. Mollaret HH, Mollaret C. Melibiose fermentation in the genus Yersinia and its importance in the diagnosis of the varieties of Y. pestis. Bull Soc Pathol Exot Filiales. 1965;58(2):154–6.
38. Zhou D, Tong Z, Song Y, Han Y, Pei D, Pang X, Zhai J, Li M, Cui B, Qi Z, et al. Genetics of metabolic variations between *Yersinia pestis* biovars and the proposal of a new biovar, microtus. J Bacteriol. 2004;186(15):5147–52.
39. Li Y, Cui Y, Hauck Y, Platonov ME, Dai E, Song Y, Guo Z, Pourcel C, Dentovskaya SV, Anisimov AP, et al. Genotyping and phylogenetic analysis of *Yersinia pestis* by MLVA: insights into the worldwide expansion of Central Asia plague foci. PLoS One. 2009;4(6):e6000.
40. Motin VL, Georgescu AM, Elliott JM, Hu P, Worsham PL, Ott LL, Slezak TR, Sokhansanj BA, Regala WM, Brubaker RR, et al. Genetic variability of Yersinia pestis isolates as pre-

dicted by PCR-based IS100 genotyping and analysis of structural genes encoding glycerol-3-phosphate dehydrogenase (glpD). J Bacteriol. 2002;184(4):1019–27.
41. Hai R, Yu DZ, Wei JC, Xia LX, Shi XM, Zhang ZK, Zhang EM. Molecular biological characteristics and genetic significance of Yersinia pestis in China. Zhonghua Liu Xing Bing Xue Za Zhi. 2004;25(6):509–13.
42. Zhou D, Han Y, Song Y, Huang P, Yang R. Comparative and evolutionary genomics of Yersinia pestis. Microbes Infect. 2004;6(13):1226–34.
43. Golubov A, Neubauer H, Nolting C, Heesemann J, Rakin A. Structural organization of the pFra virulence-associated plasmid of rhamnose-positive Yersinia pestis. Infect Immun. 2004;72(10):5613–21.
44. Platonova ME, Evseevaa VV, Efremenkob DV, Afanas'evc MV, Verzhutskic DV, Kuznetsovab IV, Shestopalovc MY, Dentovskayaa SV, Kulichenkob AN, Balakhonovc SV, et al. Intraspecies classification of rhamnose-positive Yersinia pestis strains from natural plague foci of Mongolia. Mol Genet Microbiol Virol. 2015;30(1):24–9.
45. Cui Y, Li Y, Gorge O, Platonov ME, Yan Y, Guo Z, Pourcel C, Dentovskaya SV, Balakhonov SV, Wang X, et al. Insight into microevolution of *Yersinia pestis* by clustered regularly interspaced short palindromic repeats. PLoS One. 2008;3(7):e2652.
46. Cornelis GR, Wolf-Watz H. The Yersinia Yop virulon: a bacterial system for subverting eukaryotic cells. Mol Microbiol. 1997;23(5):861–7.
47. Hinnebusch BJ, Rudolph AE, Cherepanov P, Dixon JE, Schwan TG, Forsberg A. Role of Yersinia murine toxin in survival of *Yersinia pestis* in the midgut of the flea vector. Science. 2002;296(5568):733–5.
48. Dong X, YU D. Plasmids in Yersinia pestis: functions and their role in epidemiology. Yu Fang Yi Xue Qing Bao Za Zhi. 1994;10(3):138–44.
49. Filippov AA, Solodovnikov NS, Kookleva LM, Protsenko OA. Plasmid content in Yersinia pestis strains of different origin. FEMS Microbiol Lett. 1990;55(1-2):45–8.
50. Dong XQ, Lindler LE, Chu MC. Complete DNA sequence and analysis of an emerging cryptic plasmid isolated from *Yersinia pestis*. Plasmid. 2000;43(2):144–8.
51. Dong X, Ye F, Peng H. Geographic distribution and feature of Yersinia pestis plasmid isolated from Yunnan province. Zhonghua Liu Xing Bing Xue Za Zhi. 2001;22(5):344–7.
52. Chu MC, Dong XQ, Zhou X, Garon CF. A cryptic 19-kilobase plasmid associated with U.S. isolates of Yersinia pestis: a dimer of the 9.5-kilobase plasmid. AmJTrop Med Hyg. 1998;59(5):679–86.
53. Song Y, Tong Z, Wang J, Wang L, Guo Z, Han Y, Zhang J, Pei D, Zhou D, Qin H, et al. Complete genome sequence of *Yersinia pestis* strain 91001, an isolate avirulent to humans. DNA Res. 2004;11(3):179–97.
54. Xu F, Yang Z, Li L, Zhao F. An analysis of plasmid profiles of Yersinia pestis isolated from natural plague foci in Xinjiang, China (in Chinese). End Dis Bull. 1997;12(1):12–5.
55. Leal NC, de Almeida AM, Ferreira LC. Plasmid composition and virulence-associated factors of Yersinia pestis isolates from a plague outbreak at the Paraiba State, Brazil. Rev Inst Med Trop Sao Paulo. 1989;31(5):295–300.
56. Cavalcanti YV, Leal NC, De Almeida AM. Typing of *Yersinia pestis* isolates from the state of Ceara, Brazil. Lett Appl Microbiol. 2002;35(6):543–7.
57. Galimand M, Guiyoule A, Gerbaud G, Rasoamanana B, Chanteau S, Carniel E, Courvalin P. Multidrug resistance in *Yersinia pestis* mediated by a transferable plasmid. N Engl J Med. 1997;337(10):677–80.
58. Guiyoule A, Gerbaud G, Buchrieser C, Galimand M, Rahalison L, Chanteau S, Courvalin P, Carniel E. Transferable plasmid-mediated resistance to streptomycin in a clinical isolate of *Yersinia pestis*. Emerg Infect Dis. 2001;7(1):43–8.
59. Guiyoule A, Grimont F, Iteman I, Grimont PA, Lefevre M, Carniel E. Plague pandemics investigated by ribotyping of *Yersinia pestis* strains. J Clin Microbiol. 1994;32(3):634–41.

60. Pei D, Pang X, Song Y, Zhai J, Chen Z, Liu H, Guo Z, Wang J, Yang R. Fluorescent amplified fragment length polymorphism for genotyping Yersinia pestis. Chi J End. 2004;23(3):210–4.
61. Guiyoule A, Rasoamanana B, Buchrieser C, Michel P, Chanteau S, Carniel E. Recent emergence of new variants of Yersinia pestis in Madagascar. J Clin Microbiol. 1997;35(11):2826–33.
62. Lucier TS, Brubaker RR. Determination of genome size, macrorestriction pattern polymorphism, and nonpigmentation-specific deletion in Yersinia pestis by pulsed-field gel electrophoresis. J Bacteriol. 1992;174(7):2078–86.
63. Drancourt M, Roux V, Dang LV, Tran-Hung L, Castex D, Chenal-Francisque V, Ogata H, Fournier P-E, Crubézy E, Raoult D. Genotyping, orientalis-like Yersinia pestis, and plague pandemics. Emerg Infect Dis. 2004;10(9):1585–92.
64. Portnoy DA, Falkow S. Virulence-associated plasmids from Yersinia enterocolitica and Yersinia pestis. J Bacteriol. 1981;148(3):877–83.
65. Protsenko OA, Filippov AA, Kutyrev VV. Integration of the plasmid encoding the synthesis of capsular antigen and murine toxin into Yersinia pestis chromosome. Microb Pathog. 1991;11(2):123–8.
66. Simonet M, Riot B, Fortineau N, Berche P. Invasion production by Yersinia pestis is abolished by insertion of an IS200-like element within the inv gene. Infect Immun. 1996;64(1):375–9.
67. Odaert M, Devalckenaere A, Trieu-Cuot P, Simonet M. Molecular characterization of IS1541 insertions in the genome of *Yersinia pestis*. J Bacteriol. 1998;180(1):178–81.
68. McDonough KA, Hare JM. Homology with a repeated *Yersinia pestis* DNA sequence IS100 correlates with pesticin sensitivity in *Yersinia pseudotuberculosis*. J Bacteriol. 1997;179(6):2081–5.
69. Klevytska AM, Price LB, Schupp JM, Worsham PL, Wong J, Keim P. Identification and characterization of variable-number tandem repeats in the *Yersinia pestis* genome. J Clin Microbiol. 2001;39(9):3179–85.
70. Adair DM, Worsham PL, Hill KK, Klevytska AM, Jackson PJ, Friedlander AM, Keim P. Diversity in a variable-number tandem repeat from Yersinia pestis. J Clin Microbiol. 2000;38(4):1516–9.
71. Huang F, Yu D, Hai R, Cai H. Study on the application of random amplified polymorphic DNA in Yersinia pestis genotyping. Zhonghua Liu Xing Bing Xue Za Zhi. 2000;21(6):424–6.
72. Yu DZ, Hai R, Dong XQ, Li M, Xia LX, Shi XM, Wei JC, Cui BZ, Wang P, Sun LZ. Genetic analysis of Yersinia pestis strains isolated in China. Zhonghua Liu Xing Bing Xue Za Zhi. 2003;24(11):1005–9.
73. Smith CL, Condemine G. New approaches for physical mapping of small genomes. J Bacteriol. 1990;172(3):1167–72.
74. Arbeit RD, Arthur M, Dunn R, Kim C, Selander RK, Goldstein R. Resolution of recent evolutionary divergence among Escherichia coli from related lineages: the application of pulsed field electrophoresis to molecular epidemiology. J Infect Dis. 1990;161(2):230–5.
75. Bercovier H, Alonso JM, Bentaiba ZN, Brault J, Mollaret HH. Contribution to the definition and the taxonomy of Yersinia enterocolitica. Contrib Microbiol Immunol. 1979;5:12–22.
76. Filippov AA, Oleinikov PV, Motin VL, Protsenko OA, Smirnov GB. Sequencing of two Yersinia pestis IS elements, IS285 and IS100. Contrib Microbiol Immunol. 1995;13:306–9.
77. Rakin A, Heesemann J. The established Yersinia pestis biovars are characterized by typical patterns of I-CeuI restriction fragment length polymorphism. Mol Gen Mikrobiol Virusol. 1995;3:26–9.
78. Huang XZ, Chu MC, Engelthaler DM, Lindler LE. Genotyping of a homogeneous group of Yersinia pestis strains isolated in the United States. J Clin Microbiol. 2002;40(4):1164–73.
79. Grimont F, Grimont PA. Ribosomal ribonucleic acid gene restriction patterns as potential taxonomic tools. Ann Inst Pasteur Microbiol. 1986;137B(2):165–75.

80. Stull TL, LiPuma JJ, Edlind TD. A broad-spectrum probe for molecular epidemiology of bacteria: ribosomal RNA. J Infect Dis. 1988;157(2):280–6.
81. Grimont F, Chevrier D, Grimont PA, Lefevre M, Guesdon JL. Acetylaminofluorene-labelled ribosomal RNA for use in molecular epidemiology and taxonomy. Res Microbiol. 1989;140(7):447–54.
82. Parkhill J, Wren BW, Thomson NR, Titball RW, Holden MT, Prentice MB, Sebaihia M, James KD, Churcher C, Mungall KL, et al. Genome sequence of *Yersinia pestis*, the causative agent of plague. Nature. 2001;413(6855):523–7.
83. Deng W, Burland V, Plunkett 3rd G, Boutin A, Mayhew GF, Liss P, Perna NT, Rose DJ, Mau B, Zhou S, et al. Genome sequence of Yersinia pestis KIM. J Bacteriol. 2002;184(16): 4601–11.
84. Grif K, Dierich MP, Much P, Hofer E, Allerberger F. Identifying and subtyping species of dangerous pathogens by automated ribotyping. Diagn Microbiol Infect Dis. 2003;47:313–20.
85. Mahillon J, Chandler M. Insertion sequences. Microbiol Mol Biol Rev. 1998;62(3):725–74.
86. Yang R, Wang J, Lin W, Li Y, Guo Z. Amplification of rDNA from Escherichia coli as probes for bacterial ribotyping. Zhonghua Liu Xing Bing Xue Za Zhi. 1992;13(Suppl2):156–9.
87. Leclercq AJ, Torrea G, Chenal-Francisque V, Carniel E. 3 IS-RFLP: a powerful tool for geographical clustering of global isolates of Yersinia pestis. Adv Exp Med Biol. 2007;603:322–6.
88. Lindsay JA. Typing pathogenic bugs on the net. Mol Med Today. 2000;6:100.
89. Suntsova VV, Suntsova NI. Ecological aspects of evolution of the plague microbe Yersinia pestis and the genesis of natural foci. Biol Bull. 2000;27(6):541–52.
90. Rachman C, Kabadjova P, Valcheva R, Prevost H, Dousset X. Identification of Carnobacterium species by restriction fragment length polymorphism of the 16S-23S rRNA gene intergenic spacer region and species-specific PCR. Appl Environ Microbiol. 2004;70(8):4468–77.
91. Catry B, Baele M, Opsomer G, de Kruif A, Decostere A, Haesebrouck F. tRNA-intergenic spacer PCR for the identification of Pasteurella and Mannheimia spp. Vet Microbiol. 2004;98(3–4):251–60.
92. Vitorino L, Ze-Ze L, Sousa A, Bacellar F, Tenreiro R. rRNA intergenic spacer regions for phylogenetic analysis of Rickettsia species. Ann N Y Acad Sci. 2003;990:726–33.
93. Song Y, Liu C, Molitoris D, Tomzynski TJ, Mc Teague M, Read E, Finegold SM. Use of 16S-23S rRNA spacer-region (SR)-PCR for identification of intestinal clostridia. Syst Appl Microbiol. 2002;25(4):528–35.
94. Rachman CN, Kabadjova P, Prevost H, Dousset X. Identification of Lactobacillus alimentarius and Lactobacillus farciminis with 16S–23S rDNA intergenic spacer region polymorphism and PCR amplification using species-specific oligonucleotide. J Appl Microbiol. 2003;95(6):1207–16.
95. Tannock GW, Tilsala-Timisjarvi A, Rodtong S, Ng J, Munro K, Alatossava T. Identification of Lactobacillus isolates from the gastrointestinal tract, silage, and yoghurt by 16S–23S rRNA gene intergenic spacer region sequence comparisons. Appl Environ Microbiol. 1999;65(9):4264–7.
96. Zhou D, Han Y, Song Y, Tong Z, Wang J, Guo Z, Pei D, Pang X, Zhai J, Li M, et al. DNA microarray analysis of genome dynamics in *Yersinia pestis*: insights into bacterial genome microevolution and niche adaptation. J Bacteriol. 2004;186(15):5138–46.
97. Raoult D, Aboudharam G, Crubezy E, Larrouy G, Ludes B, Drancourt M. Molecular identification by "suicide PCR" of Yersinia pestis as the agent of medieval black death. Proc Natl Acad Sci U S A. 2000;97(23):12800–3.
98. Drancourt M, Aboudharam G, Signoli M, Dutour O, Raoult D. Detection of 400-year-old *Yersinia pestis* DNA in human dental pulp: an approach to the diagnosis of ancient septicemia. Proc Natl Acad Sci U S A. 1998;95(21):12637–40.
99. Gilbert MT, Cuccui J, White W, Lynnerup N, Titball RW, Cooper A, Prentice MB. Absence of Yersinia pestis-specific DNA in human teeth from five European excavations of putative plague victims. Microbiology. 2004;150(Pt 2):341–54.

100. Wood J, DeWitte-Avina S. Was the Black Death yersinial plague? Lancet Infect Dis. 2004;4(8):485.
101. Wood J, DeWitte-Avina S. Was the Black Death yersinial plague? Lancet Infect Dis. 2003;3(6):327–8. discussion 328.
102. Prentice MB, Gilbert T, Cooper A. Was the Black Death caused by *Yersinia pestis*? Lancet Infect Dis. 2004;4(2):72.
103. Wood JW, Ferrell RJ, Dewitte-Avina SN. The temporal dynamics of the fourteenth-century Black Death: new evidence from English ecclesiastical records. Hum Biol. 2003;75(4):427–48.
104. Raoult D, Drancourt M. Cause of Black Death. Lancet Infect Dis. 2002;2(8):459.
105. Pepinski W, Janica J, Aleksandrowicz-Bukin M, Skawronska M, Koc-Zorawska E, Niemcunowicz-Janica A. Population genetics of 10 short tandem repeat (STR) loci in a population sample of the ethnic group of Polish Tatars living in the Podlasie area (Northeastern Poland). Folia Morphol (Warsz). 2004;63(2):249–52.
106. Kondopoulou H, Kouvatsi A, Triantaphyllidis C. Forensic evaluation of 10 STRs and two minisatellite loci in the Greek population. Forensic Sci Int. 2001;124(2-3):228–30.
107. Parra A, Fernandez-Llario P, Tato A, Larrasa J, Garcia A, Alonso JM, Hermoso de Mendoza M, Hermoso de Mendoza J. Epidemiology of Mycobacterium bovis infections of pigs and wild boars using a molecular approach. Vet Microbiol. 2003;97(1-2):123–33.
108. Banu S, Gordon SV, Palmer S, Islam MR, Ahmed S, Alam KM, Cole ST, Brosch R. Genotypic analysis of Mycobacterium tuberculosis in Bangladesh and prevalence of the Beijing strain. J Clin Microbiol. 2004;42(2):674–82.
109. Crawford JT. Genotyping in contact investigations: a CDC perspective. Int J Tuberc Lung Dis. 2003;7(12 Suppl 3):S453–7.
110. Sabat A, Krzyszton-Russjan J, Strzalka W, Filipek R, Kosowska K, Hryniewicz W, Travis J, Potempa J. New method for typing Staphylococcus aureus strains: multiple-locus variable-number tandem repeat analysis of polymorphism and genetic relationships of clinical isolates. J Clin Microbiol. 2003;41(4):1801–4.
111. Savine E, Warren RM, van der Spuy GD, Beyers N, van Helden PD, Locht C, Supply P. Stability of variable-number tandem repeats of mycobacterial interspersed repetitive units from 12 loci in serial isolates of Mycobacterium tuberculosis. J Clin Microbiol. 2002;40(12):4561–6.
112. Kim W, Hong YP, Yoo JH, Lee WB, Choi CS, Chung SI. Genetic relationships of Bacillus anthracis and closely related species based on variable-number tandem repeat analysis and BOX-PCR genomic fingerprinting. FEMS Microbiol Lett. 2002;207(1):21–7.
113. Chanchaem W, Palittapongarnpim P. A variable number of tandem repeats result in polymorphic alpha -isopropylmalate synthase in Mycobacterium tuberculosis. Tuberculosis (Edinb). 2002;82(1):1–6.
114. van Belkum A, Scherer S, van Leeuwen W, Willemse D, van Alphen L, Verbrugh H. Variable number of tandem repeats in clinical strains of Haemophilus influenzae. Infect Immun. 1997;65(12):5017–27.
115. Le Fleche P, Hauck Y, Onteniente L, Prieur A, Denoeud F, Ramisse V, Sylvestre P, Benson G, Ramisse F, Vergnaud G. A tandem repeats database for bacterial genomes: application to the genotyping of Yersinia pestis and Bacillus anthracis. BMC Microbiol. 2001;1(1):2.
116. Pourcel C, Andre-Mazeaud F, Neubauer H, Ramisse F, Vergnaud G. Tandem repeats analysis for the high resolution phylogenetic analysis of *Yersinia pestis*. BMC Microbiol. 2004;4(1):22.
117. Denoeud F, Vergnaud G. Identification of polymorphic tandem repeats by direct comparison of genome sequence from different bacterial strains: a web-based resource. BMC Bioinform. 2004;5(1):4.
118. Li Y, Dai E, Cui Y, Li M, Zhang Y, Wu M, Zhou D, Guo Z, Dai X, Cui B, et al. Different region analysis for genotyping Yersinia pestis isolates from China. PLoS One. 2008;3(5):e2166.

119. Morelli G, Song Y, Mazzoni CJ, Eppinger M, Roumagnac P, Wagner DM, Feldkamp M, Kusecek B, Vogler AJ, Li Y, et al. Yersinia pestis genome sequencing identifies patterns of global phylogenetic diversity. Nat Genet. 2010;42(12):1140–3.
120. Cui Y, Yu C, Yan Y, Li D, Li Y, Jombart T, Weinert LA, Wang Z, Guo Z, Xu L, et al. Historical variations in mutation rate in an epidemic pathogen, Yersinia pestis. Proc Natl Acad Sci U S A. 2013;110(2):577–82.
121. Comas I, Homolka S, Niemann S, Gagneux S. Genotyping of genetically monomorphic bacteria: DNA sequencing in Mycobacterium tuberculosis highlights the limitations of current methodologies. PLoS One. 2009;4(11):e7815.
122. Li Y, Cui Y, Cui B, Yan Y, Yang X, Wang H, Qi Z, Zhang Q, Xiao X, Guo Z, et al. Features of variable number of tandem repeats in Yersinia pestis and the development of a hierarchical genotyping scheme. PLoS One. 2013;8(6):e66567.
123. Cong L, Ran FA, Cox D, Lin S, Barretto R, Habib N, Hsu PD, Wu X, Jiang W, Marraffini LA, et al. Multiplex genome engineering using CRISPR/Cas systems. Science. 2013;339(6121):819–23.
124. Pennisi E. The CRISPR craze. Science. 2013;341(6148):833–6.
125. Mali P, Yang L, Esvelt KM, Aach J, Guell M, DiCarlo JE, Norville JE, Church GM. RNA-guided human genome engineering via Cas9. Science. 2013;339(6121):823–6.
126. Horvath P, Barrangou R. CRISPR/Cas, the immune system of bacteria and archaea. Science. 2010;327(5962):167–70.
127. Deveau H, Garneau JE, Moineau S. CRISPR/Cas system and its role in phage-bacteria interactions. Annu Rev Microbiol. 2010;64:475–93.
128. Marraffini LA, Sontheimer EJ. CRISPR interference: RNA-directed adaptive immunity in bacteria and archaea. Nat Rev Genet. 2010;11(3):181–90.
129. Horvath P, Romero DA, Coute-Monvoisin AC, Richards M, Deveau H, Moineau S, Boyaval P, Fremaux C, Barrangou R. Diversity, activity, and evolution of CRISPR loci in Streptococcus thermophilus. J Bacteriol. 2008;190(4):1401–12.
130. Garneau JE, Dupuis ME, Villion M, Romero DA, Barrangou R, Boyaval P, Fremaux C, Horvath P, Magadan AH, Moineau S. The CRISPR/Cas bacterial immune system cleaves bacteriophage and plasmid DNA. Nature. 2010;468(7320):67–71.
131. Deveau H, Barrangou R, Garneau JE, Labonte J, Fremaux C, Boyaval P, Romero DA, Horvath P, Moineau S. Phage response to CRISPR-encoded resistance in Streptococcus thermophilus. J Bacteriol. 2008;190(4):1390–400.
132. Barrangou R, Fremaux C, Deveau H, Richards M, Boyaval P, Moineau S, Romero DA, Horvath P. CRISPR provides acquired resistance against viruses in prokaryotes. Science. 2007;315(5819):1709–12.
133. Jansen R, Embden JD, Gaastra W, Schouls LM. Identification of genes that are associated with DNA repeats in prokaryotes. Mol Microbiol. 2002;43(6):1565–75.
134. Pourcel C, Salvignol G, Vergnaud G. CRISPR elements in Yersinia pestis acquire new repeats by preferential uptake of bacteriophage DNA, and provide additional tools for evolutionary studies. Microbiology. 2005;151(Pt 3):653–63.
135. Mojica FJ, Diez-Villasenor C, Garcia-Martinez J, Soria E. Intervening sequences of regularly spaced prokaryotic repeats derive from foreign genetic elements. J Mol Evol. 2005;60(2):174–82.
136. Goyal M, Saunders NA, van Embden JD, Young DB, Shaw RJ. Differentiation of Mycobacterium tuberculosis isolates by spoligotyping and IS6110 restriction fragment length polymorphism. J Clin Microbiol. 1997;35(3):647–51.
137. Liu F, Barrangou R, Gerner-Smidt P, Ribot EM, Knabel SJ, Dudley EG. Novel virulence gene and clustered regularly interspaced short palindromic repeat (CRISPR) multilocus sequence typing scheme for subtyping of the major serovars of Salmonella enterica subsp. enterica. Appl Environ Microbiol. 2011;77(6):1946–56.

138. Bachmann NL, Petty NK, Ben Zakour NL, Szubert JM, Savill J, Beatson SA. Genome analysis and CRISPR typing of Salmonella enterica serovar Virchow. BMC Genomics. 2014;15:389.
139. Kovanen SM, Kivisto RI, Rossi M, Hanninen ML. A combination of MLST and CRISPR typing reveals dominant Campylobacter jejuni types in organically farmed laying hens. J Appl Microbiol. 2014;117(1):249–57.
140. Vergnaud G, Li Y, Gorge O, Cui Y, Song Y, Zhou D, Grissa I, Dentovskaya SV, Platonov ME, Rakin A, et al. Analysis of the three Yersinia pestis CRISPR loci provides new tools for phylogenetic studies and possibly for the investigation of ancient DNA. Adv Exp Med Biol. 2007;603:327–38.
141. Riehm JM, Vergnaud G, Kiefer D, Damdindorj T, Dashdavaa O, Khurelsukh T, Zoller L, Wolfel R, Le Fleche P, Scholz HC. Yersinia pestis lineages in Mongolia. PLoS One. 2012;7(2):e30624.
142. Grissa I, Vergnaud G, Pourcel C. CRISPRFinder: a web tool to identify clustered regularly interspaced short palindromic repeats. Nucleic Acids Res. 2007. doi:10.1093/nar/gkm1360.
143. Grissa I, Vergnaud G, Pourcel C. CRISPRcompar: a website to compare clustered regularly interspaced short palindromic repeats. Nucleic Acids Res. 2008;36:W145–8.
144. Ochman H, Moran NA. Genes lost and genes found: evolution of bacterial pathogenesis and symbiosis. Science. 2001;292(5519):1096–9.
145. Hacker J, Hentschel U, Dobrindt U. Prokaryotic chromosomes and disease. Science. 2003;301(5634):790–3.
146. Dai E, Tong Z, Wang X, Li M, Cui B, Dai R, Zhou D, Pei D, Song Y, Zhang J, et al. Identification of different regions among strains of Yersinia pestis by suppression subtractive hybridization. Res Microbiol. 2005;156(7):785–9.
147. Platonov ME, Evseeva VV, Svetoch TE, Efremenko DV, Kuznetsova IV, Dentovskaia SV, Kulichenko AN, Anisimov AP. The phylogeography of the Yersinia pestis vole strains isolated from the natural foci of caucasian region. Mol Gen Mikrobiol Virusol. 2012;3:18–21.
148. Roumagnac P, Weill FX, Dolecek C, Baker S, Brisse S, Chinh NT, Le TA, Acosta CJ, Farrar J, Dougan G, et al. Evolutionary history of Salmonella typhi. Science. 2006;314(5803): 1301–4.
149. Pearson T, Busch JD, Ravel J, Read TD, Rhoton SD, U'Ren JM, Simonson TS, Kachur SM, Leadem RR, Cardon ML, et al. Phylogenetic discovery bias in Bacillus anthracis using single-nucleotide polymorphisms from whole-genome sequencing. Proc Natl Acad Sci U S A. 2004;101(37):13536–41.
150. Alland D, Whittam TS, Murray MB, Cave MD, Hazbon MH, Dix K, Kokoris M, Duesterhoeft A, Eisen JA, Fraser CM, et al. Modeling bacterial evolution with comparative-genome-based marker systems: application to Mycobacterium tuberculosis evolution and pathogenesis. J Bacteriol. 2003;185(11):3392–9.
151. Harbeck M, Seifert L, Hansch S, Wagner DM, Birdsell D, Parise KL, Wiechmann I, Grupe G, Thomas A, Keim P, et al. Yersinia pestis DNA from skeletal remains from the 6(th) century AD reveals insights into Justinianic Plague. PLoS Pathog. 2013;9(5):e1003349.
152. Wagner DM, Klunk J, Harbeck M, Devault A, Waglechner N, Sahl JW, Enk J, Birdsell DN, Kuch M, Lumibao C, et al. Yersinia pestis and the plague of Justinian 541–543 AD: a genomic analysis. Lancet Infect Dis. 2014;14(4):319–26.
153. Rasmussen S, Allentoft ME, Nielsen K, Orlando L, Sikora M, Sjogren KG, Pedersen AG, Schubert M, Van Dam A, Kapel CM, et al. Early divergent strains of Yersinia pestis in Eurasia 5,000 years Ago. Cell. 2015;163(3):571–82.
154. Wood F. The Silk Road: two thousand years in the heart of Asia. 1st ed. Berkeley: University of California Press; 2002.
155. Yang B. Horses, silver, and cowries: Yunnan in global perspective. J World Hist. 2004;15(3):281–322.
156. Morelli G, Song Y, Mazzoni CJ, Eppinger MPR, Wagner DM, Feldkamp M, Kusecek B, Vogler AJ, Li Y et al. Plague out of China, again and again. Nat Genet 2010;In Press.
157. Prentice MB, Rahalison L. Plague. Lancet. 2007;369(9568):1196–207.

158. Abath FG, Almeida AM, Ferreira LC. Electrophoretic characterisation of the outer membrane proteins of Yersinia pestis isolated in north-east Brazil. Epidemiol Infect. 1989;103(3):595–602.
159. Abath FG, Ferreira LC. Comparative studies of Yersinia pestis outer membrane isolation techniques and their potential use in plaque epidemiology. Rev Inst Med Trop Sao Paulo. 1990;32(2):78–83.
160. Song Y, Guo Z, Zhang M, Yang R, Zhao M, Cong X, Zhang C. Cellular fatty acids analysis of Yersinia pestis strains in China. Zhonghua Wei Sheng Wu Yu Mian Yi Xue Za Zhi. 2002.
161. Tan Y, Wu M, Liu H, Dong X, Guo Z, Song Z, Li Y, Cui Y, Song Y, Du Z, et al. Cellular fatty acids as chemical markers for differentiation of Yersinia pestis and Yersinia pseudotuberculosis. Lett Appl Microbiol. 2010;50(1):104–11.
162. Singhal N, Kumar M, Kanaujia PK, Virdi JS. MALDI-TOF mass spectrometry: an emerging technology for microbial identification and diagnosis. Front Microbiol. 2015;6:791.
163. Wahl KL, Wunschel SC, Jarman KH, Valentine NB, Petersen CE, Kingsley MT, Zartolas KA, Saenz AJ. Analysis of microbial mixtures by matrix-assisted laser desorption/ionization time-of-flight mass spectrometry. Anal Chem. 2002;74(24):6191–9.
164. Liu H, Du Z, Wang J, Yang R. Universal sample preparation method for characterization of bacteria by matrix-assisted laser desorption ionization-time of flight mass spectrometry. Appl Environ Microbiol. 2007;73(6):1899–907.
165. Ayyadurai S, Flaudrops C, Raoult D, Drancourt M. Rapid identification and typing of Yersinia pestis and other Yersinia species by matrix-assisted laser desorption/ionization time-of-flight (MALDI-TOF) mass spectrometry. BMC Microbiol. 2010;10:285.

Chapter 4
Physiology of *Yersinia pestis*

Robert R. Brubaker

Abstract This chapter outlines the physiology of *Yersinia pestis* with emphasis on identifying unique functions required for tissue invasion and acute disease. These activities are opposed to often incompatible processes expressed by very closely related *Yersinia pseudotuberculosis*, which causes localized gastrointestinal infection. Gain of new information in *Y. pestis* entailed lateral transfer of plasminogen activator and anti-phagocytic capsular antigen via the plasmids pPCP and pMT, respectively, and derepression of the pigmentation locus facilitating colonization of the flea vector. The ability to survive in austere natural environments became unnecessary following mastery of the closed flea–rodent–flea life cycle permitting concomitant chromosomal degeneration (large and small deletions, additions, inversions, translocations, transposon inserts, and single base substitutions causing nonsense and missense mutations). Consequently, modern *Y. pestis* lacks a functional pentose–phosphate pathway, glyoxylate bypass, and is unable to directly catabolize L-aspartate and close metabolic derivatives directly via the tricarboxylic acid cycle. The missing gene products accounting for these and numerous other metabolic lesions are now well-established. This group includes formyltetrahydrofolate deformylase (PurU) required for synthesis of glycine. This deficiency is associated with a dramatic ability of *Y. pestis* to catabolize L-serine, required by the host to initiate methylation of DNA (necessary in turn to initiate successful innate immune processes leading to delayed-type hypersensitivity).

Keywords Bacterial structure • Bacterial function • Intermediary metabolism • Anabolism • Catabolism

---

R.R. Brubaker (✉)
Department of Microbiology, Immunology, and Molecular Genetics,
Chandler Medical Center, The University of Kentucky, MS 415, 300 Rose Street,
Lexington, KY 40536-0298, USA
e-mail: robert.brubaker@uky.edu

R. Yang, A. Anisimov (eds.), *Yersinia pestis: Retrospective and Perspective*,
Advances in Experimental Medicine and Biology 918,
DOI 10.1007/978-94-024-0890-4_4

4.1 Introduction

A large number of mammalian species including man are vulnerable to *Yersinia pestis*, the causative agent of plague. This organism causes both severe mortality in typical hosts (over 50 % for the bubonic and 90 % for the pneumonic forms, respectively) and high infectivity (about 10 bacteria by fleabite and a 100–1000-fold more by aerosol). This lethal combination generated three worldwide epidemics termed the Justinian Plague of the mid-sixth century (about 50 million deaths), the Black Death initiated during the mid-fourteenth century (between 75 and 200 million deaths), and the Third Pandemic throughout the late nineteenth and early twentieth century (approximately 12 million deaths). Concerns that the earlier pandemics were actually caused by distinct etiologies have now been resolved by demonstration of *Y. pestis*-specific DNA sequences in human skeletal remains from ancient plague cemeteries [1, 2]. Consequently, it is now generally accepted by microbiologists and historians alike that plague is the most devastating epidemic bacterial disease experienced by mankind.

The physiological basis for this extraordinary pathogenicity remains uncertain although it is established that wild-type cells of *Y. pestis* resist all forms of innate immunity including containment within granulomatous lesions, the hallmark of delayed-type hypersensitivity [3]. Nevertheless, impressive progress has been made in resolving the recent evolution of *Y. pestis* from very closely related but relatively benign enteropathogenic *Yersinia pseudotuberculosis*. Plague emerged less than 10,000 years ago [4] and, as judged by a rooted phylogenetic tree based upon single nucleotide polymorphisms [5],divergence has proceeded to the present day giving rise to specific lineages or molecular groups (Fig. 4.1) analogous to the classical biovars of Devignat [6]. The brevity of this evolutionary process only permitted change by acquisition of new genes by lateral transfer and modification or loss of existing genes. The former include determinants encoded by new plasmids or high pathogenicity islands that facilitate invasion of tissue or survival in the flea vector. The latter entail structural or regulatory modifications of similar shared genetic elements or elimination of primarily chromosomally encoded activities no longer required by *Y. pestis* in its new closed life cycle (but essential to *Y. pseudotuberculosis* for expression of chronic enteropathogenic disease or survival for extended periods in austere natural environments). As described in the next sections, the progression occurring from *Y. pseudotuberculosis* until divergence of the 0.PE4 molecular group is straightforward but often tumultuous. Curiously, members of this lineage and its progenitors possess most or all of the known virulence determinants shared by contemporary *Y. pestis* [5, 7]. However, these strains only promote disease in murid rodents as opposed to isolates from the next 0.ANT1 molecular group and all subsequent lineages (Fig. 4.1). These later strains cause plague in non-murid rodents as well as a variety of other mammalian species including primates [7, 8].

Little is known about the genetic change(s) that account for this impressive late increase in invasiveness and virulence. The still extant *Y. pestis* strains lacking this critical alteration were assigned to the subspecies *Microtus* and are considered to be

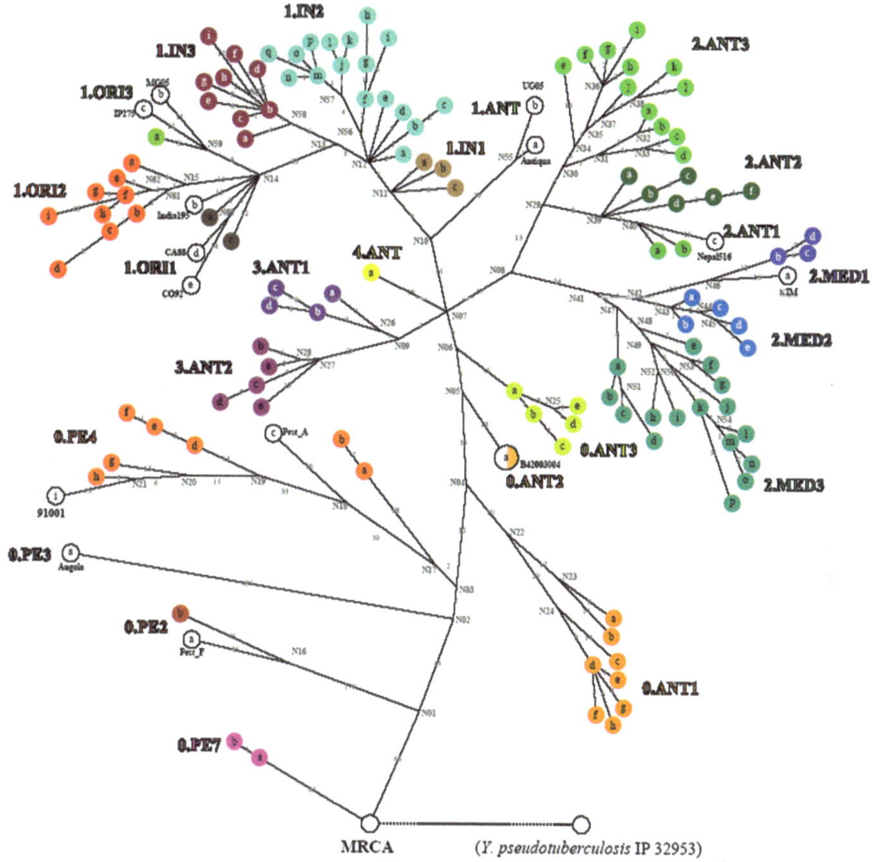

Fig. 4.1 The evolution of *Yersinia pestis*; change from the enzootic to epidemic phenotype occurred after divergence of the 0.PE4 but before emergence of the 0.ANT1 lineages, respectively [5]

"enzootic, non-main, or pestoides" isolates [7, 8]. The purpose of this chapter is to define the structure and function of contemporary highly pathogenic "epidemic or main" *Y. pestis* emphasizing both unique genetic information acquired by recent lateral transfer and the physiological consequences of extensive chromosomal loss on fitness in vitro.

4.2 Phenotypes

Y. pseudotuberculosis is a member of the family Enterobacteriaceae known to possess fully functional and tightly regulated Embden–Meyerhof (glycolytic), pentose–phosphate, Entner–Doudoroff, and tricarboxylic acid (TCA) cycle or pathways plus a complete set of respiratory cytochromes. These organisms are also capable of growth, at least under aerobic conditions, at room temperature (26 °C) in minimal medium containing only inorganic salts and a fermentable source of carbon. Growth at 26 °C but not host temperature (37 °C) also induces production of complex lipopolysaccharide (LPS) containing unusual and highly antigenic terminal hydrophobic dideoxyhexoses [9, 10] that favor survival within harsh or austere natural environments. Genes promoting production of this full LPS are downregulated at 37 °C in *Y. pseudotuberculosis* and are largely cryptic in *Y. pestis* [7]. Nevertheless, a comparison of salient portions of the two genomes demonstrated that *Y. pseudotuberculosis* serotype O:1b (encompassing the dideoxyhexose paratose) was the progenitor of *Y. pestis* [11].

Total DNA of *Y. pestis* was assayed under stringent conditions of association and found to have 9 %, 23 %, and 83 % homology with that of *Escherichia coli*, *Yersinia enterocolitica*, and *Y. pseudotuberculosis*, respectively [12]. This relationship predicted that the amount of DNA found to be specific for *Y. pestis* would be minimal and the results of subsequent sequencing revealed that only the two plasmids of the species and 32 chromosomal genes are unique [13]. In contrast, about 13 % of all functional genes within *Y. pseudotuberculosis* are now inactive in *Y. pestis* [13] due to chromosomal degeneration caused by inversions and translocations, small additions and deletions, missense mutations, and rampant proliferation of IS elements. The latter in *Y. pseudotuberculosis* exist as a combined total of 20 copies of IS*100*, IS*1541*, IS*1661*, and IS*285*, whereas this number was about sixfold higher in epidemic *Y. pestis* [13–15]. The consequences of this evident random loss extend to all areas of intermediary metabolism as well as the integrity and effectiveness of the envelope. The resulting physiological lesions are obviously not detrimental to infectivity and virulence but do prevent long-term survival in limited natural environments and thus serve to contain *Y pestis* within a closed flea to rodent to flea life cycle. This phenotype is atypical in that the plague bacillus is both lethal to the flea vector due to blockage of its foregut and also causes death of the mammalian host, thereby assuring that freshly infected resident fleas will depart in search of new victims. A response of this nature is consistent with that of a newly evolved parasite as contrasted with more highly adapted pathogens that tend to minimize damage to their hosts [16].

All three pathogenic yersiniae share an approximately 70-kb plasmid termed pCD (calcium dependence) in *Y. pestis* and pYV (yersiniae virulence) in the enteropathogenic species. pCD mediates a low-calcium response (LCR) at 37 °C but not 26 °C defined as bacteriostasis accompanied by upregulation of a pCD-encoded type 3 secretion system (T3SS) during cultivation without Ca^{2+} versus the occurrence of vegetative growth and downregulation of the T3SS in media containing ≥2.5 mM Ca^{2+} [17]. It is significant that this concentration of Ca^{2+} is identical to that of mammalian plasma and that, in enriched Ca^{2+}-deficient media, the LCR of *Y. pestis* is exacerbated [18] by addition of that concentration of Mg^{2+} (20 mM) present in mammalian cytoplasm [19]. This relationship was defined by the early discoveries that oxalated solid medium containing added 20 mM Mg^{2+} is selective at 37 °C for mutants lacking pCD [20] and that the same concentration (in the absence of added Ca^{2+}) also upregulated components [21] later found to be encoded by pCD that mediate the T3SS. These products include virulence effectors termed Yops (yersiniae outer proteins) [22] and activities that promote their translocation (Lcr^+). Ca^{2+} is dominant over Mg^{2+} in that pCD-encoded gene products remain suppressed in vitro if the former is present at concentrations of ≥2.5 mM regardless of the molarity of added Mg^{2+} [23]. These nutritional aspects of the LCR in *Y. pestis* are similar to those observed for *Y. pseudotuberculosis*, although the onset of bacteriostasis is typically less pronounced in this species. In contrast, cells of most strains of *Y. enterocolitica* continue slow logarithmic growth at 37 °C in Ca^{2+}-deficient liquid cultures and form colonies on solid media containing magnesium oxalate. These findings emphasize that temperature and calcium are important factors that affect structure and function of wild-type yersiniae in various fashion depending upon the species.

4.2.1 Temperature

Chromosomally encoded YmoA imposes downregulation at 26° by the pCD/pYV-encoded transcriptional activator LcrF (VirF in *Y. enterocolitica*). Thermal release is probably accomplished by melting salient DNA bends and results in induction of T3SS virulence effecters and machinery [24]. Other regulatory systems, of course, account for expression of the numerous additional temperature-dependent functions of yersiniae [25].

4.2.2 Ca^{2+}

Downregulation of pCD/pYV-encoded gene products by Ca^{2+} is accomplished by the negative regulator YopD and its chaperone LcrH [26]; mutational loss of these genes results in a Ca^{2+}-blind phenotype defined as an inability to multiply at 37 °C regardless of the presence of sufficient Ca^{2+} to ensure growth of wild-type yersiniae [17]. Ca^{2+} also closes physiological gates comprised of YopN–SycN–YscB–TyeA [27, 28] and LcrG-LcrV (V antigen) [29]. The Ca^{2+}-blind phenotype is also conferred by mutational loss of YopN and LcrG, whereas mutants lacking *lcrV*, encoding a major protective antigen [21, 30], are Ca^{2+}-independent. Although Ca^{2+} downregulates the LCR in vitro, this inhibition is blocked in Ca^{2+}-rich mammalian plasma or lymph where contact of yersiniae with the surface of professional and nonprofessional phagocytes promotes translocation of Yops into host cell cytoplasm [31, 32]. This ability to prevent Ca^{2+} from downregulating the LCR in vivo may reflect competition at the bacterial surface between the free cation and a specific receptor. This possibility was strengthened by studies of YscF, a small pCD-encoded protein that polymerizes to form the external needle of the T3SS injectisome. This activity contains aspartyl residues which, when replaced with neutral amino acids (unable to bind calcium), yielded Ca^{2+}-blind mutants suggesting a role as sensor [33]. This observation is consistent with the finding that bacteriostatic Ca^{2+}-starved Lcr$^+$ *Y. pestis* bound significantly more carrier-free $^{45}Ca^{2+}$ at 37 °C than did growing mutants cured of pCD [34].

Results of further analysis suggest that the injection of virulence effectors can also occur following secretion of translocators and their associated Yops across the inner and outer membrane prior to impending host cell contact. These complexes then assemble and associate with pores of the host target cell plasma membrane where the translocators mediate entry of their Yops into cytoplasm [35, 36]. As mentioned below, this concept concerns other variables, especially the concentration of Mg^{2+} and proteolysis caused by the plague plasminogen activator.

The physiological consequences of Ca^{2+} deprivation and the reason why the cation serves as a temperature-dependent nutritional factor have received additional study. As subsequently shown, static Ca^{2+}-starved Lcr$^+$ cells of *Y. pestis* are enlarged and contain nucleoids in the form of axial filaments. This morphology is typical of Enterobacteriaceae in stationary phase or otherwise starved for some essential nutrient and reflects a reduced adenylate energy charge inconsistent with cell division. This proved to be the case with Lcr$^+$ yersiniae as shown by using a Ca^{2+}-deficient but otherwise highly enriched synthetic medium where temperature shift from 26 to 37 °C permitted completion of ongoing rounds of DNA replication but blocked the initiation of new rounds [18]. These findings suggested that Ca^{2+} starvation promotes a lesion in bioenergetics associated with expression of the LCR and the results of further work showed that a combination of Na^+, Cl^-, and L-glutamic acid was necessary to assure bacteriostasis [37]. The pH of chemically defined Ca^{2+}-deficient culture media (supplied with K^+ to maintain cationic equivalents) also influenced growth of Lcr$^+$ cells and cultures containing added NaCl and L-glutamate at pH 7.5 prompted almost immediate shutoff of cell division. In contrast, essentially full-scale growth plus complete expression of all tested pCD-encoded gene products occurred at pH 6.5 in the presence of added L-glutamate but not NaCl [38]; it is probably significant that this environment mimics Na^+- and Cl^--deficient host cell cytoplasm [19]. Other Ca^{2+}-deficient combinations that nurtured essentially full-scale growth were pH 5.5 with added Na^+ and L-glutamate, where expression of pCD-encoded functions was minimal, and pH 8.0 with added Na^+, where NADH:ubiquinone oxidoreductase (Na^+-NQR) is active as a primary electrogenic sodium pump [39]. It is of interest in this context that Ca^{2+}-blind *lcrG* and *yopN*, but not *yopD* or *lcrH*, mutants were also rescued from temperature-sensitivity in the three permissive environments [40].

These findings suggest that temperature-dependent nutritional-dependence upon Ca^{2+} is a nutritional artifact prompted by the ubiquitous presence of Na^+ (and Cl^-) in media prepared from natural sources, use of sodium oxalate to chelate Ca^{2+}, and the common practice of adding NaOH to adjust the pH of chemically defined media. The combination of Na^+ plus L-glutamate without added Ca^{2+} proved to be especially inhibitory at neutral pH, possibly because Na^+ serves as a porter for its uptake via a major (inner membrane) glutamate:sodium symporter (GltS), thereby enabling Na^+ to enter bacterial cytoplasm. It is well-known that Na^+ is a cytoplasmic poison and its accumulation within Ca^{2+}-starved Lcr$^+$ yersiniae may account for the observed reduced adenylate charge associated with bacteriostasis. Additional metabolic lesions specifically associated with catabolism of L-glutamate/L-aspartate in Lcr$^+$ cells of *Y. pestis* are considered below.

4.2.3 Y. Pestis-*Specific Functions*

As already noted, the ability of *Y. pestis* to cause acute invasive disease is caused in part by laterally transferred determinants that promote invasion of tissues. Major examples are the ~10-kb plasmid pPCP (pesticin, coagulase, and plasminogen activator) and the ~100-kb plasmid pMT (murine toxin) or pFra (fraction 1) [41]. In addition, epidemic strains possess a 102-kb chromosomal pigmentation region (Pgm$^+$) containing genes within a high pathogenicity island [42] for siderophore (yersiniabactin or Ybt) biosynthesis, necessary for high-affinity uptake of Fe^{3+} [43–45]. In addition, this sequence contains genes encoding biofilm synthesis and regulatory functions that result in the binding of hemin at room temperature (Hms$^+$), a function accounting for the classical "pigmented" phenotype of Jackson and Burrows [46].

4.2.3.1 pPCP

This small plasmid harbors only three structural genes [47] but it is critical for the invasiveness of *Y. pestis*. It encodes a bacteriocin termed pesticin discovered by Ben-Gurian and Hertman [48], its immunity protein, a "fibrinolytic activity" first reported by Madison [49], and the "coagulase" of Jawetz and Meyer [50]. Hall and Brubaker [51] showed that pesticin converts target bacteria to osmotically stable spheroplasts (Fig. 4.2) and additional study showed that the bacteriocin displays muramidase activity [52]. Furthermore, Beesley et al. [53] demonstrated that both the "fibrinolytic" and "coagulase" functions co-purified and required processing by the host to initiate activity as evidenced by inhibition of the fibrinolysin with a specific plasminogen activator inhibitor (ε-aminocaproic acid) on fibrin plates containing plasminogen or by destroying the plasminogen therein by heating. These findings first demonstrated that the fibrinolytic activity elaborated by *Y. pestis* is a plasminogen activator (Pla). In parallel studies, the consequences of mutational loss of pPCP-encoded pesticin and linked Pla [41] on virulence were evaluated in mice [54]. The results showed that Pla was of little importance following intravenous injection but reduced lethality by a million-fold when the bacteria were administered subcutaneously (Table 4.1). This finding established Pla as a major tissue invasion of epidemic *Yersinia pestis*. Essentially identical results were reported independently by others [55] and this group first established that Pla was a member of the omptin protease family [55]. Pla is now known to directly or indirectly activate a number of additional host activities that favor invasion of host tissues

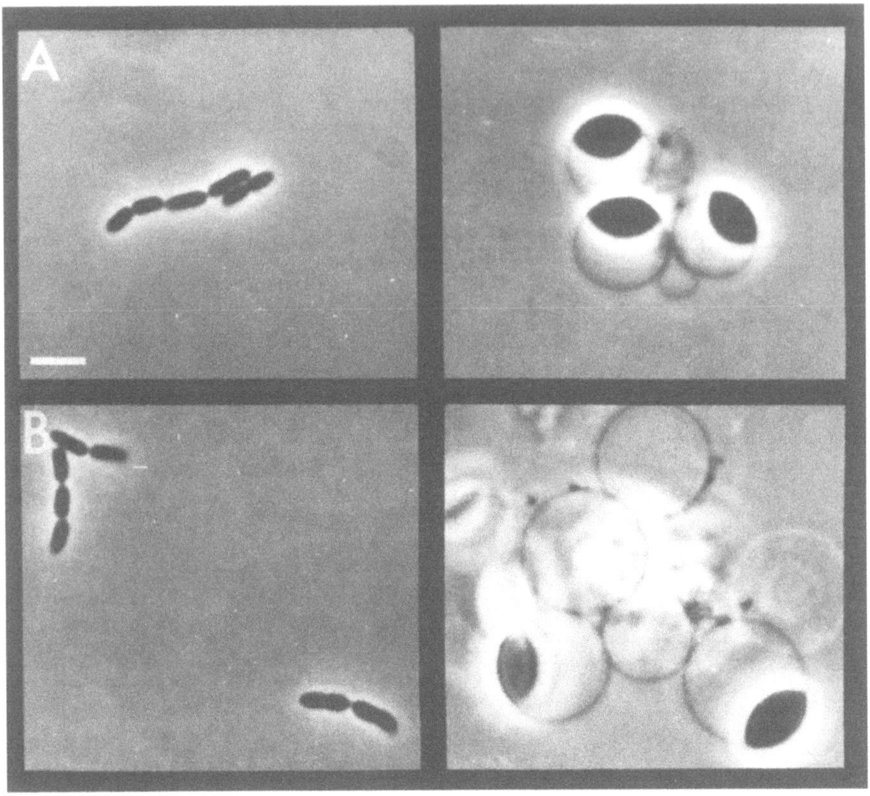

Fig. 4.2 Morphology of cells of *Yersinia enterocolitica* strain P76M after cultivation for 4 h (**a**) and 6 h (**b**) without pesticin (*left*) or with 40 U of pesticin per ml of MOPS-buffered defined medium (*right*) [51]

(Fig. 4.3). Further discussion of these reactions, that account for coagulation and deposition of fibrin are beyond the scope of this chapter and the reader is referred to the excellent review of Korhonen et al. for this information [56]. Curiously, Pla was not required for tissue dissemination in a primitive enzootic isolate normally lacking pPCP but otherwise sharing unique sequences of *Y. pseudotuberculosis* that possibly contribute to the innate invasiveness of this species [57].

Another interesting feature of Pla is that it degrades undelivered pCD-encoded Yops, thereby preventing expression of these virulence effectors in vitro [58–60]. Pla itself is first processed and then affixed in the outer membrane in association with rough LPS [56] where it contacts and hydrolyzes all known secreted pCD-encoded gene products except LcrV. This same type of degradation by cells of wild-type *Y. pestis* obviously does not always occur during growth in vivo (where Yops are normally delivered upon host cell contact, probably by the same mechanisms defined for the enteropathogenic yersiniae).

Table 4.1 Influence of pCD/pYV, pCDC, and Pgm on 50 % lethal doses of *Yersinia pestis* and *Yersinia pseudotuberculosis* by intravenous and peripheral (intraperitoneal and subcutaneous) routes of injection in control mice and mice receiving sufficient injected iron (40 µg) to saturate serum transferrin[a]

Phenotype	Injected Fe^{3+} (40 µg/mouse)[b]	Route of injection: Intravenous	Intraperitoneal	Subcutaneous
Y. pestis				
pCD⁺, pPCP⁺, Pgm⁺	0	8.1×10^0	9.9×10^0	6.1×10^0
	+	9.8×10^0	9.8×10^0	5.9×10^0
pCD⁻, pPCP⁺, Pgm⁺	0	$>10^7$	$>10^7$	$>10^7$
	+	$>10^7$	$>10^7$	$>10^7$
pCD⁺, pPCP⁻, Pgm⁺	0	7.1×10^1	3.8×10^5	$>10^7$
	+	2.3×10^1	1.4×10^1	$>10^7$
pCD⁺, pPCP⁺, Pgm⁻	0	1.5×10^1	$>10^7$	$>10^7$
	+	4.3×10^1	$>10^7$	$>10^7$
Y. pseudotuberculosis				
pYV⁺	0	3.9×10^1	2.9×10^4	1.1×10^4
	+	3.1×10^1	2.0×10^2	1.4×10^2
pYV⁻	0	$>10^7$	$>10^7$	$>10^7$
	+	$>10^7$	$>10^7$	$>10^7$

[a]Data from references [54, 97], and unpublished observations (R. R. Brubaker)

4.2.3.2 pMT

This large plasmid encodes proteinaceous capsular antigen fraction 1 (Caf1), which is essential for full virulence in the guinea pig but dispensable in the mouse [61] and probably man [62]. In addition, pMT harbors the structural gene for murine toxin. This phospholipase is highly toxic for murids but essentially innocuous in other mammals. Murine toxin is, however, required for plague bacilli to colonize the flea [63]. The remaining genes located on pMT are plebeian and unnecessary for growth in vitro or colonization of either the host or vector.

4.2.3.3 Pigmentation

The term "pigmentation" applies to the ability of *Y. pestis* to produce a biofilm [64] at 26 °C, but not 37 °C, that promotes absorption of certain exogenous small planar molecules, especially hemin [46] or the dye Congo red [65], by bacteria growing as colonies on agar surfaces. Reports describing this discovery also demonstrated that injected iron restored full virulence of Pgm⁻ mutants in mice infected by peripheral routes [66], although it is now established that administration of sufficient iron to saturate serum transferrin also inhibits innate immunity, as occurs with *yopE* mutants not known to be defective in iron transport [67]. The Hms⁺ phenotype is important for biofilm formation in fleas, thereby facilitating (blockage-dependent) transmission to mammals and maintenance of yersiniae within the foregut of the

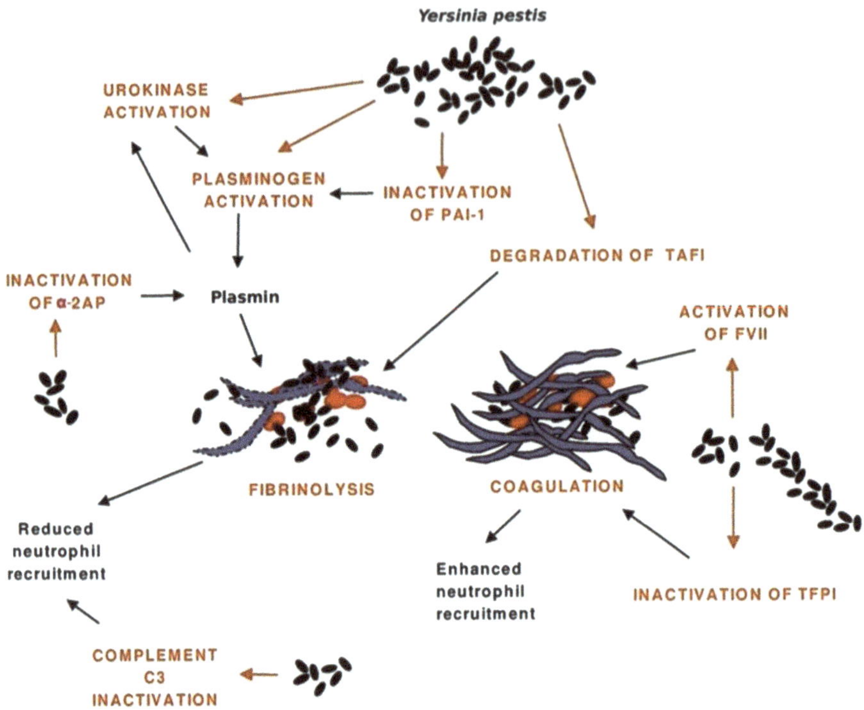

Fig. 4.3 Pla enhances fibrinolysis directly in *Yersinia pestis* by activating plasminogen to plasmin and indirectly by inactivating the antiprotease plasminogen activator inhibitor-1 (PAI-1), by activating the precursor of the plasminogen activator urokinase, and by inactivating the plasmin inhibitor α_2-antiplasmin (α_2AP). Fibrin degradation allows bacterial dissemination and reduces engagement of neutrophils at the infection foci, which is also affected by the Pla-mediated cleavage of the C3 complement protein. Pla favors coagulation by activating the precursor of the enzyme factor VII (FVII) and by inactivating the anticoagulant tissue factor pathway inhibitor (TFPI). Fibrin traps bacteria and decreases their dissemination, and it also increases leukocyte engagement [56]

flea vector [68, 69]. The Hms$^+$ property has no role in causing disease in the mammalian host [70–72]. In contrast, the ability to produce and transport Ybt via its specific receptor (Psn) is, like Pla, essential for invasion of the host following injection by peripheral routes (subcutaneous or intraperitoneal) but of no consequence if the organisms are administrated intravenously (Table 4.1). Furthermore, patterns of attenuation were observed in the bubonic plague model for mutants missing the entire Pgm$^+$ sequence that were essentially identical to those of mutants specifically lacking Psn (or its attendant ABC transporter) [73, 74]. Considered together, these findings indicate that Ybt per se is essential by the subcutaneous route, critical by the intranasal route (with virulence differences in biosynthetic vs transport mutants), and irrelevant by the intravenous route.

It is of interest in this context that Psn required for high-affinity uptake of iron also serves as the receptor of pesticin [75]. This relationship was exploited to show that the mutation rate from Pgm$^+$ to Pgm$^-$ was 10^{-5} [76], a high value caused, at least in part, by reciprocal recombination of flanking *IS*100 elements causing deletion of the whole 102-kb sequence [44]. Additional genes are encoded within the Pgm region that may also influence virulence; an interesting prospect is RipA (required for intracellular proliferation) [77, 78]. Perry, Bobrov, and Fetherston [79] recently reviewed the mechanics, regulation, and roles in virulence of Pgm$^+$-dependent and Pgm$^+$-independent transport systems for iron and other biologically important transition metals.

4.2.3.4 Selective Losses

The ability to disseminate within the host and thus cause acute disease (or colonize the flea) is sometimes incompatible with the chronic enteropathogenic habit (or maintenance of equipment necessary for persistence in soil and water). *Y. pestis* has therefore undergone selection for a number of loss mutations that eliminate activities favoring chronic disease or survival in natural environments. Some important examples are loss of the major pYV-encoded outer membrane adhesin (YadA) of the enteropathogenic yersiniae [80] and maximal inflammation caused by YopJ, although a truncated version of this protein remains [81]. Similarly, loss by *Y. pestis* of the urease in the enteropathogenic yersiniae favored colonization of the flea [82].

4.3 Structure

Thin sections of *Y. pestis* dividing at 37 °C in vitro (Fig. 4.4a) revealed normal gram-negative anatomy including typical irregular lobate nucleoids, ribosome-rich cytosol, and envelopes comprising the inner membrane, periplasmic space, and outer membrane [83]. In contrast, static Ca^{2+}-starved Lcr$^+$ cells (Fig. 4.4b) became enlarged (but not filamentous) and contained nucleoids in the form of long axial filaments [23, 83]. Those genes of *Y. pseudotuberculosis* that regulate or encode flagella and LPS O-group structure are only expressed at 26 °C and are either absent or cryptic in *Y. pestis*. Plague bacilli are therefore always nonmotile and morphologically rough at room temperature, although at 37 °C colonies exhibit a smooth appearance because of the temperature-dependent production of Caf1[25]. The lipid A moiety of LPS in both *Y. pestis* and *Y. pseudotuberculosis* has also undergone temperature-dependent modification promoting its hexa-acetylation at 26 °C but only tetra-acetylation at 37 °C. This feature, as well as other modifications of the outer membrane, lessens inflammation, thereby minimizing the effectiveness of innate immunity [84–86]. Another example of this phenomenon already discussed is the destruction of Yops in the absence of host cell contact by Pla located in the outer membrane.

Cells of *Y. pestis* and *Y. pseudotuberculosis* (but not more distantly related *Y. enterocolitica*) produce a host temperature-dependent [25] periplasmic catalase-

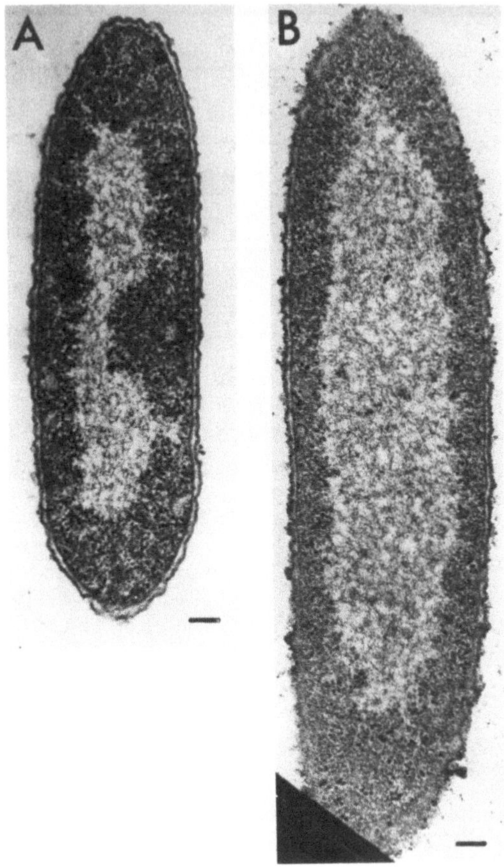

Fig. 4.4 Thin sections of Lcr+ cells of *Yersinia pestis* strain EV76 grown at 37 °C in synthetic medium in the (**a**) presence of 2.5 mM Ca^{2+} and (**b**) absence of added Ca^{2+}; bars,0.1 μm [83]

peroxidase now termed KatY. The raison d'être of KatY is unknown other than it is distinct from vegetative catalase (KatA), produced in great abundance at 37 °C in enriched media containing Fe^{3+}, but not essential for expression of virulence [87, 88]. Putative changes of inner membrane function associated with starvation for Ca^{2+} were already noted and are considered below in more detail.

4.4 Function

The unique latterly acquired elements of epidemic *Y. pestis* discussed above are shared with most enzootic strains and thus do not account for the broad host range of the former. Similarly, they have no known role in reducing the ability to compete in natural environments and thus do not account for the reduced metabolic fitness observed upon comparison of *Y. pestis* with *Y. pseudotuberculosis*. These differences in fitness are now generally recognized to occur as a consequence of the unavoidable chromosomal degeneration permitted by adaptation of *Y. pestis* to the two enriched environments of the mammalian host and flea vector. As mentioned above, the consequent phenotypic loss is extensive (13 %) and probably almost always random except for changes driven by selective pressure to escape restrictions imposed by the enteropathogenic habit. The possibility still exists, however, that regulatory changes imposed by loss of one or more structural genes might directly or indirectly account for the increased host range of epidemic *Y. pestis*.

4.4.1 Intermediary Metabolism

Traditionally, *Y. pestis* was assumed to grow optimally at room temperature and, regardless of temperature, always more slowly than *Y. pseudotuberculosis*. These notions are fallacious. Cells of both species exhibit minimum doubling times of ~70 min at 37 °C in chemically defined medium designed to compensate for salient gene loss in *Y. pestis* [38].

4.4.1.1 Catabolism

Like *Y. pseudotuberculosis*, cells of *Y. pestis* exhibit a complete Embden–Meyerh pathway but lack a functional pentose–phosphate pathway (or hexose-monophosphate shunt) due to loss of glucose 6-phosphate dehydrogenase (*zwf*) [89, 90]. D-glucose facilitates rapid growth but promotes accumulation of excessive acidic end products in all but the most vigorously aerated media. This characteristic, which likely contributes to morbidity during terminal septicemia, is readily bypassed in vitro by use of D-gluconate that is catabolized by the Entner–Doudoroff pathway. Cells of *Y. pestis* contain a complete TCA cycle and readily oxidize the 4C and 5C intermediates of this pathway. The glyoxylate bypass in *Y. pestis* is expressed constitutively due to loss of the IclR transcriptional repressor via frameshift mutation [91]. This mutation has no known physiological relevance and likely occurred as a random event.

Only those naturally occurring amino acids used as sources of energy or nitrogen by *E. coli* were catabolized by cells of *Y. pestis* and *Y. pseudotuberculosis* (Table 4.2) [92]. The resulting rates of destruction fell into two groups (Fig. 4.5). The first consisted of L-serine, glycine, L-threonine, and L-alanine that are all converted to pyruvate prior to entering the TCA cycle (as acetyl-Co A). This initial step occurs directly from L-serine by its deaminase (SdaA) or from L-alanine by its racemase (DadX) and deaminase (DadX), respectively, or indirectly by prior conversion of glycine to L-serine by serine hydroxymethyltransferase (GlyA) or of L-threonine to glycine by threonine dehydrogenase (Tdh) and then 2-amino-3-ketobutyrate CoA ligase (Kbl). These amino acids, especially L-serine, were generally destroyed more rapidly by cells of *Y. pestis* than those of *Y. pseudotuberculosis*. The very high and evidently unregulated SdaA activity in *Y. pestis* (243.8 nmols min mg dry weight $^{-1}$)

Table 4.2 Destruction of readily catabolized amino acids at pH 7.0 by resting Lcr[+] cells of *Yersinia pestis* strain EV76 and *Yersinia pseudotuberculosis* strain PB1 after cultivation at 37 °C in enriched Ca^{2+}-deficient medium[a]

	Rate of destruction (nmol min mg dry wt^{-1})		
Amino acid	*Y. pestis*	*Y. pseudotuberculosis*	R[b]
L-Serine	243.8	53.1	4.59
Glycine	18.1	5.8	3.12
L-Threonine	7,7	10.1	0.76
L-Alanine	12.8	9.3	1.37
L-Asparagine[c]	6.4	46.6	0.14
L-Aspartic acid	0.0	31.0	–
L-Glutamine	0.0	2.2	–
L-Glutamic acid[c]	2.5	10.2	0.25
L-Proline[c]	11.9	11.3	1.05

[a]Data from Dreyfus and Brubaker [92]
[b]R = ratio, where values for *Y. pestis* were set at unity
[c]Amino acids that promoted secretion of L-aspartic acid

is surprising possibly correlated with the broad host range of epidemic isolates and considered again in the subsequent discussion of glycine auxotrophy.

The second group of amino acids is those that were destroyed more rapidly by Ca^{2+}-starved cells of *Y. pseudotuberculosis* (L-asparagine, L-aspartate, L-glutamine, L-glutamate, and L-proline). Results of early work showed that growth of *Y. pestis* was greatly stimulated by CO_2 later found to be physiologically fixed by phosphoenolpyruvic carboxylase (Ppc) and phosphoenolpyruvate carboxykinase (Pck) into oxaloacetate [93]. This finding suggested that the oxaloacetate pool is low during growth in vitro and in need of replenishment by fixation of CO_2. Results of further study indicated that this reduction in pool size resulted as a consequence of defective catabolism of L-aspartic acid due to mutational loss of an almost ubiquitous prokaryote enzyme termed aspartate ammonium-lyase or aspartase (AspA) [92]. This enzyme hydrolyzes L-aspartate to oxaloacetate. As an obvious consequence of this loss, Ca^{2+}-starved cells of *Y. pestis* excrete metabolic L-aspartate following uptake of L-asparagine, L-proline, and L-glutamate, which remain in metabolic equilibrium in AspA[+] *Y. pseudotuberculosis* where they undergo complete oxidation to CO_2, NH_4^+, and H_2O (Table 4.2). The block imposed by loss of AspA in Ca^{2+}-starved *Y. pestis* is especially insidious in the case of L-glutamate which, like L-serine, serves as a favored source of carbon and energy. That is, the consequential loss of metabolic carbon as L-aspartate facilitates reduction of the adenylate energy charge, thereby causing bacteriostasis. Note that the amino acid change in AspA (and *zwf*) reflects a missense mutation that was not predicted to affect activity; similar apparently innocuous alterations in other proteins could have comparable effects that cannot be identified by genome annotation.

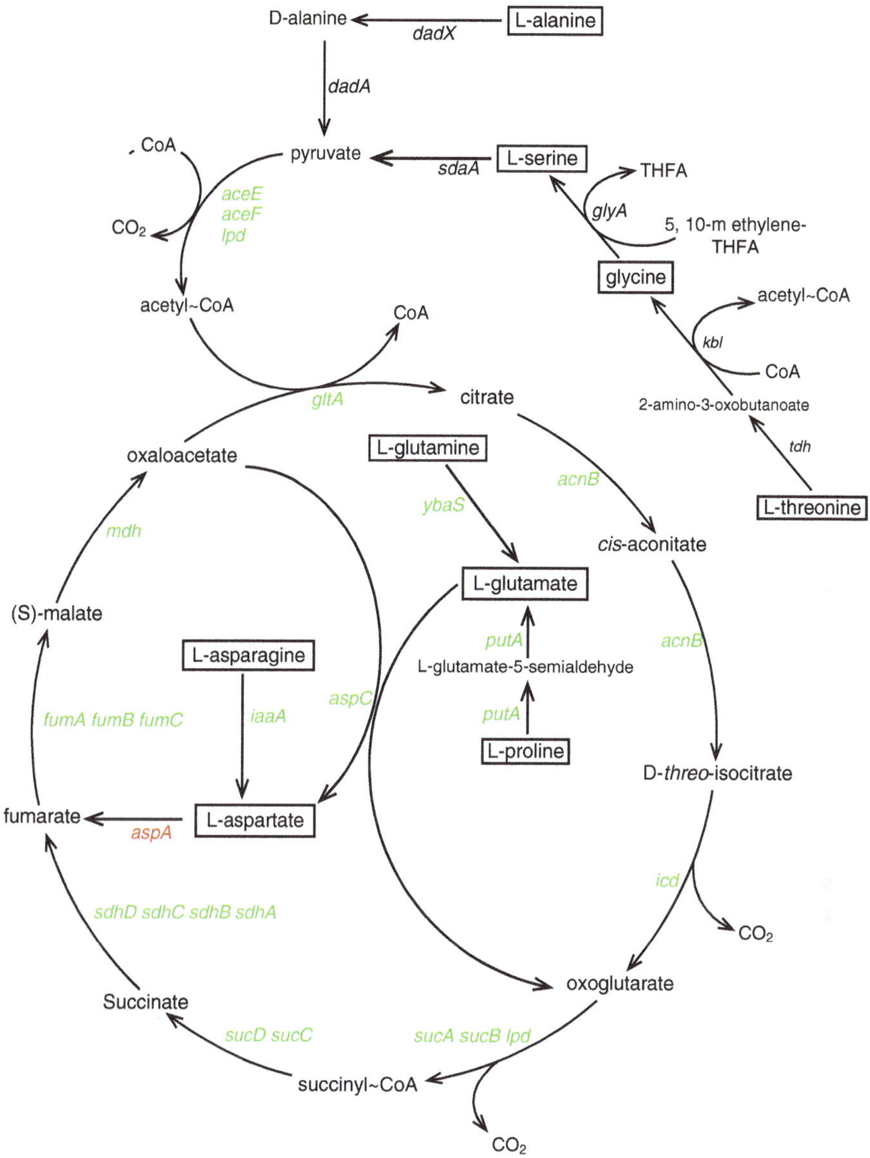

Fig. 4.5 Entrance of readily catabolized naturally occurring amino acids (*shown within boxes*) into the TCA cycle via acetyl-CoA from pyruvate or fumarate from L-aspartate. The structural genes encoding these portals and all other depicted enzymes are functional in *Y. pseudotuberculosis* and *Y. pestis* (*shown in green*) except for aspartase in *Y. pestis* (*shown in red*). Serine deaminase (*sdaA*) is especially active in *Y. pestis* (*heavy arrow*) accounting for the pronounced catabolism of L-threonine, glycine, and, especially, L-serine), whereas the absence of aspartase markedly inhibits the destruction of L-aspartate and metabolically related amino acids as shown in Table 4.2

4.4.1.2 Anabolism

Processes of macromolecular synthesis tend to be conserved in all forms of life and no unique mechanisms of this nature have been reported for *Y. pestis*. Considerable diversity, however, exists in routes of small molecule production, interaction, and regulation although these procedures in yersiniae generally resemble those known to occur in *E. coli*. Unique to most if not all epidemic strains of *Y. pestis* are nutritional requirements for L-cysteine, L-phenylalanine, L-methionine, and either glycine or L-threonine.

1. *L-cysteine*. This requirement is caused by a frameshift in the gene encoding 3'-phospho-adenylylsulfate reductase (CysH) that catalyzes the formation of sulfite ions (SO_3^{-2}) from sulfate ions (SO_4^{-2}). It is commonplace to fulfill this requirement in chemically defined solid medium by addition of more stable thiosulfate ions ($S_2O_3^{-2}$) that also provide expedient reducing potential.
2. *L-phenylalanine*. The nutritional requirement for L-phenylalanine reflects loss of bifunctional chorismate mutase/prephenate dehydratase (PheA) by frameshift due to introduction of an IS element (amino acid codon 253). Curiously, this event only blocks L-phenylalanine synthesis by isolates of the 1.ORI lineage (Fig. 4.1) and a different insert (amino acid codon 311) causing a larger deletion exists in certain L-phenylalanine-independent strains of earlier lineages.
3. *L-methionine*. Loss of *O*-succinylhomoserine (thiol)-lyase/*O*-succinylhomoserine lyase (MetB) by frameshift accounts for this nutritional requirement. The enzyme utilizes L-cysteine (already a nutritional requirement) and *O*-succinyl-L-homoserine to form L-cystathionine, long known to be capable of replacing L-methionine as an essential nutritional factor [94].
4. *Glycine (or L-threonine)*. The nature of this requirement was perplexing because the most common glycine auxotroph of *E. coli* can also utilize L-serine for growth. This replacement in plague bacilli, however, is not effective, although L-threonine is known to yield glycine in these organisms by catabolic cleavage. The genes encoding all anabolic enzymes required for de novo synthesis of L-serine and its dehydroxylation to glycine are functional in *Y. pestis* as judged by genomic annotation or direct enzymatic assay. Solution of this question, therefore, required an explanation of why GlyA is evidently inactive in vitro. One possibility was that the observed high rate of L-serine degradation via SdaA in *Y. pestis* was sufficient to successfully compete with GlyA, although the possibility also existed that the lesion in question actually promoted dramatic SdaA upregulation per se.

This problem was resolved upon realization that an insertion element causes the loss of formyltetrahydrofolate deformylase (PurU) in *Y. pestis* [15] and that this enzyme in *E. coli* is necessary to maintain physiological levels of uncharged (deformylated) tetrahydrofolate within the bacterial cell [95]. As in the case of *E. coli* [96], PurU is necessary in *Y. pestis* to ensure the presence of sufficient tetrahydrofolate to catalyze the GlyA-dependent synthesis of glycine from L-serine (V. L. Motin, personal communication). Furthermore, restoration of the PurU$^+$ phenotype reduced the marked ability of epidemic *Y. pestis* to destroy L-serine to a level similar to that of *Y. pseudotuberculosis* (R. R. Brubaker, unpublished observations). The physiological significance of this putative regulatory role of PurU is uncertain but it is of considerable interest that *purU* was first introduced into the *Y. pestis* genome shortly before appearance of the 0.ANT1 lineage (a time close to the 1st plague pandemic) and remained fixed in the population thereafter (R. Yang, personal communication).

References

1. Bos KI, Schuenemann VJ, Golding GB, Burbano HA, Waglechner N, Coombes BK, McPhee JB, DeWitte SN, Meyer M, Schmedes S, et al. A draft genome of *Yersinia pestis* from victims of the black death. Nature. 2011;478:506–10. doi:10.1038/nature10549.
2. Harbeck M, Seifert L, Hänsch S, Wagner DM, Birdsell D, Parise KL, Wiechmann I, Grupe G, Thomas A, Keim P, et al. *Yersinia pestis* DNA from skeletal remains from the 6(th) century AD reveals insights into justinianic plague. PLoS Pathog. 2013;9(5):e1003349. doi:10.1371/journal.ppat.1003349.
3. Nakajima R, Motin VL, Brubaker RR. Suppression of cytokines in mice by protein A-V antigen fusion peptide and restoration of synthesis by active immunization. Infect Immun. 1995;63(8):3021–9.
4. Achtman M, Morelli G, Zhu P, Wirth T, Diehl I, Kusecek B, Vogler AJ, Wagner DM, Allender CJ, Easterday WR, et al. Microevolution and history of the plague bacillus, Yersinia pestis. Proc Natl Acad Sci U S A. 2004;101(51):17837–42.
5. Morelli G, Song Y, Mazzoni CJ, Eppinger M, Roumagnac P, Wagner DM, Feldkamp M, Kusecek B, Vogler AJ, Li Y, et al. *Yersinia pestis* genome sequencing identifies patterns of global phylogenetic diversity. Nat Genet. 2010;42(12):1140–3. doi:10.1038/ng.1705.
6. Devignat R. Varietes de l'espece Pasteurella pestis. Nouvelle hypothese. Bull WHO. 1951;4 SRC – GoogleScholar:247–63.
7. Brubaker RR. Yersinia pestis. In: Tang Y-W, Liu D, Schwartzman JD, Sussman M, Poxton IR, editors. Molecular medical microbiology, vol. 3. 2nd ed. San Diego: Academic Press/Elsevier; 2015. p. 1845–65.
8. Anisimov AP, Lindler LE, Pier GB. Intraspecific diversity of Yersinia pestis. Clin Microbiol Rev. 2004;17(2):434–64.
9. Davies DAL. Dideoxysugars of Pasteurella pseudotuberculosis-specific polysaccharides, and the occurrence of ascarylose. Nature. 1961;191 SRC – GoogleScholar:43–4.
10. Samuelsson K, Lindberg B, Brubaker RR. Structure of O-specific side chains of lipopolysaccharides from Yersinia pseudotuberculosis. J Bacteriol. 1974;117(3):1010–6.
11. Skurnik M, Peippo A, Ervela E. Characterization of the O-antigen gene clusters of *Yersinia pseudotuberculosis* and the cryptic O-antigen gene cluster of *Yersinia pestis* shows that the plague bacillus is most closely related to and has evolved from *Y. pseudotuberculosis* serotype O:1b. Mol Microbiol. 2000;37:316–30. doi:10.1046/j.1365-2958.2000.01993.x.
12. Moore RL, Brubaker RR. Hybridization of deoxyribonucleotide sequences of *Yersinia enterocolitica* and other selected members of Enterobacteriaceae. Int J Syst Bact. 1975;25:336–9.
13. Chain PSG, Carniel E, Larimer FW, Lamerdin J, Stoutland PO, Regala WM, Georgescu AM, Vergez LM, Land ML, Motin VL, et al. Insights into the evolution of Yersinia pestis through whole-genome comparison with Yersinia pseudotuberculosis. Proc Natl Acad Sci U S A. 2004;101(38):13826–31.
14. Chain PSG, Hu P, Malfatti SA, Radnedge L, Larimer FW, Vergez LM, Worsham P, Chu MC, Andersen GL. Complete genome sequence of *Yersinia pestis* strains Antiqua and Nepal516: evidence of gene reduction in an emerging pathogen. J Bacteriol. 2006;188(12):4453–63.

15. Parkhill J, Wren BW, Thomson NR, Titball RW, Holden MT, Prentice MB, Sebaihia M, James KD, Churcher C, Mungall KL, et al. Genome sequence of Yersinia pestis, the causative agent of plague. Nature. 2001;413(6855):523–7.
16. Smith T. Parasitism and disease. Princeton: Princeton University Press; 1934.
17. Goguen JD, Yother J, Straley SC. Genetic analysis of the low calcium response in *Yersinia pestis* mu d1(Ap *lac*) insertion mutants. J Bacteriol. 1984;160(3):842–8.
18. Zahorchak RJ, Charnetzky WT, Little RV, Brubaker RR. Consequences of Ca^{2+} deficiency on macromolecular synthesis and adenylate energy charge in *Yersinia pestis*. J Bacteriol. 1979;39:792–9.
19. Kugelmass NL, Charles C. Biochemistry of blood in health and disease. Springfield: Thomas Inc; 1959.
20. Higuchi K, Smith JL. Studies on the nutrition and physiology of Pasteurella pestis: VI. A differential plating medium for the estimation of the mutation rate to avirulence. J Bacteriol. 1961;81 SRC – GoogleScholar:605–8.
21. Lawton WD, Erdman RL, Surgalla MJ. Biosynthesis and purification of V and W antigen in *Pasteurella pestis*. J Immunol. 1963;91 SRC – GoogleScholar:179–84.
22. Straley SC, Bowmer WS. Virulence genes regulated at the transcriptional level by Ca^{2+} in *Yersinia pestis* include structural genes for outer membrane proteins. Infect Immun. 1986;51:445–54.
23. Brubaker RR, Surgalla MJ. The effect of Ca^{++} and Mg^{++} on lysis, growth, and production of virulence antigens by *Pasteurella pestis*. J Infect Dis. 1964;114:13–25.
24. Cornelis GR, Sluiters C, Delor I, Geib D, Kaniga K. Lambert de Rouvroit C, Sory MP, Vanootegham JC, Michiels T: *ymoA*, a *Yersinia enterocolitica* chromosomal gene modulating the expression of virulence functions. Mol Microbiol. 1991;5:1023–34.
25. Crumpton MY, Davies DAL. An antigenic analysis of *Pasteurella pestis* by diffusion of antigens and antibodies in agar. Proc R Soc Lond Ser B. 1956;145:109–34.
26. Francis MS, Lloyd SA, Wolf-Watz H. The type III secretion chaperone LcrH co-operates with YopD to establish a negative, regulatory loop for control of Yop synthesis in *Yersinia pseudotuberculosis*. Mol Microbiol. 2001;42:1075–93.
27. Cheng LW, Kay O, Schneewind O. Regulated secretion of YopN by the type III machinery of *Yersinia enterocolitica*. J Bacteriol. 2001;183(18):5293–301.
28. Ferracci F, Schubot FD, Waugh DS, Plano GV. Selection and characterization of *Yersinia pestis* YopN mutants that constitutively block Yop secretion. Mol Microbiol. 2005;57(4): 970–87.
29. Matson JS, Nilles ML. LcrG-LcrV interaction is required for control of Yops secretion in *Yersinia pestis*. J Bacteriol. 2001;183(17):5082–91.
30. Motin VL, Nakajima R, Smirnov GB, Brubaker RR. Passive immunity to yersiniae mediated by anti-recombinant V antigen and protein A-V antigen fusion peptide. Infect Immun. 1994;62(10):4192–201.
31. Cornelis GR, Wolf-Watz H. The Yersinia Yop virulon: a bacterial system for subverting eukaryotic cells. Mol Microbiol. 1997;23(5):861–7.
32. Rosqvist R, Magnusson KE, Wolf-Watz H. Target cell contact triggers expression and polarized transfer of *Yersinia* YopE cytotoxin into mammalian cells. EMBO J. 1994;13(4):964–72.
33. Torruellas J, Jackson MW, Pennock JW, Plano GV. The *Yersinia pestis* type III secretion needle plays a role in the regulation of Yop secretion. Mol Microbiol. 2005;57(6):1719–33.
34. Perry RD, Brubaker RR. Transport of Ca^{2+} by *Yersinia pestis*. J Bacteriol. 1987;169(10): 4861–4.
35. Edgren T, Forsberg A, Rosqvist R, Wolf-Watz H. Type III secretion in *Yersinia*: injectisome or not? PLoS Pathog. 2012;8(5), e1002669.
36. Akopyan K, Edgren T, Wang-Edgren H, Rosqvist R, Fahlgren A, Wolf-Watz H, Fallman M. Translocation of surface-localized effectors in type III secretion. Proc Natl Acad Sci U S A. 2011;108(4):1639–44.
37. Fowler JM, Brubaker RR. Physiological basis of the low calcium response in *Yersinia pestis*. Infect Immun. 1994;62(12):5234–41.

38. Brubaker RR. Influence of Na^+, dicarboxylic amino acids, and pH in modulating the low-calcium response of *Yersinia pestis*. Infect Immun. 2005;73(8):4743–52.
39. Reyes-Prieto A, Barquera B, Juárez O. Origin and evolution of the sodium -pumping NADH: ubiquinone oxidoreductase. PLoS One. 2014;9(5):e96696.
40. Fowler JM, Wulff CR, Straley SC, Brubaker RR. Growth of calcium-blind mutants of *Yersinia pestis* at 37 degrees C in permissive Ca^{2+}-deficient environments. Microbiology. 2009;155(8): 2509–21.
41. Ferber DM, Brubaker RR. Plasmids in *Yersinia pestis*. Infect Immun. 1981;31(2):839–41.
42. Buchrieser C, Prentice M, Carniel E. The 102-kilobase unstable region of *Yersinia pestis* comprises a high-pathogenicity island linked to a pigmentation segment which undergoes internal rearrangement. J Bacteriol. 1998;180(9):2321–9.
43. Heesemann J. Chromosomal-encoded siderophores are required for mouse virulence of enteropathogenic *Yersinia* species. FEMS Microbiol Lett. 1987;48:229–33.
44. Fetherston JD, Perry RD. The pigmentation locus of *Yersinia pestis* KIM6+ is flanked by an insertion sequence and includes the structural genes for pesticin sensitivity and HMWP2. Mol Microbiol. 1994;13(4):697–708.
45. Perry RD, Fetherston JD. Yersiniabactin iron uptake: mechanisms and role in *Yersinia pestis* pathogenesis. Microbes Infect. 2011;13(10):808–17. doi:10.1016/j.micinf.2011.1004.1008.
46. Jackson S, Burrows TW. The pigmentation of *Pasteurella pestis* on a defined medium containing haemin. Br J Exp Pathol. 1956;37:570–6.
47. Hu P, Elliott J, McCready P, Skowronski E, Garnes J, Kobayashi A, Brubaker RR, Garcia E. Structural organization of virulence-associated plasmids of *Yersinia pestis*. J Bacteriol. 1998;180(19):5192–202.
48. Ben-Gurion R, Hertman I. Bacteriocin-like material produced by *Pasteurella pestis*. J Gen Microbiol. 1958;19:289–97.
49. Madison RR. Fibrinolytic specificity of *Bacillus pestis*. Proc Soc Exp Biol Med. 1936;34:301–2.
50. Jawetz E, Meyer KF. Studies on plague immunity in experimental animals. II. Some of the immunity mechanism in bubonic plague. J Immunol. 1944;49:15–30.
51. Hall PJ, Brubaker RR. Pesticin-dependent generation of osmotically stable spheroplast-like structures. J Bacteriol. 1978;136(2):786–9.
52. Vollmer W, Pilsl H, Hantke K, Höltje JV, Braun V. Pesticin displays muramidase activity. J Bacteriol. 1997;179(5):1580–3.
53. Beesley ED, Brubaker RR, Janssen WA, Surgalla MJ. Pesticins. III. Expression of coagulase and mechanism of fibrinolysis. J Bacteriol. 1967;94:19–26.
54. Brubaker RR, Beesley ED, Surgalla MJ. *Pasteurella pestis*: role of pesticin I and iron in experimental plague. Science. 1965;149:422–4.
55. Sodeinde OA, Subrahmanyam YV, Stark K, Quan T, Bao Y, Goguen JD. A surface protease and the invasive character of plague. Science. 1992;258:1004–7.
56. Korhonen TK, Haiko J, Laakkonen L, Järvinen HM, Westerlund-Wikström B. Fibrinolytic and coagulative activities of *Yersinia pestis*. Front Cell Infect Microbiol. 2013;3:35. doi:10.3389/fmicb.2015.00063.
57. Garcia E, Worsham P, Bearden S, Malfatti S, Lang D, Larimer F, Lindler L, Chain P. Pestoides F, an atypical *Yersinia pestis* strain from the former Soviet Union. Adv Exp Med Biol. 2007;603:17–22.
58. Sample AK, Fowler JM, Brubaker RR. Modulation of the low-calcium response in *Yersinia pestis* via plasmid-plasmid interaction. Microb Pathog. 1987;2(6):443–53.
59. Sample AK, Brubaker RR. Post-translational regulation of Lcr plasmid-mediated peptides in pesticinogenic Yersinia pestis. Microb Pathog. 1987;3(4):239–48.
60. Mehigh RJ, Braubaker RR. Major stable peptides of *Yersinia pestis* synthesized during the low-calcium response. Infect Immun. 1993;61(1):13–22.
61. Burrows TW. Virulence of Pasteurella pestis. Nature. 1957;179:1246–7.
62. Winter CC, Cherry WB, Moody MD. An unusual strain of *Pasteurella pestis* isolated from a fatal case of human plague. Bull WHO. 1960;23:408–9.

63. Hinnebusch BJ, Rudolph AE, Cherepanov P, Dixon JE, Schwan TG, Forsberg A. Role of *Yersinia* murine toxin in survival of *Yersinia pestis* in the midgut of the flea vector. Science. 2002;296(5568):733–5.
64. Darby C, Hsu JW, Ghori N, Falkow S. *Caenorhabditis elegans*: plague bacteria biofilm blocks food intake. Nature. 2002;417(6886):243–4.
65. Surgalla MJ, Beesley ED. Congo red-agar plating medium for detecting pigmentation in *Pasteurella pestis*. Appl Microbiol. 1969;18(5):834–7.
66. Jackson S, Burrows TW. The virulence enhancing effect of iron on non-pigmented mutants of virulent strains of *Pasteurella pestis*. Br J Exp Pathol. 1956;37:577–83.
67. Mehigh RJ, Sample AK, Brubaker RR. Expression of the low calcium response in Yersinia pestis. Microb Pathog. 1989;6(3):203–17.
68. Hinnebusch BJ, Perry RD, Schwan TG. Role of the Yersinia pestis hemin storage (hms) locus in the transmission of plague by fleas. Science (New York, NY). 1996;273(5273):367–70.
69. Jarrett CO, Deak E, Isherwood KE, Oyston PC, Fischer ER, Whitney AR, Kobayashi SD, DeLeo FR, Hinnebusch BJ. Transmission of *Yersinia pestis* from an infectious biofilm in the flea vector. J Infect Dis. 2004;190:783–92.
70. Lillard JW, Bearden SW, Fetherston JD, Perry RD. The haemin storage (Hms+) phenotype of *Yersinia pestis* is not essential for the pathogenesis of bubonic plague in mammals. Microbiology. 1999;145:197–209.
71. Abu Khweek A, Fetherston JD, Perry RD. Analysis of HmsH and its role in plague biofilm formation. Microbiology. 2010;156:1424–38. doi:10.1099/mic.1420.036640-036640.
72. Bobrov AG, Kirillina O, Ryjenkov DA, Waters CM, Price PA, Fetherston JD, Mack D, Goldman WE, Gomelsky M, Perry RD. Systematic analysis of cyclic di-GMP signalling enzymes and their role in biofilm formation and virulence in *Yersinia pestis*. Mol Microbiol. 2011;79(2):533–51. doi:10.1111/j.1365-2958.
73. Fetherston JD, Kirillina O, Bobrov AG, Paulley JT, Perry RD. The yersiniabactin transport system is critical for the pathogenesis of bubonic and pneumonic plague. Infect Immun. 2010;78(5):2045–52.
74. Bobrov AG, Kirillina O, Fetherston JD, Miller MC, Burlison JA, Perry RD. The *Yersinia pestis* HmsCDE regulatory system is essential for blockage of the oriental rat flea (Xenopsylla cheopis), a classic plague vector. Mol Microbiol. 2014;17(4):947–59.
75. Bearden SW, Fetherston JD, Perry RD. Genetic organization of the yersiniabactin biosynthetic region and construction of avirulent mutants in *Yersinia pestis*. Infect Immun. 1997;65(5): 1659–68.
76. Brubaker RR. Mutation rate to nonpigmentation in *Pasteurella pestis*. J Bacteriol. 1970; 98:1404–6.
77. Pujol C, Grabenstein JP, Perry RD, Bliska JB. Replication of *Yersinia pestis* in interferon gamma-activated macrophages requires *ripA*, a gene encoded in the pigmentation locus. Proc Natl Acad Sci U S A. 2005;102(36):12909–14.
78. Torres R, Swift RV, Chim N, Wheatley N, Lan B, Atwood BR, Pujol C, Sankaran B, Bliska JB, Amaro RE, et al. Biochemical, structural and molecular dynamics analyses of the potential virulence factor RipA from *Yersinia pestis*. PLoS One. 2011;6(9):e25084. doi:10.1371/journal.pone.0025084.
79. Perry RD, Bobbrov AG, Fetherston JD. The role of transition metal transporters for iron, zinc, manganese, and copper in the pathogenesis of *Yersinia pestis*. Metallomics. 2015; (In press). doi:10.1039/C1034MT00332B.
80. Skurnik M, Wolf-Watz H. Analysis of the *yopA* gene encoding the Yop1 virulence determinants of *Yersinia* spp. Mol Microbiol. 1989;3(4):517–29.
81. Zauberman A, Cohen S, Mamroud E, Flashner Y, Tidhar A, Ber R, Elhanany E, Shafferman A, Velan B. Interaction of *Yersinia pestis* with macrophages: limitations in YopJ-dependent apoptosis. Infect Immun. 2006;74(6):3239–50.
82. Chouikha I, Hinnebusch BJ. Silencing urease: a key evolutionary step that facilitated the adaptation of *Yersinia pestis* to the flea-borne transmission route. Proc Natl Acad Sci U S A. 2014;111:18709–14. doi:10.1073/pnas.1413209111.

83. Hall PJ, Yang GC, Little RV, Brubaker RR. Effect of Ca^{2+} on morphology and division of *Yersinia pestis*. Infect Immun. 1974;9(6):1105–13.
84. Matsuura M. Structural modifications of bacterial lipopolysaccharide that Facilitate gram-negative bacteria evasion of host innate immunity. Front Immunol. 2013;24(4):109. doi:10.3389/fimmu.
85. Rebeil R, Ernst RK, Jarrett CO, Adams KN, Miller SI, Hinnebusch BJ. Characterization of late acyltransferase genes of *Yersinia pestis* and their role in temperature-dependent lipid A variation. J Bacteriol. 2006;188:1381–8.
86. Vladimer GI, Weng D, Paquette SWM, Vanaja SK, Rathinam VAK, Aune MH, Conlon JE, Burbage JJ, Proulx MK, Liu Q, et al. The NLRP12 inflammasome recognizes *Yersinia pestis*. Immunity. 2012;37(1):96–107.
87. Garcia E, Nedialkov YA, Elliott J, Motin VL, Brubaker RR. Molecular characterization of KatY (antigen 5), a thermoregulated chromosomally encoded catalase-peroxidase of Yersinia pestis. J Bacteriol. 1999;181(10):3114–22.
88. Han Y, Geng J, Qiu Y, Guo Z, Zhou D, Bi Y, Du Z, Song Y, Wang X, Tan Y, et al. Physiological and regulatory characterization of KatA and KatY in *Yersinia pestis*. DNA Cell Biol. 2008;27(8):453–62.
89. Mortlock RP. Gluconate metabolism of *Pasteurella pestis*. J Bacteriol. 1962;84:53–9.
90. Mortlock RP, Brubaker RR. Glucose 6-phosphate dehydrogenase and 6-phosphogluconate dehydrogenase activities of Pasteurella pestis. J Bacteriol. 1962;84(5):1122–3.
91. Sebbane F, Jarrett CO, Linkenhoker JR, Hinnebusch BJ. Evaluation of the role of constitutive isocitrate lyase activity in *Yersinia pestis* infection of the flea vector and mammalian host. Infect Immun. 2004;72(12):7334–7.
92. Dreyfus LA, Brubaker RR. Consequences of aspartase deficiency in *Yersinia pestis*. J Bacteriol. 1978;136(2):757–64.
93. Baugh CL, Lanham JW, Surgalla MJ. Effects of bicarbonate on growth of *Pasteurella pestis*. II. Carbon dioxide fixation into oxalacetate by cell-free extracts. J Bacteriol. 1964;88:553–8.
94. Englesberg E. The irreversibility of L-methionine synthesis from cysteine in Pasteurella pestis. J Bacteriol. 1952;63:675–80.
95. Nagy PL, Marolewski A, Benkovic SJ, Zalkin H. Formyltetrahydrofolate hydrolase, a regulatory enzyme that functions to balance pools of tetrahydrofolate and one-carbon tetrahydrofolate adducts in *Escherichia coli*. J Bacteriol. 1995;177:1292–8.
96. Nagy PL, McCorkle GM, Zalkin H. *purU*, a source of formate for purT-dependent phosphoribosyl-N-formylglycinamide synthesis. J Bacteriol. 1993;175:7066–73.
97. Une T, Brubaker RR. In vivo comparison of avirulent Vwa^- and Pgm^- or Pst^r phenotypes of yersiniae. Infect Immun. 1984;43(3):895–900.

Chapter 5
Ecology of *Yersinia pestis* and the Epidemiology of Plague

Vladimir M. Dubyanskiy and Aidyn B. Yeszhanov

Abstract This chapter summarizes information about the natural foci of plague in the world. We describe the location, main hosts, and vectors of *Yersinia pestis*. The ecological features of the hosts and vectors of plague are listed, including predators – birds and mammals and their role in the epizootic. The epizootic process in plague and the factors affecting the dynamics of epizootic activity of natural foci of *Y. pestis* are described in detail. The mathematical models of the epizootic process in plague and predictive models are briefly described. The most comprehensive list of the hosts and vectors of *Y. pestis* in the world is presented as well.

Keywords Epizootic activity • Yersinia pestis • Focus • Foci • Model • Host • Vector

5.1 Distribution of Plague Foci and Unanswered Questions

In this section, we describe the distribution of natural foci of plague around the world, briefly presenting the landscapes, carriers, and vectors of the plague microbe *Yersinia pestis*.

V.M. Dubyanskiy (✉)
Stavropol Research Anti-Plague Institute, Stavropol, Russian Federation
e-mail: dvmplague@rambler.ru; dvmplague@gmail.com

A.B. Yeszhanov
M. Aikimbayev Kazakh Scientific Center of Quarantine and Zoonotic Diseases, Almaty, Kazakhstan

R. Yang, A. Anisimov (eds.), *Yersinia pestis: Retrospective and Perspective*, Advances in Experimental Medicine and Biology 918,
DOI 10.1007/978-94-024-0890-4_5

5.1.1 Brief Description of the Focal Areas

Plague is one of the most dangerous epidemic diseases. During the three pandemics (in VI, XIV, and XIX–XX c. AD), it has claimed the lives of about 170 million people [1]. The morbidity of people is largely due to the epizootic activity of plague foci [2, 3].

Natural plague foci are located approximately at the level of 49° northern latitude in Eastern Europe and 34° south latitude in Africa, 51° northern latitude in North America and East Asia, and 37° south latitude in South America. Conditionally, this space can be called "the plague area."

Natural plague foci are located in areas with a continental climate, the arid steppes and deserts, with insufficient rainfall amount for the development of woody vegetation. In the area of high-altitude steppes, within this zone the alpine foci of plague have been formed. In general, in Africa, Eurasia, and North and South America, plague foci confined to the arid region of the deserts, semideserts, dry plains savannas, steppes, and mountain meadows with their characteristic communities of warm-blooded animals and their ectoparasites. Such foci can be considered primary. The primary resistant natural foci of plague are adjacent secondary foci, emerged in populations of synanthropic rodents in the port cities and wet-tropical mainland areas, inhabited by rats of different species [30].

Unfortunately, detailed maps of the focal areas exist only in some countries. Modern approximate and far not complete location of the natural plague foci is shown in Fig. 5.1.

The modern-day situation is described in Tables 5.1 and 5.2. Despite the absence of a pandemic, the number of plague cases in the world is currently>2000 per year,

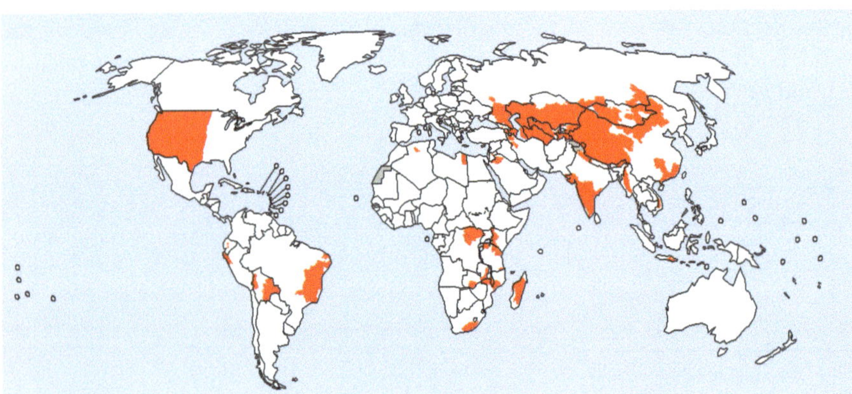

Fig. 5.1 Global distribution of natural plague foci as of March 2016 (▬ Areas (First administrative level representation) with potential plague natural foci based on the historical data and current information. Source: WHO/PED, as of 15 March 2016. The boundaries and names shown and the designations used on this map do not imply the expression of any opinion whatsoever on the part of the World Health Organization concerning the legal status of any country, territory, city or area or of its authorities, or concerning the delimitation of its frontiers or boundaries. Dotted lines on maps represent approximate border lines for which there may not yet be full agreement. ©WHO 2016. All rights reserved)

Table 5.1 Reports of human plague cases (and related mortalities) in countries: 2002–2011

Continents and countries	2002	2003	2004	2005	2006	2007	2008	2009	2010	2011
Total in the world	**1782**	**2319 (1)**	**2303 (169)**	**1895 (136)**	**2271 (179)**	**2418 (155)**	**2686 (136)**	**985 (72)**	**31 (1)**	**335 (56)**
Africa	**1672**	**2091**	**2256 (156)**	**1855 (134)**	**2225 (176)**	**2310 (150)**	**2649 (132)**	**940 (67)**	**12 (1)**	**333 (56)**
Algeria		11(0)	0	0	0	0	4(1)	0		
Zambia			0	0	0	425(2)	34(0)	0		
Congo	798(?)	1092(?)	1042(58)	1434(99)	1789(119)	966(47)	1962(52)	618(27)		
Libya			0	0	0	0	0	5(1)		23(7)
Madagascar	658	933	1214(98)	421(35)	412(51)	583(71)	535(71)	289(38)	5(1)	310(49)
Tanzania	19(0)		0	0	0	59(1)	74(4)	2(0)	7(0)	
Uganda	60(0)	24(0)	0	0	24(6)	277(29)	40(4)	26(1)		
Malawi	92(0)									
Mozambique	45(0)	31(0)								
Asia	**99**	**29(1)**	**36 (12)**	**16 (2)**	**4**	**75 (3)**	**6 (3)**	**12 (3)**	**0**	**0**
India	16(0)		8(3)	0	0	0	0	0		
Indonesia	1(0)	2(0)	7(0)	11(0)	4(0)	71(1)	3(0)	0		
China	68(0)	13(0)	21(9)	5(2)	0	2(1)	2(2)	12(3)		
Mongolia	6(0)	10(0)	0	0	0	2(1)	1(1)	0		
Kazakhstan		4(1)								
Vietnam	8(0)									
America	**11**	**199**	**11 (1)**	**24**	**42 (3)**	**33 (2)**	**31 (1)**	**33 (2)**	**19**	**2 (0)**
Peru	9(0)	198(0)	8(0)	16(0)	25(1)	26(0)	28(1)	25(0)	17(0)	
USA	2(0)	1(0)	3(1)	8(0)	17(2)	7(2)	3(0)	8(2)	2(0)	2(0)

Table 5.2 Reports of human plague cases (and related mortalities) in countries: 2012–2015 (by [4] corrected by Dubyanskiy V.M.)

Continents and countries	2012	2013	2014	2015
Total in the world	**427 (81)**	**772 (130)**	**622 (130)**	**321 (77)**
Africa	**416 (81)**	**743 (126)**	**597 (125)**	**301 (71)**
Congo	131 (15)	55 (5)	78 (12)	18 (5)
Madagascar	256 (60)	675 (118)	482 (112)	275 (63)
Tanzania	7 (0)	0 (0)	31 (1)	5 (0)
Uganda	22 (6)	13 (3)	6 (0)	3 (0)
Asia	**1 (0)**	**1 (1)**	**5 (3)**	**4 (2)**
China	1 (0)	0 (0)	3 (3)	0 (0)
Mongolia	0 (0)	0 (0)	1 (0)	3 (2)
Russian Federation	0 (0)	0 (0)	1 (0)	1 (0)
Kyrgyzstan	0 (0)	1 (1)	0 (0)	0 (0)
America	**10 (0)**	**28 (3)**	**20 (2)**	**16 (4)**
Bolivia	0 (0)	0 (0)	2 (1)	0 (0)
Peru	6 (0)	24 (2)	8 (1)	0 (0)
USA	4 (0)	4 (1)	10 (0)	16 (4)

with most occurring in connection with natural plague foci. These are presented in the following sections.

5.1.2 Distribution of Plague Foci

Please see the last part of this chapter for detail.

5.1.3 Unanswered Questions

The main question that remains unanswered is whether there is an interepizootic period. Among the many hypotheses formulated to answer this question is the one proposing that the epizootic process proceeds as a chain of acute sporadic cases. This hypothesis of a "nomadic" plague suggests that the epizootic process enters a territory when conditions are optimal for the development of an epizootic number of hosts and vectors. Plague microbes may be preserved in microfoci, which serve as optimal habitats consistently populated by hosts and vectors and thereby continuously maintain the epizootic process. The gradual passage of the pathogen is ensured by the ability of plague microbes to persist for long periods of time in the bodies of their arthropod (fleas and ticks) vectors. This offers a means of preservation of the plague in the interepizootic period, the circulation of atypical strains, and the extended transmission of plague by predatory mammals and birds [5–12]. Other hypotheses regarding plague transmission derive from the studies of Yermilov et al. [13], Soldatkin and Rudenchik [14], Dyatlov et al. [15], Popov [16], Sludskyi et al. [17],

Sagymbek et al. [18], Karimov and Neronov [1], and many others. They assume that the plague pathogen can persist in soil, in corpses, in the excrement of fleas, and in a symbiotic relationship with soil microorganisms, with the bacterium temporarily existing in a saprophytic form, and describe the possibility of repeated acts of speciation. An evaluation of these hypotheses is beyond the scope of this chapter, but the knowledge on which they are based is essential.

Another question that remains largely unstudied is the change in the plague microbe at different phases of the epizootic process in connection with the dynamics of the number of hosts and vectors at different plague foci. In pre-Balkhash natural foci, strains of fraction-1-negative plague have been periodically detected. However, fraction-1 negativity is practically absent from natural foci in Caspian sandy regions.

Whether foci can be poly-hostal or are mono-hostal is also a matter of long-standing debate. According to supporters of mono-hostal foci, the main host of the plague is sufficient for the long-term functioning of the epizootic. In the 1940s, Fenyuk and, later, Rall reported on mono-hostal foci of plague. However, since then a number of researchers have advocated the poly-hostal theory of foci, in which the integration of an abundance of different species of rodents living in a natural focus plays a major role in the maintenance of a plague enzootic, regardless of the number and ecological features of those species.

Conversely, Burdelov [32] refuted the division of plague hosts into main and secondary ones. Given the contacts between the different rodent species in the burrows of the great gerbil and in the presence of numerous fleas, it was argued that an epizootic cannot flow in a population comprising one species without being influenced by other species. Circulation of the plague microbe was proposed to be ensured by three processes: preservation of the pathogen, its accumulation, and its dissemination. Burdelov divided hosts and vectors into "preservers," in which the plague microbe persists for a certain time, "hoarders" in the body of which plague microbes can increase in number, and "disseminators" of infection, i.e., animal and bird species that can effectively carry fleas.

Existing uncertainty regarding the natural foci of plague hosts has affected the development and efficiency of analytical and predictive models of epizootic processes and the establishment of preventive measures.

5.2 Ecological Characteristics of *Yersinia pestis* Hosts and Vectors

5.2.1 Hosts

According to recent data, over 300 species of mammals and three species of birds serve as plague hosts [1, 19]. Among mammals, only one [1] species, the opossum *Monodelphis domestica*, belongs to the subclass marsupials (Metatheria). All other mammalian hosts of *Y. pestis* belong to the subclass of placental animals (Eutheria).

The 341 species of mammalian hosts can be classified in seven taxonomic orders: rodents (Rodentia), 273 species; predatory carnivores (Carnivora), 30 species; lagomorphs (Lagomorpha), 16 species; insectivores (Insectivora), 10 species; even-toed ungulates (Artiodactyla), 6 species; primates (Primates), 4 species; and hyrax (Hyracoidea), 1 species. Wild birds, mostly commonly wheatears (*Oenanthe* sp.), also can be infected by plague (3 known species).

5.2.1.1 Animal Species

The different host animal species differ in their contribution to maintaining the circulation of *Y. pestis* in natural foci. In the Soviet and Russian literature on this subject, most authors adhere to the separation of main, secondary(additional), and random hosts.

According to Sludskyi [19], the **main host** of the plague is a background species with conservative (obligate) use of certain local areas for digging burrows within the habitat. These burrows are relatively complex and deep and suited for long-term (several years) use. As such, they are perfect hygro-thermostats for enormous numbers of fleas [107], the exception being vole-type foci. Additional characteristics of the main host are its leading position in both the biocenosis and the natural focus as well as its wide distribution throughout the territories of the foci.

According to Emelyanov [21], the main hosts of the plague are not limited to rodents but include any species inhabiting mountain steppe, steppe, and desert ecosystems in plague foci. The main host may also be characterized by persistently high levels of epizootic contact, whether achieved by high population numbers or by high mobility and migration [22]. However, the computer simulation studies of Dubyanskiy [23] showed that the epizootic process can also function within a population main hosts at low abundance. This feature is one of the hallmarks distinguishing main from secondary or random hosts.

The settlements of main hosts should be relatively continuous over large areas or, if they consist of mosaics, interconnected.

Among the pathogenic features of the main hosts, the most important are: the occurrence of massive and long-term bacteremia, the presence of lingering forms of plague, and the generalization of infection, which in some cases enables the long-term preservation of the plague microbe despite epizootic decay. Other important features are the relatively high sensitivity to infection of the young of some host species, the predominance of septic forms of the disease in these animals, and the heterogeneity of a population, that is, the presence of sensitive and resistant individuals. Exceptions to the latter can be found in the Caucasus and in the northwest Caspian region, where there are natural foci of plague with consistently highly sensitive hosts, i.e., gerbils of the *Meriones* genus (Vinogradov, Persian, Asia Minor, Tamarisk) [24, 25]. Fleas are the most specific of the parasites involved in plague transmission. They provide active dissemination of the pathogen during warm seasons and its preservation during cold seasons. Among the fleas preying on the main host, at least one or two species are involved in plague transmission.

Additional or secondary hosts are mammalian species that contribute to the legitimate circulation of the pathogen in nature and are therefore important (but less than main hosts) in the sustainability and long-term preservation of the plague microbe in natural ecosystems [26]. For example, in places where the main host is a large gerbil, the Libyan jird may be a secondary host. However, for ecological and pathogenic reasons, secondary hosts are not capable of providing constant circulation of the pathogen in natural foci.

Secondary hosts increase the reliability of mechanisms of infection, as under certain conditions they may harbor the pathogen in their populations without the participation of the main host and therefore, at least temporarily, assume its function [27]. In the case of the disappearance of the main host from a focus, it might be replaced by another species that previously served as the secondary host [30]. Moreover, in some foci the same species can act as the main and as the secondary hosts of the plague microbe.

According to Sludskyi [19, 20], **random hosts** are animal species that are rarely infected by plague in nature, for both ecological and physiological reasons, such as sporadic settlements, low population density, the strict specificity of their fleas, the lack of effective vectors, and high resistance to *Yersinia pestis*. If they are infected, they will become ill but will not play a role in the development of the epizootic.

5.2.1.2 The Role of Predatory Birds and Mammals in Natural Foci of Plague

Tarasov [28] was one of the first to recognize the important role of predators in natural foci of plague. He wrote that, according to the dietary specialization of certain predators and the numbers of species, the abundance of different groups of rodents could be determined. He also proposed a technique to identify a plague epizootic based on the accumulation of predatory birds.

Demidova [29] showed that the role of predators is the removal of dead and sick animals, which reduces the number that are infected but also spreads the infection (infected fleas and animals corpses) over long distances. The mechanism of infected animal distribution by birds of prey was described by Dubyanskiy [156]. One route is when predators bring the infected, flea-bearing rodents they have caught into their nests.

Small mustelids that can penetrate the burrows of their prey (e.g., great gerbils) can be attacked by the same fleas that parasitize their victims and thereby serve as additional hosts of *Y. pestis*-infected fleas. The predators then spread the fleas across the hunting area and often become infected with plague themselves. Our field observations in 2014, which included radiometric tracking of great gerbils, showed another mechanism of spread linked with predators. After an attack by the marbled polecat (*Vormela peregusna*), the gerbils moved from colony to colony for a few days, presumably actively exchanging fleas in the many visited colonies. They then disappeared and could not be found over an area of 4 km, having perhaps moved to a new territory, further spreading the infection.

5.2.2 *Vectors of* Y. pestis

General information on the hosts and vectors of *Y. pestis* can be found in Sludskyi [19, 20], Karimova, and Neronov [1], and Kozlov and Sultanov [30]. In natural foci of plague, only fleas are its main vectors. Under natural conditions, the plague pathogen has been detected in 280 species and subspecies of fleas belonging to 62 genera [19]. Besides fleas, other species of bloodsucking arthropods (*Argasidae*, *Gamasidae*, *Ixodidae*, *Anoplura*, *Heteroptera*, etc.; see Table 5.8) also harbor the bacterium but they do not play a significant role in plague epizootics and epidemics.

As with mammalian and bird hosts, different species of fleas differ in their importance in maintaining the circulation of the plague microbe in natural foci. Similar to hosts, vectors may be main, secondary (additional), or random. The main species of bloodsucking arthropods that define it as a vector, i.e., the specific parasite of the main host, are life cycle, number, and the ability to transmit the pathogen and to allow for its continuous transmission from one host to another. Kunitskii [31] was the first to describe the criteria of the main vectors of plague:

1. Parasitism in major hosts
2. The presence of the imago in the population throughout the year
3. A sufficient number of transmissions (and therefore abundance) of the pathogen to the imago during the epizootic season, both annually and over the long term
4. In seasons critical for flea existence (e.g., in deserts and semideserts, the hot and dry period at the end of summer), the maintenance of this level at least in optimal habitats
5. A high level of vital activity of the imago during seasons in which there is an abundance of both highly sensitive individuals of the main hosts and contacts between individuals from different burrows
6. A high degree of adaptation of the plague pathogen to the vector (microbe intensively reproduces and is protractedly preserved in the flea intestine)
7. Relative durability during the interepizootic season

The main features of plague vectors [19, 159] are summarized in Table 5.3.

Petrov [22] considered the term "vector" highly inaccurate, because it describes the essence of the transmission process only partially. This view was based on the fact that the total residence time of the plague microbe in fleas is much higher than

Table 5.3 Important features and advantages of the main vector of the plague microbe

Dominant species	Transmission of the pathogen at bacteremia of any intensity
	Permanent "immunosuppressive pressure" on the host
Constant presence of the imago in the population	Stability
	Stable abundance
Constant migration	Stability of intrapopulational and interpopulational epizootic connections
Frequent feeding	Stable level of pathogen transmission

in warm-blooded animals and the need for long-term preservation of the pathogen in fleas during periods when the epizootic rapidly declines or stops completely (e.g., in marmots, during their annual 9-month hibernation). In the above-described classification of Burdelov [32], fleas play dual roles, acting as conservators and accumulators of the plague microbe.

The habitat for the plague microbe is the digestive tract of fleas, which is divided into front ectodermal (esophagus with a suction pump and pre-stomach), middle endodermal (midgut or stomach), and rear ectodermal (large and small intestine, rectal ampoule, rectum) segments. The flea esophagus consists of a thin tube with a muscular expansion and a suction pump at its anterior and posterior ends, respectively. Food traveling down the esophagus first reaches the conical pre-stomach, which consists of an exterior cuticle inside and an interior well-developed circular and longitudinal musculature. Chitin needles within the pre-stomach tear and chop ingested red blood cells. The suction pump is surrounded by an unclosed dorsal chitin ring, from which muscle bundles travel in the dorsolateral direction and regulate the volume of the pump. The relatively homogeneous midgut is made up of secretory suction and regeneration cells. The posterior large and small intestine have strong plicate cuticles. Water and salts are absorbed from the feces mostly via rectal papillae in the ampoule.

Although the survival of *Y. pestis* in the flea was initially thought to be limited by the digestive tract, [161] and Vashchenok [33] demonstrated the penetration of the pathogen in the midgut epithelial cells and in the cytoplasm of digestive cells. The bacterium causes separation of the epithelial cells of the insect stomach, with subsequent penetration into the resulting intercellular space. Breaching of the epithelial cell membrane is followed by infiltration of the cytoplasm and the destruction of intracellular organelles [34, 35].

5.3 Characteristics of the Epizootic Process in Plague Foci

According to Korenberg [36], the plague epizootic process relies on the continuous interaction between populations of the pathogen and its natural hosts. In earlier work, Rall [55] divided the epizootic process in nature into development, flow, and extinction phases, followed by an interepizootic period during which infection persists in fleas and in rodents infected by latent forms of plague. The exacerbation and development of an epizootic under living conditions unfavorable for rodents are regular and necessary steps within the larger chain of events that characterize a plague enzootic.

An epizootic can be divided into four phases. Phase 1, accumulation of the plague agent, is characterized by sporadic cases or local epizootics in which the causative agent is in a highly virulent form. The intensive (acute or "spilled") phase 2 of an epizootic is characterized by (a) simultaneous detection of the pathogen in different sites within the enzootic area (continuous field), (b) flow (i.e., a local, strong epizootic) followed by its step-by-step spatial development (extension), and (c) independent epizootic spots in different localized areas. During phase 3, epizootic attenuation, there is a high number of infected fleas, but their rodent hosts are immune to the pathogen, which has developed atypical (avirulent, weak virulent) strains. In the fourth and final interepizootic phase, plague epizootic manifestations alternate with periods of complete quiescence of ≥ 3–5 years duration within a number of natural foci [19].

The epizootic process can be examined in detail using the example of the Central Asian desert natural focus, where the plague host is the great gerbil (*Rhombomys opimus*; first described by Lichtenstein in 1823). The infection of a single individual of the great gerbil by the causative agent of plague in spring marks the beginning of the epizootic and reflects the overwintering of pestilential fleas, mostly of the genus *Xenopsylla* [37–39]. Although some rodents may be infected in winter, this seems to be rare [10]. With the springtime infection of rodents, exacerbated by the presence of individuals infected in winter, a spring-summer epizootic is initiated. According to Soldatkin [39], passivation of the plague microbe requires 7–12 days (10 days average), during which infection develops in the great gerbil (3–4 days) and its fleas become capable of transmission, a process involving blockage by the bacterium of the flea stomach and referred to as block formation (6–7 days). At low temperatures, block formation may be delayed for over 2–3 months, whereas at high temperature it can occur within 3 days. Passage of the fleas to other animals results in their infection, in a chain reaction, although data on the number of colonies of animals infected in spring per unit territory are lacking.

According to some researchers, block formation by the fleas that parasitize great gerbils is not an absolute requirement for passing of the plague microbe or the development of a massive preagonal bacteremia [40, 41]. This implies that the frequency of causative agent transmission may be higher than determined quantitatively based on the frequency and speed of block formation and preagonal bacteremia. This should be taken into account when interpreting the results of plague modeling.

In the first half of spring, great gerbils are highly mobile, associated with the collapse of wintering populations and the formation of mating pairs, each of which occupies a separate burrow [10, 42]. Thus, after passing, the fleas are carried into a few neighboring burrows [10, 39]. These subsequently infected burrows are located in groups (typically 2–5) set around the burrow containing the first sick animals [43].

At the same time, the infected fleas may be transported over a distance of 1–2 km, which expands the range of plague spots or, following the perpetuation of this process, 2–3 new plague spot [39]. Usually, about a third of animals in a given plague spot are infected, and roughly half of them will die from the infection. These plague spots, typically a few hundred meters in diameter [43], are therefore considered as the smallest structural unit of the epizootic [44], and they provide a simultaneously occurring miniature of all the phases of epizootic development. Thus, acutely developing infection of the rodents can be seen in some of the burrows, whereas in others the process has reached its end and the rodents have developed immunity. Some burrows are unimpaired by the epizootic and therefore provide a reserve for its later, further development. The unevenness of epizootic flow in plague spots and the constant movement of the epizootic throughout the territory are well-recognized features of the process. Studies of the spatial structure (focality) and the epizootic process [7, 10, 43, 45, 46] have shown that plague spots of plague-infected burrows constitute a group of neighboring colonies where both great gerbils that are infected and have recovered from plague and infected ectoparasites reside. These groups are irregularly located and separated from other such groups by statistically significant distances. The accumulation of plague spots forms an epizootic spot, measuring several kilometers in diameter and considered an intermediate structure of the epizootic, whereas its large structure ranges over tens of kilometers, as is the case in areas of "spilled" plague epizootics [43].

In the second half of spring, some animals have already started breeding and are mostly sedentary [47]. However, with the growth of young animals, the gerbils visit neighboring burrows and the plague spot therefore expands. Before the appearance of the second brood, females nursing the first brood leave the burrow and move to another, traveling hundreds of meters. The males are even more mobile, usually visiting females in several burrows [10]. At the same time, fleas of the genus *Xenopsylla* feed intensively on their hosts, often changing hosts [48]. The transmission speed of the plague microbe may increase as temperature conditions and the frequency of flea feeding become optimal for block formation, resulting in the further extension of the plague spot. According to Soldatkin [39], the number of infected fleas is typically much higher than that required for the infection of gerbils in a plague spot. This redundancy in the number of infected fleas together with the flea exchange rate between gerbils in the colony contributes to the high reliability of the epizootic process. Even an initially unsuccessful transmission of *Y. pestis* will, following repeated exposure of the animal, eventually lead to its infection. In places of epizootic development, 90–100 % of the gerbils may be infected by the bacterium [49].

In summer, the major factor changing the structure of the epizootic is the high mobility of the gerbils, as they increase their search for food after the drying up of spring forage plants. Young animals continue their resettlement, migrating over distances up to 18 km [10]. Infected animals will also be redistributed during these territorial migrations. However, in summer the old fleas die off and the small number of newly infected fleas does not provide an adequate level of plague microbe transmission. Therefore, the number of infected burrows will increase only slightly during June and in July–August will rapidly decline, a process enhanced by the acquisition of immunity after the spring-summer epizootic.

According to Rudenchik and Soldatkin [50], despite the small number of newly infected fleas in summer, conditions favor the dispersion of the pathogen within a given territory. Regular migrations of an entire gerbil family differ from the springtime visits of animals to nearby colonies. The frequency and long distances of the former (over a distance of hundreds of meters) contribute to the wide-ranging spread of the pathogen. When the number of fleas is small, this process can go on invisibly and lead to the dispersion of only a small number of infected animals; at higher flea abundances, the result will be a diffused plague epizootic [51].

For *Xenopsylla* fleas, autumn marks the beginning of the breeding period of wintering populations, which ends with the onset of winter, when the gerbils become inactive [39]. Conditions supporting epizootic expansion are the entry into the process of fleas of the genus *Coptopsylla*, whose hatching starts in autumn. These fleas are efficient vectors of the plague microbe, but they feed less often and the longer blocking formation time reduces the speed of plague transmission.

Also in autumn, as the great gerbils begin to store food, they form wintering groups such that their mobility again increases [47], which in turn increases the number of infected burrows. Adjacent burrows are predominantly infected, but in spring expansion of the plague spots begins anew. According to Soldatkin [39], the conditions for passing in autumn are similar to those in spring. In some cases, however, because of the delayed transmission of the infecting agent and the briefness of the autumn season, the plague spots will not reach their springtime size and may be limited to 3–8 burrows. In many burrows containing gerbils with *Y. pestis* bacteremia, despite the presence of several (1–10) infected burrows by the end of autumn, climate conditions will not favor block formation and therefore plague transmission. Transmission in winter is further hindered by the low mobility of the gerbils and the almost complete cessation of contacts between burrows. But with the arrival of fleas of the genus *Xenopsylla* in an inactive condition and their persistence in the burrow throughout the winter, transmission of the pathogen the following spring is ensured.

A mandatory element in the development of a plague epizootic is territorial displacement of the vector, which occurs in two ways. The first is transmission of the plague microbe to neighboring burrows, which according to Soldatkin [39], is 100–150 m per 10 days. The second relates to the distant transport of the microbe, over distances of 1–2 km but up to 10–18 km [10, 39]. Naiden et al. [52] traced the development of a plague epizootic in Central Kyzylkum in 1961–1966 and found an average speed of distribution of as much as 50 km per year. In the northern pre-Aral region, the annual range of movement of the plague microbe reached 15 km [53]. More detailed field studies by the same author, conducted at the pre-Ustyurt plague focus in the Ushkan camp, showed a maximum speed of the epizootic spots of 1–2 km per month.

During the interepizootic period, the preservation of plague spots and their slow increase in "epizootic-free" territories occur. At an as-yet-undefined critical threshold of foci accumulation, the plague epizootic spreads, initially according to the principle of a branched-chain reaction [51] and later as an acute and spilled epizootic, resulting in the sudden coverage of large areas. Also of importance in an epizootic is that previously unoccupied burrows can be visited by animals of two or three neighboring family groups [26, 42]. This provides all the prerequisites for long-term parasitic contact between great gerbils and between them and other rodents. According to Burdelov et al. [26], unoccupied burrows are the nodal points of epizootic contact.

The abundances of rodents and fleas are also important determinants of epizootic contact. The plague microbe may be found less frequently or not at all in areas with a maximum abundance of great gerbils, whereas an epizootic may be ongoing in adjacent parts of the settlements with low or intermediate numbers of the rodent [100, 125, 129]. High rodent numbers imply increased long-range migration, which supports distribution of the plague microbe. Furthermore, an epizootic may bypass densely populated areas of great gerbils [52], reflecting the lack of places of free exchange with ectoparasites and/or a deficit of free ectoparasites [26]. Thus, in the absence of sporadic cases in rodents, the later emergence of a spilled epizootic is prevented [54].

Even at low numbers, great gerbils may migrate further, from one uninhabited burrow to another, thereby intensifying ectoparasite exchange in unoccupied burrows and increasing the so-called dynamic density [55]. Thus, despite the low number of hosts, the level of epizootic contact is probably sufficient to allow their sporadic infection. The dynamics of an epizootic on the Bakanas plain between 1949 and 1995 are described in Fig. 5.2. As shown in the figure, the intensity of the epizootic can be described by the percentages of infected sectors, i.e., those where epizootic manifestations have been recorded in any form (e.g., in terms of infected or recovered animals or infected ectoparasites). It can also be seen that over the long term the four phases of epizootic occur cyclically [56, 57], although recurrence is disordered. This reflects the varying number of hosts, the weak expression in some foci, and the complete absence in others [58, 60].

For epizootics involving great gerbils, the first phase has been poorly studied and remains controversial. The second phase typically begins with a high and relatively constant background number of the rodents [10, 44] and with a high number of *Xenopsylla* fleas. At this stage, epizootics may also include other species of fleas, which promotes infection and transmission. Disease is most often acute, usually involving bacteremia, and a high percentage of rodents and fleas in the population are infected.

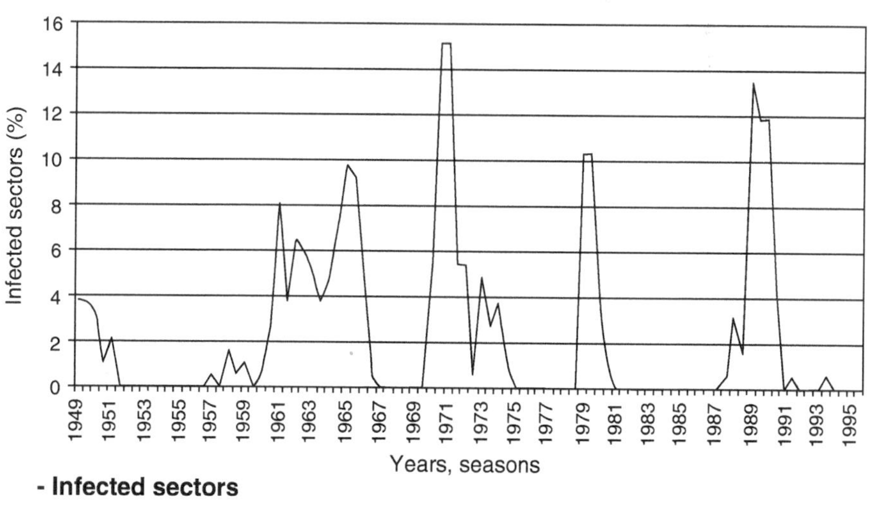

Fig. 5.2 Infestation of the sectors (in percentages) in epizootic on the Bakanas Plain between 1949 and 1995

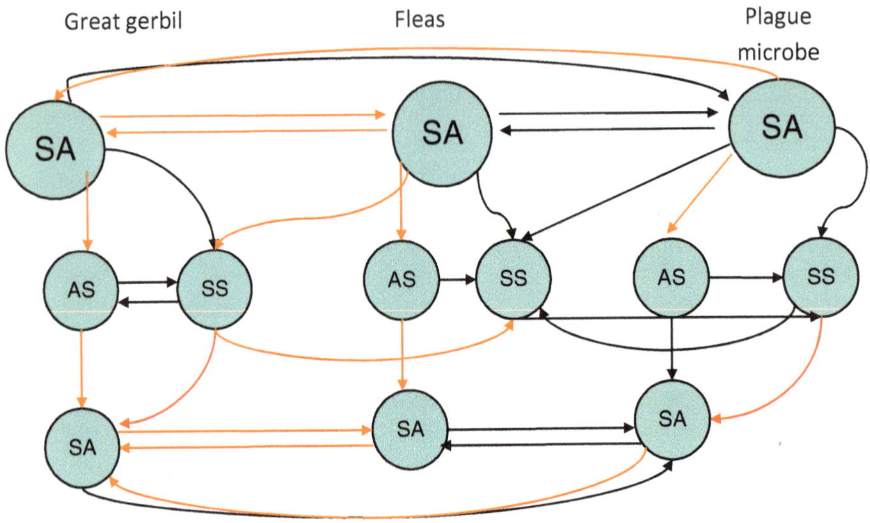

Fig. 5.3 Scheme of the connections between the numbers of members of the plague epizootic triad (host, vector, microbe) [58]. SA, SS, and AS, growth in the number of plague epizootic triad (host, vector, microbe) components from spring to autumn (SA), spring to spring (SS), and autumn to spring (AS), respectively. Red lines, negative link; black lines, positive link

In the period of epizootic attenuation (phase 3), there is a decrease in the number of weak and chronically infected animals in the burrows and a reduction in the proportion of recently infected fleas. Atypical (weakly virulent, phage-infected) *Y. pestis* strains are isolated during this phase [10].

The interepizootic period (phase 4) is also controversial and is the least-studied phase (see Sect. 5.1.3). The phases of the plague epizootic have been linked to changes in the level of pathogen virulence. The most virulent strains of *Y. pestis* are thought to circulate at the beginning or during the peak of the epizootic, whereas in the attenuation phase virulence decreases. Figure 5.3 presents a highly simplified scheme of the epizootic process.

5.4 Factors Influencing *Y. pestis* and Plague Dynamics

5.4.1 Environmental and Ecological Parameters

Plague microbe transmission varies from year to year and depends on both biotic (ecological) and abiotic (environmental: humidity, soil chemistry, etc.) factors [19, 58].

5.4.1.1 Abiotic Factors

Spatial and geophysical factors account for variations in the conditions of hosts, vectors, and the responsible microbe. These factors act indirectly, through the hydrometeorological conditions that impact biological processes [16, 58], but direct effects cannot be excluded. Cyclicality, linked with the features of solar activity, is inherent to the dynamics of host and vector numbers and to epizootics. There are also geographically varying cycles, including the zonality of atmospheric processes impacting biological relationships. Lavrovskii [57] pointed out the connection between the cyclical patterns within natural plague foci and solar and climatic factors. Dubyanskiy [58] evaluated the complex effect of local abiotic factors on epizootic activity within plague foci (see the above-described example of the Central Asian desert focus with the great gerbil as the main host).

Climatic factors are significant determinants of great gerbil populations and their fleas [59–61, 156]. Among the members of the plague triad (host, vector, microbe), the host plays the primary role in the epizootic process, given that parasites are vitally dependent on the host and because expression of the microbe-induced pathogenic effect requires a receptive, sensitive host. Weather and climate conditions exert indirect influences on the food supply and other intermediates. In the case of the great gerbil and its life cycle, there is a very close dependence on weather conditions. Consequently, cold winds increase the susceptibility of the gerbils to infection and thereby increase of epizootic activity. The nature of the weather in the plague foci in spring and autumn is that cooling typically lasts 2–4 days, which alternates with periods of warming. Flooding of the burrows, from meltwater during thawing in autumn, winter, and spring, has a disastrous impact on great gerbil populations.

Changes in external conditions have an even greater impact on the poikilothermic flea vectors. However, the fleas' hosts offer better protection than granted to the great gerbil by its burrows. This would explain the delay in the dying off of old fleas in the deepest burrows in summer.

Whether other factors, such as electric and magnetic fields and spatial changes in the atmospheric charge of the atmosphere, also affect the epizootic process is unclear. Dubyanskiy and Bogatyrev [62] found a delayed (16–20 days) positive dependence of plague epizootic intensification with increased solar (sunspots speakers) and geomagnetic activities. Laboratory experiments using static electricity fields demonstrated a stable change in the viscosity of the blood of rodents, which undoubtedly would affect both the feeding process of fleas and block formation. Changes in the properties of the plague microbe infecting the fleas in response to a negative electric field include the appearance of large amounts of Pgm^-, Ca^{++}-independent colonies [58]. Serzhanov associated the cyclicity of the plague epizootic process with changes in the hydrothermal coefficient (HTC), an index of the combined effect of temperature and precipitation on fleas [63]. In Central Asian desert foci, conditions close to the optimum for plague development are characterized by HTC values > 0.08–0.09 for southern desert subzones, 0.10–0.15 for the transitional band, and 0.15–0.20 for the northern subzone. In the latter two, these conditions occur during a 7–8-year period and in the southern subzone for anywhere between 2 and 3 years and up to 10 years. In years with a high HTC and an active epizootic process, abiotic factors are of less importance. During the interepizootic period, which as a rule is characterized by low HTC values, the ability to preserve elementary foci, especially at a low groundwater levels, is important [63].

5.4.1.2 Biotic Factors

The complexity of the biocenotic structure of the focus is one of the many factors influencing the character of the epizootic and circulation of the pathogen in nature. It is determined by the various hosts, which differ from each other in their ecological features, their vulnerability to the plague microbe, disease pathogenesis following infection [64], the species composition and ecological characteristics of the vectors, and the nature of the biocenotic connections.

Among the biotic factors that directly affect the epizootic are the number and dynamic density of rodent hosts [65] and their settlement types [10, 66]; the number, migration, and alimentary activity of the fleas, especially of the *Xenopsylla* genus; and the characteristics of the epizootics in the previous and current season and those of the different strains of the plague microbe.

The abundance and activity of the hosts and vectors are of particular importance. Within a focus, the number of great gerbils can vary greatly, from a few tens of individuals per hectare to near extinction. In summer, despite the decreased number of fleas, an epizootic can develop in areas where the immunity of the gerbils is weakened [10]. Conversely, plague epizootics can fade after a few months because of the strengthened immunity of the great gerbils, even if other favorable conditions are present [67]. Studies of blood groups conducted by Ergaliev et al. [157] showed dynamic changes in the proportions of animal sensitive and resistant to the plague bacterium. Indeed, a change in population susceptibility is one of the basic mechanisms regulating the epizootic process. Also, in some years, summer and autumn breeding can differ in their intensity; in the southern part of the focus, breeding can occur even in winter [60].

Suleymenov [19, 160] summarized the impact of abiotic and biotic factors on the epizootic process (Table 5.4).

Factors affecting natural foci in the Palearctic have been studied by Karimova and Neronov [1] using principal components analysis (PCA). In that study, the four main components, with a total dispersion of 86.8 %, described a variety of environmental conditions that shape the dynamics of the epizootic. The most important factor, temperature conditions, explained 34.5 % of the dispersions of the system and was followed by wind (23 %), which was thoroughly investigated by Dubyanskiy and coworkers [2, 61, 68]. They found that the frequency of occurrence of wind of separate rhumbs was the most important aspect, followed by the power of the wind. This component was negatively associated with the height above sea level of the

Table 5.4 The main factors influencing the epizootic and their functions

Factor	Effect/function
Climatic conditions of the previous year (average monthly temperature and precipitation)	Spring abundance of the main host and main vector
	Physiological condition of the main host and main vector
	Number and physiological condition of secondary hosts and vectors
Climatic conditions of the current year (average monthly temperatures and precipitation)	Physiological condition of the main host
	Intensity of host breeding
	Dynamics of the main host and main vector
	Dynamics of secondary hosts and vectors
Main host species	Intensity, speed, development, and distribution of the epizootic
Main host settlement type	Direction and speed of distribution of the epizootic
Main host settlement density	Intensity and speed of distribution of the epizootic
Seasonal host abundance (spring, summer and autumn)	Determinant of the beginning, speed, intensity, and extent of the epizootic
Age and sex structure of the main host population	Level of interpopulation contacts
Intensity of reproduction	Dynamics of host numbers
Immunological structure of the main host population	Character of the epizootic
Population of secondary hosts	Dynamics of the epizootic
Population of predatory mammals	Speed of distribution and the direction of the epizootic Direction of the process
Main host numbers	Level and speed of intraspecific and interspecific epizootic contacts
	Intensity and extent of the epizootic
	Level of resistance of the main and secondary hosts, including the dynamics of acute and chronic infectious processes
Number of secondary hosts	Degree and speed of intraspecific and interspecific epizootic contacts
	Intensity and extent of the epizootic
	Level of host resistance
	Dynamics of acute and chronic infectious processes

focus, i.e., the wind is stronger in open spaces. The third and fourth components identified in the PCA were radiation (14.9%), which was related to temperature conditions, and rainfall from September to April, respectively.

Dubyanskiy analyzed > 500 biotic and abiotic factors to identify those most important for the functioning of the epizootic process in different parts of the Central Asian desert known to harbor natural foci. The results are summarized in Table 5.5.

Table 5.5 Biotic and abiotic factors influencing the epizootic process in the Central Asian desert

Abiotic factor	Hydrothermal coefficient[a]
	Wolf-Volfer number
	Increase in the Wolf-Volfer number from summer to summer
	Amount of monthly precipitation[a]
	Average speed of the southeast wind in March–May, the northeast wind in September–November, the western wind in March–August, September–November, the northern wind in September and November, the southern wind in September–November, the northwest wind in June and November, the southwest wind in June–August
	The product of the frequency of the eastern wind and the sum of the air temperature sum in June–August[a]
	Total humidity between April and September[a]
	Frequency of eastern winds in March and November, southeast winds in June and November, northeast winds in March–May, southwest winds in March–May and September–November, and southern winds in September–November
	Air humidity in May[a]
	Sum of the monthly average temperatures in October–November[a]
	Total number of days in June–November with an eastern type of atmospheric circulation and in March–May and June–August with a meridional type of atmospheric circulation
	Annual index of geomagnetic activity per year
Biotic	Total number of spring and autumn infestation points (%)
	Increase in these points from spring to autumn (%)
	Number of great gerbils/burrow in spring, during two springs, in autumn, two autumns, and spring and autumn
	Increase in the number of fleas/great gerbil burrow from spring to autumn
	Intensity of great gerbil breeding in June and
	May (maximum from average decade values)
	Number of fleas/great gerbil burrow in autumn
	Increase (%) in the number of great gerbils/burrow from spring to autumn
	Proportion of flea females with large eggs in May (maximum from average decade values)
	Captivity of tamarisk jird in spring and in autumn (%)
	Captivity of murine species in autumn (%)

[a]These factors are generalized; specific monthly values vary depending on the geographic location of the focus and the status of the members of the epizootic triad

5.5 Modeling Plague Epizootic and Epidemic Events

5.5.1 Analytical Models

The impact of specific factors on the distinct components of the epizootic process has been evaluated in several models, for example, in that of Keeling and Gilligan [69], who used multiple regression analyses. Holt et al. [70] used a specially designed geographic information system spatial-temporal model (Maxent) and was thus able to identify ecological niches favorable for the plague epizootic in California and to preliminarily predict epizootics with respect to climate change. Stenseth et al. [71], Snall et al. [72], and Kausrud et al. [73] calculated the average temperature increase needed to stimulate the activity of plague foci. The spatial heterogeneity of Karginsk meso-foci was shown to have been determined mainly by the confinement of epizootics to glacial and proluvial quaternary depositions of the second ice age associated with the Middle Pleistocene (113–104 thousand years ago) [74]. Studies of the plague bacterium have revealed its ability to fill available ecological niches [74]. In a new and original approach, by Davis et al. [75], *Y. pestis* was shown to spread in great gerbil settlements in accordance with the principles of percolation theory.

SEIR [susceptible (stage of infection), exposed (incubation period), infected (stage of infection development), and removal (postinfection immunity)] models, which simulate epidemic or epizootic processes and are based on cellular automata [76, 77, 78], have bypassed the challenges of directly observing an epizootic [79] while allowing its nuances in nature to be taken into account. Soldatkin and Rudenchik showed that the epizootic process is feedback-inhibited. One such mechanism is entry of the pathogen into already infected burrows, which leads to the loss of transmission. This type of inhibition is linearly dependent on the density of burrows in the foci, measured in terms of the number of plague pathogen-free burrows adjacent to those that are infected. Another type of "braking" is the gradual depletion of the population of susceptible animals in the infected burrows, such that subsequent passage of the pathogen in these burrows increasingly falls on resistant individuals, which impedes the progress of the epizootic. The speed of movement of the epizootic between colonies averages 3.3–3.5 intervals (maximum = 19) per month.

Survival of the disease chain, that is, the development of the epizootic over the course of a year in foci of pestilential burrows, has been modeled based on a probability of 0.999. The results showed that the minimum number of disease chains is between 8 and 17, depending on the initial conditions. Changing the input data of the model confirmed the hypothesis that explosive-intensive epizootics can be attributed to the delayed extinction of overwintered fleas.

Long-term studies [80] have shown that great gerbil settlements tend to be ordered, rather than diffuse. Dubyanskiy et al. [160] used SEIR models and cellular automata to show that the spatial structure of great gerbil settlements is quantitatively related to the intensity of the plague epizootic. In the more ordered settlements of great gerbils, epizootics were shown to occur at a much lower level (sixfold; $p < 0.01$) of epizootic contact [defined as any ecological connection (topical, trophic, etc.) between animals, in which plague transmission from donor to recipient occurs at a certain rate [23, 81]] than in diffuse settlements (Figs. 5.4 and 5.5). This means that transmission of the plague microbe in these settlements requires fewer infected fleas and fewer rodents. In areas with a high abundance of great gerbils, the structure of the settlement is one of the factors that shape epizootic flow (see Sect. 5.3) [82–88].

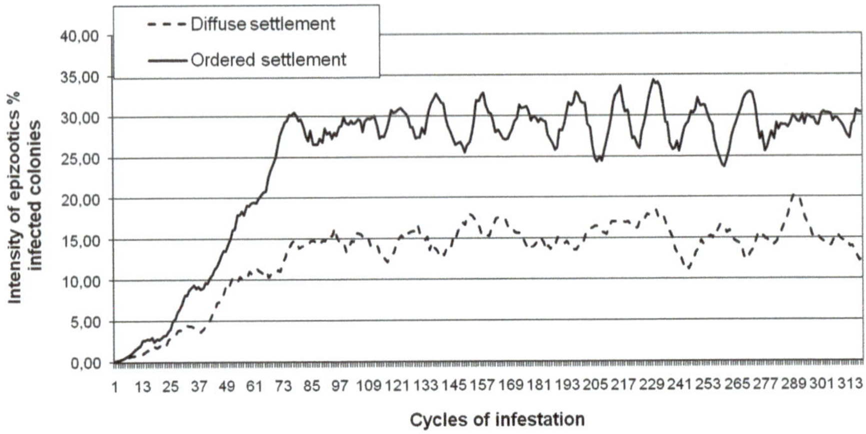

Fig. 5.4 The intensity of the epizootic process in diffuse (1.33 colonies/ha) and orderly (0.44 colonies/ha) great gerbil settlements

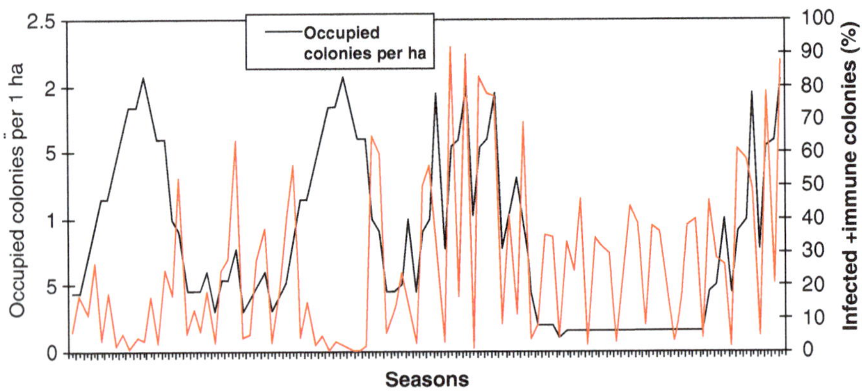

Fig. 5.5 Densities of occupied colonies over an area of 1 hectare and the relative intensity (%) of the (modeled) epizootic process

An analysis of the spatial characteristics enabling the transmission of plague infection in the different types of settlements determined that the lower limit of the number of great gerbil burrows providing short-term (2–3 months) transmission ranged from 25 to 400, depending on the environmental conditions and the burrow density [87, 88]. Similar studies of the plague epizootic process over the long term, based on permanent flow for 25 years, demonstrated that the epizootic process can continue uninterrupted at very low densities of occupied burrows (0.1 burrows per ha), that is, even when great gerbil numbers are depressed.

5.5.2 *Predictive Models*

In a one-factor simulation model, Soldatkin and Rudenchik [89, 90] predicted epizootic activity based on the number of fleas that had ingested the isotope-labeled blood of a gerbil. Other models have addressed the roles of air temperature in March and total summer precipitation [91]; the Wolf number, which measures the number of sunspots [92], on the number of great gerbils over a 2-year period [158]; and the density of colonies inhabited by great gerbils for the two preceding years [93, 94]. However, experience has shown that these models are inadequate because they do not cover all the major influencing factors [87, 88].

Using the cybernetic and statistical techniques developed by Genese [95], Kunitskii et al. [96] created a system that combines two factors: the number of hosts and the number of vectors of the plague microbe. This model, which uses historical data, recognized 85 % of the cases in the Southern pre-Balkhash. Other authors have used multiple regression analyses to predict the number of hosts [97–99], but in those cases the utility of this approach was limited by the inaccuracy of the initial observations of the plague foci in terms of functional connections and other relevant parameters.

Models based on the theory of immunocomputing [100, 101] showed good correspondence with historical data, but whether they can be used to predict epizootics in real time was not determined.

Models based on differential equations or Markov chains have a number of significant drawbacks. They can be used only for well-mixed populations with close arrangements of the hosts [102], a condition that is not always satisfied in nature. Grabowski [103] noted that: (a) Well-known mathematical models based on differential equations impose very rigid requirements on the original information. This, however, points out the fact that the volume and accuracy of data collected during the monitoring of natural foci of plague are often inadequate for the correct use of this type of model. (b) Differential equations are used in the simulation of space and assume that all of the events unfold from a single point. Examples include regression models, polynomial models of epizootic activity and predictors, and others in which the analysis is performed on the averaged data in the hierarchy: the sector of the primary area, the primary area, the landscape-epizootiological area (LER), or on autonomous foci (Fadeev and Serdobintseva 1983; [2, 100]).

The prediction method proposed by Dubyanskiy et al. [2], based on the heterogeneous sequential statistical recognition procedure (HSSRP) [104], has thus far had the best results in studies of plague foci where the main host is a great gerbil. HSSRP links methods based on Bayes' theorem with the sequential analysis of Wald. Among the 55 predictions made using this approach, 85.4 % were true, 12.7 % were indeterminate, and 1.8 % were false [58].

Climate change is expanding both the areas and the species of *Y. pestis* hosts and vectors [1, 105], making spatially predictive models more important. A detailed spatial prediction of potential focal territories was provided by Karimova and Neronov [1] using discriminant analysis of orographic (altitude), climatic (number of days with frost per month, sum of rainfall per month, average monthly temperatures, maximum temperatures per month, number of days with precipitation, cloudiness per month, solar radiation per month), landscape (types), and biotic (number of hosts, presence/absence of plague epizootic) factors. The authors predicted that new foci and focal territories will eventually arise in the Palearctic, in the northern pre-Caspian, on the border of the current plague area in Kazakhstan, in southern Turkmenistan, in northeastern Iran and northern Afghanistan, and in the desert regions of Mongolia and China.

In literature searches (domestic and foreign publications) and long-term investigations conducted in natural and anthropogenic foci of plague in Africa, Eurasia, and North and South America, at least 340 species of mammals and birds vulnerable to infection by *Yersinia pestis*, the causative agent of plague, have been identified. These are listed in Table 5.6.

Table 5.6 Plague hosts and the geographic region(s) where they are found, described with respect to a list of vertebrates among the world's fauna ([20], corrected by Dubyanskiy V.M., Yeszhanov A.B.)

Number	Host	Focus/foci
	Infraclass Metatheria	
	Blood line Didelphidae	
1	*Monodelphis domestica* L.	Brasilia
	Infraclass Euteria	
	Order Insectivora	
2	*Crocidura olivieri* Less.	Senegal
3	*Crocidura suaveolens* Pall.	Uralian Prairie, Ural-Emba* (Kazakhstan), Talas (Kyrgyzstan), Volga-Ural sandy areas (Russia, Kazakhstan), Caspian sands (Russia)
4	*Diplomesodon pulchellum* Licht.	Volga-Ural sandy areas (Russia, Kazakhstan), pre-Aral Karakum, Zaaralsky, pre-Balkhash (Kazakhstan), Karakum (Turkmenistan)
5	*Hemicentetes nigriceps* Gunther	Madagascar
6	*Hemiechinus auritus** Gmelin	Uralian Prairie (Kazakhstan)
7	*Hylomys suillus** Muller	Southeast Asia
8	*Sorex* sp.	Central Caucasian (Russia)
9	*Sylvisorex gemmeus* Thom.	Congo
10	*Suncus murinus* L.	India, Cambodia, Vietnam, Java
11	*Neomys fodiens* Pennant	Transcaucasian highland (Armenia, Georgia)
	Order: **Primates**	
12	*Macaca mulatta* Zimm.	India
13	*Macaca mulatta* Zimm	India
14	*Semnopithecus entellus* Dufr.	Central India
	Order: **Scandentia**	
15	*Tupaia glis* Diard	Vietnam
	Order: Lagomorpha	
16	*Ochotona alpina* Pall.	Altay mountains (Russia)

(continued)

Table 5.6 (continued)

Number	Host	Focus/foci
17	*Ochotona daurica* Pall.	Baikal Prairie (Russia), Tuva (Mongun-Taiga) mountains (Russia), Altay mountains (Russia), Chentij and Khangai (Mongolia), Manchuria (China)
18	*Ochotona pallasi pricei* Gray	Tuva highland (Russia), Altay mountains (Russia), Chentij, Khangai (Mongolia), Gobi Altai (Mongolia)
19	*Lepus californicus* Gray	USA
20	*Lepus capensis* L.	South Africa
21	*Lepus europaeus* L.	North pre-Aral, Ural-Emba* (Kazakhstan), Argentina (acclimatized)
22	*Lepus saxatilis* Cuvier	Zimbabwe
23	*Lepus tolai* Pall.	Karakum (Turkmenistan), Kyzylkum (Kazakhstan, Uzbekistan, Turkmenistan), Muyunkum (Kazakhstan), Mangyshlak* (Kazakhstan), and pre-Ustyurt* (Kazakhstan), Ural-Emba* (Kazakhstan), pre-Balkhash (Kazakhstan) desert, Tyanshanskandy highland (Kyrgyzstan, Kazakhstan), Altay mountains (Russia)
24	*Oryctolagus cuniculus* L.	England
25	*Sylvilagus brasiliensis* andinus	Peru
26	*Sylvilagus audubonii* Baird	USA
27	*Sylvilagus bachmani* Water	USA
	Sylvilagus brasiliensis L.	Argentina, Bolivia
29	*Sylvilagus caudatus*	Peru
30	*Sylvilagus brasiliensis* gibsoni	Bolivia
31	*Sylvilagus nuttallii* Bachm.	USA
	Order: Rodentia (Family Sciuridae)	
32	*Callosciurus erythraeus* Patt.	Yunnan (China), Southeast Asia
33	*Dremomys rufigenis* Blanford	Southeast Asia
34	*Funambulus palmarum* L	Pakistan, Sri Lanka
35	*Funambulus pennanti* Wrought	India
36	*Glaucomys sabrinus* Shaw	USA, Canada
37	*Menetes berdmorei* *Blyth	Vietnam
38	*Paraxerus cepapi* A. Smith	Zimbabwe
39	*Sciurus niger* L.	USA
40	*Sciurus stramineus* Geoffr.	South Loya (Ecuador), Lankoves (Peru)
41	*Spermophilopsis leptodactylus* Licht.	Kyzylkum (Kazakhstan, Uzbekistan, Turkmenistan), Taukum (Kazakhstan), pre-Balkhash (Kazakhstan) deserts
42	*Tamiasciurus douglasi* Bachm.	California (USA)
43	*Tamiops macclellandi* *Hors.	Vietnam
44	*Xerus erythropus* Geoffr.	Senegal
45	*Xerus inauris* Zimm.	South Africa

(continued)

Table 5.6 (continued)

Number	Host	Focus/foci
46	*Tamias* quadrivittatus Say.	USA
47	*Tamias minimus* Bach,	USA
48	*Tamias speciosus* Merriam	USA
49	*Tamias townsendii* Bach.	USA
50	*Tamias umbrinus* J. Allen	USA
51	*Marmota baibacina* Kastsch	Tien-Shan highland (Kazakhstan, Kyrgyzstan), Altay mountains (Russia), ridges of this mountain system in China
52	*Marmota caudata* Geoffr.	Alay highland (Kyrgyzstan), Talas (Kyrgyzstan), Gissar highland (Tajikistan), Boro-Horo Ridge (Tianshan, China)
53	*Marmota himalayana* Hods.	China
54	*Marmota flaviventris* Aud. et Bachm.	USA, Canada
55	*Marmota sibirica* Radde	Baikal Prairie (Russia), Tuva highland (Russia), Mongolia, Inner Mongolia (China)
56	*Cynomys gunnisoni* Baird.	USA
57	*Cynomys leucurus* Merriam	USA
58	*Cynomys ludovicianus* Ord.	USA
59	*Cynomys mexicanus* Merr.	USA, Mexico
60	*Cynomys parvidens* All.	USA
61	*Spermophilus (Citellus) alaschanicus* Bachner	Ningxia Hui (China)
62	*Spermophilus armatus* Kenn.	USA
63	*Spermophilus beecheyi* Rich.	USA
64	*Spermophilus beldingi* Merr.	USA
65	*Spermophilus columbianus* Ord.	USA
66	*Spermophilus dauricus* Br.	Baikal Prairie, East Mongolia (Russia), Manchuria (Northeastern China)
67	*Spermophilus erythrogenys* Br.	Pre-Balkhash, Betpak-Dala (Kazakhstan), pre-Alakol low mountains (Kazakhstan), Southeast Mongolia, Inner Mongolia (China)
68	*Spermophilus fulvus* Licht.	Volga-Ural sandy areas (Russia, Kazakhstan), Volga-Ural prairie (Kazakhstan), Uralian(Ural-Uil) prairie (Kazakhstan), Ural-Emba (Kazakhstan), pre-Ustyurt (Kazakhstan), Ustyurt (Kazakhstan, Uzbekistan, Turkmenistan), Mangyshlak (Kazakhstan), Kyzylkum (Kazakhstan, Uzbekistan, Turkmenistan), Karakum (Turkmenistan), pre-Aral Karakum (Kazakhstan), North pre-Aral (Kazakhstan), Zaaralsky (Kazakhstan), Muyunkum (Kazakhstan), Betpak-Dala (Kazakhstan) Deserts
69	*Spermophilus mollis* idahoensis	USA
70	*Spermophilus lateralis* Say	USA

(continued)

Table 5.6 (continued)

Number	Host	Focus/foci
71	*Spermophilus leucurus*	USA
72	*Spermophilus major* Pall. 1778	Uralian Prairie (Kazakhstan)
73	*Spermophilus mexicanus*	USA
74	*Spermophilus musicus* Menetrie	Central Caucasian (Russia)
75	*Spermophilus pygmaeus* Pall.	Terek-Sunzha (Russia), Dagestan plains and foothills (Russia), Caspian Northwest prairie (Russia), Caspian sands (Russia), Volga-Ural prairie (Russia, Kazakhstan), Volga-Ural sandy (Russia, Kazakhstan), Uralian (Ural-Uil) prairie (Kazakhstan), Ural-Emba (Kazakhstan), pre-Ustyurt (Kazakhstan), Ustyurt (Kazakhstan, Uzbekistan, Turkmenistan), North pre-Aral (Kazakhstan), pre-Aral Karakum (Kazakhstan) deserts
76	*Spermophilus relictus* Kaschk.	Tien-Shan (Kyrgyzstan, Kazakhstan), Talas (Kazakhstan)
77	*Spermophilus richardsonii* Sab	USA, Canada
78	*Spermophilus (Xerospermophilus) spilosoma*	USA
79	*Spermophilus townsendii* Bach.	USA
80	*Spermophilus (Ictidomys) tridecemlineatus*	USA
81	*Spermophilus undulatus* Pall.	Altay mountains (Russia), Tuva highlands (Russia), Chentij and Khangai (Mongolia), China
82	*Spermophilus (Otospermophilus) variegatus* Erxl.	USA
83	*Spermophilus (Urocitellus) washingtoni* Howell	USA
	Family: Geomyidae	
84	*Thomomys bottae* Eud. et Cerv.	USA
85	*Thomomys fossor*	USA
	Family: Heteromyidae	
86	*Dipodomys ordii* Wood	USA
87	*Heteromys anomalus* Thom.	Venezuela
	Family: Pedetidae	
88	*Pedetes cafer* Pall.	South Africa
89	*Pedetes capensis*[a]	Zimbabwe
	Family: Cricetidae	
90	*Akodon arviculoides* Wagner	Brazil
91	*Akodon dolores* Thom.	Argentina
92	*Akodon mollis* Thom.	Ecuador, Peru

(continued)

Table 5.6 (continued)

Number	Host	Focus/foci
93	*Akodon orophilus* Osgood	Peru
94	*Allocricetulus eversmanni* Brandt	Volga-Ural (Russia, Kazakhstan), Uralian (Ural-Uil) prairie (Kazakhstan), Ural-Emba, pre-Ustyurt, Northern pre-Aral*, pre-Aral Karakum desert (Kazakhstan), Khangai (Mongolia)
95	*Calomys (Hesperomys) bimaculatus* Waterh.	Bolivia
96	*Calomys callosus* Rengr.	Brazil
97	*Calomys fecundus* Thom.	Bolivia
98	*Calomys laucha* Desm.	Argentina
99	*Calomys murillus* Thom.	Argentina
100	*Calomys tener* Winge	Brazil
101	*Calomys venustus* Thom.	Bolivia
102	*Cricetus cricetus* L.	Uralian Prairie (Kazakhstan), Dzhungarian* highland (Kazakhstan)
103	*Cricetulus barabensis* Pall.	Baikal prairie (Russia), North China
104	*Cricetulus migratorius* Pall.	Dagestan (Russia), South Caucasus (Armenia, Georgia), Gissar(Tajikistan), Tien-Shan (Kazakhstan, Kyrgyzstan), Talas (Kyrgyzstan, Kazakhstan), Dzhungarian* highlands, Volga-Ural (Russia, Kazakhstan) and Uralian (Ural-Uil) prairies (Kazakhstan), Ural-Emba, pre-Ustyurt, Ustyurt, North pre-Aral, Zaaralsky, Mangyshlak, pre-Aral Karakum deserts (Kazakhstan), Caspian (Russia) and Volga-Ural sandy (Russia, Kazakhstan), pre-Alakol low mountains (Kazakhstan)
105	*Cricetulus triton* Wint.	North China
106	*Eligmodontia moreni* Thom.	Argentina
107	*Eligmodontia hirtipes* Thom.	Argentina
108	*Graomys (Phyllotis) cachinus*	Argentina
109	*Graomys chacoensis*	Argentina
110	*Graomys griseoflavus* Waterh.	Argentina, Bolivia
111	*Graomys medius*	Argentina
112	*Holochillus balnearum* Thom.	Argentina
113	*Holochillus brasiliensis* Desmarest	Brazil
114	*Holochillus sciureus* Wagn.	Brazil
115	*Mesocricetus brandti* Nehr.	Transcaucasian highland Armenia, Georgia
116	*Mystromys albicaudatus*	USA
117	*Neotoma albigula* Hart.	USA
118	*Neotoma cinerea* Ord.	USA
119	*Neotoma desertorum* Merr.	USA
120	*Neotoma fuscipes* Baird.	USA
121	*Neotoma intermedia* Rhoads.	USA

(continued)

Table 5.6 (continued)

Number	Host	Focus/foci
122	*Neotoma lepida* Thom.	USA
123	*Neotoma micropus* Baird	USA
124	*Onychomys leucogaster* W.-N.	USA
125	*Onychomys torridus* Coues	USA
126	*Oryzomys andinus* Org.	Peru
127	*Oryzomys arenalis* Thom.	Peru
128	*Oryzomys flavescens* Waterh.	Argentina, Bolivia
129	*Oryzomys intermedius* Leche	Brazil
130	*Oryzomys palustris* Harl.	USA
131	*Oryzomys phaeopus* Thom.	Ecuador
132	*Oryzomys pyrrhorhinus* Vied.	Brazil
133	*Oryzomys stolzmanni* Thom.	Peru
134	*Oryzomys subflavus* Wagner	Brazil
135	*Oryzomys xanthaeolus* Thom.	Peru, Ecuador
136	*Oryzomys x. baroni* All.	Ecuador
137	*Oxymycterus paramensis* Thom.	Bolivia
138	*Peromyscus boylii* Baird	USA
139	*Peromyscus*	USA
140	*Peromyscus maniculatus* Wagn.	USA
141	*Peromyscus truei* Shuf.	USA
142	*Phodopus sungorus* Pall.	Altay mountains (Russia)
143	*Phyllotis amicus* Thom.	Peru
144	*Phyllotis darwini* Thom.	Argentina
145	*Phyllotis fruticolus* Anth.	Ecuador
146	*Phyllotis wolfsohni* Thom.	Bolivia
147	*Rhipidomys equatorius* Thom.	Peru
148	*Rhipidomys leucodactylus* Tsch.	Bolivia
149	*Sigmodon hirsutus* Burm.	Venezuela
150	*Sigmodon peruanus* All.	Peru, Ecuador
151	*Sigmodon puna*	Ecuador
152	*Zygodontomys lasiurus* Lund	Brazil
153	*Zygodontomys pixuna*	Brazil
154	*Myospalax psilurus* Miln.-Edw.	China
155	*Alticola argentatus (roylei)* Severtz.	Alay (Kyrgyzstan), Tien-Shan (Kazakhstan, Kyrgyzstan), Talas (Kyrgyzstan), Gissar (Tajikistan) highland, Khangai (Mongolia)
156	*Alticola strelzovi* Kastsch.	Altai and Tuva mountains (Russia)
157	*Arvicola terrestris* L.	South Caucasus (Armenia, Georgia) and Dagestan (Russia) highland, Ural-Emba (Kazakhstan)

(continued)

Table 5.6 (continued)

Number	Host	Focus/foci
158	*Clethrionomys frater* Thom.	Talas* (Kazakhstan), Dzhungarian* (Kazakhstan) highlands
159	*Ellobius lutescens* Thom.	Kurdistan (Iran)
160	*Ellobius talpinus* Pall.	Alay (Kyrgyzstan) highlands, Volga-Ural (Russia, Kazakhstan) and Uralian*(Kazakhstan) prairie, Volga-Ural (Russia, Kazakhstan) sands, Zaaralsky, Mangyshlak (Kazakhstan), Kyzylkum (Kazakhstan, Uzbekistan, Turkmenistan), Karakum (Turkmenistan) deserts
161	*Eothenomys melanogaster* Milne-Edw.	South China
162	*Eothenomys miletus*	Northwestern Yunnan (China)
163	*Lagurus curtatus* Cope	USA
164	*Lagurus lagurus* Pall.	Volga-Ural (Russia, Kazakhstan) and Uralian (Kazakhstan) prairies, Volga-Ural sands (Russia, Kazakhstan)
165	*Lasiopodomys brandtii* Radde	Baikal prairie (Russia), Khangai, Chentij (Mongolia), Jilin Gol Plateau (Inner Mongolia)
166	*Microtus arvalis* Pall.	South Caucasus (Armenia, Georgia), Dagestan (Russia), Central Caucasus (Russia) highland, Volga-Ural(Russia, Kazakhstan) and Uralian (Kazakhstan) prairies, Volga-Ural sands (Russia, Kazakhstan)
167	*Microtus daghestanicus* Schidlowski	Dagestan highland (Russia)
168	*Microtus californicus* Peal	California (USA)
169	*Microtus carruthersi* Thom.	Gissar highlands (Tajikistan), Talas* highlands (Uzbekistan, Kazakhstan)
170	*Microtus gregalis* Pall.	Alay (Kyrgyzstan), Tien-Shan (Kazakhstan, Kyrgyzstan), Altay mountains* (Russia) highlands, Baikal (Russia) prairie, Tuva (Mongun-Taiga) (Russia), Khangai (Mongolia)
171	*Microtus (Chionomys) gud* Satun.	Transcaucasian highland (Armenia, Georgia)
172	*Microtus fortis* Buchn.	China
173	*Microtus fuscus*	Tibet
174	*Microtus kirgisorum (ilaeus)* Ognev	Tien-Shan highland (Kyrgyzstan, Kazakhstan), Talas, Dzhungarian* highlands (Kazakhstan)
175	*Microtus (montanus)* manus	USA
176	*Microtus minutus*	South China
177	*Microtus montanus* Peal	USA
178	*Microtus (Chionomys) nivalis* Mart.	Transcaucasian highland (Armenia, Georgia)

(continued)

Table 5.6 (continued)

Number	Host	Focus/foci
179	*Microtus socialis* Pall	Low pre-Araksinsk mountains (Armenia, Azerbaijan), Transcaucasian highland (Armenia, Georgia), Dagestan valleys and foothills (Russia), Volga-Ural, Uralian prairie (Kazakhstan), pre-Ustyurt, Ustyurt, North pre-Aral, pre-Aral Karakum, Taukum, pre-Balkhash desert (Kazakhstan), Caspian sands (Russia), Dzhungarian* highlands (Kazakhstan)
180	*Microtus townsendii* Bachm.	USA
181	*Ondatra zibethicus* L.	Caspian sands (Russia), pre-Balkhash desert* (Kazakhstan)
182	*Desmodillus auricularis* Smith.	South Africa
183	*Gerbillus gerbillus* Olivier	Mauritania
184	*Gerbillus nanus* Blanford	Mauritania
185	*Gerbillus (Gerbillurus) paeba* Smith.	South Africa
186	*Meriones blackleri (tristrami)* Thom.	Low pre-Araksinsk mountains (Armenia, Azerbaijan), South Transcaucasian valleys and foothills (Azerbaijan, Georgia), Kurds-Iranian (Iran)
187	*Meriones erythrourus (libycus)* Gray	Pre-Ustyurt (Kazakhstan), Ustyurt (Kazakhstan, Uzbekistan, Turkmenistan), Ural-Emba (Kazakhstan), North pre-Aral (Kazakhstan), Zaaralsky (Kazakhstan), Mangyshlak (Kazakhstan), pre-Aral Karakum (Kazakhstan), Karakum (Turkmenistan), Kopet Dag (Turkmenistan), Kyzylkum (Kazakhstan, Uzbekistan, Turkmenistan), Muyunkum (Kazakhstan), Taukum (Kazakhstan), pre-Balkhash (Kazakhstan), Betpak-Dala (Kazakhstan) desert, Alakol mountains (Kazakhstan), Ili intermountains (Kazakhstan), South Transcaucasian plain-piedmont (Azerbaijan, Georgia), Kurds-Iranian mountain prairie, Iran-Afghan low mountains, desert, Syrian-Mesopotamian desert (Libya*)
188	*Meriones meridianus* Pall.	Caspian Northwest (Russia), Volga-Ural (Russia, Kazakhstan), Uralian (Kazakhstan) Prairie, pre-Araksinsk low mountains (Armenia, Azerbaijan), Caspian (Russia), Volga-Ural (Russia, Kazakhstan), Ural-Emba (Kazakhstan), pre-Ustyurt (Kazakhstan), Ustyurt (Kazakhstan, Uzbekistan, Turkmenistan), North pre-Aral (Kazakhstan), Zaaralsky (Kazakhstan), Mangyshlak (Kazakhstan), pre-Aral Karakum (Kazakhstan), Karakum (Turkmenistan), Kyzylkum (Kazakhstan, Uzbekistan, Turkmenistan), Muyunkum (Kazakhstan), Taukum (Kazakhstan), pre-Balkhash (Kazakhstan), Ili intermountains (Kazakhstan), Betpak-Dala (Kazakhstan) Desert (Southeast Mongolia), Inner Mongolia (China) sands

(continued)

Table 5.6 (continued)

Number	Host	Focus/foci
189	*Meriones persicus* Blanf.	Low pre-Araksinsk mountains (Armenia, Azerbaijan), Kurds-Iranian (Iran)
190	*Meriones shawi* Duv., 1842	Kurds-Iranian (Iran)
191	*Meriones tamariscinus* Pall.	Dagestan foothills-flats (Russia), Volga-Ural (Russia, Kazakhstan), Uralian (Kazakhstan) Prairie, Caspian (Russia), Volga-Ural (Russia, Kazakhstan) sands, Ural-Emba (Kazakhstan), pre-Ustyurt (Kazakhstan), Ustyurt (Kazakhstan, Uzbekistan, Turkmenistan), Mangyshlak (Kazakhstan), pre-Aral Karakum (Kazakhstan), North pre-Aral (Kazakhstan), Zaaralsky (Kazakhstan), Betpak-Dala (Kazakhstan), Karakum (Turkmenistan), Kyzylkum (Kazakhstan, Uzbekistan, Turkmenistan), Muyunkum (Kazakhstan), Taukum (Kazakhstan), pre-Balkhash (Kazakhstan) desert, pre-Alakol low mountains (Kazakhstan)
192	*Meriones unguiculatus* M.-Edw.	Gobi Altai (Mongolia), Inner Mongolia, Manchuria (China)
193	*Meriones vinogradovi* Heptn.	Pre-Araksinsk low mountains (Armenia, Azerbaijan), Kurds-Iranian (Iran)
194	*Psammomys obesus* Cretzschmar	Africa
195	*Rhombomys opimus* Licht.	Uralian prairie (Kazakhstan), Ural-Emba (Kazakhstan), pre-Ustyurt (Kazakhstan), Ustyurt (Kazakhstan, Uzbekistan, Turkmenistan), North pre-Aral (Kazakhstan), Zaaralsky (Kazakhstan), Mangyshlak (Kazakhstan), pre-Aral Karakum (Kazakhstan), Karakum (Turkmenistan), Kopet Dag (Turkmenistan), Kyzylkum (Kazakhstan, Uzbekistan, Turkmenistan), Muyunkum (Kazakhstan), Taukum (Kazakhstan), pre-Balkhash (Kazakhstan), Betpak-Dala (Kazakhstan), pre-Alakol low mountains
196	*Tatera afra* Gray	Africa (Kazakhstan), Iran-Afghan low mountains and desert, Gobi Altai (Mongolia), Northwest China
197	*Tatera brantsi* Smith.	South Africa
198	*Tatera indica* Hardw.	Kurds-Iranian (Iran) mountains and prairie, Syrian-Mesopotamian (Syria) desert, and North, Central, and South India
199	*Tatera leucogaster* Peters	Africa
200	*Tatera lobengulae* de Vint.	South Africa
201	*Tatera nigrita* Hatt.	Africa
202	*Tatera robusta* Creter	Tanzania
203	*Tatera schinzi* Noack.	Africa
204	*Tatera valida* Bosage	Africa
	Family: **Muridae**	
205	*Acomys cahirinus* Desm.	Egypt
206	*Apodemus agrarius* Pall.	Caspian sands, Manchuria (China)
207	*Apodemus chevrieri*	Northwestern Yunnan (China)

(continued)

Table 5.6 (continued)

Number	Host	Focus/foci
208	*Apodemus speciosus* Temminck	Northwestern Yunnan (China)
209	*Apodemus sylvaticus* L.	Caspian Northwest prairie (Russia), Uralian (Ural-Uil) Prairie (Kazakhstan) low pre-Alakol mountains (Kazakhstan), Terek-Sunzha (Russia), Central Caucasus (Russia), Talas (Kyrgyzstan, Kazakhstan), Dzhungarian* (Kazakhstan), Alay (Kyrgyzstan), South Caucasus (Armenia, Georgia), Gissar (Tajikistan) highlands
210	*Dasymys incomtus* Sund.	Africa
211	*Dendromus haymani* Hatt.	Africa
212	*Dendromus insignis* Thom.	Africa
213	*Dendromus melanotis* Smith	Congo
214	*Dendromus mesomelas* Brants	Congo
215	*Dendromus mystacalis* Heuglin	Congo
216	*Hybomys univittatus* Peters	Congo
217	*Lemniscomys griselda* Thom.	Senegal, Kenya*
218	*Lemniscomys striatus* L.	Kenya, Tanzania*, Congo
219	*Lophuromys aguilus* Dollm.	Africa
220	*Lophuromys flavopunctatus* Thom.	Tanzania, Congo
221	*Lophuromys sikapusi** Temminck	Africa
222	*Malacotrix typicus* Smith.	Africa
223	*Micromys minutus* Pall.	China
224	*Mus booduga* Gray.	India
225	*Mus bufo* Thom.	Congo
226	*Mus cervicolor* Hodg.	Southeast Asia
228	*Mus (Leggada) minutoides* A. Smith	Congo
229	*Mus musculoides* Temm.	Africa
230	*Mus musculus* L.	Dagestan plains and foothills (Russia), Volga-Ural (Russia, Kazakhstan), Uralian (Kazakhstan) prairie, Central Caucasus (Russia) highlands, Ural-Emba, pre-Ustyurt, North pre-Aral, Zaaralsky, Mangyshlak, pre-Aral Karakum (Kazakhstan), Karakum (Turkmenistan), Kyzylkum (Kazakhstan, Uzbekistan, Turkmenistan), Muyunkum (Kazakhstan) deserts, Volga-Ural (Russia, Kazakhstan), Caspian (Russia) sands, pre-Alakol (Kazakhstan), pre-Araksinsk (Armenia, Azerbaijan), South Caucasus flat-foothill mountain (Azerbaijan, Georgia), South Caucasus highlands (Armenia, Georgia), Sarydzhaz highlands (Kazakhstan), Inner Mongolia, Northeast China, Vietnam, South America

(continued)

Table 5.6 (continued)

Number	Host	Focus/foci
231	*Mus platythrix* Bennett	India, Vietnam*
232	*Mus triton* Thom.	Africa
233	*Mylomys cuninghamei* Thom.	Africa
234	*Rhabdomys pumilio* Sparrm.	Kenya, Zimbabwe
235	*Steatomys pratensis* Pet.	Africa
236	*Aethomys chrysophilus* de Winton	Africa
237	*Aethomys hindei* Thom.	Africa
238	*Aethomys namaquensis* A. Smith	Africa
239	*Arvicanthis abyssinicus* Rupp.	Kenya, Tanzania, Congo
240	*Arvicanthis niloticus* Desm.	Tanzania, Senegal, Kenya
241	*Bandicota bengalensis* Gray	South India, Burma
242	*Bandicota indica* Bech.	South India, Southeast Asia
243	*Bandicota (bengalensis) gracilis* Nehr.	India, Sri Lanka
244	*Bandicota malabarica* Shaw.	Sri Lanka
245	*Bandicota savilei* Thom.	Vietnam
246	*Berylmys bowersi* Anderson	Vietnam
247	*Cricetomys gambianus* Waterh.	Tanzania, Senegal, Congo
248	*Dipodomys* sp.	USA
249	*Golunda ellioti* Gray	India
250	*Grammomys dolichurus* Smith	Congo, Tanzania*, Kenya*
251	*Grammomys drays* Thom.	Africa
252	*Leopoldamys edwardsi* Thom.	Southeast Asia
253	*Mastomys (Praomys) coucha*	Africa
254	*Mastomys natalensis* Smith	Kenya, Tanzania, Mozambique, Congo, Senegal
255	*Millardia meltada* Gray.	North and Central India
256	*Nesokia indica* Gray	India
257	*Niviventer niviventer* Hodg.	Vietnam
258	*Oenomys hypoxanthus* Pucheran	Congo
259	*Otomys angoniensis* Wrought.	Kenya, Tanzania
260	*Otomys denti* Thom.	Congo, Tanzania
261	*Otomys irroratus* Br.	Africa
262	*Otomys tropicalis* Wrought.	Africa
263	*Otomys unisulcatus* Cuvier et Geoffr.	Africa
264	*Paratomys brantsi* Smith.	Africa
265	*Pelomys campanae* Huet.	Africa
266	*Pelomys fallax* Pet.	Zambia, Tanzania*

(continued)

Table 5.6 (continued)

Number	Host	Focus/foci
267	*Rattus andersoni* Thom.	China
268	*Rattus argentiventer* Robinson et Kloss	Southeast Asia
269	*Rattus blanfordi* Thom.	India
270	*Rattus concolor* Blyth.	Java
271	*Rattus exulans* Peale	Java, Vietnam
272	*Rattus flavipectus* Milne-Edwards	Yunnan, Guangdong, Fyuan (South China), Vietnam
273	*Rattus griseiventer* Bouh.	Java
274	*Rattus hawaiiensis* Stone	Hawaiian Islands
275	*Rattus kajzeri* Noack	Africa
276	*Rattus losea* Swinh.	Southeast Asia
277	*Rattus molliculus** Robinson et Kloss	Vietnam
278	*Rattus nitidus* Hodyson	Northwestern Yunnan (China), Vietnam
279	*Rattus norvegicus* Berk.	Volga-Ural (Russia, Kazakhstan), Caspian (Russia) sands, Africa, India, Northeast and Southeast China, South America
280	*Rattus rattus* L.	Africa, Madagascar, Hawaiian Islands, India, Southeast Asia, China, South America
281	*Rattus sladeni (koratensis)* Kloss	Southeast Asia
	Family: **Gliridae**	
282	*Dryomys nitedula* Pall	South Caucasus (Armenia, Georgia), Gissar highlands, Sarydzhaz * (Kazakhstan), Talas* (Kazakhstan), Dzhungarian* (Kazakhstan) highlands
	Family: **Dipodidae**	
283	*Allactaga elater* Licht.	Volga-Ural (Russia, Kazakhstan), Trans-Ural (Kazakhstan) prairie, Ural-Emba, pre-Ustyurt, Ustyurt, North pre-Aral, Zaaralsky, Mangyshlak, pre-Aral Karakum (Kazakhstan), Kyzylkum (Kazakhstan, Uzbekistan, Turkmenistan), Muyunkum (Kazakhstan), Karakum (Turkmenistan) Desert, Volga-Ural (Russia, Kazakhstan), Caspian (Russia) sands, pre-Araksinsk low mountains (Armenia, Azerbaijan), Iran
284	*Allactaga jaculus (major)* Pall.	Caspian sands (Russia), Volga-Ural (Russia, Kazakhstan), Uralian* prairie (Kazakhstan), Ural-Emba,* pre-Ustyurt,*Ustyurt,* North pre-Aral, pre-Aral Karakum (Kazakhstan)
285	*Allactaga severtzovi* Vinogr.	Zaaralsky (Aryskum-Daryalik Takyr)*, pre-Aral Karakum* (Kazakhstan) deserts
286	*Allactaga sibirica (saltator)* Forster	Baikal prairie (Russia), Tuva highlands (Russia), Altay mountains (Russia), Ustyurt*, pre-Aral Karakum*(Kazakhstan) desert, Sarydzhaz* highlands (Kazakhstan), Inner Mongolia (China)
287	*Allactagulus acontion* Pall.	Uralian* prairie (Kazakhstan), Ural-Emba,* pre-Ustyurt,* Ustyurt,* North pre-Aral,* Kyzylkum, Taukum *(Kazakhstan), Karakum (Turkmenistan) desert

(continued)

Table 5.6 (continued)

Number	Host	Focus/foci
288	*Dipus sagitta* Pall.	Ural-Emba, North pre-Aral*, pre-Aral Karakum* (Kazakhstan), Kyzylkum (Kazakhstan, Uzbekistan, Turkmenistan), pre-Balkhash (Kazakhstan), Karakum (Turkmenistan) deserts, Volga-Ural sands (Russia, Kazakhstan), Inner Mongolia (China)
289	*Eremodipus lichtensteini* Vinogr.	Kyzylkum (Kazakhstan, Uzbekistan, Turkmenistan), pre-Balkhash (Kazakhstan) desert
290	*Paradipus ctenodactylus* Vinogr.	Karakum (Turkmenistan), Kyzylkum (Kazakhstan, Uzbekistan, Turkmenistan) deserts
291	*Pygerethmus* Pygerethmus platyurus Licht.	Ural-Emba*, Pre-Ustyurt*, Mangyshlak deserts (Kazakhstan)
292	*Scirtopoda telum* Licht.	Caspian sands (Russia), Caspian Northwest (Russia), Uralian* prairie (Kazakhstan), Ural-Emba*, pre-Ustyurt*, Ustyurt*, North pre-Aral, Zaaralsky, Mangyshlak*, pre-Aral Karakum*, Kyzylkum, Muyunkum, Taukum, pre-Balkhash deserts (Kazakhstan)
	Family: **Caviidae**	
293	*Cavia aperea* Erxl.	Brazil, Ecuador
294	*Cavia pamparum* Thom.	Argentina
295	*Cavia porcella* L.	Ecuador
296	*Cavia tschudii* Osg.	Peru, Ecuador
297	*Galea musteloides* Burm.	Argentina, Bolivia
298	*Galea spixii* Wagl.	Brazil
299	*Kerodon rupestris* Wied.	Brazil
300	*Microcavia australis* Geoffr. et D. Orb.	Argentina
	Family: **Dasyproctidae**	
301	*Dasyprocta variegata*	Bolivia
	Family: **Chinchillidae**	
302	*Lagostomus maximus* Thom.	Argentina
	Family: **Echimyidae**	
303	*Cercomys cunicularius* Thom.	Brazil
304	*Cercomys inermis* Pict.	Brazil
	Order: **Carnivora**	
	Family: **Canidae**	
305	*Canis familiaris* L.	Africa, India*, Peru*
306	*Canis latrans* Say.	USA
307	*Vulpes corsac* L.	Emba*, pre-Ustyurt*, Ustyurt*, North pre-Aral*, Zaaralsky*, pre-Aral Karakum*, Kyzylkum* (Kazakhstan), Karakum (Turkmenistan) deserts
308	*Vulpes vulpes* L.	Volga-Ural prairie (Russia), Volga-Ural sands* (Kazakhstan), Ural-Emba*, pre-Ustyurt*, North pre-Aral*, Zaaralsky*, Mangyshlak*, pre-Aral Karakum*, Kyzylkum* (Kazakhstan) deserts, Tien-Shan highlands (Kazakhstan, Kyrgyzstan)

(continued)

Table 5.6 (continued)

Number	Host	Focus/foci
309	*Urocyon cinereoargenteus**	USA
	Family: Procyonidae	
310	*Procyon lotor*	USA
	Family: Mustelidae	
311	*Martes americana*	USA
312	*Meles meles* L.	Zaaralsky*, pre-Aral Karakum* deserts (Kazakhstan), Tien-Shan highlands (Kazakhstan, Kyrgyzstan), USA*
313	*Mustela altaica* Pall.	Baikal prairie (Russia), Iran
314	*Mustela erminea*	Tien-Shan highlands (Kazakhstan, Kyrgyzstan), North pre-Aral deserts* (Kazakhstan), Talas* and Dzhungarian* highland (Kazakhstan), Altay mountains* (Russia), Hissar* (Tajikistan) highlands
315	*Mustela eversmanni* Less.	Caspian Northwest (Russia), Volga-Ural (Kazakhstan), Baikal prairie (Russia), Ural-Emba*, pre-Ustyurt*, Ustyurt*, North pre-Aral, Zaaralsky, Mangyshlak, pre-Aral Karakum, pre-Balkhash, Muyunkum*, Betpak-Dala*(Kazakhstan), Kyzylkum (Kazakhstan, Uzbekistan, Turkmenistan) deserts, Tien-Shan (Kazakhstan, Kyrgyzstan), Talas* highland (Kazakhstan), Tuva (Russia), Altay mountains (Russia)
316	*Mustela furo*	South Africa
317	*Mustela nigripes*	USA
318	*Mustela nivalis* L.	Uralian prairie* (Kazakhstan), Ustyurt*, North pre-Aral*, Zaaralsky*, Mangyshlak*, pre-Aral Karakum*, Kyzylkum Muyunkum*, pre-Balkhash (Kazakhstan), Karakum (Turkmenistan) deserts, South Caucasus (Armenia, Georgia), Gissar (Tajikistan), Sarydzhaz *(Kazakhstan) highlands, China
319	*Mustela sibirica* Pall.	Baikal prairie (Russia)
320	*Mustela putorius* L.	Odessa, 1911 (Russia)
321	*Taxidea taxus*	USA
322	*Vormela peregusna* Gueld.	North pre-Aral*, Zaaralsky*, Mangyshlak*, pre-Aral Karakum* (Kazakhstan), Kyzylkum (Kazakhstan, Uzbekistan, Turkmenistan), Muyunkum* (Kazakhstan), Karakum (Turkmenistan) deserts
	Family: Viverridae	
323	*Cynictis* penicillata Cuvier	Africa
324	*Herpestes auropunctatus*	Hawaiian Islands
325	*Herpestes javanicus* E. Geoffr.	Vietnam
326	*Mangusta ichneumon* L.	India
327	*Paradoxurus hermaphrodites* Pal.	Vietnam
328	*Helogale parvula*	Kenya

(continued)

Table 5.6 (continued)

Number	Host	Focus/foci
329	*Suricata* suricatta Erhl.	India
	Family: Mephitidae	
330	*Mephitis mephitis*	USA
	Family: Felidae	
331	*Felis catus* L.	Zaaralsky* (Kazakhstan), Karakum (Turkmenistan), Kyzylkum (Uzbekistan, Turkmenistan) deserts, Sarydzhaz highlands* (Kazakhstan), China, South Africa, USA
332	*Felis (Puma) concolor*	USA
333	*Felis libyca* Forst.	Uralian (Ural-Uil)* prairie (Kazakhstan), Zaaralsky*, pre-Balkhash* deserts (Kazakhstan), pre-Alakol low mountains* (Kazakhstan)
334	*Felis rufa (Lynx rufus* Schreb.)	USA
	Order: Hyracoidea	
335	*Procavia capensis* L.	Africa
	Order: Artiodactyla	
	Family: Suidae	
336	*Sus scrofa* L	USA
	Family: Camelidae	
337	*Camelus bactrianus* L.	Volga-Ural sands (Russia), Mangyshlak*, pre-Aral Karakum* (Kazakhstan), Kyzylkum (Kazakhstan, Uzbekistan) deserts
338	*Camelus dromedarius* L.	Karakum desert (Turkmenistan)
	Family: Bovidae	
339	*Saiga tatarica* L.	Mangyshlak, Muyunkum*, Betpak-Dala* (Kazakhstan) deserts
340	*Bubalus* sp.*	Africa
341	Domestic goat (*Capra aegagrus hircus*)*[b]	Libya
	Class: Aves	
	Order: Passeriformes	
342	*Eremophila alpestris* L	Mongolia
343	*Oenanthe isabellina* Temm.	Volga-Ural prairie, Volga-Ural sands (Russia, Kazakhstan), Uralian* prairie (Kazakhstan), Mangyshlak* (Kazakhstan), Kyzylkum* (Kazakhstan, Uzbekistan, Turkmenistan) desert, Altay mountains* (Russia), Hissar* (Tajikistan) highlands, Mongolia
344	*Podoces panderi* Fisch.	Karakum (Turkmenistan), Kyzylkum*(Kazakhstan) desert

Species described as plague hosts as a result of identification of the antibodies in their organism to Fr I, without isolation of the causative agent of plague, marked with*

[a]The last species as a plague host described by Neronov et al. [105]. Experts on mammalian taxonomy recognize only one species in this family: *Pedetes cafer* or *P. capensis*, but whether there are one or two species is unclear

[b]There is a reported case of six people infected by plague during cutting of two domestic goats; one of the six had antibodies to fraction-I (Christie et al., 1980)

Arthropods (fleas, ticks, lice) infected with *Y. pestis*, the causative agent of the plague, in nature have been reported in the literature: [1, 19, 20, 107, 119]. They are listed in Tables 5.7 and 5.8. The names of the arthropods are given in accordance with the source citation.

Table 5.7 Species and subspecies of fleas infected with *Y. pestis* in nature as reported in the literature

	Species	Location	Host
1	*Acropsylla episema*	India	The available literature lacks relevant information
2	*Adoratopsylla (Tritopsylla) intermedia copha*	South America	*Opossums Didelphis* sp.
3	*Amalareus dissimilis dagestanicus*	Caucasus	*Microtus arvalis*
4	*A. dissimilis* ssp. n.	Eurasia	Cricetidae
5	*A. penicilliger*	Eurasia	Voles *Arvicola terrestris*, *Clethrionomys frater*; *Microtus gregalis* and other murine rodents
6	*Amphalius runatus*	Eurasia	*Ochotona pricei* and other pika species. In the mountain Altai focus, *Y. pestis* was isolated from fleas and was obtained from Daurian pikas, long-tailed ground squirrel, and jerboas and from the entrances of Altai marmot burrows. In Mongolia, the flea is found on marmots long-tailed and Daurian ground squirrels, Brandt's vole, and terrestrial predators
7	*Amphipsylla anceps*	Eurasia	*Cricetulus migratorius* and other murine rodent species
8	*A. asiatica*	Eurasia (Tianshan highlands)	*Microtus gregalis*
9	*A. kuznetzovi*	Eurasia	Voles *Alticola* sp., *Microtus* sp.
10	*A. montana*	Eurasia	(*Alticola argentatus*)
11	*A. phaiomydis*	Gissar Ridge	*Microtus carruthersi*; other murine rodents
12	*A. primaris*	Eurasia	Voles *Alticola strelzovi*, *Lasiopodomys brandti*, pikas (*Ochotona pricei*)
13	*A. rossica*	Caucasus	*Microtus arvalis*; other murine rodent species

(continued)

Table 5.7 (continued)

	Species	Location	Host
14	*A. schelkovnikovi*	Eurasia	*Cricetulus migratorius*; other murine rodents
15	*Anomiopsylla hiemalis*	North America	Hamsters *Neotoma* sp.
16	*A. nudatus*		Cricetidae
17	*Atyphloceras multidentatus*	North America	Hamsters *Peromyscus* sp.
18	*Callopsylla caspia*	Transcaucasian, Dagestan, Gissar, Talas	*Microtus arvalis*, *M. carruthersi*, *Alticola argentatus*, other murine rodents
19	*C. dolabris*	Eurasia	Marmots *Marmota* sp. (Xinjiang, China)
20	*C. saxatilis*	Caucasus	Cricetidae
21	*Catallagia decipiens*	North America	Hamsters *Lagurus* sp., *Peromyscus* sp.
22	*C. sculleni*		Cricetidae
23	*C. wymani*	North America	Voles *Microtus* sp., hamsters *Peromyscus* sp.
24	*Cediopsylla spillmanni*	South America	American rabbits *Sylvilagus* sp.
25	*Ceratophyllus borealis*	Caucasus (Armenia)	Birds (Aves)
26	*C. styx avicitelli*	Eurasia	Birds (Aves)
27	*Chaetopsylla homoea*	Eurasia	Carnivorous mammals (Carnivora); rodents
28	*Chiastopsylla numae*	Africa	Muridae
29	*C. rossi*	South Africa	Gerbils *Tatera* sp., *Otomys* sp.
30	*Citellophilus lebedevi*	Tianshan, Alay, and Gissar Ridge	Marmots *Marmota baibacina* and *M. caudata*
31	*C. relicticola*	Tianshan	*Spermophilus relictus*
32	*C. tesquorum tesquorum*	Eurasia	Gophers *Spermophilus* (*Citellus*) sp.
33	*C. t. altaicus*	Mountain Altai	(*M. baibacina*) (*S. undulatus*)
34	*C. t. ciscaucasicus*	Caucasus	(*S. pygmaeus*)
35	*C. t. mongolicus*	Eurasia	Sciuridae (marmots gophers)
36	*C. t. transvolgensis*	North Caspian	(*S. pygmaeus*)
37	*C. trispinus*	Eurasia	(*Spermophilus fulvus*) (*S. erythrogenys*) gophers gerbils *Rhombomys opimus*, *Meriones* sp.
38	*C. ullus*	Eurasia	
39	*Coptopsylla bairamaliensis*	Eurasia	Gerbils *Rhombomys opimus*, *Meriones* sp.
40	*C. caucasica*	Eurasia	Gerbils, *Meriones* sp.
41	*C. lamellifer*	Eurasia	*Rhombomys opimus*, *Meriones* sp.
42	*C. olgae*	Eurasia	(*Rhombomys opimus*)
43	*Craneopsylla minerva wolffhuegeli*		Cricetidae
44	*Ctenocephalides canis*	Africa	Carnivorous mammals

(continued)

Table 5.7 (continued)

	Species	Location	Host
45	*C. felisi*	India	
46	*C. f. strongylus*		
47	*C. f. orientalis*		
48	*Ctenophthalmus bogatschevi*	Caucasus (Azerbaijan)	Cricetidae
49	*C. breviatus*	Eurasia	*Spermophilus pygmaeus* and other rodents
50	*C. cabirus*	Africa	*Arvicanthis* sp. (Muridae)
51	*C. calceatus cabirus*	Africa	Cricetidae
52	*C. congener*	Eurasia	Mouse *Apodemus* sp., voles *Microtus* sp.
53	*C. dolichus*	Eurasia	Gerbils *Meriones* sp., *Rhombomys opimus*
54	*C. golovi*	Eurasia	*Spermophilus musicus*, voles *Arvicola terrestris*, *Microtus arvalis* and other murine rodents
55	*C. intermedius*	Eurasia	(*Microtus arvalis*)
56	*C. iranus iranus*	Caucasus (Azerbaijan)	Cricetidae
57	*C. orientalis*	Eurasia	*Spermophilus musicus*, hamster voles, and other rodents
58	*C. phyris*	Africa	(*Arvicanthis* sp., *Lemniscomys* sp., *Otomys* sp.)
59	*C. pollex*	Eurasia	(*Spermophilus pygmaeus*) (*Ct. migratorius*)
60	*C. quadratus*	Eurasia	Muridae
61	*C. secundus*	Eurasia	Voles *Microtus* sp.
62	*C. teres*	Caucasus	*Microtus arvalis*
63	*C. wagneri*	Caucasus	Voles *Arvicola terrestris*, *Microtus* sp.
64	*C. wladimiri*	Caucasus	*Microtus arvalis*
65	*Ctenophyllus hirticrus*	Eurasia	Pikas, *Ochotona* sp.
66	*Dactylopsylla (Foxella) ignota*	North America	Rodents of the family Geomyidae
67	*Delostichus talis*	South America	*Cavia* sp.
68	*Diamanus (Oropsylla) montanus*	North America	Gophers *Spermophilus* (*Citellus*) sp.
69	*Dinopsyllus brachypectus*	Madagascar	Gray and black rats
70	*D. ellobius*	Africa	Mouse, *Rhabdomys* sp., gerbils *Tatera* sp.
71	*D. lypusus*	Africa	Muridae (*Arvicanthis* sp.)
72	*Echidnophaga gallinacea*	Africa, North America	Insectivora

(continued)

Table 5.7 (continued)

	Species	Location	Host
73	*E. oschanini*	Eurasia	Gerbils (*Rhombomys opimus*, *Meriones erythrourus*)
74	*Epitedia stanfordi*	North America	Cricetidae
75	*E. wenmanni wenmanni*	North America	Cricetidae
76	*E. w. testor*	North America	Cricetidae
77	*Euhoplopsyllus andensis*	South America	Lagomorpha
78	*E. glacialis affinis*	North America	Lagomorpha
79	*E. manconis*	South America	Lagomorpha
80	*Foxella ignota*	North America	*Thomomys*
81	*Frontopsylla caucasica*	Caucasian Ridge	Voles *Microtus arvalis*
82	*F. elata*	Eurasia	Voles *Microtus* sp. mouse *Apodemus* sp., hamsters *Cricetulus* sp., pikas *Ochotona* sp.
83	*F. elatoides*	Eurasia	(*Spermophilus undulatus*)
84	*F. frontalis*	Eurasia	Birds (*Oenanthe* sp.), different species of rodents
85	*F. glabra vara*	Gissar Ridge	(*Microtus carruthersi*)
86	*F. hetera*	Eurasia	*Ochotona* sp. in mountain Altai foci, found on all animal species
87	*F. luculenta*	Eurasia	*Ochotona* sp., gophers *Spermophilus* sp., voles *Lasiopodomys brandti*, other rodent species
88	*F. ornata*	Eurasia	Voles *Clethrionomys* sp., *Microtus* sp., *Apodemus* sp., other murine rodent species
89	*F. protera*	Eurasia	Voles *Alticola* sp., *Microtus* sp.
90	*F. semura*	Eurasia	Gophers *Spermophilus* sp.
91	*F. spadix spadix*	Eurasia	Sciuridae
92	*F. wagneri*	Eurasia	Dipodidae, other rodent species
93	*Hectopsylla eskeyi*	South America	*Cavia* sp.
94	*H. suarezi*	South America	*Cavia* sp.
95	*Hiphiopsylla lippa*	Africa	Muridae
96	*Hoplopsyllus andensis*	South America	Rabbits, *Sylvilagus* sp.
97	*H. anomalus*	North America	Gophers *Spermophilus* sp. (Lagomorpha)
98	*H. glacialis affinis*	North America	Rabbits, *Sylvilagus* sp.
99	*H. manconis*	South America	Rabbits, *Sylvilagus* sp.
100	*Hystrichopsylla dippiei truncata*	North America	Cricetidae
101	*H. linsdalei*	North America	Voles *Microtus* sp.
102	*H. talpae*	Eurasia	Moles (*Talpa* sp.), voles *Microtus* sp., and other rodent species

(continued)

Table 5.7 (continued)

	Species	Location	Host
103	*Leptopsylla aethiopica aethiopica*	Africa	Muridae
104	*L. (Pectinoctenus) nemorosa*		*Apodemus sylvaticus*. In Gissar Ridge also *Microtus carruthersi*, *Cricetulus migratorius*
105	*L. nana*	Eurasia	Voles *Alticola* sp., *Chionomys* sp., *Microtus* sp., *Cricetulus* sp.
106	*L. segnis*	Eurasia	*Mus musculus*; in China, several species of rodents
107	*L. pavlovskii*	Eurasia	*Phodopus sungorus* and other murine rodent species
108	*L. tashcenbergi*	Eurasia	Mouse *Apodemus* sp., other rodent species and insectivores
109	*Listropsylla dorippae*	Africa	Gerbils, *Tatera* sp.
110	*Malaraeus bitterrootensis*	North America	Cricetidae
111	*M. sinomus*	North America	Cricetidae
112	*M. telchinus*	North America	Voles *Microtus* sp., hamsters *Peromyscus* sp.
113	*Megabothris abantis*	North America	Cricetidae
114	*M. clantoni clantoni*	North America	*Lagurus* sp., *Peromyscus* sp.
115	*M. c. johnsoni*	North America	Cricetidae
116	*M. exilis*	North America	Cricetidae, Hamsters *Onychomys* sp.
117	*M. eumolpi*	North America	Cricetidae, *Tamias* sp.
118	*M. turbidus*	Eurasia	Voles *Microtus arvalis*, *Clethrionomys* sp., *Apodemus* sp.
119	*M. wagneri*	North America	Cricetidae, *Lagurus* sp., *Peromyscus* sp.
120	*Meringis shannoni*	North America	*Lagurus* sp. and other rodents
121	*Mesopsylla apscheronica*	Eurasia (Transcaucasus)	*Allactaga* sp.
122	*M. eucta*	Eurasia	Jerboas, *Alactagulus* sp., *Allactaga* sp., *Scirtopoda* sp., Libyan jird (*Meriones erythrourus*)
123	*M. hebes*	Eurasia	Jerboas, *Allactaga* sp.; gerbils, *Meriones* sp.
124	*M. lenis*	Eurasia	Jerboas, *Allactaga* sp., Libyan jird (*Meriones erythrourus*)
125	*M. tuschkan*	Eurasia	Jerboas, *Allactaga* sp., etc.
126	*Monopsyllus (Ceratophyllus) anisus*	Eurasia	*Rattus* sp.
127	*Neopsylla abagaitui*	Eurasia	(*Spermophilus dauricus*)
128	*N. bidentatiformis*	Eurasia	gophers *Spermophilus* sp., *Cricetulus* sp., *Rattus* sp., and other rodent species

(continued)

Table 5.7 (continued)

	Species	Location	Host
129	*N. galea*	Eurasia	(*Lasiopodomys brandti*) (*Cricetulus barabensis*)
130	*N. hongyangensis*	Eurasia	
131	*N. inopina*	North America	Sciuridae
132	*N. mana*	Eurasia	Voles *Lasiopodomys brandti*, gophers *Spermophilus undulatus*, and other rodent species
133	*N. meridiana*	Eurasia (Tianshan highlands)	
134	*N. pleskei*	Eurasia	Voles *Microtus* sp., *Cricetulus migratorius*, and other murine rodents
135	*N. setosa*	Eurasia	Sciuridae; gophers *Spermophilus* sp.
136	*N. specialis*	Eurasia	Muridae
137	*N. teratura*	Eurasia	*Cricetulus migratorius* and other murine rodents
138	*Neotyphloceras rosenbergi*	South America	*Didelphis* sp.
139	*Nosopsyllus aralis*	Eurasia	Jirds *Meriones* sp.
140	*N. arcotus*	India	
141	*N. consimilis*	Eurasia	Voles *Microtus arvalis* and other murine rodent species
142	*N. fasciatus*	Eurasia and India	*Rattus* sp.
143	*N. fidus*	Eurasia	*Mus musculus* and other Muridae species
144	*N. iranus*	Eurasia	Jirds *Meriones* sp.
145	*N. laeviceps*	Eurasia	*Meriones* sp. (*Ct. migratorius*)
146	*N. londiniensis*	Brazil, Ecuador	Muridae
147	*N. mokrzeckyi*	Eurasia	*Mus musculus* and other murine rodent species
148	*N. monstrosus*	Eurasia	Jirds *Meriones* sp.
149	*N. nicanus*	Eurasia	Muridae
150	*N. nilgiriensis*	South Asia	Muridae
151	*N. punjabensis*	South Asia	Muridae
152	*N. tersus*	Eurasia	Great gerbil, *Rhombomys opimus*
153	*N. turkmenicus*	Eurasia	Jirds *Meriones* sp.
154	*Odontopsyllus sp.*	South America	Rabbits, *Sylvilagus* sp.
155	*Ophthalmopsylla kasakiensis*	Eurasia	(*Dipus sagitta*)
156	*O. kiritschenkoi*	Eurasia	Dipodidae, Cricetidae
157	*O. kukusckini*	Eurasia	Murine rodent species
158	*O. praefecta*	Eurasia	Dipodidae; Hurhinsk meso-foci (Mongolia); China from the fur of *S. alaschanicus*

(continued)

Table 5.7 (continued)

	Species	Location	Host
159	*O. volgensis*	Eurasia	Jerboas, *Alactagulus* sp., *Allactaga* sp., *Dipus* sp.
160	*Opisocrostis bruneri*	North America	Prairie dogs *Cynomys* sp.
161	*O. hirsutus*	North America	
162	*O. labis*	North America	*Cynomys* sp.
163	*O. tuberculatus*	North America	*Cynomys* sp.
164	*Opisodasis keeni nesiotus*	North America	Voles *Microtus* sp., hamsters *Peromyscus* sp., *Reithrodontomys* sp.
165	*O. k. keeni*	North America	Cricetidae
166	*Orchopeas howardii howardii*	North America	Cricetidae
167	*O. leucopus*	North America	Hamsters *Neotoma* sp.
168	*O. neotomae*	North America	
169	*O. sexdentatus*	North America	*Neotoma* sp., *Peromyscus* sp.
170	*Oropsylla alaskensis*	Eurasia	Gophers *Spermophilus dauricus*, *S. undu*
171	*O. bruneri*	North America	
172	*O. hirsuta*	North America	Prairie dogs *Cynomys ludovicianus*
173	*O. idahoensis*	North America	Gophers *Spermophilus* sp., *Otospermophilus* sp.
174	*O. ilovaiskii*	Eurasia	Gophers *Spermophilus* sp.
175	*O. rupestris*	North America	Gophers *Spermophilus* sp., *Otospermophilus* sp.
176	*O. silantiewi*	Eurasia	Marmots *Marmota* sp. In Mongolia, on susliks, Brandt's vole, pikas, terrestrial predators
177	*Panallius galeanus*	South America	*Cavia* sp.
176	*Paradoxopsyllus curvispinus*	Eurasia	Muridae
177	*P. custodies*	Eurasia	Muridae
178	*P. dashidorzhii*	Eurasia	*Ochotona* sp.; in the Altay mountains, pikas, voles, and other small animals; in Mongolia, long-tailed ground squirrel, narrow-skulled vole
179	*P. integer*	Eurasia	*Ochotona* sp., *Rattus* sp.
180	*P. kalabukhovi*	Eurasia	*Ochotona* sp.
181	*P. repandus*	Eurasia	Gerbils, *Meriones* sp., *Rhombomys opimus*
182	*P. scorodumovi*	Eurasia	Gerbils, *Ochotona* sp.
183	*P. teretifrons*	Eurasia	Gerbils, *Meriones* sp., *Rhombomys* sp.
184	*Paramonopsyllus skaloni*	Eurasia	Pikas, *Ochotona* sp.
185	*Paraneopsylla ioffi*	Eurasia	*Ochotona* sp., voles *Alticola* sp.

(continued)

Table 5.7 (continued)

	Species	Location	Host
186	*Peromyscopsylla bidentata*	Caucasus(Azerbaijan)	Cricetidae
187	*P. hesperomys adelpha*	North America	Hamsters *Peromyscus* sp.
188	*Phalacropsylla alios*	North America	Cricetidae
189	*Pleochaetis dolens*	South America	*Cavia* sp., hamsters *Oryzomys* sp., rabbits, *Sylvilagus* sp.
190	*P. equatoris*	South America	Hamsters *Akodon* sp., *Oryzomys* sp., rabbits, *Sylvilagus* sp.
191	*P. sibynus*	North America	Cricetidae
192	*Plocopsylla hector*	South America	Hamsters *Thomasomys* sp.
193	*Polygenis bohlsi*	South America	*Cavia* sp., hamsters *Oryzomys* sp. (*Monodelphis domestica*)
194	*P. brachinus*	South America	Hamsters *Akodon* sp., *Oryzomys* sp.
195	*P. byturus*	South America	*Cavia* sp., hamsters *Graomys* sp.
196	*P. gwyni*	North and South America	
197	*P. litargus*	South America	Hamsters *Oryzomys* sp., squirrels *Sciurus* sp.
198	*P. platensis cisandinus*	South America	*Cavia* sp.
199	*P. tripus*	South America	Hamsters *Oryzomys* sp., rats, *Rattus* sp.
200	*Pulex irritans*	Eurasia, South America	*Rattus* sp., marmots *Marmota* sp., predatory mammals, animals, human houses
201	*P. similans*	North America	Sciuridae
202	*Rhadinopsylla altaica*	Eurasia	*Ochotona* sp.
203	*R. altifrons*	Gissar Ridge	*Microtus carruthersi*
204	*R. angusta*	Eurasia	Voles *Microtus* sp. (*M. gregalis*, Tien-Shan highlands); *Clethrionomys* sp., gophers *Spermophilus* sp., other rodent species
205	*R. bivirgis*	Eurasia	Jirds *Meriones* sp. and other rodents
206	*R. cedestis*	Eurasia (Transcaucasus, Betpak-Dala)	Gerbils, *Meriones* sp., great gerbil *Rhombomys opimus*
207	*R. dahurica*	Eurasia	*Ochotona* sp., gophers *Spermophilus* sp., *Microtus* sp., and other rodent species, steppe polecat
208	*R. dives*	Eurasia	*Cricetulus* sp., *Apodemus* sp., gophers *Spermophilus* sp.
209	*R. insolita*	Eurasia	*Cricetulus* sp., *Apodemus* sp., and other small rodents
210	*R. integella*	Eurasia	*Microtus arvalis* and other small rodents

(continued)

Table 5.7 (continued)

	Species	Location	Host
211	*R. li*	Eurasia	Marmots *Marmota* sp., gophers *Spermophilus* sp., *Cricetulus* sp., and other small rodents
212	*R. rothschildi*	Eurasia	*Lasiopodomys brandti* and other small rodents and carnivorous mammals
213	*R. socia*	Eurasia	Jirds *Meriones* sp.
214	*R. tenella*	Eurasia	*Cricetulus* sp., *Apodemus* sp.
215	*R. ucrainica*	Eurasia (Transcaucasus)	Voles *Microtus socialis*, gerbils *Meriones* sp.
216	*Rhopalopsyllus* sp.	South America	Edentata
217	*R. casicus*	Peru	
218	*Rostropsylla daca*	Eurasia	*Spermophilopsis leptodactylus*
219	*Sphinctopsylla mars*	South America	Hamsters *Hesperomys* sp.
220	*Stenischia humilis*	Eurasia	Cricetidae
221	*Stenistomera (Miochaeta) alpina*	Eurasia	Cricetidae
222	*S. macrodactyla*	North America	Hamsters *Peromyscus* sp.
223	*Stenoponia conspecta*	Eurasia	Gerbils, great (*Rhombomys opimus*), Libyan *Meriones erythrourus*
224	*S. ivanovi*	Eurasia	*Microtus arvalis*
225	*S. tripectinata insperata*	Eurasia	Jirds *Meriones* sp.
226	*S. vlasovi*	Eurasia	Jirds *Meriones* sp., *Pallasiomys* sp.
227	*Stivalius ahalae*	India, Java	Muridae
228	*S. aporus*	India	
229	*S. cognatus*	Java	Muridae
230	*Synopsyllus estradei*	Madagascar	Muridae
231	*S. fonque*	Madagascar	Muridae
232	*Synosternus cleopatrae*	Africa	Cricetidae
233	*S. longispinus*	Eurasia	Insectivora, Carnivora, Rodentia
234	*S. pallidus*	Eurasia, Africa	Insectivores, carnivorous mammals, rodent
235	*Thrassis acamantis*	North America	Sciuridae (gophers)
236	*T. arizonensis*	North America	Sciuridae (gophers)
237	*T. bacchi bacchi*	North America	Gophers *Spermophilus* sp.
238	*T. bacchi johnsoni*	North America	*Lagurus* sp., hamsters *Peromyscus* sp.
239	*T. focus*	North America	Hamsters *Onychomys* sp.
240	*T. francisi*	North America	Sciuridae
241	*T. gladiolis*	North America	Sciuridae
242	*T. howelli*	North America	
243	*T. pandorae*	North America	Gophers *Spermophilus* sp.
244	*T. petiolatus*	North America	Sciuridae

(continued)

Table 5.7 (continued)

	Species	Location	Host
245	*T. stanfordi*	North America	*Marmota* sp.
246	*Tiamastus cavicola*	South America	Cricetidae; *Cavia* sp.
247	*Tunga penetrans*	Africa	
248	*Xenopsylla astia*	India	Muridae
249	*X. brasiliensis*	Africa, India, South America	Muridae
250	*X. buxtoni*	Eurasia	Jirds *Meriones* sp.
251	*X. cheopis*	Eurasia, South America	*Rattus* sp.
252	*X. conformis*	Eurasia	*Meriones* sp.
253	*X. eridos*	Africa	Hamsters *Otomys* sp.
254	*X. gerbilli caspica*	Eurasia	Gerbils *Rhombomys opimus*, *Meriones* sp.
255	*X. gerbilli gerbilli*	Eurasia	*Rhombomys opimus*, *Meriones* sp.
256	*X. g. minax*	Eurasia	*Rhombomys opimus*, *Meriones* sp.
257	*X. hirsuta*	Africa	Cricetidae
258	*X. hirtipes*	Eurasia	Gerbils *Meriones* sp., *Rhombomys opimus*
259	*X. magdalenae*	Eurasia	Cricetidae
260	*X. nubica*	Africa	Cricetidae
261	*X. nuttalli*	Eurasia	Great gerbil *Rhombomys opimus*
262	*X. philoxera*	Africa	Cricetidae
263	*X. phyllomae*	Africa	Cricetidae
264	*X. piriei*	Africa	Cricetidae
265	*X. skrjabini*	Eurasia	*Rhombomys opimus*
266	*X. versuta*	Africa	*Rhabdomys* sp.
267	*X. vexabilis hawaiiensis*	Hawaiian Islands	*Rattus hawaiiensis*, *R. rattus*
268	*Xiphiopsylla lippa*	Africa	*Lophuromys* sp.
269	*Wagnerina longicaudata*	Eurasia	*Ochotona* sp., voles *Alticola* sp.

(Sludskyi, 2014)

Table 5.8 Ticks and lice in nature infected with plague ([19, 20], corrected by Dubyanskiy V.M., Yeszhanov A.B.)

	Species
Ticks	
Argasidae	*Ornithodoros alactagalis*, *O. tartakovskyi*
Gamasides	*Eulaelaps kolpakovae*; In Mongolia, *Y. pestis* isolated from *Haemogamasus manschuricus* and *Hg. kitanoi*; *Hg. nidi* (Transcaucasian highlands), *Laelaps algericus* Hirst.
Ixodidae	*Dermacentor nuttalli*, *Haemaphysalis numidiana turanica*, *Hyalomma asiaticum*, *Ixodes crenulatus*, *I. persulcatus*, *Rhipicephalus schulzei*, *R. pumilio*
Lice	*Hoplopleura acanthopus* (in Baikal parasite on 29 mammalian species). *Linognathoides (Neohaematopinus) laeviusculus*, *L. palearcticus*, *L. p. tarbagani*, *Pediculus humanus*, *P. corporis*

Supplementary Details for "5.1.2 Distribution of Plague Foci"

5.1.2.1 Africa

5.1.2.1.1 North Africa (Group Countries)

Natural plague foci in North Africa have hardly been studied, although cases of plague in humans and camels in Libya, Tunisia, and Egypt have been described.
Hosts: *Acomys cahirinus*. The plague microbe was isolated in 1927–1928 [107].
Vectors: Verified data are lacking.

5.1.2.1.2 Mauritania, Western Sahara

Hosts: *Psammomys obesus*, with an ecology similar to that of *Rhombomys opimus, Xerus erythropus, Mus musculus, Acomys cahirinus, Meriones libycus, Gerbillus pyramidum, G. gerbillus, G. agag, G. nanus. Dipodillus campestris, Pachyuromys duprasi*, and *Jaculus jaculus* [108, 109]
Vectors: *Xenopsylla ramesis, Synosternus cleopatrae, X. nubica* [1]

5.1.2.1.3 Algeria

Hosts: *Rattus rattus, Rattus norvegicus*, gerbils of the genus *Gerbillus* and *Meriones*
Vectors: *Xenopsylla cheopis*

5.1.2.1.4 Democratic Republic of the Congo

Plague foci along the border with Tanzania:
Landscapes: Savanna
Hosts: *Mastomys natalensis, Rattus rattus*
Vectors: *Xenopsylla brasiliensis, X. cheopis* [30]

5.1.2.1.5 Uganda

Northern and western provinces:
Landscapes: The available literature lacks relevant information.
Hosts: *Mastomys natalensis*, *Rattus rattus*
Vectors: *Xenopsylla cheopis, Ctenocephalides* spp. [30, 110]

5.1.2.1.6 Kenya

Landscapes: The available literature lacks relevant information.
Hosts: *Arvicanthis niloticus, Mastomys natalensis*
Vectors: *Xenopsylla cheopis, X. brasiliensis* [30, 111]

5.1.2.1.7 Tanzania

The slopes of Mount Kilimanjaro and the area southeast of Lake Victoria:
Landscapes: 900–2500 m above sea level, forests, pastures
Hosts: *Rattus rattus alexandrinus, Tatera robusta* (Cretzschmar)
Vectors: *Ctenophthalmus eximius, Nosopsyllus incisus* [20, 30, 112, 113]
Epizootic area west of Lake Victoria borders plague foci in the Democratic Republic of the Congo:
Landscapes: The available literature lacks relevant information.
The southeast epizootic area borders the plague focus in Kenya:
Landscapes: The available literature lacks relevant information.
Hosts: *Lemniscomys griselda* Thom., *Lemniscomys striatus L.* [20]
Both the area in the vicinity of Lake Rukk and the interstream area between the Greater Rvakh and Ruhundzhi share a focus with that in Malawi and Zambia:
Landscapes: The available literature lacks relevant information.
Hosts (for all foci in Tanzania): *Rattus rattus alexandrines, Tatera robusta* (Cretzschmar), *Pelomys fallax* (Peters), *Lemniscomys griselda* (Thomas), *Lemniscomys striatus* [19]

5.1.2.1.8 Namibia

Foci in the interstream area between Kunene and Okavango, on the Angola border:
Landscapes: Desert savanna
Hosts: *Tatera schinzi, Desmodillus auricularis* [30]
Landscapes: The available literature lacks relevant information.
Foci on the southern half of the Damaraland Plateau and the southern slopes of the Kamala Plateau:
Landscapes: The available literature lacks relevant information.
Hosts: *Tatera schinzi, Desmodillus auricularis*
Vectors (for all foci in Namibia): *X. philoxera*, *X. brasiliensis*, and *Dinopsyllus ellobius* [30]
Foci on the Namaland Plateau
Landscapes: Sandy desert
Hosts: *Tatera schinzi, Desmodillus auricularis* [30]
Vectors: The available literature lacks relevant information.

5.1.2.1.9 Botswana

Foci east and west of the Okavango Delta and on the borders of Botswana, Namibia, and southern Angola:
Landscapes: Desert
Hosts: *Tatera schinzi, Desmodillus auricularis* [30]
Vectors: The available literature lacks relevant information.
Foci are also present in the south, just outside South Africa.

5.1.2.1.10 Malawi

Vectors: The available literature lacks relevant information.

5.1.2.1.11 Zambia

Focus of plague in eastern Zambia within the plateau; a single focus in Tanzania:
Hosts: *Tatera valida*
Vectors: *Xenopsylla philoxera, X. hipponax* [30]

5.1.2.1.12 Zimbabwe

Foci in most parts of the country:
Landscapes: Desert, savanna
Hosts: *Tatera leucogaster, Aethomys chrysophilus* [30]
Vectors: *Xenopsylla philoxera, X. brasiliensis* [114]

5.1.2.1.13 South Africa

Focal areas in the Great Karoo:
Hosts: *Desmodillus auricularis, Rhabdomys pumilio*
Vectors: *Xenopsylla pirici, X. Brasiliensis, Dinopsyllus ellobius* ([30, 111]; Plague Control Guidelines for South Africa (http://www.nicd.ac.za/assets/files/National_Plague_Control_Guidelines.pdf))
Foci between the Great Escarpment and the Orange River:
Hosts: *Desmodillus auricularis, Tatera brantsi*
Vectors: *Xenopsylla pirici, X. Brasiliensis, Dinopsyllus ellobius* ([30]; Plague Control Guidelines for South Africa (http://www.nicd.ac.za/assets/files/National_Plague_Control_Guidelines.pdf))
Focal areas located in the interfluves of the Orange River and the Vaal River, extending until the Drakensberg mountains in the east:
Landscapes: Semideserts, deserts

Hosts: *Tatera brantsii, T. schinzi*
Focal area on the northeastern slopes of the Witwatersrand Ridge:
Vectors (for all foci of South Africa): *X. philoxera, X. brasiliensis*, and *Dinopsyllus ellobius* ([30]; http://www.who.int/csr/resources/publications/plague/WHO_CDS_CSR_EDC_99_2_EN/en/)

5.1.2.1.14 Madagascar

Focal areas on the high plateau:
Landscapes: Madagascar savanna
Hosts: *Rattus rattus*
Vectors: *Synopsylla fonquernii, Xenopsylla cheopis* [30]

5.1.2.2 Southeast Asia

5.1.1.2.1 India

In northern India, foci on the Indo-Gangetic Plain:
Landscapes: The available literature lacks relevant information.
Hosts: *Tatera indica, Bandicota indica*
Vectors: *Xenopsylla astia, X. cheopis* [30]
In central India, foci in Madhya Pradesh:
Vectors: The available literature lacks relevant information.
Hosts: *Tatera indica, Bandicota indica*
Vectors: *Xenopsylla astia, X. cheopis* [30]
Foci in southern India:
Vectors: The available literature lacks relevant information.
Hosts: *Tatera indica, Bandicota indica*
Vectors: *Xenopsylla astia, X. cheopis* [30]

5.1.1.2.2 Nepal

Vectors: The available literature lacks relevant information.
Hosts: *Rattus rattus*
Vectors: *Xenopsylla cheopis* [115]

5.1.1.2.3 Burma

Foci in the central uplands, between the Irrawaddy and Sittwe Rivers:
Vectors: The available literature lacks relevant information.
Hosts: *Rattus rattus, Bandicota bengalensis*
Vectors: *Xenopsylla cheopis* [116]

5.1.1.2.4 Thailand

Foci on the Khorat Plateau and in the northern and western parts of central Thailand:
Vectors: The available literature lacks relevant information.
Hosts: *Rattus concolor, R. rattus*
Vectors: *Xenopsylla cheopis* [30]

5.1.1.2.5 Vietnam

Landscapes: Coastal lowland plains
Hosts: *Rattus rattus, R. norvegicus, R. indica*
Vectors: *Xenopsylla cheopis* [30]

5.1.1.2.6 Indonesia

Foci on Java
Landscapes: Hillsides, highlands, plateaus with rice fields
Hosts: *Rattus concolor, R. norvegicus*
Vectors: *Xenopsylla cheopis, Stivalius cognatus* [30]

5.1.2.3 Countries of the Former USSR (Part of English Foci Names by Anisimov et al. [117], 2015; Zhou et al. [118])

5.1.2.3.1 Pre-Araks

Natural foci (07) (it is the registration number of focus according to the Russian system of count) in the low mountains:
Landscapes: Deserts, low mountains, midlands
Hosts: *Meriones vinogradovi, M. persicus, M. tristrami, M. dahli, Allactaga elater, Microtus socialis, Mus musculus* [119]
Vectors: *Xenopsylla conformis conformis, Nosopsyllus iranus iranus, Pulex irritans* [15]

5.1.2.3.2 Transcaucasian Valleys and Foothills

Natural foci in the Kur-Arak lowlands:
Landscapes: Dry steppes, semideserts, deserts
Hosts: *Meriones libycus, M. tristrami, Mus musculus* [15]
Vectors: *Xenopsylla conformis conformis, Nosopsyllus laeviceps laeviceps* [15]

5.1.2.3.3 Transcaucasian Highland

Group of autonomous natural foci (04–06):
Landscapes: Mountainous steppe, steppe, subalpine and alpine meadows
Hosts: *Microtus arvalis, Microtus socialis, Mus musculus* [25]
Vectors: *Callopsylla caspia, Nosopsyllus consimilis* [15]

5.1.2.3.4 Dagestan Highland

Natural focus (39) located at 2,000–3,000 m above sea level:
Landscapes: Alpine meadows
Hosts: *Microtus arvalis, Cricetulus migratorius, Arvicola terrestris* [112, 121]
Vectors: *Callopsylla caspia, Amalaraeus dissimilis dagestanicus* [15, 122]

5.1.2.3.5 Central Caucasian

Natural focus of plague (1) at Mount Elbrus:
Landscapes: Mountain steppes, alpine and subalpine meadows
Hosts: *Spermophilus musicus, Microtus arvalis, Sylvaemus uralensis, Mus musculus* [15, 119]
Vectors: *Citellophilus tesquorum caucasicus, Neopsylla setosa* [15, 123]

Dagestan valleys and foothills: Natural foci of plague (03) in the Caspian Plain within Dagestan
Landscapes: Deserts, semideserts
Hosts: *Spermophilus pygmaeus, Meriones tamariscinus, Mus musculus* [15, 124]
Vectors: *Neopsylla setosa, Citellophilus tesquorum caucasicus* [15]

Terek-Sunzha natural focus of plague (02) near the Terek and Sunzha Rivers:
Landscapes: Desert steppes
Hosts: *Spermophilus pygmaeus* [15, 119]
Vectors: *Neopsylla setosa, Citellophilus tesquorum* [125]

5.1.2.3.6 Pre-Caspian sandy natural foci (43) in sandy regions near the Caspian Sea in the Republics of Kalmykia, Dagestan, and Stavropol:
Landscapes: Semideserts
Hosts: *Meriones meridianus, Spermophilus pygmaeus* [126]
Vectors: *Nosopsyllus laeviceps, Xenopsylla conformis conformis* [127]

5.1.2.3.7 Pre-Caspian northwestern natural foci of plague (14) in the Astrakhan region, Kalmykia, Northwest Caspian steppe:
Landscapes: Semideserts
Hosts: *Spermophilus pygmaeus, Meriones meridianus, Mus musculus* [119]
Vectors: *Citellophilus tesquorum, Neopsilla setosa* [1]

5.1.2.3.8 Volga-Ural steppe natural foci of plague (15) in the northern Volga and Ural interstream, from the Caspian lowland to the Common Syrt:
Landscapes: Steppe, semideserts
Hosts: *Spermophilus pygmaeus pygmaeus* [128]
Vectors: *Neopsilla setosa*, *Citellophilus tesquorum transvolgensis* [128]

5.1.2.3.9 Volga-Ural sandy natural foci of plague (16) in the southern part of the Caspian Depression, in the Ural and the Volga interstream
Landscapes: Sandy
Hosts: *M. meridianus, M. tamariscinus, Spermophilus fulvus, Mus musculus*
Vectors: *X. conformis, Nosopsyllus laeviceps* [128]

5.1.2.3.10 Trans-UralTrans-Ural natural foci of plague (17) east of the Ural River in West Kazakhstan, Aktobe and Atyrau regions
Landscapes: Steppes, semideserts
Hosts: *S. pygmaeus, Rhombomys opimus* [128]
Vectors: *N. setosa, C. tesquorum* [128]

5.1.2.3.11 Central Asian Desert

5.1.2.3.11.1 Ural-Emba autonomous plague focus (18) in the Ural-Emba desert between the Ural and Emba Rivers
Landscapes: Semideserts, river floodplains
Hosts: *Rhombomys opimus, M. tamariscinus, M. meridianus* [128]
Vectors: *X. skrjabini, N. laeviceps, C. lamellifer* [128]

5.1.2.3.11.2 Pre-Ustyurt autonomous desert plague foci (19) between the Emba River, Ustyurt Plateau, and Caspian Sea
Landscapes: Semideserts, deserts
Hosts: *Rhombomys opimus, M. lybicus, M. meridianus* [128]
Vectors: *X. skrjabini, C. lamellifer, N. laeviceps* [128]

5.1.2.3.11.3 Mangyshlak autonomous plague focus (23) in the Mangyshlak Desert and in Mangistau region covering an area of the Buzachi Peninsula that includes the plains and mountains of the Mangyshlak region
Landscapes: Deserts
Hosts: *Rhombomys opimus, M. lybicus*
Vectors: *X. skrjabini, X. nuttalii* [128]

5.1.2.3.11.4 Ustyurt autonomous plague focus (20) on the Ustyurt Plateau of the Ustyurt Desert between the Mangyshlak region to the west and the Aral Sea to the east in the territories of the Republics of Kazakhstan and Uzbekistan
Landscapes: Deserts
Hosts: *Rhombomys opimus, C. migratorius* [128]
Vectors: *X. skrjabini, X. nuttalii, X. g. caspica* [128]

5.1.2.3.11.5 North-Pre-Aral autonomous plague focus (21) in the northwest pre-Aral region including Large and Small Badger Sands in the North Pre-Aral Desert

Landscapes: Deserts
Hosts: *Rhombomys opimus, M. lybicus, M. meridianus* [128]
Vectors: *X. skrjabini, C. lamellifer, N. laeviceps* [128]

5.1.2.3.11.6 Pre-Aral-Karakum autonomous focus (24) in the Pre-Aral Karakum Desert and the Aktobe, Kyzylorda, and Karaganda regions of Kazakhstan north-east of the Aral Sea

Landscapes: Sandy, clayey, gravelly deserts
Hosts: *Rhombomys opimus, M. lybicus, M. meridianus*
Vectors: *X. skrjabini, C.* lamellifer, *N. laeviceps* [128]

5.1.2.3.11.7 Trans-Aral autonomous focus (22) in the Aryskum-Daryalyktakyr Desert and theKyzylorda and Karaganda regions of Kazakhstan

Landscapes: Sandy deserts
Hosts: *Rhombomys opimus, M. lybicus, M. meridianus*
Vectors: *X. gerbilli, X. skrjabini* [128]

5.1.2.3.11.8 Karakum autonomous focus (25) in Karakum Desert, Turkmenistan

Landscapes: Sandy deserts
Hosts: *Rhombomys opimus, M. lybicus, M. meridianus*
Vectors: *X.* gerbilli, *X. hirtipes* [1]

5.1.2.3.11.9 Kopet Dag autonomous desert focus (26) in Southwestern Kopet Dag

Landscapes: Desertified lowlands
Hosts: *Rhombomys opimus, M. lybicus*
Vectors: *X. conformis, N. laeviceps* [1]

5.1.2.3.11.10 Kyzylkum autonomous focus (27) in the Kyzylkum Desert within Kazakhstan, Uzbekistan, and Turkmenistan's eastern outskirts

Landscapes: Sandy deserts
Hosts: *Rhombomys opimus, M. lybicus, M. meridianus*
Vectors: *X. gerbilli, X. hirtipes, X. skrjabini* [128]

5.1.2.3.11.11 Muyunkum autonomous focus (28) in the northern Muyunkum Desert, in the subzone within the territories of the Zhambyl and South Kazakhstan regions

Landscapes: Deserts
Hosts: *Rhombomys opimus, M. lybicus, M. meridianus*
Vectors: *X. g. minax, X. conformis* [128]

5.1.2.3.11.12 Taukum autonomous focus (29) in the Taukum Desert within the territory of Almaty and Zhambyl between the Ili River in the north and the Chu-Ili Mountains in the south

Landscapes: Deserts
Hosts: *Rhombomys opimus, M. lybicus, M. tamariscinus*
Vectors: *X. skrjabini, X. hirtipes, X. g. minax* [128]

5.1.2.3.11.13 Pre-Balkhash autonomous focus (30) in the Balkhash Desert within the territory of Almaty in the interfluves of the Ili-Karatal, Aksu, and Lepsy Rivers
Landscapes: Deserts
Hosts: *Rhombomys opimus, M. meridianus*
Vectors: *X. g. minax, X. hirtipes, X. skrjabini* [128]

5.1.2.3.11.14 Betpak-Dala autonomous focus (42) in the Betpakdala Desert west of the Sarysu River and probably extending until Balkhash Lake in the east within Zhambyl and South Kazakhstan
Landscapes: Deserts
Hosts: *Rhombomys opimus, M. lybicus*
Vectors: *X. g. minax, X. conformis* [128]

5.1.2.3.11.15 In the Pre-Alakol Lowlands, an autonomous focus (45) in the eastern Pre-Alakol and Dzungarian Gate,wwith a difference in height of 400–900 m above sea level
Landscapes: Deserts
Hosts: *Rhombomys opimus, M. meridianus, M. lybicus*
Vectors: *X. g. minax, X. skrjabini, X. conformis* [128]

5.1.2.3.11.16 In the Ili Intermountains: an autonomous focus (46) in the Ili valley from the Kapshagai Reservoir in the west to the border with China in the east and from the lowlands of the southern Jungar Ridge in the north to the foothills of Ketmen, Turaigyr, and Trans-Ili Alatau in the south
Landscapes: Deserts, desert lowlands
Hosts: *Rhombomys opimus, M. meridianus*
Vectors: *X. g. minax, X. hirtipes, X. skrjabini* [128]

5.1.2.3.12 Talas: a natural focus (40) in Kyrgyzstan on the northern macroslope of the Talas Ridge from Manas City in the west until the Kolba Ridge in the east extending north to the slopes of the Kyrgyz Ridge (including the Kazakh portion)
Landscapes: Mountain steppes, meadow steppes, alpine meadows 1600–3700 m above sea level
Hosts: *Marmota caudata*
Vectors: *Citellophilus lebedevi, Pulex irritans* [128]

5.1.2.3.13 Gissar: a natural focus (34) on the northern macroslope of the Gissar Ridge (Tajikistan)
Landscapes: Tree and shrub belts, subalpine meadows
Hosts: *Microtus juldaschi, Marmota caudata* [119]
Vectors: *Callopsylla caspia, Frontopsylla glabravara* [17]

5.1.2.3.14 Sarydzhaz natural focus (foci) of plague in the Tian Shan Mountains in the Sarydzhaz Mountains: an autonomous focus (31) within the Almaty region of Kazakhstan and the Issyk-Kul region of Kyrgyzstan on the slopes of Sary-Djaz on the northeastern tip of the Terskey Alatau Ridge
Landscapes: Midlands, highland areas
Hosts: *Marmota baibacina*
Vectors: *Oropsylla silantiewi, Rhadinopsylla liventricosa, Citellophilus lebedevi* [128]

5.1.2.3.15 Upper Naryn autonomous focus (32) within Kyrgyzstan
Landscapes: Midlands, mountainous areas
Hosts: *Marmota baibacina*
Vectors: *Oropsylla silantiewi, Rhadinopsylla liventricosa, Citellophilus lebedevi* [128]

5.1.2.3.16 Aksaiautonomous focus (33), within Kyrgyzstan
Landscapes: Midlands, mountainous areas
Hosts: *Marmota baibacina*
Vectors: *Oropsylla silantiewi, Rhadinopsylla liventricosa, Citellophilus lebedevi* [128]

5.1.2.3.17 Alai natural focus (35) located on the slopes of the Alai Ridge (Kyrgyzstan)
Landscapes: Mountain steppes and subalpine and alpine meadows 2800–5000 m above sea level
Hosts: *Marmota caudata, Alticola argentatus* [129]
Vectors: *Oropsylla silantiewi crassa, Citellophilus lebedevi princeps* [129]

5.1.2.3.18 Mountain-Altai(or Saylyugem, Mongolia), focus (36) in the mountain range of Saylyugem, Kurai, South Chu
Landscapes: Mountain steppes, alpine meadows 2000–2500 m above sea level
Hosts: *Ochotona pallasi, O. alpine*[1]
Vectors: *Paradoxopsyllus scorodumovi, Rhadinopsylla dahurica dahurica* [1]

5.1.2.3.19 Tuva (Mongun-Taiga) focus (37) in the Republic of Tyva in the south adjacent to the Hanhiro-Turgensk natural focus (Mongolia):
Landscapes: Mountain steppes 1650–2550 m above sea level
Hosts: *Spermophilus undulatus, Ochotona pallasi* [119]
Vectors: *Citellophilus tesquorum altaicus, Amphalius runatus* [1]

5.1.2.3.20 Trans-Baikal natural plague focus (38) in the trans-Baikal steppe; (Russia) and foci in Mongolia (Mongolia-Daurian) and China (Barginsky or Hulun-Byrsk focus):
Landscapes: Steppe
Hosts: *Spermophilus dauricus, Marmota sibirica* [130]
Vectors: *Citellophilus tesquorum sungaris, Oropsylla silatiewi* [119, 131]

5.1.2.3.21 Dzungarian Highland natural focus (44) of mixed type (vole-ground squirrel-marmot) in the Dzungarian Highlands (Dzungarian Alatau) wwith a presumably larger part of the focus in China on the Boro-Horo Ridge
Landscapes: Subalpine and alpine meadows
Hosts: *Microtus kirgizorum, Spermophilus undulates*
Vectors: *C. tesquorum, Ct. arvalis, C. ullus, C. assimilis, F. elata* [120]

5.1.2.4 Mongolia

5.1.2.4.1 In Khan-Khukhey natural focus in the Altai Mountain system on the Khan-Khukhey Ridge
Landscapes: Mountain steppes 2500–2800 m above sea level
Hosts: *Marmota sibirica, Spermophilus undulatus* [124, 183]
Vectors: Specific fleas

5.1.2.4.2 KhuKh-Serkh-Munkh-Khairkhan natural focus in the central part of the Mongolian Altai
Landscapes: Subalpine and alpine meadows
Hosts: *Marmota baibacina, Marmota sibirica* [1]
Vectors: *Oropsylla silantiewi*

5.1.2.4.3 Gobi Altai Mountains natural focus 3000–3500 m above sea level
Landscapes: Mountain steppes
Hosts: *Marmota sibirica, Ochotona pallasi, Spermophilus undulates* [133, 134]
Vectors: *Oropsylla silantiewi, Paramonopsyllus scalonae, Citellophilus tesquorum altaicus* [127]

5.1.2.4.4 Changay Mountain Steppe focus in central Mongolia, in the area of the Khangai Highlands
Landscapes: Dry plains and mountain steppe 1200–3000 m above sea level
Hosts: *Marmota sibirica, Spermophilus undulatus* [136, 137]
Vectors: *Oropsylla silantiewi, Citellophilus tesquorum sungaris* [1], [137]

5.1.2.4.5 Khentey natural focus south and southeast of the highlands
Landscapes: Steppe 1500–2000 m above sea level
Hosts: *Marmota sibirica, Spermophilus undulates* [128, 138].
Vectors: *Oropsylla silantiewi* [1]

5.1.2.4.6 Mongolian Altai Plain natural focus in the Altaic Gobi
Landscapes: Gravelly desert
Hosts: *Rhombomys opimus, Meriones meridianus* [128, 139]
Vectors: *Xenopsylla conformis conformis, X. skrjabini* [140]

5.1.2.4.7 In Gurvan Saikhan, a natural focus in the South Gobi Gurvan Saikhan Mountains
Landscapes: Desert steppes 2400–2800 m above sea level
Hosts: *Ochotona pallasi, O. daurica, Meriones unguiculatus* [138, 141]
Vectors: *Amphalius runatus, Ctenophyllus hirticrus* [141]

5.1.2.4.8 On the South Gobi Plain, a natural focus probably connected with that of Inner Mongolia
Landscapes: Gravelly desert
Hosts: *Rhombomys opimus, Meriones meridianus* [1]
Vectors: *Xenopsylla skrjabini, X. conformis conformis* [1]

5.1.2.5 China

5.1.2.5.1 Plague focus in the Dzungarian Alatau
Landscapes: Mountain steppes, subalpine meadows
Hosts: *Marmota baibacina, Spermophilus undulates* [142]
Vectors: *Oropsylla silantiewi, Citellophilus tesquorum altaicus, C. tesquorum dzetysuensis* [131]

5.1.2.5.2 Inner Mongolia, a natural focus in Dzamin-Uden, Erlyan
Landscapes: Semideserts, deserts
Hosts: *Meriones unguiculatus, Ochotona daurica, O. pallasi* [131, 136, 143]
Vectors: *Xenopsylla conformis conformis, Nosopsyllus laeviceps kuzenkovi, Neopsylla pleskei orientalis* [131, 140]

5.1.2.5.3 Xilingol Grassland, a natural focus east of Inner Mongolia
Landscapes: Steppe, desert 1100–1300 m above sea level
Hosts: *Lasiopodomys brandtii, Spermophilus dauricus* [131]
Vectors: *Frontopsylla luculenta luculenta, Amphipsalta primaries mitis, Neopsylla pleskei orientalis* [128]

5.1.2.5.4 In Sungari-Lyaoh (Manchuria), a natural focus in northeast China
Landscapes: Forest, wet steppes
Hosts: *Spermophilus dauricus, Rattus norvegicus* [131]
Vectors: *Citellophilus tesquorum sungarus, Xenopsylla cheopis* [131]

5.1.2.5.5 Gansu-Ningxia Hui (Ordos, Shaanxi): Natural focus in the Loess Plateau
Landscapes: Dry steppes 1500–2950 m above sea level
Hosts: *Spermophilus alashanicus, Ochotona daurica* [131]
Vectors: *Neopsylla abagaitui, Citellophilus tesquorum mongolicus* [131]

5.1.2.5.6 Qinghai-Tibet natural focus in the Tibetan Plateau
Landscapes: Subalpine and alpine meadows 2700–5450 m above sea level
Hosts: *Marmota himalayana, Allactaga sibirica* [131]
Vectors: *Oropsylla silantiewi, Callopsylla dolabris* [131]

5.1.2.6 *Near East*

5.1.2.6.1 Yemen

Hosts: *Rattus rattus, Meriones* sp.
Vectors: The available literature lacks relevant information.

5.1.2.6.2 Saudi Arabia

Saudi-Yemeni: Natural focus on the desert plains and uplands along the Saudi Arabia-Yemen border:
Landscapes: Up to a height of 1000 m above sea level, rocky and sandy deserts, desert savannas, shrub-tree xerophilic communities
Possible hosts: *Gerbillus nanus, G. cheesmani, Meriones rex, Rattus rattus, Mus musculus, Acomys cahirinus, Jaculus jaculus* [144]

5.1.2.6.3 In the Lebanese mountains and on the steppe, foci near the Syrian coast and the mountainous parts of Lebanon and northern Israel
Landscapes: Forb-grass mountain steppes
Hosts: Gerbils of the genus *Meriones*

5.1.2.6.4 In the Syrian-Mesopotamian Desert: natural foci in Upper and Lower Mesopotamia, the Syrian Plateau, and Kuwait
Landscapes: Desert flood plains of the Tigris and Euphrates Rivers and surrounding desert plateau
Hosts: *Meriones libycus, Tatera indica, Nesokia indica, Rattus rattus, Rattus norvegicus, Mus musculus*. Perhaps: *Gerbillus cheesmani, G. dasyurus, G. nanus, Meriones crassus* [144]
Vectors: *Xenopsylla buxtoni, X. astia, Stenoponia tripectinata insperata*; plague transmission from rodents to humans via the human flea *Pulex irritans* [145]

5.1.2.6.5 Natural foci in the semidesert and desert of southern Turkey, northern Syria
Landscapes: rocky ephemeral-wormwood deserts
Hosts: *Mesocricetus brandti, Microtus socialis, Tatera indica, Meriones vinogradovi, M. libycus, M. tristrami, Nesokia indica* [146]
Vectors: *Xenopsylla buxtoni*

5.1.2.6.6 Kurds-Iranian mountains and steppe: Natural focus in the western Iranian plateau, southeastern Turkey, northeastern Iraq [1]
Landscapes: Mountain plateaus, intermountain valleys 1000–2000 m above sea level, covered with herb-grass steppes
Hosts: *Meriones persicus, M. libycus, M. vinogradovi, M. tristrami* [145], *Allactaga elater, Mesocricetus brandti, Ellobius lutescens* [147, 148]
Vectors: *Xenopsylla buxtoni, X. conformis conformis, Nosopsyllus iranus iranus, Stenoponia tripectinata insperata* [147, 148], *Pulex irritans* [149], and others [1]

5.1.2.6.7 Natural focus in the Armenian-Anatolian mountains and steppe, previously considered part of the Kurdish-Iranian focus
Landscapes: Steppe, mountain steppe
Hosts: *Meriones libycus, Spermophilus xanthoprymnus* [1, 150]
Vectors: Same as those of the Kurdish-Iranian focus

5.1.2.6.8 In the Iran-Afghan Desert lowlands, a natural focus in the central low and northern mountains of Iran, Afghanistan's western plateau
Landscapes: Semideserts, deserts of Turan and Southwest Asia, and possibly on the outskirts of the Central Asian desert focus [151]:
Hosts: *Rhombomys opimus, Meriones libycus, M. meridianus, Mus musculus* [1]
Vectors: *Xenopsylla gerbilli, X. nuttalli, X. conformis conformis, Nosopsyllus laeviceps* [1]

5.1.2.6.9 Along the South Afghan flats, a natural focus in the sandy desert of Registan, sands in the lower reaches of the Helmand River, and the clay-stony desert Dasht-Markokhim
Hosts: *Rhombomys opimus, Meriones persicus, M. libycus, Mus musculus* [152]
Vectors: *Xenopsylla cheopis, X. gerbilli, X. conformis conformis* [153]

5.1.2.6.10 Natural foci in the Afghan-Pakistani midland desert-steppe of central, eastern Afghanistan and the western regions of Pakistan
Landscapes: Subtropical dry mountain steppes, acacia woodlands
Hosts: *Meriones libycus, M. persicus, Marmota caudata, Spermophilus fulvus, Mus musculus, Rattus rattus* [152]
Vectors: *Xenopsylla cheopis, X. gerbilli, X. conformis conformis* [153]

5.1.2.6.11 Natural focus in the Hindu Kush Highlands [150]
Landscapes: Alpine herb-grass meadows
Hosts: *Marmota caudata, Dryomysnitedula, Microtus juldashi, Meriones persicus, Mus musculus*
Vectors: *Xenopsylla cheopis, Citellophilus lebedevi, Frontopsylla protera*
A focus in Northern Afghanistan: within the range of *Rhombomys opimus* [154]

5.1.2.7 North America

5.1.2.7.1 In California and Oregon, foci, between the ridges of the Sierra Nevada and the coast, and the San Joaquin and Sacramento Rivers
Landscapes: The available literature lacks relevant information.
Hosts: *Spermophilus beecheyi*
Vectors: *Diamanus montanus, Hoplopsyllus anomalus* [30]

5.1.2.7.2 In Oregon, Idaho, Montana, and Wyoming, in theBlue Cascade Mountains, Bitter Root Ridge, and eastern slopes of Moustache Ridge
Hosts: *Citellus lateralis, Citellus richardsoni* [30]
Nevada and Utah: Foci in the basin of the Great Salt Lake
Landscapes: Desert
Hosts: *Neotoma lepida, Cynomys parvidens*

5.1.2.7.3 In New Mexico, Arizona, and Utah, foci in the Colorado Plateau and Llano Estacado (entering the territory of Mexico)
Landscapes: Deserts and steppes
Hosts: *Cynomys ludovicianus, C. gunnisoni*

5.1.2.7.4 In California, foci in the mountainous area of San Bruno
Hosts: *Microtus californicus, Peromyscus maniculatus* [30]
Vectors (for the whole territory): *Oropsylla montana*, O. *idahoensis*, *Ceratophyllus ciliatus*, *Orchopeas nepos* (Smith et al., 2009)

5.1.2.8 South America

5.1.2.8.1 Argentina

5.1.2.8.1.1 Foci in the provinces of Santiago del Estero, Salta, Catamarca
Landscapes: Rainforests
Hosts: *Microcavia australis, Galeamus teloides* [30]

5.1.2.8.1.2 Foci in the provinces of Cordoba, San Juan, Mendoza, La Pampa, and Rio Negro, on the eastern slopes of the Andes
Landscapes: Alpine steppe
Hosts: *Microcavia australis, Galeamus teloides*
Focus in Buenos Aires Province
Landscapes: Steppe
Hosts: *Microcavia australis* [30]
Vectors: The available literature lacks relevant information.

5.1.2.8.2 Bolivia

5.1.2.8.2.1 Foci in the departments of Chuquisaca, Santa Cruz, Tarija, probably connected with the northern foci of Argentina
Landscapes: The available literature lacks relevant information.
Hosts: Genus *Graomys*, *Hesperomys*
Vectors: *Puligenis biturdus, P. bochisi* [30]

5.1.2.8.3 Brazil

5.1.2.8.3.1 Enzootic areas in Sinara, Pernambuco, Paraiba, Alagoas, Bahia, Minas Gerais, Teresopolis
Landscapes: The available literature lacks relevant information.
Hosts: *Cavia aperea, C. spixii, Oryzomys subflavus, Zygedon tomispixuna*
Vectors: *Poligenis bohsli, P. tripus* [30, 155]

5.1.2.8.4 Ecuador

5.1.2.8.4.1 Natural focus in Manabi, western Ecuador
Hosts: *Sigmodon hispidus*
Vectors: *X. cheopis* [30]

5.1.2.8.4.2 Natural foci in Chimborazo and Tungurahua, central Ecuador
Landscapes: Alpine steppes
Hosts: *Sylvilagus brasiliensis*
Vectors: *Pleochaetis dolens, Hoplopsyllus manconis* [30]

5.1.2.8.4.3 Natural foci in Loho and El Oro, southern Ecuador
Landscapes: The available literature lacks relevant information.
Hosts: *Sciurus stramineus, Akodon mollis*
Vectors: *Poligenis litardus, Pulex irritans* [30]

5.1.2.8.5 Peru

5.1.2.8.5.1 In northern Peru: possibly the same focus as that in Ecuadorian Loho and El Oro Provinces [30]
Hosts: *Sciurus stramineus, Akodon mollis*

5.1.2.8.5.2 In southern Peru, a possible focus in the Pune Desert [30]
Vectors (for all foci in Peru): *Hectopsylla* sp., *Tiamastus cavicola* (http://www.who.int/csr/resources/publications/plague/WHO_CDS_CSR_EDC_99_2_EN/en/)

5.1.2.8.6 Venezuela

5.1.2.8.6.1 Nature foci in the states of Miranda and Aragua
Hosts: *Sigmodon hispidus, Heteromys anomalus* [30]
Vectors: The available literature lacks relevant information.

5.1.2.9 Hawaiian Islands

Possible foci of plague in the districts of Hamakura and Hilo:
Landscapes: The available literature lacks relevant information.
Hosts: *Rattus hawaiiensis*
Vectors: *Xenopsylla vexabilis* [30]

References

1. Karimova T.Yu., Neronov V.M. The nature plague foci of Palaearctic. Moscow. Nauka. 2007. P. 199.
2. Dubyanskiy MA, Kenzhebayev A, Stepanov VM, Asenov GA, Dubyanskaya LD. Prediction of plague epizootic activity in the Pre-Aral and Kyzylkum. Nukus of Karakalpakstan. 1992. p. 240.
3. Rivkus YZ, Naumov AV, Khotko NI, Geldyev A. Epidemiology and prevention of plague. Ashgabat. "Magaryf"; 1992. p. 239.
4. Bertherat E. Plague around the world 2010–2015.Weekly epidemiological record. WHO.# 8. 2016;91:89–104.
5. Ageyev VS, Burdelov LA, Serzhan OS. Slowdown mechanisms of circulation of the plague pathogen associated with fleas - poikilothermic hosts of this infection. Quarantine and Zoonotic Diseases in Kazakhstan. Almaty, 2003;2(8):131–40.
6. Akiyev AK. Question status of the mechanism of preservation of plague in inter epizootic years. (Overview). Problems of Particularly Dangerous Infections. Saratov. 1970;4(14):13–33.
7. Dubyanskiy MA. About the external signs of plague epizootic in the settlements of great gerbil in various phases of its development. Mater. scientific conf. on natural focality and plague prevention. Alma-Ata; 1963b. p. 76–8.
8. Dubyanskiy MA, Kanatov YV, Kurganov VA, Asenov G, Kenzhebayev AY, Dernovaya VF, Alibayev IA, Dubyanskaya LD, Bogatyrev SK, Tusnolobova TP. Regurgitates of birds of prey as a factor of plague microbe distribution. Ecological problems of Kazakhstan. Alma-Ata. 1990. p. 11–3.
9. Levi MI. Hypotheses explaining the epizootic process in plague. // Medical Parasitology and parasitic diseases. 1985. № 1. p. 36–42.
10. Naumov NP, Lobachyov VS, Dmitriev PP, Smirin VM. Natural foci of plague in the Pre-Aral Karakum. M. Ed. MoscowStateUniversity.1972. p. 402.
11. Stogov II, Afanasyev AG, Baranowskii SK. About the distance of transport of great gerbil fleas by Isabelline wheatear. In: Proceedings of the IV Scientific. Conf. of natural foci and plague prevention. Alma-Ata; 1965. p. 248–250.
12. Shevchenko VL. About the mechanisms of territorial spread of plague.// XII Inter-republican scientific. Practical. conf. anti-plague institutions of Central Asia and Kazakhstan on prevention of plague. Alma-Ata. 1985. p. 154–6.
13. Yermilov AP, Soldatkin IS, Rothschild EV. The problem of preservation of plague microbe in nature and prospects for further research. State and perspectives of plague prophylaxis: Abstracts at the All-Union conf. anti-plague institute "Microbe". Saratov, 1978. p. 9–12.
14. Soldatkin IS, Rudenchik Yu V. Epizootic process in natural foci of plague (audit of concept). Ecology of sapronoses pathogens. M. 1988. p. 117–31.
15. Dyatlov AI, Antonenko AD, Grizhebovskii GM, Labunets NF. Natural focality of plague in the Caucasus. Stavropol; 2001. p. 345.

16. Popov NV. Discrete - the main feature of space-time manifestation of plague outbreaks in gopher type foci. Saratov: Publishing house of Saratov University; 2002. p. 192.
17. Sludskyi AA, Derlyatko KI, Golovko EI, Ageyev VS. Gissar natural foci of plague. Saratov: Publishing house of Saratov University; 2003. p. 248.
18. Sagymbek UA, Aykimbaev AM, Meka-Mechenko TV, Rapoport LP. Selected questions of epizootiology of plague in the example of Moyinkum autonomous foci, Almaty. 2003; p. 204.
19. Sludskyi AA. Epizootology of plague (review of studies and hypotheses). Deposited in Russian Institute for Scientific and Technical Information of Russian Academy of Sciences, 08.11.2014, № 231, 2014), Saratov; 2014a. p. 313.
20. Sludskyi AA. The list of the world vertebrate fauna – the hosts of the plague pathogen. Problems of Particularly Dangerous Infections, vol 3. 2014b. p. 42–51.
21. Emelyanov PF. Rodents – edificators biocenosis as the main hosts of plague. Especially dangerous infections in Caucasus: Thes.report. Stavropol, 1987.p. 300–2.
22. Petrov VS. USSR natural foci of plague (typing, structure, genesis): abstract. thesis ....doc. biol.science. - Almaty, 1968, p. 40.
23. Dubyanskiy VM. Computer modeling of the epizootic situation with the use of remote sensing in the epidemiological surveillance of plague (on the example of Central Asian natural focus).abstract. thesis ....doctor. Biol Sciences. Moscow, 2015. p. 47.
24. Dyatlov AI. Evolutionary aspects in natural focality of plague. Stavropol: Bk. Publishing House; 1989. p. 197.
25. Eygelis Yu K. Rodents of Eastern Transcaucasia and the problem of rehabilitation of local plague foci. Saratov: Publishing House of Saratov University; 1980. p. 262.
26. Burdelov LA, Burdelov AS, Bondar EP, ZubovVV, Maslennikova ZP, Rudenchik NF. Use of burrows by great gerbil (Rhombomys opimus, Rodentia, Cricetidae) and epizootic significance its uninhabited colonies in the Central Asian foci of plague. Zool J. 1984. LXIII. V. 12. p. 1848–58.
27. Hrustselevsky VP. Biocenotic factors of natural focality of plague in Central Asia and the Trans-Baikal. Doctoral thesis abstract, Saratov; 1974. p. 52.
28. Tarasov PP. About the significance of predatory birds in the Khangay plague foci // Mat. Ir-sk. Anti-plague Inst. 1949. V.VII. p. 126–9.
29. Demidova EK. About the role of terrestrial and avian predators in the spread of plague // Proc. rep. conf. Irkutsk. Anti-plague in-te. Ulan-Ude; 1958;3. p. 41–2.
30. Kozlov MP, Sultanov GV. Plague. In: Natural focality, epizootology.V.3. Makhachkala; 2000. p. 303.
31. Kunitskii VN. Ecological and geographical sketch of fleas of south-west Azerbaijan in relation to their value in the natural foci of plague. Abstract thesis PhD. Almaty; 1966. p. 23.
32. Burdelov LA. Hostal and functional structure of the Central Asian desert plague focus (on the example of the Pre-Aral area): Abstract. thesis.... doctor. biol. science – Saratov, 1991 (All-Union. Anti-plague. Research Institute "Microbe"). p. 42.
33. Vashchenok VS. Fleas – vectors of human and animal diseases – L.: Nauka, 1988. – p. 163.
34. Konnov NP. Ultra structural and functional analysis of Y. pestis and its relationship with the organism of fleas: abstract.thesis doc. of biol.science. Saratov; 1990; p. 59.
35. Kutyrev VV, Kononov NP, Volkov YP. The causative agent of plague: ultrastructure and localization in the vector. M. Medicine; 2007. p. 224.
36. Korenberg EI. The natural focality of infections: current problems and perspectives of the research. J Zool. 2010;89(1):5–17.
37. Kunitskii VN, Gauzshtein DM, Gauzshtein LD, Kudinova TP, Kunitsa NK, Pavlova AE, Rasin BV, Filimonov VI, Shpes EM. About the structure and some quantitative indexes of plague epizootic in the great gerbil populations in southern Pre-Balkhash. Message 1.Activity of epizootic in Bosugen area in 1963–64.// Mater. V scientific conf. anti-plague institutions in Central Asia and Kazakhstan, devoted to 50th anniversary of the Great October Socialist Revolution. Alma-Ata; 1967. p. 30–2.

38. Sviridov GG. To modeling of some elements of plague epizootic in the Ili-Karatal interfluve. PhD M.S. Spec.14780 (epidemiology). Alma-Ata; 1969. p. 142.
39. Soldatkin IS. Enzootic of the plague as a self-regulating process. Author.diss. dr. of biol. sciences. 780. Saratov; 1968. p. 51.
40. Vashchenok VS. The role of fleas (Siphonaptera) in epizootology of plague. Parasitology. 1999;33(3):198–209.
41. Suleimenov BM. Transmission of plague by "unblocked" fleas. – Abstract. doctor. med.sciences: 14.00.30. Almaty. 1995. p. 48.
42. Dubrovsky YA. Gerbils and natural focality of cutaneous leishmaniasis. Moscow: Nauka; 1978. p. 183.
43. Rothschild EV. The spatial structure of the natural plague focus and methods of its study.// M. Ed. Moscow State University.1978. p.192.
44. Chumakova IV, Kozlov MP. Causative agent of plague as an element of epizootic process. Stavropol. 2008. p. 247.
45. Kunitskii VN, Gauzshtein DM, Dubovitskii NM, Rasin BV, Shpes EM, Novikov GS, Larionov GM, Bogatyreva LM, Savelov Yu V. About the spatial structure of the plague focal areas in the lower reaches of Ili. Materials of VIII scientific.conf. anti-plague institutions in Central Asia and Kazakhstan. Alma-Ata; 1974. p. 175–7.
46. Rivkus YZ, Mitropolskii OV, Urmanov RA, Belyaeva SI. Development features of plague epizooties among Kyzylkum rodents. Fauna and ecology of rodents. 1985;16:5–106.
47. Kucheruk VV. Kulik IL, Dubrovsky YA. Great gerbil as a vital form of the desert. In the book: Fauna and ecology of rodents. M. MSU. 1972; p. 5–70.
48. Gauzshtein DM, Kunitskii VN, Kunitskaya NT, Filimonov VI. About time of stay of the great gerbil fleas on the host's body in natural conditions. In: Proceedings of the IV scientific conference of natural foci and plague prevention. Alma-Ata; 1965. p. 66–8.
49. Kanatov YV, Lobachyov VS. Misaleva OS. Efficiency of bacteriological and serological diagnosis of plague in great gerbils in the North Pre-Aral region. In: Serological methods of diagnosis in epizootic survey of natural foci of plague. Saratov; 1975. p. 133–7.
50. Rudenchik YV, Soldatkin IS. Seasonal changes in the mobility of the great gerbils and spread of the plague epizooties in the Northern Kyzylkum. Problems of Particularly Dangerous Infections. 1969;1:34–9.
51. Naiden PE. Some aspects of the spatial development of plague epizooties.// In the book. Especially dangerous infections in the Caucasus. Mater.II scientific.conf. on epidemiology, epizootiology, prevention of especially dangerous infections of anti-plague institutions of Caucasus. Stavropol. 1970;1:174–6.
52. Naiden PE, Dyatlov AI, Melnikov IF, Breer VD. To the question of the spatial development of plague epizootic in Kyzylkum. Problems of Particularly Dangerous Infections. 1969;1:26–33.
53. Rothschild EV. Some spatial features of natural focality of plague in the northern Aral region. Problems of Particularly Dangerous Infections № 6. Saratov; 1969. p. 13–21.
54. Burdelov LA, Warsawskiy BS. To the question about the relationship between micro-focal and migration forms of existence of plague microbe in northern Pre-Aral – Proc: epidemiology and epizootology of plague. Saratov. 1980. p. 14–8.
55. Rall YM. Natural focality and plague epizootology. M: Medicine. 1965. p. 363.
56. Burdelov AS. About cyclical changes in the number of great gerbils and epizootic in their populations. In the book. In: Proceedings of the Central Asian Research Anti-Plague Institute. Alma-Ata. 1959. № 5.p. 179–87.
57. Lavrovskii AA. About epizootic cycling in natural foci of plague and the reasons causing it. Problems of Particularly Dangerous Infections.№ 1. Saratov; 1969. p. 3–10.
58. Dubyanskiy MA, Epizootic plague prediction on the example of the Central Asian desert focus. Diss. Dr. B. S. 14.00.30, Alma-Ata; 1983. p.352.

59. Dubyanskaya LD. Ecological factors that determine the seasonal changes in the number of the great gerbil (for example of six geographic populations).Author.PhD.B.S.: 03.00.16. Almaty; 1987. p. 21.
60. Dubyanskiy VM. Experience of system analysis of the dynamics of spring number of great gerbils in Central Kyzylkum.Author. PhD B.S. 03.00.08. Almaty; 2007. p. 26.
61. Dubyanskiy MA. Zoning of Central Asian natural plague foci based on epizootic activity communications of its parts with periodicity of atmospheric circulation types. Book: Problems of natural foci of plague. Abstracts IV to the Soviet-Mongolian experts anti-plague conference. institutions. Irkutsk; 1980. №1. p. 56–7.
62. Dubyanskiy MA, Bogatyrev SK. To the questions about the mechanism of the impact of solar activity on intensity of plague epizooties. In: Proceedings X conference antiplague institutions of Central Asia and Kazakhstan. Alma-Ata; 1979, № 1. p. 185–7.
63. Serzhanov OS, Aubakirov SA, Ageyev VS. About possible role of the hydrogeological conditions of the existence of plague enzootic during interepizootic period in the Central Asian desert focus // condition and prospects. proph. plague. Saratov; 1978. p. 17–20.
64. Petrov VS, Shmuter MF. Features of plague epizooties in natural foci of different types // Tr. Central Asian scientific anti-plague institutions. 1958; 4. p. 3–21.
65. Soldatkin IS, Rudenchik YV. Some questions of plague enzootic as a form of existence of self-regulating system "rodent – flea – pathogen". Fauna and ecology of rodents. V. 10. Moscow State University. 1971. p. 5–29.
66. Dubyanskiy MA. Ecological structure of settlements of the great gerbil in Pre-Aral Karakum PhD B. S. 097. Alma-Ata; 1970. p.23.
67. Marin SN, Kamnev PI, Trofimov AS. Some of the mechanisms of plague epizootic self-regulation among great gerbils. In the book: epidemiology and epizootiology of plague. Saratov; 1980. p. 8–14.
68. Dubyanskiy MA, Bogatyrev SK, Yermilov AP, Dubyanskya LD, Bogatyrev LM. About the correlation between the regime of winds and epizootic activity in some parts of the Central Asian desert plague focus. Book: Mater. VIII scientific conf. anti-plague. Institutions of Central Asia and Kazakhstan. Alma-Ata; 1974. p. 168–9.
69. Keeling MJ, Gilligan CA. Bubonic plague: a metapopulation model of a zoonosis. Proc R Soc London Ser B-Biol Sci. 2000;267(1458):2219–30.
70. Holt AC, Salkeld DJ, Fritz CL, Tucker JR, Gong P. Spatial analysis of plague in California: niche modeling predictions of the current distribution and potential response to climate change. // International Journal of Health Geographics 2009;8:38. doi:10.1186/1476-072X-8-38. The electronic version of this article is the complete one and can be found online at: http://www.ij-healthgeographics.com/content/8/1/38
71. Stenseth NC, Samia NI, Viljugrein H, Kausrud KL, Begon M, Davis S, Leirs H, Dubyanskiy VM, Esper J, Ageyev VS, Klassovskiy, Nikolay L, Pole SB, Chan KS. Plague dynamics are driven by climate variation. Proc Natl Acad Sci U S A. 2006; 103(35):13110–15. Published online 2006 August 21. doi:10.1073/pnas.0602447103.
72. Snall T, O'Hara RB, Ray C, Collinge SK. Climate-driven spatial dynamics of plague among prairie dog colonies. Am Nat. 2008;171:238–48.
73. Kausrud KL, Viljugrein H, Frigessi A, Begon M, Davis S, Leirs H, Dubyanskiy V, Stenseth NC. Climatically-driven synchrony of gerbil populations allows large-scale plague outbreaks. In: Proceedings of the Royal Society of London, Series B. 2007;274 p. 1963–9.
74. Kol NA, Kalush Yu A, Rostovtsev MG, Chuldum AF, Mamash EA. The use of geographic information technologies for the analysis of spatial dynamics of Karginsky meso-foci of Tuva natural focus of plague. Interexpo Geo-Siberia. 2008. №2. V. 3.
75. Davis S, Trapman P, Leirs H, Begon M, Heesterbeek JAP. The abundance threshold for plague as a critical percolation phenomenon. Nature. 2008;454:634–7.
76. Rudenchik YV. Study of epizootic process in plague in the imitation statistical model. Author. diss. of Dr. Biol Sci. 14.00.30. Saratov; 1979. p. 48.

77. Soldatkin IS, Rodnikovskii VB, Rudenchik YV. Experience of statistical modeling of epizootic process in plague. Zool J. 1973;52(5):751–6.
78. La Force FM, Acharya IL, Stott G, Brachman PS, Kaufman AF, Clapp RF, Shah NK. Clinical and epidemiological observations of plague outbreak in Nepal. Bulletin of the World Health Organisation. February 1971;45(6):693–706.
79. Soldatkin IS, Fenyuk BK, Possibilities and field of use of natural situations plague epizootic process modeling. Rodents and their ectoparasites. Saratov: Saratov.un-ty; 1968. p. 439–45.
80. Wilschut LI, Laudisoit A, Hughes NK, Addink EA, de Jong SM, Heesterbeek HAP, Reijniers J, Eagle S, Dubyanskiy VM, Begon M. Spatial distribution patterns of plague hosts: point pattern analysis of the burrows of great gerbils in Kazakhstan .J Biogeo. 2015; 42(7). doi:10.1111/jbi.12534.
81. Kuznetsov AA. Principles of quantifying of transmissible epizootic process by plague. Mater. scient. conf. dedicated to the 100th anniversary of the Anti-Plague Service of Russia. Saratov. 1997;1. p. 84.
82. Platonov ME, Evseeva VV, Efremenko DV, Afanas'ev MV, Verzhutski DB, Kuznetsova IV, M. Yu S, Dentovskaya SV, Kulichenko AN, Balakhonov SV, Anisimov AP. Intraspecies classification of rhamnose-positive Yersinia pestis strains from natural plague foci of Mongolia. Mol Genet Microbiol Virol. 2015;30(1):24–9.
83. Warsawskiy S N, Kazakevitch V P, Lavrovskii A A, Nekipelov N V. The geography of natural foci of plague in Central Asia (West, Southwest and South-Eastern part of the MPR). Problems of Particularly Dangerous Infections. 1975;3/4(43/44). p. 5–14.
84. Klein JM, Poulet AR, Simonkovich E. Observations ecologiques dans une zone epizootique de pesteen Mauritanie. 1. Leslongeurseten particulier Gerbillus gerbillus Oliver, 1981 (Rodentia, Gerbillinae) //Cah. ORSTOM. Ser. Entomol. med. et parasitol. 1975a;13(1):13–28.
85. Baltazard M, Seydian B. Enquetesur les conditions de la peste on Moyen-Orient. Ibid.1960a;23(2/3):157–67.
86. Dubyanskiy MA, Dubyanskaya LD, Bogatyrev SK. The relationship between the amount of atmospheric condensation and the probability of human infection with the plague in the natural focus. ZHMEI. 1992b. № 9–10. p. 42–6.
87. Zarhidze VA, Zuychenko NA, Chelnokov VN, Obrikas RG. Some features of plague epizootic in Chilmamedkum in 1965–1966. Problems of Particularly Dangerous Infections. 1971;2:112–7.
88. Fadeyev GS, Serdobintseva TA. Mathematical prediction of seasonal changes in the number of great gerbil. Rodents, Mater. IV All-union meeting – L. Science, 1983. p. 464–5.
89. Soldatkin IS, Rudenchik Yu.V. About prediction of epizootic situations in the southern deserts. Report. Irkutsk Anti-Plague Institute: Kyzyl. 1966. p. 98–9.
90. Soldatkin IS, Rudenchik YV. The use of mathematical models in the study of natural foci of disease.Book: natural-focal anthropozoonoses. Omsk. 1976. p. 36–7.
91. Rothschild EV, Yermilov AP, Danilenko ID, Postnikov GB. Long-term dynamics of plague epizootic among the great gerbils in the North-East Caspian and link with the weather. Problems of especially dangerous infections, Saratov; 1970 № 6, p. 120–31.
92. Popov NV. Development of main principles for predicting of the plague epizooties in the settlements of small gopher in natural focus of the North-West Pre-Caspian.Author.PhD B.S. Saratov. 1977. p. 19.
93. Davis S, Begon M, De Bruyn L, et al. Predictive thresholds for plague in Kazakhstan. Science. 2004;304(5671):736–8.
94. Davis S, Leirs H, Viljugrein H, et al. Empirical assessment of a threshold model for sylvatic plague. J R Soc Interface. 2007;4:649–57.
95. Genes VS. Some simple methods of cybernetic processing of data of diagnostic and physiological studies. M Sci. 1967. p. 207.
96. Kunitskii VN, Gauzshtein DM, Kunitskaya NT, Shpes EM. The informational content of value of the density of great gerbils populations and their fleas in relation in respect of identification of epizootic phase of the cycle. Book: Ecology and medical meaning of the gerbils of fauna of the USSR. M., 1977. p. 327–8.

97. Dubyanskiy MA, Dubyanskaya LD. Experience of use of multiple correlation method to predict the number of the great gerbil. Book L IV inter-agency meeting on phenological prediction. L. 1977; p. 169–71.
98. Okulova NM. Reproduction and mortality in the population of red-backed vole, and the main factors affecting these processes. Zool J. 1975;54(11):1703–14.
99. Sergeyev GE. The use of multiple correlation for the prediction of pests. In the book: Labor of the All-Union Scientific Research Institute of Plant Protection. L. 1970. p. 238–46.
100. Sokolova SP, Kuzmina EA, Abdullina VZ. Monitoring of especially dangerous infections (example of plague). Math Biol Bioinform. 2007;2(1):82–97.
101. Tarakanov A, Sokolova S, Abramov B, DubyanskiyV. Comprehensive assessment of plague epizootic as a result of the recognition of the model. Scientific and technical, economic, industrial "Tauar." 1999. p. 51–4.
102. Fu SC, Milne G. Epidemic modelling using cellular automata. In: Proceedings of the 1st Australian Conference on Artificial Life (ACAL.03), Canberra, December 2003. p. 43–57.
103. Grabowskii VI. 1995. Cellular automatons as simple models of complex systems. Successes of modern biology. 115(4), p. 412–418.
104. Gubler EV. Computational methods of analysis and detection of pathological processes. – L .: Medical, Leningrad Branch, 1978. p. 296.
105. Burdelov LA, Dubyanskiy VM, Meka-Mechenko VG, Semenko OV, Sadovskaya VP. About the reasons of recent expansion of the great gerbil area (Rhombomys opimus Licht) in Kazakhstan./ Zoological and hunting management studies in Kazakhstan and neighboring countries. Mat. international scient. conf. Almaty. 2012. p. 69–73.
106. Neronov VM, Malchazova SM, Tikonov VS. Regional geography of plague. M; 1991. p. 232
107. Rall Yu. M. Rodents and Natural focality . M. Medgiz. 1960. p. 224.
108. Warsawskiy SN, Kazakevitch VP, Lavrovskii AA. Natural focality of plague in the northern and western Africa. Problems of Particularly Dangerous Infections. 1971:3(19) p. 149–159.
109. Klein JM, Simonkovich E, Alonso JM, Baranton G. Observations ecologiquesdansune zone epizootique de pesteenMauritanie. 2. Les puces de rongeurs (Insecta, Siphonaptera). Ibid. 1975b; 13(1):29–39.
110. Borchert JN, Mach JJ, Linder TJ. Invasive rats and bubonic plague in northwest Uganda. Managing Vertebrate Invasive Species. In: Proceedings of an International Symposium. USDA/APHIS/WS, National Wildlife Research Center, Fort Collins, CO. 2007. p. 283–93.
111. Roberts JI. The relationship of field rodents to plague in Kenya. J Hyg. 1939;39(03):334–44.
112. Ziva MH, Matee MI, Hang'ombe BM, Lyamuya EF, Kilonzo BS. Plague in Tanzania: an overview. Tanzan J Health Res. 2013;15(4):252–8. http://dx.doi.org/10.4314/thrb.v15i4.7
113. Laudisoit A, Neerincx S, Makundi RH, Leiers H, Krasnov BR. Are local endemicity and ecological characteristics of vectors and reservoirs related? A case study in north-east Tanzania. Curr Zool. 2009;55(3):200–11.
114. Poland JD, Dennis DT. Plague manual/ Epidemiology, Distribution, Surveillance and Control. Treatment of Plague. 1999. WHO/CDS/EDC/99.2.p. 55–134.
115. LaForce FM, Acharya IL, Stott G, Brachman PS, Kaufman AF, Clapp RF, Shah NK .Clinical and epidemiological observations on an outbreak of plague in Nepal.// Bulletin of the World Health Organisation. February 1971; 45(6):693–706.
116. Barnes A. Vector-borne diseases control, Burma. Assignment report, 27 December 1981 – 5; March 1982.WHO. p. 30.
117. Anisimov AP, Lindler LE, Pier GB. Intraspecific diversity of Yersinia pestis. Clin Microbiol Rev. 2004;17(2):434–64.
118. Zhou D, Han Y, Song Y, Huang P, Yang R. Comparative and evolutionary genomics of *Yersinia pestis*. Microbes Infect. 2004;6:1226–34.
119. Natural foci of plague of the Caucasus, Pre-Caspian, Central Asia and Siberia.M. Medicine. 2004. p. 192.

120. Kazakov VP, Dyatlov AI. About the spatial structure and landscape confinement of outbreaks in the Dagestan mountainous plague foci. Prevention of natural focal infections. Stavropol. 1983; p. 76–8.
121. Kazakov VP, Nurmagomedov NM. Epizooty of plague among gray hamsters in mountainous Dagestan. Diseases with natural focality in the Caucasus. Stavropol. 1982; p. 66–7.
122. Kazakova TI, Zemelman BM, Labunets NF. About block formation and the infecting ability of some fleas of common voles from Dagestan mountain plague focus. Diseases with natural focality in the Caucasus. Stavropol; 1982. p. 70–2.
123. Suchkov EG, Naiden PE, Rozanova GN et al. Zoning of Central Caucasus natural focus of plague on the basis of some features of allocated here strains of plague microbe // Diseases with natural focality in the Caucasus. Stavropol; 1982. p. 127–8.
124. Popov NV, Warsawskiy BS. The activity of natural foci of plague in the CIS in 1978–1992. Problems of Particularly Dangerous Infections. 1993;3(73):3–15.
125. Dobronravov VP, Abdurakhmanov GA. About detection of plague epizootic among small gophers in the Malgobek district of the Chechen-Ingush ASSR. Problems of Particularly Dangerous Infections.1971;1. p. 167–8.
126. Toporkov VP, Podsvirin AV, Yashkuljv KB. Ecological and epidemiological monitoring of predictors of extreme epidemic situations in the natural foci of plague in the North-Western Caspian region. Elista; 1999. p. 126.
127. Popov NV, Rogatkin AA. Kozlov TA, Bukayeva IN. Cyclical of plague epizootic in the North and North-West of the Caspian region and its determining factors. Astrakhan; 1999. p. 112.
128. Atlas of especially dangerous infections spread in the Republic of Kazakhstan, edited by prof. Burdelov L A, Almaty; 2012. p. 232 (Rus, Kaz.).
129. Ergeshbayev MB, Seryakov VA, Sadykov AD, Burdelov LA. Some features of plague epizootic from 1988–1989 in Western Alai. Organisation in plague surveillance and measures for its prevention. Mater. interstate. scientific.-practical. conf. Alma-Ata; 1992. Part 2, p. 291–3.
130. Golubinskii EP, Zhovyi IF, Lemesheva LB. About the plague in Siberia. Irkutsk: Irkutsk University Publishing House; 1987. p. 244.
131. The atlas of Plague and its environments in the People's Republic of China. Beijing: Science Press. 2000. p. 206.
132. Lipayev VM, Busoedova NM, Derevich SM et al. The main results of ten-year (1967–1976) epidemiological survey of the North-Western part of Mongolia. Epidemiology and prophylaxis of especially-dangerous infections in Mongolia and the USSR. Ulaanbaatar; 1977. p. 38–40.
133. Balakhonov SV, Batsuh D, Adyasuren Z, Tserenbuuzhid N. Spectrum of plasmids of Y. pestis strains from different mesofoci of plague of Mongolian Altai. Organisation in plague surveillance and measures for its prevention. Materials interstate.scientific-practical. conf. Alma-Ata. 1992; Part 1:p. 82–4.
134. Munhtuyaa O, Schur N, Sarah M. Application of serological diagnosis in the natural foci of Mongolia. Epidemiology and Prevention of especially-dangerous infections in Mongolia and the USSR: Ulaanbaatar; 1977. p. 71–4.
135. Munhtumur D, Oyuunchimeg D, Dorj S. Fleas of mammals and birds as prime cause of emergence of the natural foci of plague in the Mongolian Altai // Natural conditions, history and culture of Western Mongolia and adjacent regions: Materials of the VII International. conf. (19–23 Sep., 2005), Kyzyl. Tuv IKOPRSBRAS. 2005; Vol 1, p. 204–7
136. Nekipelov NV. Epizootology of plague in the Mongolian People's Republic. Izv.Irkutsk.state: Anti-Plague.Research Institute of Siberia and Far East; 1959. V. 22. p. 108–243.
137. Otgonsuren L, Galbadrah D, Purevzhal C. Current condition of Khangay natural foci of plague. Problems of natural foci of plague. Abstracts.IV Soviet-Mongolian Conf. Antiplague professionals. Agencies. Irkutsk. 1980. Part 1. p. 68–9.
138. Batsuh D, Dembrel Zh, Dembrel B, Batbold Zh. Current condition of natural foci of plague of Mongolian People's Republic. Nature focality of plague in the Mongolian People's Republic. Irkutsk. 1988;3–5.

139. Tarasov MP. Rodents of south-eastern part of the Mongolian Altai and adjacent Gobi // Mat. Irkutsk anti-plague research institute of Siberia and the Far East. 1958; 19, p. 60–71.
140. Kiefer M, Krumpal M, Cendsuren N et al. Cheklist, distribution and bibliography of Mongolian Siphonaptera. Erforshc. Biol. Ress. MVR (Halle). 1984. N. 4. p. 91–123.
141. Chinbold L. Natural foci of plague in the Gurvan-Saikhan mountains. International and national aspects of surveillance in plague. Irkutsk. 1975. p. 28–30.
142. Fenyuk BK. Notes about natural foci of plague and anti-plague work in the People's Republic of China. ZHMEI. 1959, № 10, p. 8–16.
143. Barkov IP, Bazunov LP, Shiryaeva A. About plague among the Mongolian gerbil and midday jirds in the Central Asian foci. News of Irkutsk anti-plague research Institute of Siberia and Far East. 1959;20:117–9.
144. Lavrovskii AA, Kazakevich VP, Warsawskiy SN. Natural foci of plague in the Front and Southwest Asia. Problems of Particularly Dangerous Infections. 1973;2(30):9–22.
145. Baltazard M, Bamanyar M, Mostachfi P et al. Rechercessur la peste in Iran. Ibid. 1960b;23(2/3). p. 141–55.
146. Misonne X. Mammiferes de la Turquiesud-oriental et du Nord de la Syrie. Mammalia. 1957;21(1):53–67.
147. Baltazard M, Bamanyar M, Mofidi C, Seydian B. Le foyer de peste du Kurdistan. Bull WHO. 1952; 5(4). p. 441–72.
148. Karimi Y. Decouverte d'un nouveau mesofoyer de pestesauvagedans I' Azarbaidjan oriental de I'Iran. Bull Soc Pathol Exot. 1980;73(1):28–35.
149. Bahmanyar M. Plague epidemics in Iran and their control. WHO. 1968, BD/PL/68.32. p. 1–7.
150. Warsawskiy SN, Kazakevitch VP. Biocenotic structure and landscape features of the foreign plague foci in the Front and Southwest Asia. Bul Mosc Soc Nat Dep Biol. 1984;89(1):13–20.
151. Kozlov MP, Sultanov GV. Epidemic manifestations of plague in the past and the present. Makhachkala, Dagest. Vol. publ. 1993. p. 336.
152. Neronov VM, Arsenyev LP. Zoogeographical analysis of Afghanistan rodent fauna // Modern problems of zoogeography.M: Science. 1980. p. 254–72.
153. Arsenyev LP, Neronov VM. Materials to the fauna of fleas of Afghanistan (Siphonaptera) // Parasitology. 1982; 16(4):306–314.
154. Akiev AK. About the great gerbil in the northern Afghanistan // rodents and their ectoparasites (ecology, epizootic value, control). Saratov: Publishing house of Saratov University; 1968. p. 422–24.
155. Almeida A. Rodent and vector surveillance in Brazil. Interregional meeting on prevention and control of plague. Antananarivo, Madagascar 1–11. April 2006. p. 36.
156. Dubyanskiy MA, Dubyanskaya LD, Nekrasova LE, Bakanurskaya TL. One more time about the importance of raptors as indicators of plague epizootic. Natural. foc. microbiology. and proph. zoonoses. Saratov; 1989. p. 40–43.
157. Yergaliev KH, Pole SB, Stepanov VM. Blood groups of the great gerbil (Rhombomys opimus Licht.) as supposed indicators of resistant populations to infection. // Genetics. - V. 26. - №1. - 1990. pp. 103–8.
158. Klassovkii LN, Kunitskii VN, Gauzshtein DM, Burdelov AS, Aykimbaev MA, Dubovitskii NM, Novikov GS, Rasin BV, Savelov YV. To the question about short-term prediction of the epizootic situation in the Ili-Karatal interfluves. Book: condition and perspectives of plague preventing. Abstracts of the all-union conference. Saratov. 1978. p. 69–71.
159. Suleimenov BM. The mechanism of plague epizootic. Almaty. 2004. p. 236.
160. Dubyanskiy VM, Burdelov LA, Barkley JL. Introduction to the computer modeling of the plague epizootic process. 07/2012; ISBN: 978-953-51-0685-2 In book: Automation, Chapter: 8, Publisher: In Tech, p. 149–70.
161. Shedrin VI. Morphological and histochemical data on blood digestion in some species of fleas – vectors of plague: PhD thesis abstract – Saratov; 1974. p. 16.

Chapter 6
Genome and Evolution of *Yersinia pestis*

Yujun Cui and Yajun Song

Abstract This chapter summarizes researches on genome and evolution features of *Yersinia pestis*, the young pathogen that evolved from *Y. pseudotuberculosis* at least 5000 years ago. *Y. pestis* is a highly clonal bacterial species with closed pan-genome. Comparative genomic analysis revealed that genome of *Y. pestis* experienced highly frequent rearrangement and genome decay events during the evolution. The genealogy of *Y. pestis* includes five major branches, and four of them seemed raised from a "big bang" node that is associated with the Black Death. Although whole genome-wide variation of *Y. pestis* reflected a neutral evolutionary process, the branch length in the genealogical tree revealed over dispersion, which was supposedly caused by varied historical molecular clock that is associated with demographical effect by alternate cycles of enzootic disease and epizootic disease in sylvatic plague foci. In recent years, palaeomicrobiology researches on victims of the Black Death, and Justinian's plague verified that two historical pandemics were indeed caused by *Y. pestis*, but the etiological lineages might be extinct today.

Keywords *Y. pestis* • Genome • Evolution • Molecular clock • Population diversity

6.1 Genome of *Yersinia pestis*

6.1.1 General Features of the Y. pestis *Chromosome*

Genomic research provides an unprecedented opportunity for us to probe into the pathogenicity and evolution of *Y. pestis*. The first *Y. pestis* genome, that of strain CO92, was decoded in 2001 [1]. Strain CO92 belongs to biovar Orientalis, which was isolated from a fatal human case of primary pneumonic plague that was contracted from an infected cat in Colorado, USA, in 1992 [2]. The genome was

Y. Cui (✉) • Y. Song
Beijing Institute of Microbiology and Epidemiology, No. Dongdajie, Fengtai, Beijing 100071, China
e-mail: cuiyujun.new@gmail.com; songyajun88@gmail.com

R. Yang, A. Anisimov (eds.), *Yersinia pestis: Retrospective and Perspective*, Advances in Experimental Medicine and Biology 918,
DOI 10.1007/978-94-024-0890-4_6

Table 6.1 General features of the chromosomes of *Y. pestis* strains 91001, CO92, and KIM

Accession number	AE017042	AL590842	AE009952
Strain	91001	CO92	KIM10
Length (bp)	4,595,065	4,653,728	4,600,755
(G+C)%	47.65 %	47.64 %	47.64 %
ORF	4037	4012	4198
Pseudogenes	141	149	54
Coding density	81.6 %	83.8 %	86 %
Average gene length (bp)	966	998	940
rRNA operons	7×(16S-23S-5S)+5S	6×(16S-23S-5S)	7×(16S-23S-5S)
tRNAs	72	70	73
IS*100*	30 intact	44 intact	35 intact
IS*1541*	43 intact	62 intact	49 intact
	2 disrupted by IS*100*	2 disrupted by IS*100*	3 disrupted by IS*100*
	3 partial	2 partial	6 partial
IS*1661*	7 intact	7 intact	8 intact
	1 partial	2 partial	2 partial
IS*285*	23 intact	21 intact	19 intact

assembled from 94,881 end sequences (giving about 9.6-fold coverage) that were derived from shotgun libraries, and it was completed using extensive gap-filling procedures. The CO92 consists of one 4.65-Mb chromosome and three plasmids, namely, pCD1 (70.3 kb), pMT1 (96.2 kb), and pPCP1 (9.6 kb).

The most striking large-scale features in the CO92 chromosome are the anomalies in GC bias (G-C/G+C). Unlike most other genomes with small but detectable bias toward G on the leading strand of the bidirectional replication fork, no clear GC-skew was seen in the CO92 genome. These anomalies might be caused by the very recent acquisition of DNA (such as prophages or genomic islands) or by the inversion or translocation of blocks of DNA [1]. Polymerase chain reactions flanking the possible translocated blocks revealed that genomic rearrangements occurred not only in different *Y. pestis* strains but also in different CO92 subcultures. These results demonstrate that the *Y. pestis* genome is extremely fluid, while ribotyping or pulsed-field gel electrophoresis typing results might be distorted by frequent intragenomic recombination events in vitro [3].

Right after the release of the CO92 genome, scientists decoded the genomes of two other representative *Y. pestis* strains. One was strain KIM10+, a biovar Medievalis derivative that is widely used in research laboratories around the world [4]. The other one was strain 91001, a biovar Microtus isolate, which has a median lethal dose of 23.2 colony-forming units (CFU) when administered subcutaneously to mice, while a subcutaneous challenge of 1.5×10^7 CFU challenging caused neither bubonic nor pneumonic plague in humans [5]. Table 6.1 lists the general features of the chromosomes of these three strains, which are quite similar to each other. An intriguing finding is that the copy numbers of insertion sequence (IS) elements in *Y. pestis* are much higher than in its ancestor, *Y. pseudotuberculosis* [6]. There are four different kinds of IS elements in *Y. pestis*, IS*100*, IS*285*, IS*1661*, and IS*1541*, which account for about 3.7 % of the chromosome. The unusual abundance of IS elements contributes to the frequent genomic rearrangements in the *Y. pestis* chromosome [7].

Another unique feature in the *Y. pestis* chromosome is the large number of pseudogenes, which might be caused by IS interruptions, deletions, frameshift mutations, or nonsense mutations [7]. A handful of genes essential for the enteropathogenicity of *Y. pseudotuberculosis* have been inactivated via different mechanisms. This kind of genome decay would offer *Y. pestis* fitness advantages for a new lifestyle as a systemic pathogen of mammals and an insect pathogen (see below) [1, 8].

6.1.2 Pan-Genome of Y. **pestis**

The pan-genome of a bacterial species describes the full complement of genes in its sequenced genomes [9]. The pan-genome includes the core genome that contains genes present in all strains, the accessory genome that contains genes present in at least two strains, and unique genes that are specific to single strains. Depending on the genomes included in the analysis, the pan-genome of a species can have large variations.

In a comparative genomic analysis that involved 133 genomes, the *Y. pestis* core genome, consisting of segments present in all the genomes, was 3.53 Mb long with 3450 annotated genes [10], while 1.92 Mb of accessory genome (1700 annotated genes) were only present in some isolates, of which 421 kb were within plasmids. Thus, the *Y. pestis* pan-genome (core genome plus accessory genome) was estimated to be 5.45 Mb. Figure 6.1 illustrates the plots of the *Y. pestis* pan-genome *vs.* the number of accumulating genomes, which reveals a platform for the size of pan-genome. Through horizontal gene transfer (HGT), many bacteria have a so-called open pan-genome. For example, the core genome of 20 strains of *Escherichia coli* contained only 2000 genes, whereas 20,000 genes were included in the pan-genome [11]. Similarly, it was predicted that the genome content of *Streptococcus agalactiae* will continue growing even after hundreds of genomes have been sequenced [12]. However, for asexual bacteria, such as *Salmonella enterica* serovar Typhi [13], clones of *Staphylococcus aureus* [14, 15], and the endosymbiont *Buchnera* [16], the genome is closed, and any acquisition of foreign genes seems to be largely restricted to bacteriophages and plasmids. The pan-genome of *Y. pestis* is also closed, with a size that is only slightly larger than the average size of the sequenced genomes [10].

Plasmids are a very important part of the accessory genome in *Y. pestis*. Typical *Y. pestis* strains have three plasmids: pCD1, which encodes a type III secretion system; pMT1, which encodes the *Yersinia* murine toxin (Ymt) and the fraction 1 capsular protein; and pPCP1, which encodes plasminogen activator protein Pla [8].

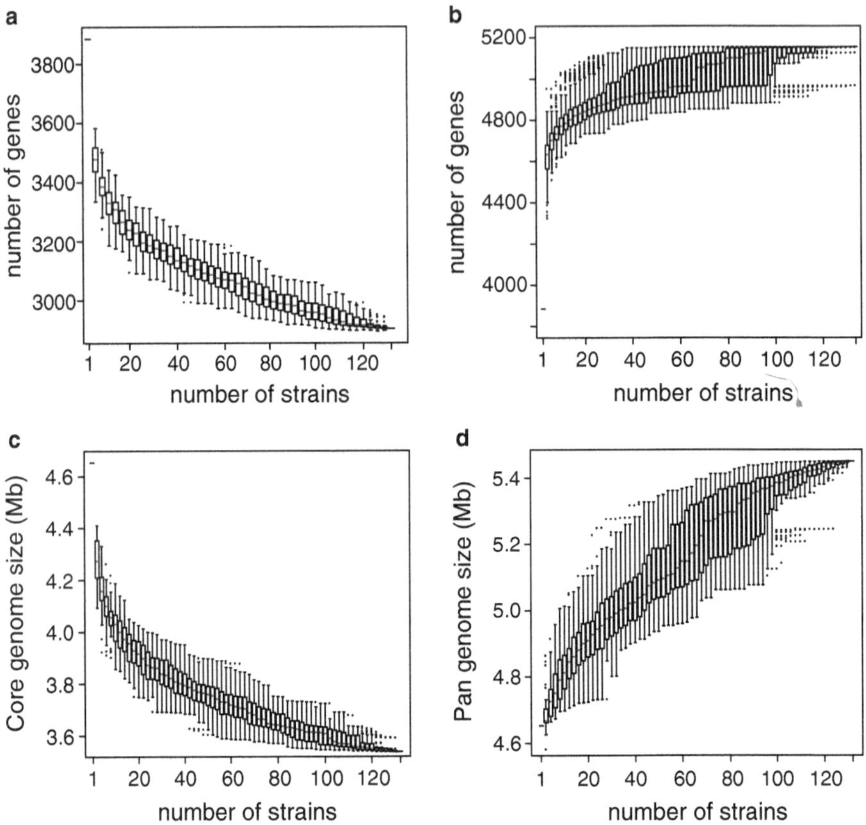

Fig. 6.1 Size estimates of the core genome (**a**, **c**) and pan-genome (**b**, **d**) of *Y. pestis* versus the number of genomic sequences. Size estimates are indicated by *box* and whisker plots, in which the box encloses the second and third quartile (25–75 %) of the data. The median is indicated by a thin *red line* within each box. The whiskers indicate the top and bottom quartiles, except for outliers, which are indicated by dots. (**a** and **b**) Numbers of genes included in the core genome and pan-genome. (**c** and **d**) Sizes of the core genome and pan-genome

Plasmid pCD1 is common to *Y. pestis*, *Y. pseudotuberculosis*, and *Y. enterocolitica*, while *Y. pestis* is believed to have acquired the other two plasmids via HGT. In addition to the three classical plasmids, other plasmids have also been reported in various strains. Strain 91001 harbors a 21.7-kb cryptic plasmid, pCRY1, which might encode a type four secretion system [5]. Another 5.9-kb cryptic plasmid, pYC, has been found in more than 230 *Y. pestis* isolates from China [17]. The deep-rooted pestoides isolate Angola harbors a 114.6-kb pMT1-pPCP1 chimeric plasmid [18]. Strain Java 9, a biovar Orientalis strain that was isolated in 1957 from a dead Javanese rat, carries two novel plasmids termed pJARS35 (35.0 kb) and pJARS36 (36.1 kb), which might be responsible for high-level arsenic resistance [19]. Notably, the first multidrug-resistant isolate of *Y. pestis* (strain IP275) was identified in Madagascar in 1995, and it contains a self-transmissible plasmid (pIP1202, 113.3 kb) that confers resistance to many of the antimicrobials that are recommended for plague treatment and prophylaxis [20]. With the rapid growth in the number of sequenced *Y. pestis* genomes, more diverse plasmid profiles will be identified in the near future.

Other than plasmids, the only confirmed genomic island that might have been acquired after the earliest nodes in the evolution of *Y. pestis* is a filamentous bacteriophage [10]. This phage, YpfPhi, forms filamentous particles that are secreted by *Y. pestis* cells into the culture supernatant without affecting bacterial growth or lysing the cells [21]. In *Y. pestis*, YpfPhi is present in isolates from the three main phylogenetic branches, suggesting that it was acquired shortly after the emergence of *Y. pestis*. However, YpfPhi is only systematically present in strains from subbranch 1.ORI. In the other subbranches, YpfPhi is only detected in some isolates [22]. In Antigua and Medievalis strains, YpfPhi forms an unstable episome, whereas it is stably integrated into the chromosome in Orientalis isolates. Deletion of the YpfPhi genome does not affect the ability of *Y. pestis* to colonize and block the flea proventriculus, but it results in an alteration of *Y. pestis* pathogenicity in mice [21].

6.2 Ancient DNA Genomics of *Y. pestis*

Historically, three great pandemics of plague claimed millions of lives around the world. The first great pandemic (541–544 AD) was known as Justinian's Plague. The second one, the Black Death, which occurred in the middle of the fourteenth century, killed an estimated 17–28 million Europeans. The third one, Modern Plague, which reached Hong Kong from Yunnan Province in 1894, spread worldwide, and it established stable enzootic foci on every major inhabited continent except Australia [23]. Since Alexandre Yersin isolated Gram-negative bacilli from a bubonic plague patient in Hong Kong at the very early stage of the third pandemic, nobody questioned the fact that *Y. pestis* is the causative agent of the third plague pandemic. However, debates about which etiologic agents caused the first two pandemics have lasted for years [24, 25]. In 1998, Raoult et al. successfully amplified *Y. pestis*-specific *pla* fragments from dental pulp DNA isolated from sixteenth- and eighteenth-century French graves of persons thought to have died of plague [26]. Several years later, they obtained *Y. pestis*-specific amplicons from victims of Justinian's Plague [27]. Increasing numbers of ancient DNA studies now support *Y. pestis* as the causative agent of the first two pandemics [28].

In 2011, Bos et al. described the first complete genome sequence of *Y. pestis* from Black Death victims [29]. First, they extracted bacterial DNA from five teeth of plague victims taken from a burial pit in London, which was established at the height of Black Death in 1348. Then, they used custom Agilent DNA capture arrays to enrich *Y. pestis* DNA amid a background of host and environmental DNA and performed Illumina sequencing. They finally assembled draft genomes with about 28× coverage, which provided valuable data for phylogenomic analysis [29]. Later, using a similar strategy, an international team successfully generated two draft *Y.*

pestis genomes from two Justinian's Plague victims from a cemetery in Bavaria, Germany [30].

More recently, an archaeology group screened 89 billion raw DNA sequence reads that were obtained from the teeth of 101 Bronze Age individuals from Europe and Asia, and they found that seven individuals carried sequences resembling *Y. pestis*. Then, they performed in-depth sequencing for the seven individuals, and assembled genomes to an average coverage of 0.14–29.5×, with 12 %–95 % of the positions in the genome covered at least once [31]. This study revealed *Y. pestis* infection in the Bronze Age for the first time, and it pushes the divergence dates for the early branching of *Y. pestis* back to 5783 years ago. However, as they did not include any enrichment procedure, five out of the seven draft genomes have a coverage that is less than one.

6.3 Evolution of *Y. pestis*

6.3.1 On the Origin of Y. pestis

Although *Y. pestis* has been proved to cause human infections as early as the Bronze Age [31], it was not identified as the cause of plague by microbiologists until 1894 [32]. Since then, the plague bacillus was transferred from one genus to another until it was finally placed in the genus *Yersinia* in 1944 [33]. During more than half a century of research, people realized that *Y. pestis* and *Y. pseudotuberculosis* are highly homologous and distinct from other bacterial species in terms of their DNA sequences [34]. According to DNA-DNA hybridization results (more than 90 % DNA relatedness), Bercovier et al. suggested that *Y. pseudotuberculosis* and *Y. pestis* should be treated as two separate subspecies of the same species, and they suggested that they be named *Y. pseudotuberculosis* subsp. *pseudotuberculosis* and *Y. pseudotuberculosis* subsp. *pestis*, respectively [35]. The sequence data obtained from rDNA sequences also supported the close relationship between these two organisms, which have fully identical 16S rDNA sequences and only a single base difference within a highly variable region of 23S rDNA [36]. Although satisfying the common criterion of being the same species based on the similarity of their DNA sequences [37], *Y. pestis* and *Y. pseudotuberculosis* are still listed as different species because of differences in their virulence, transmission route, as well as historical importance to human culture.

To deduce the origin and phylogenetic history of *Y. pestis*, Achtman et al. collected 76 *Y. pestis* strains, which were isolated from around the globe and that represent all three biovars, 13 strains of *Y. enterocolitica*, and 12 strains of *Y. pseudotuberculosis* of various serotypes, and they analyzed the fragments of five housekeeping genes (*thrA*, *trpE*, *glnA*, *tmk*, and *dmsA*) and one lipopolysaccharide (LPS) synthesis-associated gene (*manB*). No differences were observed across all six gene fragments for all trains of *Y. pestis* used in the study. Three genes revealed identical alleles between *Y. pestis* and some strains of *Y. pseudotuberculosis*.

Furthermore, the genetic distances between the alleles in *Y. pestis* and *Y. pseudotuberculosis* were within the range seen in *Y. pseudotuberculosis* alone. These results indicate that *Y. pestis* is a genetically monomorphic clone of *Y. pseudotuberculosis* that differentiated so recently that only few sequence polymorphisms have accumulated [38]. By sequencing housekeeping genes in subsequent research, Achtman et al. showed that groups of atypical *Y. pestis* strains, including pestoides strains in Central Asia and the Microtus strain in China, also have fully identical alleles with all three classical *Y. pestis* biovars, despite their different biochemical characteristics [39]. This study further defines the range of *Y. pestis* species and supports the view that all *Y. pestis* strains evolved from *Y. pseudotuberculosis*.

However, there are still arguments over whether the sequences of housekeeping genes represent the full genetic characteristics of *Y. pseudotuberculosis* and *Y. pestis* and, thus, whether these species could constitute a sister taxon. To confirm the hypothesis that *Y. pestis* is a clone of *Y. pseudotuberculosis*, Cui et al. compared the genomes of 133 *Y. pestis* strains and four *Y. pseudotuberculosis* strains. The four *Y. pseudotuberculosis* genomes share the same single nucleotide polymorphisms (SNPs) with *Y. pestis*, and a highly robust maximum-likelihood tree confirmed that all 133 *Y. pestis* genomes exhibited limited diversity and were descended from *Y. pseudotuberculosis* [10]. With rapid advancements in sequencing technology, more *Yersinia* genomes were deciphered subsequently. A more comprehensive study analyzed 241 *Yersinia* strains that belonged to 18 different species. A phylogeny based on core genes revealed that *Y. pestis* strains exhibited very limited diversity with other *Yersinia* species and that the lineages of *Y. pestis* and *Y. pseudotuberculosis* formed a tight group, with the nonpathogenic *Y. similis* appearing ancestral to these species on the tree. These whole genome-based studies provided further evidence that *Y. pestis* is a genetically monomorphic clone of the more diverse species, *Y. pseudotuberculosis*.

Although robust evidence at the genetic level supports the view that *Y. pestis* originated from *Y. pseudotuberculosis*, the natural process of this transition is still little known. The enormous differences between their virulence, transmission route, and survival niche make it difficult to speculate how the niche jumping occurred. One possible evolutionary scenario is that *Y. pseudotuberculosis* acquired necessary genetic factors, such as the pMT1 and pPCP1 plasmids, from the resident gut microbiota during co-colonization of the rodent gastrointestinal tract or the flea midgut and occasionally causes fatal septicemia in animals stressed by cold, famine, or capture [38, 40]. As marmots are known to be the major host of *Y. pestis* in many natural plague foci, some former Soviet Union researchers speculate that the evolution from *Y. pseudotuberculosis* to *Y. pestis* might be associated with the living habits of this rodent species. Before winter, marmots need to seal their burrow entries using a mixture of soil, twigs, and, sometimes, their feces, to prepare for hibernation. The twigs or sharp stones in the soil might cause wounds on their lips, which might have enabled *Y. pseudotuberculosis* in the feces to enter into the blood, which was a suitable environment for its further evolution and eventually shaped the new species. Nonetheless, the scenario of the origin of *Y. pestis* still needs to be explored by archaeology and laboratory studies.

6.3.2 Genealogy of Y. pestis

Y. pestis is so monomorphic that no sequence variations have been observed in its housekeeping genes [38, 39]. Therefore, before the population genomics era, various molecular techniques that targeted multiple loci of whole genomes were employed in attempts to determine the genealogy and evolutionary history of *Y. pestis*. These techniques include restriction fragment length polymorphisms of the locations of ISs [38], gain and loss of genomic segments [41–43], the diversity of clustered regularly interspaced short palindromic repeats (CRISPRs) [44–46], and multiple locus variable number tandem repeat analyses (MLVAs) [47–49]. These efforts proved the linkage between the molecular genotypes that were determined by molecular methods and the traditional biovars, as well as the linkage between genotypes and the geographical distributions of the isolates. However, because these molecular markers are either unreliable, are homoplasies, lack adequate resolution, or are affected by phylogenetic discovery bias, they are not ideally suited for the evolutionary reconstruction of this genetically monomorphic organism [50]. Generally, strains that are closely related genetically could be accurately attributed to the same cluster/genotype by the above methods; however, the phylogenetic relationship across deep branches of an evolutionary tree is usually unstable, and different methods yield contradictory results.

Using 76 synonymous SNPs that were found in 3250 orthologous coding sequences from the first three *Y. pestis* genomes, Achtman et al. first described a robust three-branch population structure of *Y. pestis* [39]. These SNPs defined a tree in which an ancestral branch, branch 0, gave rise to branches 1 and 2, with the ancestral status of the synonymous SNPs defined according to their status in the genome of *Y. pseudotuberculosis* (IP32953). Since then, many new branches and subpopulations have been recognized with an increasing number of analyzed genomes; however, the three-branch structure was proven to be the backbone genealogy of *Y. pestis*, and it was only slightly modified by subsequent studies. According to this backbone, Achtman et al. designed a nomenclature system for *Y. pestis*, which was widely used by subsequent typing and evolution studies. This nomenclature system combined SNP-defined molecular relatedness and the traditional biovar designations. The populations were designated according to the branch plus an abbreviation for the biovar (abbreviations: ANT, Antiqua; PE, pestoides (including Microtus isolates); MED, Medievalis; ORI, Orientalis), followed by a number for sequential populations (see Chap. 3 for details).

To increase the discriminative power of the system, Achtman et al. combined different multi-locus molecular methods, including synonymous SNP, MLVA, and IS*100* typing, to investigate the microevolution of *Y. pestis*. Based on the three-branch structure, eight populations were recognized. The pestoides and biovar Microtus strains were distributed on branch 0, the nearest one to the root of the tree, and, hence, they are the ancestral populations of *Y. pestis*. Branch 1 includes most of the biovar Orientalis (1.ORI) strains and a group of Antiqua strains (1.ANT). Branch 2 includes all of the Medievalis strains (2.MED) and another group of

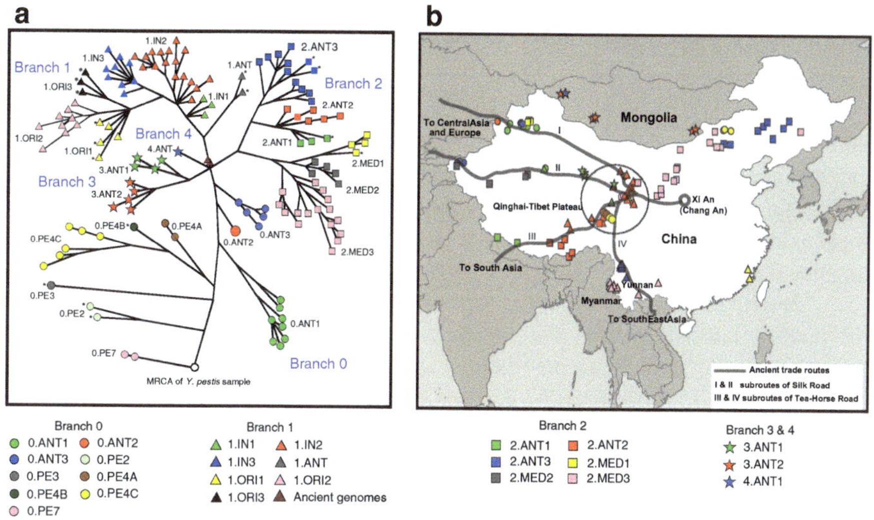

Fig. 6.2 Genealogy of *Y. pestis* constructed based on 2, 298 SNPs. (**a**) Whole genome-based MSTree of 133 strains of *Y. pestis*, using *Y. pseudotuberculosis* IP 32953 as the outgroup of most recent common ancestor (MRCA) of *Y. pestis*. Distinct symbol shapes represent different branches, while different colors indicate distinct populations within the branches. The publicly available genomes are labeled by an asterisk, and they are not included in the panel B. (**b**) Geographic origins of strains sequenced in the study. Gray lines illustrate the ancient trade routes. The circled area is the Qinghai–Tibet Plateau, where the most diverse isolates were isolated, indicating the possible origin of *Y. pestis*

Antiqua strains (2.ANT). This indicated that *Y. pestis* populations are only partially compatible with the classical biovars that are based on phenotypic characteristics, and, hence, biovars are not necessarily monophyletic and should not be used for evolutionary or taxonomic purposes. These inferences were supported by a phylogeny that used 1346 SNPs from 3349 coding sequences in 16 *Y. pestis* genomes [51].

With advancements in next-generation sequencing technology, population genomics was introduced into evolutionary studies of *Y. pestis*. Cui et al. sequenced 118 *Y. pestis* genomes from China and the vicinity and combined them with 15 previously published genomes (for a total of 133 genomes) and analyzed them for SNP calling, which identifies variable sites [10]. Almost all core SNPs within *Y. pestis* correspond to substitutions that have occurred only once and, thus, define a unique, fully parsimonious path of changes. The core genome of *Y. pestis* contains 2298 such SNPs, which were used to generate a rooted minimum spanning tree (MSTree) based on sequential changes since the most recent common ancestor (MRCA) (Fig. 6.2). The large number of new genomes revealed a novel, additional genealogical substructure, which includes branches 3 and 4 that split from branch 0. Branch 1–4 concurrently descended from the junction node N07 and formed an extensive "big bang" polytomy at the node (Fig. 6.2). The ancient genomes that deciphered from *Y. pestis* of Black Death victims [29] are also very close to this polytomy. They are located on branch 1, and they acquired one or three SNPs since diverging from node N07, indicating an extensive polytomy that gave rise to branches 1 through 4 shortly before the Black Death.

The genomes of modern isolates disclosed the phylogenetic relationships between existing populations, while the ancient DNA studies provided clues of the extinct lineages, some of which might have played important roles in human history. Using the ancient genomes obtained from the teeth of Bronze Age humans (approximately 3000 BC) across Europe and Asia, a new *Y. pestis* population (which has not been designated) that split earlier than any known lineage was identified [31]. This finding suggests that the virulent, flea-borne *Y. pestis* strain that caused the historic bubonic plague pandemics evolved from a less pathogenic *Y. pestis* lineage that infected human populations long before recorded evidence of plague outbreaks. The draft genomes of *Y. pestis* obtained from two individuals who died in the first pandemic formed a unique Justinian branch, which is interleaved between two extant groups, 0.ANT1 and 0.ANT2, and is distant from strains associated with the second and third pandemics [30]. From human remains of the last European plague outbreak, from 1722 in Marseille, France, five *Y. pestis* genomes were acquired, which represent a lineage that descended from strains from the fourteenth-century Black Death. These data suggest the existence of a previously uncharacterized historical plague focus that persisted for at least three centuries [52]. All of these ancient genomes have no known contemporary representatives; thus, their corresponding lineages likely went extinct.

6.3.3 *Genomic Decay and Neutral Evolution of* Y. pestis

Genome decay, means gene inactivation and gene loss, occurs when a gene or gene cluster is no longer useful for survival of a microbe or when a microbe attempts to adapt to a new ecological niche, which is a common theme during the evolution of bacterial pathogens [53]. Dumping unnecessary genetic elements will offer pathogens better fitness when facing new niches.

The first published *Y. pestis* genome, CO92, harbors a large number of ISs, comprising 3.7 % of the genome, far more than most other known bacteria. IS interruption and frameshift mutations have resulted in 149 pseudogenes in the CO92 genome [1]. The similar large number of pseudogenes was also observed in the following deciphered *Y. pestis* genomes. A brief comparison of the CO92 and *Y. pseudotuberculosis* IP32953 genomes identified 317 genes that were inactivated or absent from *Y. pestis*, which means that a large proportion of *Y. pseudotuberculosis* genes (13 %) are no longer functional in *Y. pestis* [6]. The accumulation of pseudogenes might promote the speciation and microevolution of *Y. pestis* [54].

For examples, *yadA* and *inv* encode the major adhesin and invasin, respectively, in *Y. pseudotuberculosis*, and they are essential for this enteropathogen to adhere to host intestinal surfaces and invade epithelial cells. Both of them are inactivated in *Y. pestis* [1]. Another important pseudogene in *Y. pestis* is *ureD*, which contains a premature stop codon that results in a malfunctioning urease operon [55]. Urease is necessary for the oral transmission of *Y. pseudotuberculosis*, while it is redundant for the flea-host-flea cycle of *Y. pestis*, and, thus, it might have been abandoned during host adaption [56].

The transcriptional regulator RcsA is known to inhibit *Yersinia* biofilm formation. *rcsA* is functional in *Y. pseudotuberculosis*, but it is a pseudogene in *Y. pestis*. Replacing the Y. pestis homologue with the functional homologue from *Y. pseudotuberculosis* abolishes biofilm formation in fleas [57]. The pseudogenization of *rcsA* is a clear case of positive selection during the evolution of *Y. pestis*.

It is well established that *Y. pestis* expresses rough LPS that lacks the O antigen, which is due to the inactivation of five genes in the O antigen gene cluster [1]. The loss of the O antigen and the expression of rough LPS are believed to facilitate Pla function and the invasiveness of *Y. pestis* [58].

For many Gram-negative pathogens, hexa-acylated LPS is able to efficiently activate Toll-like receptor 4 (TLR4) signaling to stimulate the host innate immune response. *Y. pestis* does not carry the *lpxL* gene, which encodes an enzyme that transfers a secondary laurate chain to lipid A at 37 °C and results in a tetra-acylated LPS that has low affinity for TLR4. Introducing *lpxL* into *Y. pestis* results in the formation of hexa-acylated lipid A at 37 °C, and the *lpxL* mutant is avirulent in mice [59]. Clearly, the loss of *lpxL* plays a vital role in *Y. pestis* immune evasion, and an *lpxL* knock-in mutant of the *Y. pestis* KIM strain has been proposed as a novel live vaccine against plague [60].

A comparison of the *Y. pestis* and *Y. pseudotuberculosis* genomes identified nine loci, including five genomic regions and four individual genes, which are present in *Y. pseudotuberculosis* and absent in *Y. pestis*. Deletion of R1, a region predicted to encode the methionine salvage pathway, altered the pathogenicity of *Y. pseudotuberculosis*. Interestingly, R1 is present and conserved in two *Y. pestis* strains on the ancient lineage, implying that R1 was lost as an early step of *Y. pestis* microevolution. Region R3 was also proven to be lost sequentially in the *Y. pestis* genome [61]. These gene loss events reflect the past genomic decay of *Y. pestis*, which is still evolving into an obligate intercellular pathogen with a minimized genome [62, 63].

Although it might be the results of natural selection that genome decay associated with survival or virulence of the *Y. pestis*, as described above, genomic variations that occurred during the evolutional history of *Y. pestis* seemed to reflect a neutral process, rather than Darwinian selection. *dN*/*dS*, the ratio of non-synonymous to synonymous mutation rates, is 0.91 for 2298 genome-wide SNPs and ranges from 0.79 to 1.37 for individual nodes in the phylogeny of *Y. pestis* [10]. The value of *dN*/*dS* ratios is near to 1, which is consistent with neutral evolution and provides no obvious signs for diversifying selection at the genome level. Another possible signal of diversifying selection is frequent homoplasies, as has been observed for mutations that resulted in antibiotic resistance [64] and resistance to antivirals [65]. However, only 28 SNPs were homoplastic in 133 *Y. pestis* genomes. A further, typical hallmark of diversifying selection is the existence of genes with multiple, clustered non-synonymous SNPs (nsSNPs). The distribution of the number of nsSNPs per gene did deviate markedly from theoretical expectations under neutral evolution, but this deviation rose largely because of only seven outlier genes (including two with homoplasies) that exhibited an excess of nsSNPs. Based on the combined criteria of homoplasies and/or clustered nsSNPs, only 22 of the 3450 genes in the core genome seem to be under diversifying selection (Table 6.2). Those genes did not possess obvious unifying features, except that one (YPO0348, *aspA*) is thought to be important for the virulence of *Y. pestis* [66] and two are related to motility. Therefore, these lines of evidence support the view that genetic drift, rather than Darwinian selection, played a major role in the evolutionary history of *Y. pestis*.

6.3.4 Mutation Rate and a Variable Molecular Clock

Mutation rate is important in evolutionary studies, as it is a necessary parameter for estimating the age of population-dividing events. It could be measured through experimentally evolved replicate lines from laboratory cultures or calculated according to the genetic distances and differences in isolation time among natural isolates.

Based on the frequencies of SNPs that were discovered in 81 *Y. pestis* isolates from Madagascar by denaturing high-performance liquid chromatography (dHPLC), Morelli et al. calculated the mutation rates of SNPs during the natural evolution of *Y. pestis* [51]. Because most of the SNPs discovered by dHPLC were singleton mutations, it is reasonable to assume that after the introduction of *Y. pestis* to Madagascar, demographic expansion was strong enough to result in star-shaped genealogies, i.e., those without any coalescent events. Based on this assumption, Morelli et al. built a full maximum-likelihood model to infer the yearly mutation rates. Historical records show that an initial plague epidemic reached Madagascar in 1898 and that a second wave began in 1921 [67]. Therefore, calculations were performed assuming a demographic expansion that started either in 1898 or 1921. Because the 81 Madagascar isolates fell into two distinct groups in a minimum spanning tree, the same calculations were performed for each of the two subgroups. All of the calculation results were comparable, with extreme value estimates ranging from 2.09E−09 to 2.30E−08 per site per year [51].

In their evolutionary studies of Madagascar strains, Morelli et al. found that an unusually large number of SNPs accumulated in certain branches of the *Y. pestis* phylogenetic tree, which was difficult to explain, even for the highest inferred mutation rate [51]. For example, there are seven SNPs that differentiate Madagascar nodes in the older cluster from their direct ancestors in Israel and India, which became fixed by 1926, the date of isolation of the EV76 strain that belongs to this cluster. The same problem applies to the seven other SNPs that distinguish Madagascar nodes from their descendants in Turkey, which had probably been established by the 1930s. The probability of such a rapid accumulation of seven SNPs is <0.05 under the assumption of an invariant molecular clock. Overdispersion of branch lengths was also observed in the phylogeny of *Y. pestis* based on whole genome-wide SNPs [10]. The number of SNPs that have accumulated since the

Table 6.2 Features of highly variable genes

Gene ID	Type[a]	Product	Length (bp)	# of SNPs	# of nsSNPs	# of sSNPs
YPO0348[a]	Multiple SNPs and homoplasies	Aspartate ammonia-lyase	1437	8	8	0
YPO0821	Multiple SNPs	Hypothetical protein	195	3	2	1
YPO2210[a]	Multiple SNPs and homoplasies	Hypothetical protein	129	4	2	2
YPO2387	Multiple SNPs	DNA-binding transcriptional repressor PurR	1026	5	5	0
YPO2639	Multiple SNPs	Integrase (partial)	186	2	2	0
YPO3041	Multiple SNPs	Nitrate/nitrite response regulator protein NarP	630	4	4	0
YPO3561	Multiple SNPs	Stringent starvation protein A	642	4	4	0
YPO0698	Homoplasies	Outer membrane usher protein	2481	6	5	1
YPO1376	Homoplasies	Putative cell division protein	3918	6	6	0
YPO0176	Homoplasies	Hypothetical protein	408	1	1	0
YPO0222a	Homoplasies	30S ribosomal protein S14	306	1	0	1
YPO0319	Homoplasies	Quinone oxidoreductase, NADPH dependent	984	3	1	2
YPO0412	Homoplasies	Putative ABC transporter ATP-binding protein	1584	3	1	2
YPO0440	Homoplasies	Purine nucleoside phosphorylase	720	1	1	0
YPO0725	Homoplasies	Flagellar hook protein FlgE	1242	1	1	0
YPO1571	Homoplasies	Hypothetical protein	1431	4	4	0
YPO1581	Homoplasies	Alpha-galactosidase	2127	2	1	1
YPO2299	Homoplasies	Putative methylated-DNA–protein-cysteine methyltransferase	519	2	1	1
YPO2904	Homoplasies	Putative 3-phenylpropionic acid transporter	1158	1	1	0
YPO3013	Homoplasies	Sulfate/thiosulfate transporter permease subunit	876	1	1	0
YPO3223	Homoplasies	DNA-binding transcriptional regulator Crl	402	2	2	0
YPO4107	Homoplasies	Xanthine/uracil permease family protein	1329	2	1	1

[a]Genes contain both multiple SNPs and homoplasies

MRCA varied from 96 (0.PE7) to 548 (0.PE3). Under a constant molecular clock, the root-to-tip distances were predicted to follow the Poisson distribution, that is, the values for the mean and the variance of the number of SNPs were similar if the molecular clock was invariable. However, through bootstrapping the root-to-tip distances to calculate 99 % CIs, the variance (1575; 99 % CI: 456–3970) was much higher than the mean (248; 99 % CI: 239–257) value of the number of SNPs accumulated from root-to-tip in the *Y. pestis* phylogeny.

Overdispersion of branch lengths was also observed with the H1N1 influenza virus [68] and methicillin-resistant *Staphylococcus aureus* [14, 15], of which accumulated mutations revealed liner correlation with the time of sampling. However, both of these organisms evolved more rapid than *Y. pestis* and as such, a substantial proportion of their diversity arose during the time covered by the sampling period. In contrast, only a small fraction of the diversity in *Y. pestis* isolates arose during the period of sampling time, and the branch lengths to the MRCA for individual strains of *Y. pestis* were uncorrelated with their year of isolation ($R^2=0.01$).

An explanation for the overdispersion of branch lengths is that the substitution rate varies across lineages. According to the genealogy based on a lognormal distribution of the clock rate that is inferred by the Bayesian relaxed clock model in BEAST software [69–72], the molecular clock between the slowest and fastest evolving branches revealed a nearly 40-fold difference. Using a strict clock model that assumed a constant rate across the tree, a $\log_{10}$ marginal likelihood of 109.6 was yielded with the Bayes factor approach implemented in Tracer, providing overwhelming support for a relaxed clock [70].

The drivers behind the strong clock rate variation could possibly be numerous events of diversifying selection, which led to the rapid fixation of beneficial mutations. However, we have shown in the above sections that there are no obvious signs of diversifying selection at the genome level of *Y. pestis* and, therefore, that natural selection cannot explain the extensive variation in the substitution rate. Another possibility is that many hypermutators, strains with increased mutation rates [73], are present in the analyzed datasets. However, a 0.PE3 Angola strain, the strain whose genome had the most SNPs, is not a hypermutator [18], nor did any of the genomes contain traces of epistatic mutations that can result in reversion of a transient hypermutator to wild type. Furthermore, only one hypermutator was identified among 294 *Y. pestis* strains that were isolated from different regions [51]. Thus, transient hypermutators are unlikely to have lead to the numerous rate variations observed throughout the *Y. pestis* phylogeny.

A third alternative is that the clock rate variation is due to different bacterial replication rates between epidemic spread and latent endemic disease. Sylvatic plagues consist of alternate cycles of enzootic disease and epizootic disease [74]. During the enzootic phase, *Y. pestis* is rare in fleas and hosts and can survive for long periods in a dormant state [75]. In contrast, during a typical epizootic rodent-flea cycle, *Y. pestis* is common and the infected rodents suffer from an intense bacteremia with $>10^7$ CFU/ml of blood [76, 77]. The mutational clock rate at the microsatellite loci is constant in the epizootic phase and laboratory serial-transfer experiments [78], but data are lacking on clock rates in enzootic plague. The same is true of human epidemics, except that SNPs can be fixed at exceptionally high rates during transmission to new geographic areas [51]. However, based on their epidemiological differences, lineages in epidemics and epizootics tend to experience more transmission events per unit time than lineages during long periods of intervening enzootic disease. While the mutation rate is constant per transmission event, the increased number of transmissions per unit time leads to an increase in the mutation rate during outbreaks and to an increased substitution rate over recurrent outbreaks, even in the absence of natural selection. This effect can be considered as analogous to the "generation-time effect" thought to occur in higher taxa, where species with more generations per unit time accumulate a greater number of mutations than species with a lower number of generations [79]. However, for *Y. pestis*, this hypothesis indicates that lineages related with more epidemic events should have higher clock rates, and conversely that higher clock rates are an indicator for an association with multiple epidemic disease, even in the absence of historical evidence. Nonetheless, it should be noted that while a demographic explanation explains the observed patterns well, we have not excluded alternative scenarios, such as host density changes driven by climatic factors [80, 81], which will also change the number of bacterial replication cycles.

6.3.5 Historical Spread

One of the most attractive aspects of *Y. pestis* for the scientific community is the causative agent of three worldwide pandemics, which claimed millions of deaths and dramatically influenced the development of human civilization [74]. Using population genetic methods, the age of population divergent events could be calculated, and in combination with historical documents, one could sketch the spread of *Y. pestis* during the prior millennia.

Recently, genomic studies estimated that the age of the MRCA of all known *Y. pestis* strains was 5783 years ago (95 % highest posterior density interval: 5021–7022 years ago). Concerning the fact of acquiring ancient *Y. pestis* genomes from the teeth of eight Bronze Age individuals from Russia, Poland, Armenia, and Estonia, it could be speculated that *Y. pestis* caused plague infections in the human populations of Eurasia for at least 3000 years before any historical recordings of pandemics [31]. The Bronze Age in Europe and Asia was characterized by large-scale human population movements and admixtures, which were accompanied by profound social and economic changes [82, 83] that might have facilitated the spread of *Y. pestis*. This finding also suggests that plague outbreaks began to influence human culture 5000 years ago.

However, a comparison with modern *Y. pestis* genomes indicated that the prehistory *Y. pestis* might not be capable of causing bubonic plague and lacked the ability to be transmitted via fleas because it lacked necessary genes and mutations [31].

Prior to the 1st millennium BC, *Y. pestis* evolved to acquire the ability to be transmitted by fleas and to cause bubonic plague, which led to three global pandemics and other historically recorded plagues.

Justinian's Plague (sixth to eighth centuries), which began in 541 AD, quickly spread across the Mediterranean basin into Europe, and it killed an estimated 100 million people and contributed to the demise of the Roman Empire. By analyzing ancient DNA in two independent laboratories, unambiguous evidence was obtained that confirmed the presence of *Y. pestis* DNA in human skeletal remains from multiple sixth-century skeletons from Germany, which ended the controversy over whether Justinian's Plague was caused by *Y. pestis* [84]. In a subsequent phylogenetic analysis, ancient genomes obtained from two victims of Justinian's Plague were shown to belong to a unique branch that interleaved between two extant groups, 0.ANT1 and 0.ANT2. Strains of the two 0.ANT groups were only identified in long-term natural plague foci of Xinjiang, China. Therefore, the results supported the theory that the first pandemic emerged from rodents in long-term plague foci in China, rather than Africa, as was originally proposed by Procopius [85] and that *Y. pestis* spread west to Europe along established trade routes, such as the Silk Road. Subsequent outbreaks in Europe lasted for two centuries after the initial introduction, and then the lineage of Justinian's Plague seemed to disappear, and no extant descendants have been identified thus far. There are several hypotheses regarding the evolutionary dead-end of the Justinian's Plague lineage, including a scarcity of susceptible hosts, insufficient numbers of susceptible hosts in a background of widespread population immunity, or a changing climate disturbed the vector-driven transmission. However, the real explanation for its extinction remains a mystery.

Approximately 500 years after the end of Justinian's Plague, another even more disastrous pandemic, the Black Death or Great Pestilence, killed 30–50% of Europe's population between 1347 and 1351 AD [23, 74]. The pandemic first appeared in China and then spread westward along the trade routes in Tauris on the Black Sea and eventually to Constantinople. From India, it reached the Crimea in 1347, and it was subsequently imported into Venice, Genoa, and Sicily. Many smaller outbreaks continued, occurring until the eighteenth century, which led to great changes in economic, societal, political, religious, and medical systems. The Black Death also greatly influenced the evolution of *Y. pestis*, and it shaped the foundation of the current *Y. pestis* population structure, which is characterized by the polytomy (termed N07) at the end of branch 0 [10, 29]. In the whole-genome phylogeny, the ancient genomes from Black Death victims (1348–1350, London) are located on branch 1, and they acquired one or three SNPs since diverging from node N07, indicating the close relationship between the Black Death and the big bang polytomy of *Y. pestis*.

It has generally been assumed that during the Black Death, *Y. pestis* was introduced from Asia to Europe in one or more waves, and then the founder strain established natural foci in Europe and caused subsequent outbreaks for several hundreds of years. However, Schmid et al. found a strong association between climate fluctuations and the occurrence of plague outbreaks during the Black Death, with a delay of 15 ± 1 years [86]. As climate change could destroy plague-infected rodent populations in Asia, their resident fleas would have to find a new host, e.g., humans or camel caravans that passed through the Asian plague foci, which finally led to the spread of *Y. pestis* into Europe. Therefore, Schmid et al. suggested that *Y. pestis* was continuously reimported into Europe during the second plague pandemic and not from permanently established plague foci. Subsequently, two independent ancient genome studies provided evidence that challenged Schmid et al.'s opinion. One presented five *Y. pestis* genomes from one of the last European outbreaks of plague, from 1722 AD in Marseille, France [52]. If these strains were reimported from Asia, they should have carried the ancestor population-derived mutations in branch 1, similar to other Asia populations, and according to the isolation time, the most probable ancestral population should be the 1.IN group. However, all five ancient genomes formed a lineage that descended from strains from the fourteenth-century Black Death victims, which was not identified in any extant population, proving that such a lineage was present in Europe for at least 300 years, rather than being reimported from Asia. Similarly, another ancient *Y. pestis* genotype that belonged to the fourteenth-century Black Death lineage was found in skeletons from seventeenth-century Germany [87]. Therefore, it is probable that both repeated reimportation and the long-term persistence of plague foci in Europe contributed to a second pandemic that lasted for several centuries. Similar to the Justinian's Plague lineage, the reason for the extinction of the fourteenth-century Black Death lineage is also unknown. Nonetheless, environmental changes and advancements in public health, along with the Industrial Revolution, might have played an important role in the decline of plague in Europe.

The third pandemic probably started in 1855 in Yunnan province, China, and then spread to the southern coast of China, reaching Hong Kong and Canton in 1894. Plague was spread via steamships to Africa, Australia, Europe, Hawaii, India, Japan, the Middle East, the Philippines, North America (the USA), and South America within a few years [51]. The third pandemic was caused by biovar Orientalis strains, which belong to the 1.ORI group that evolved approximately 200 years ago. There are three subgroups of 1.ORI, named 1.ORI1, 1.ORI2, and 1.ORI3, and they represent three major waves of *Y. pestis* transmission.

The 1.ORI1 subpopulation might have evolved in China and then spread to and established natural plague foci in the USA. A steamship that carried plague-infected animals/fleas that sailed from Hong Kong to San Francisco (via Hawaii) in 1899 might be responsible for the earliest import of *Y. pestis* to the USA. Plague was also imported on later occasions to Oregon, New York, and cities on the Gulf Coast (New Orleans, Pensacola, and Galveston) [88], but only the foci introduced into California and/or Oregon are thought to have persisted [89], and, thus, all extant *Y. pestis* strains in the USA seem to be derived from a single import.

The 1.ORI2 subgroup includes multiple radiated expansions, which began from China and reached Europe, South America, Africa, and Southeast Asia, and one of the branches might have returned to China after circulating in Southeast Asia. The ancestor of subgroup 1.ORI3 was transmitted to Madagascar in 1898, probably via a plague-containing ship from India [67], and local plague foci appeared after a dozen or more years, accompanied by the formation of two major clusters of 1.

ORI3. In the 1930s, the 1.ORI3 strains were transmitted to Turkey and other countries in the Middle East, but did not establish long-term foci; therefore, the 1.ORI3 group only survived and evolved in Madagascar for approximately 100 years.

The descendants of all three 1.ORI subpopulations still thrive in natural plague foci across the world, and they continue to cause sporadic cases and local outbreaks in Africa, Asia, and North and South America. Therefore, strictly, the third pandemic has not yet passed, and we are still living with it.

References

1. Parkhill J, Wren BW, Thomson NR, Titball RW, Holden MT, Prentice MB, Sebaihia M, James KD, Churcher C, Mungall KL, et al. Genome sequence of *Yersinia pestis*, the causative agent of plague. Nature. 2001;413(6855):523–7.
2. Doll JM, Zeitz PS, Ettestad P, Bucholtz AL, Davis T, Gage K. Cat-transmitted fatal pneumonic plague in a person who traveled from Colorado to Arizona. Am J Trop Med Hyg. 1994;51(1):109–14.
3. Revazishvili T, Johnson JA. Pulsed-field gel electrophoresis of *Yersinia pestis*. Methods Mol Biol. 2015;1301:115–28.
4. Deng W, Burland V, Plunkett 3rd G, Boutin A, Mayhew GF, Liss P, Perna NT, Rose DJ, Mau B, Zhou S, et al. Genome sequence of *Yersinia pestis* KIM. J Bacteriol. 2002;184(16):4601–11.
5. Song Y, Tong Z, Wang J, Wang L, Guo Z, Han Y, Zhang J, Pei D, Zhou D, Qin H, et al. Complete genome sequence of *Yersinia pestis* strain 91001, an isolate avirulent to humans. DNA Res. 2004;11(3):179–97.
6. Chain PS, Carniel E, Larimer FW, Lamerdin J, Stoutland PO, Regala WM, Georgescu AM, Vergez LM, Land ML, Motin VL, et al. Insights into the evolution of *Yersinia pestis* through whole-genome comparison with *Yersinia pseudotuberculosis*. Proc Natl Acad Sci U S A. 2004;101(38):13826–31.
7. Tong Z, Zhou D, Song Y, Zhang L, Pei D, Han Y, Pang X, Li M, Cui B, Wang J, et al. Pseudogene accumulation might promote the adaptive microevolution of *Yersinia pestis*. J Med Microbiol. 2005;54(Pt 3):259–68.
8. McNally A, Thomson NR, Reuter S, Wren BW. 'Add, stir and reduce': *Yersinia* spp. as model bacteria for pathogen evolution. Nat Rev Microbiol. 2016;14(3):177–90.
9. Medini D, Donati C, Tettelin H, Masignani V, Rappuoli R. The microbial pan-genome. Curr Opin Genet Dev. 2005;15(6):589–94.
10. Cui Y, Yu C, Yan Y, Li D, Li Y, Jombart T, Weinert LA, Wang Z, Guo Z, Xu L, et al. Historical variations in mutation rate in an epidemic pathogen, Yersinia pestis. Proc Natl Acad Sci U S A. 2013;110(2):577–82.
11. Touchon M, Hoede C, Tenaillon O, Barbe V, Baeriswyl S, Bidet P, Bingen E, Bonacorsi S, Bouchier C, Bouvet O, et al. Organised genome dynamics in the *Escherichia coli* species results in highly diverse adaptive paths. PLoS Genet. 2009;5(1):e1000344.
12. Tettelin H, Masignani V, Cieslewicz MJ, Donati C, Medini D, Ward NL, Angiuoli SV, Crabtree J, Jones AL, Durkin AS, et al. Genome analysis of multiple pathogenic isolates of *Streptococcus agalactiae*: implications for the microbial "pan-genome". Proc Natl Acad Sci U S A. 2005;102(39):13950–5.
13. Holt KE, Parkhill J, Mazzoni CJ, Roumagnac P, Weill FX, Goodhead I, Rance R, Baker S, Maskell DJ, Wain J, et al. High-throughput sequencing provides insights into genome variation and evolution in *Salmonella* Typhi. Nat Genet. 2008;40(8):987–93.

14. Harris SR, Feil EJ, Holden MT, Quail MA, Nickerson EK, Chantratita N, Gardete S, Tavares A, Day N, Lindsay JA, et al. Evolution of MRSA during hospital transmission and intercontinental spread. Science. 2010;327(5964):469–74.
15. Nubel U, Dordel J, Kurt K, Strommenger B, Westh H, Shukla SK, Zemlickova H, Leblois R, Wirth T, Jombart T, et al. A timescale for evolution, population expansion, and spatial spread of an emerging clone of methicillin-resistant *Staphylococcus aureus*. PLoS Pathog. 2010;6(4):e1000855.
16. Moran NA, McLaughlin HJ, Sorek R. The dynamics and time scale of ongoing genomic erosion in symbiotic bacteria. Science. 2009;323(5912):379–82.
17. Dong XQ, Lindler LE, Chu MC. Complete DNA sequence and analysis of an emerging cryptic plasmid isolated from *Yersinia pestis*. Plasmid. 2000;43(2):144–8.
18. Eppinger M, Worsham PL, Nikolich MP, Riley DR, Sebastian Y, Mou S, Achtman M, Lindler LE, Ravel J. Genome sequence of the deep-rooted *Yersinia pestis* strain Angola reveals new insights into the evolution and pangenome of the plague bacterium. J Bacteriol. 2010;192(6):1685–99.
19. Eppinger M, Radnedge L, Andersen G, Vietri N, Severson G, Mou S, Ravel J, Worsham PL. Novel plasmids and resistance phenotypes in *Yersinia pestis*: unique plasmid inventory of strain Java 9 mediates high levels of arsenic resistance. PLoS One. 2012;7(3):e32911.
20. Welch TJ, Fricke WF, McDermott PF, White DG, Rosso ML, Rasko DA, Mammel MK, Eppinger M, Rosovitz MJ, Wagner D, et al. Multiple antimicrobial resistance in plague: an emerging public health risk. PLoS One. 2007;2(3):e309.
21. Derbise A, Chenal-Francisque V, Pouillot F, Fayolle C, Prevost MC, Medigue C, Hinnebusch BJ, Carniel E. A horizontally acquired filamentous phage contributes to the pathogenicity of the plague bacillus. Mol Microbiol. 2007;63(4):1145–57.
22. Derbise A, Carniel E. YpfPhi: a filamentous phage acquired by Yersinia pestis. Front Microbiol. 2014;5:701.
23. Perry RD, Fetherston JD. *Yersinia pestis*–etiologic agent of plague. Clin Microbiol Rev. 1997;10(1):35–66.
24. Altschuler EL, Kariuki YM. Was the Justinian Plague caused by the 1918 flu virus? Med Hypotheses. 2009;72(2):234.
25. Zietz BP, Dunkelberg H. The history of the plague and the research on the causative agent Yersinia pestis. Int J Hyg Environ Health. 2004;207(2):165–78.
26. Drancourt M, Aboudharam G, Signoli M, Dutour O, Raoult D. Detection of 400-year-old *Yersinia pestis* DNA in human dental pulp: an approach to the diagnosis of ancient septicemia. Proc Natl Acad Sci U S A. 1998;95(21):12637–40.
27. Drancourt M, Roux V, Dang LV, Tran-Hung L, Castex D, Chenal-Francisque V, Ogata H, Fournier PE, Crubezy E, Raoult D. Genotyping, Orientalis-like *Yersinia pestis*, and plague pandemics. Emerg Infect Dis. 2004;10(9):1585–92.
28. Tran TN, Signoli M, Fozzati L, Aboudharam G, Raoult D, Drancourt M. High throughput, multiplexed pathogen detection authenticates plague waves in medieval Venice, Italy. PLoS One. 2011;6(3):e16735.
29. Bos KI, Schuenemann VJ, Golding GB, Burbano HA, Waglechner N, Coombes BK, McPhee JB, DeWitte SN, Meyer M, Schmedes S, et al. A draft genome of *Yersinia pestis* from victims of the Black Death. Nature. 2011;478(7370):506–10.
30. Wagner DM, Klunk J, Harbeck M, Devault A, Waglechner N, Sahl JW, Enk J, Birdsell DN, Kuch M, Lumibao C, et al. *Yersinia pestis* and the plague of Justinian 541–543 AD: a genomic analysis. Lancet Infect Dis. 2014;14(4):319–26.
31. Rasmussen S, Allentoft ME, Nielsen K, Orlando L, Sikora M, Sjogren KG, Pedersen AG, Schubert M, Van Dam A, Kapel CM, et al. Early divergent strains of *Yersinia pestis* in Eurasia 5,000 years ago. Cell. 2015;163(3):571–82.
32. Yersin A. La peste bubonique à Hong-Kong. Ann Inst Pasteur. 1894;2:428–30.
33. Van Loghem JJ. The classification of the plague-bacillus. Antonie Van Leeuwenhoek. 1944;10(1–2):15.

34. Moore RL, Brubaker RR. Hybridization of deoxyribonucleotide sequences of Yersinia enterocolitica and other selected members of Enterobacteriaceae. Int J Syst Bacteriol. 1975;25(4):336–9.
35. Bercovier H, Mollaret HH, Alonso JM, Braul J, Richard Fanning G, Steiger-walt AG, Brenner DJ. Intra- and interspecies relatedness of Yersinia pestis by DNA hybridization and its relationship to Yersinia pseudotuberculosis. Curr Microbiol. 1980;4:225–9.
36. Trebesius K, Harmsen D, Rakin A, Schmelz J, Heesemann J. Development of rRNA-targeted PCR and in situ hybridization with fluorescently labelled oligonucleotides for detection of *Yersinia* species. J Clin Microbiol. 1998;36(9):2557–64.
37. Wayne LG. International Committee on Systematic Bacteriology: announcement of the report of the ad hoc Committee on Reconciliation of Approaches to Bacterial Systematics. Zentralbl Bakteriol Mikrobiol Hyg A. 1988;268(4):433–4.
38. Achtman M, Zurth K, Morelli G, Torrea G, Guiyoule A, Carniel E. *Yersinia pestis*, the cause of plague, is a recently emerged clone of *Yersinia pseudotuberculosis*. Proc Natl Acad Sci U S A. 1999;96(24):14043–8.
39. Achtman M, Morelli G, Zhu P, Wirth T, Diehl I, Kusecek B, Vogler AJ, Wagner DM, Allender CJ, Easterday WR, et al. Microevolution and history of the plague bacillus, *Yersinia pestis*. Proc Natl Acad Sci U S A. 2004;101(51):17837–42.
40. Reuter S, Connor TR, Barquist L, Walker D, Feltwell T, Harris SR, Fookes M, Hall ME, Petty NK, Fuchs TM, et al. Parallel independent evolution of pathogenicity within the genus Yersinia. Proc Natl Acad Sci U S A. 2014;111(18):6768–73.
41. Zhou D, Han Y, Song Y, Huang P, Yang R. Comparative and evolutionary genomics of *Yersinia pestis*. Microbes Infect. 2004;6(13):1226–34.
42. Zhou D, Han Y, Song Y, Tong Z, Wang J, Guo Z, Pei D, Pang X, Zhai J, Li M, et al. DNA microarray analysis of genome dynamics in *Yersinia pestis*: insights into bacterial genome microevolution and niche adaptation. J Bacteriol. 2004;186(15):5138–46.
43. Li Y, Dai E, Cui Y, Li M, Zhang Y, Wu M, Zhou D, Guo Z, Dai X, Cui B, et al. Different region analysis for genotyping *Yersinia pestis* isolates from China. PLoS One. 2008;3(5):e2166.
44. Pourcel C, Salvignol G, Vergnaud G. CRISPR elements in *Yersinia pestis* acquire new repeats by preferential uptake of bacteriophage DNA, and provide additional tools for evolutionary studies. Microbiology. 2005;151(3):653–63.
45. Vergnaud G, Li Y, Gorge O, Cui Y, Song Y, Zhou D, Grissa I, Dentovskaya SV, Platonov ME, Rakin A, et al. Analysis of the three Yersinia pestis CRISPR loci provides new tools for phylogenetic studies and possibly for the investigation of ancient DNA. Adv Exp Med Biol. 2007;603:327–38.
46. Cui Y, Li Y, Gorge O, Platonov ME, Yan Y, Guo Z, Pourcel C, Dentovskaya SV, Balakhonov SV, Wang X, et al. Insight into microevolution of *Yersinia pestis* by clustered regularly interspaced short palindromic repeats. PLoS One. 2008;3(7):e2652.
47. Li Y, Cui Y, Hauck Y, Platonov ME, Dai E, Song Y, Guo Z, Pourcel C, Dentovskaya SV, Anisimov AP, et al. Genotyping and phylogenetic analysis of *Yersinia pestis* by MLVA: insights into the worldwide expansion of Central Asia plague foci. PLoS One. 2009;4(6):e6000.
48. Denoeud F, Vergnaud G. Identification of polymorphic tandem repeats by direct comparison of genome sequence from different bacterial strains: a web-based resource. BMC Bioinformatics. 2004;5(1):4.
49. Pourcel C, Andre-Mazeaud F, Neubauer H, Ramisse F, Vergnaud G. Tandem repeats analysis for the high resolution phylogenetic analysis of *Yersinia pestis*. BMC Microbiol. 2004;4(1):22.
50. Achtman M. Evolution, population structure, and phylogeography of genetically monomorphic bacterial pathogens. Annu Rev Microbiol. 2008;62:53–70.
51. Morelli G, Song Y, Mazzoni CJ, Eppinger M, Roumagnac P, Wagner DM, Feldkamp M, Kusecek B, Vogler AJ, Li Y, et al. Yersinia pestis genome sequencing identifies patterns of global phylogenetic diversity. Nat Genet. 2010;42(12):1140–3.
52. Bos KI, Herbig A, Sahl J, Waglechner N, Fourment M, Forrest SA, Klunk J, Schuenemann V, Poinar D, Kuch M, et al. Eighteenth century Yersinia pestis genomes reveal the long-term persistence of an historical plague focus. bioRxiv; 2016.

53. Cerdeno-Tarraga A, Thomson N, Parkhill J. Pathogens in decay. Nat Rev Microbiol. 2004;2(10):774–5.
54. Tong Z, Zhou D, Song Y, Zhang L, Pei D, Han Y, Pang X, Li M, Cui B, Wang J, et al. Pseudogene accumulation might promote the adaptive microevolution of *Yersinia pestis*. J Med Microbiol. 2005;54(3):259–68.
55. Sebbane F, Devalckenaere A, Foulon J, Carniel E, Simonet M. Silencing and reactivation of urease in *Yersinia pestis* is determined by one G residue at a specific position in the ureD gene. Infect Immun. 2001;69(1):170–6.
56. Chen PE, Cook C, Stewart AC, Nagarajan N, Sommer DD, Pop M, Thomason B, Thomason MP, Lentz S, Nolan N, et al. Genomic characterization of the *Yersinia* genus. Genome Biol. 2010;11(1):R1.
57. Sun YC, Hinnebusch BJ, Darby C. Experimental evidence for negative selection in the evolution of a *Yersinia pestis* pseudogene. Proc Natl Acad Sci U S A. 2008;105(23):8097–101.
58. Kukkonen M, Suomalainen M, Kyllonen P, Lahteenmaki K, Lang H, Virkola R, Helander IM, Holst O, Korhonen TK. Lack of O-antigen is essential for plasminogen activation by *Yersinia pestis* and *Salmonella enterica*. Mol Microbiol. 2004;51(1):215–25.
59. Montminy SW, Khan N, McGrath S, Walkowicz MJ, Sharp F, Conlon JE, Fukase K, Kusumoto S, Sweet C, Miyake K, et al. Virulence factors of *Yersinia pestis* are overcome by a strong lipopolysaccharide response. Nat Immunol. 2006;7(10):1066–73.
60. Sun W, Six D, Kuang X, Roland KL, Raetz CR, Curtiss 3rd R. A live attenuated strain of Yersinia pestis KIM as a vaccine against plague. Vaccine. 2011;29(16):2986–98.
61. Pouillot F, Fayolle C, Carniel E. Characterization of chromosomal regions conserved in *Yersinia pseudotuberculosis* and lost by *Yersinia pestis*. Infect Immun. 2008;76(10):4592–9.
62. Moran NA. Microbial minimalism: genome reduction in bacterial pathogens. Cell. 2002;108(5):583–6.
63. Ochman H. Genomes on the shrink. Proc Natl Acad Sci U S A. 2005;102(34):11959–60.
64. Roumagnac P, Weill FX, Dolecek C, Baker S, Brisse S, Chinh NT, Le TA, Acosta CJ, Farrar J, Dougan G, et al. Evolutionary history of Salmonella typhi. Science. 2006;314(5803):1301–4.
65. Kryazhimskiy S, Dushoff J, Bazykin GA, Plotkin JB. Prevalence of epistasis in the evolution of influenza A surface proteins. PLoS Genet. 2011;7(2):e1001301.
66. Bearden SW, Sexton C, Pare J, Fowler JM, Arvidson CG, Yerman L, Viola RE, Brubaker RR. Attenuated enzootic (pestoides) isolates of *Yersinia pestis* express active aspartase. Microbiology. 2009;155(1):198–209.
67. Brygoo ER. Epidemiologie de la peste à Madagascar. Archive de l Institut Pasteurde Madagascar. 1966;35:7–147.
68. Smith GJ, Vijaykrishna D, Bahl J, Lycett SJ, Worobey M, Pybus OG, Ma SK, Cheung CL, Raghwani J, Bhatt S, et al. Origins and evolutionary genomics of the 2009 swine-origin H1N1 influenza A epidemic. Nature. 2009;459(7250):1122–5.
69. Drummond AJ, Ho SY, Phillips MJ, Rambaut A. Relaxed phylogenetics and dating with confidence. PLoS Biol. 2006;4(5):e88.
70. Drummond AJ, Rambaut A. BEAST: Bayesian evolutionary analysis by sampling trees. BMC Evol Biol. 2007;7:214.
71. Drummond AJ, Suchard MA. Bayesian random local clocks, or one rate to rule them all. BMC Biol. 2010;8:114.
72. Drummond AJ, Suchard MA, Xie D, Rambaut A. Bayesian phylogenetics with BEAUti and the BEAST 1.7. Mol Biol Evol. 2012;29(8):1969–73.
73. Barrick JE, Yu DS, Yoon SH, Jeong H, Oh TK, Schneider D, Lenski RE, Kim JF. Genome evolution and adaptation in a long-term experiment with *Escherichia coli*. Nature. 2009;461(7268):1243–7.
74. Gage KL, Kosoy MY. Natural history of plague: perspectives from more than a century of research. Annu Rev Entomol. 2005;50:505–28.
75. Drancourt M, Houhamdi L, Raoult D. Yersinia pestis as a telluric, human ectoparasite-borne organism. Lancet Infect Dis. 2006;6(4):234–41.

76. Hinnebusch BJ. The evolution of flea-borne transmission in *Yersinia pestis*. Curr Issues Mol Biol. 2005;7(2):197–212.
77. Lorange EA, Race BL, Sebbane F, Hinnebusch BJ. Poor vector competence of fleas and the evolution of hypervirulence in *Yersinia pestis*. J Infect Dis. 2005;191(11):1907–12.
78. Girard JM, Wagner DM, Vogler AJ, Keys C, Allender CJ, Drickamer LC, Keim P. Differential plague-transmission dynamics determine *Yersinia pestis* population genetic structure on local, regional, and global scales. Proc Natl Acad Sci U S A. 2004;101(22):8408–13.
79. Laird CD, McConaughy BL, McCarthy BJ. Rate of fixation of nucleotide substitutions in evolution. Nature. 1969;224(5215):149–54.
80. Stenseth NC, Samia NI, Viljugrein H, Kausrud KL, Begon M, Davis S, Leirs H, Dubyanskiy VM, Esper J, Ageyev VS, et al. Plague dynamics are driven by climate variation. Proc Natl Acad Sci U S A. 2006;103(35):13110–5.
81. Xu L, Liu Q, Stige LC, Ben Ari T, Fang X, Chan KS, Wang S, Stenseth NC, Zhang Z. Nonlinear effect of climate on plague during the third pandemic in China. Proc Natl Acad Sci U S A. 2011;108(25):10214–9.
82. Allentoft ME, Sikora M, Sjogren KG, Rasmussen S, Rasmussen M, Stenderup J, Damgaard PB, Schroeder H, Ahlstrom T, Vinner L, et al. Population genomics of Bronze Age Eurasia. Nature. 2015;522(7555):167–72.
83. Haak W, Lazaridis I, Patterson N, Rohland N, Mallick S, Llamas B, Brandt G, Nordenfelt S, Harney E, Stewardson K, et al. Massive migration from the steppe was a source for Indo-European languages in Europe. Nature. 2015;522(7555):207–11.
84. Harbeck M, Seifert L, Hansch S, Wagner DM, Birdsell D, Parise KL, Wiechmann I, Grupe G, Thomas A, Keim P, et al. Yersinia pestis DNA from skeletal remains from the 6(th) century AD reveals insights into Justinianic Plague. PLoS Pathog. 2013;9(5):e1003349.
85. Little LK. Plague and the end of antiquity : the pandemic of 541–750. New York: Cambridge University Press; 2007.
86. Schmid BV, Buntgen U, Easterday WR, Ginzler C, Walloe L, Bramanti B, Stenseth NC. Climate-driven introduction of the Black Death and successive plague reintroductions into Europe. Proc Natl Acad Sci U S A. 2015;112(10):3020–5.
87. Seifert L, Wiechmann I, Harbeck M, Thomas A, Grupe G, Projahn M, Scholz HC, Riehm JM. Genotyping Yersinia pestis in historical plague: evidence for long-term persistence of Y. pestis in Europe from the 14th to the 17th century. PLoS One. 2016;11(1):e0145194.
88. Link VB. A history of plague in United States of America. Public Health Monogr. 1955;26:1–120.
89. Adjemian JZ, Foley P, Gage KL, Foley JE. Initiation and spread of traveling waves of plague, Yersinia pestis, in the western United States. Am J Trop Med Hyg. 2007;76(2):365–75.

Chapter 7
Pathology and Pathogenesis of *Yersinia pestis*

Zongmin Du and Xiaoyi Wang

Abstract Various types of animal models of plague have been developed, including mice, rats, guinea pigs, and nonhuman primates. Studies have indicated that rodent and nonhuman primate models of pneumonic plague closely resemble the human disease and that the pathologic changes that occur during bubonic plague are very similar in rodents, nonhuman primates, and humans. In this section, the pathological changes caused by *Y. pestis* in different animal models are described. The bacterium *Y. pestis* causes deadly plague, whereas the other two closely related enteropathogenic *Yersinia* species merely cause limited gastrointestinal manifestations. *Y. pestis* has unique virulence mechanisms that enable it to be a successful flea-borne and highly virulent pathogen. Massive gene losses and inactivation play important roles, as well as the gene acquisitions, in the evolution process of this pathogen. Here, we summarized several newly acquired features of *Y. pestis*, including the unique lipid A modification, biofilm formation ability, and loss of adhesions for enteric colonization that are realized by gene inactivation and plasminogen activator and F1 capsular that are realized by gene acquisition, which contribute to the unique transmission and pathogenesis of *Y. pestis*.

Keywords *Y. pestis* • Pathology • Pathogenesis • Molecular mechanism

Plague is a zoonotic disease, caused by *Yersinia pestis*, which mainly infects rodents, while humans are only accidental hosts. Plague infections can result from a fleabite, direct handling of infected animal tissues, ingestion of infective materials, or inhalation of aerosolized bacteria. Pneumonic plague can be spread person-to-person via respiratory droplets. Depending on the route of infection, there are three major clinical manifestations of plague: bubonic, pneumonic, and septicemic. Fleabite or subcutaneous (*s. c.*) infections usually cause bubonic plague, which is characterized by an acute, necrotizing lymphadenitis. In bubonic plague, *Y. pestis* bacteria colonize the regional lymph nodes that drain the intradermal infection site, and they multiply

Z. Du (✉) • X. Wang
Beijing Institute of Microbiology and Epidemiology, Beijing 100071, China
e-mail: zongmindu@gmail.com; wgenome@163.com

R. Yang, A. Anisimov (eds.), *Yersinia pestis: Retrospective and Perspective*,
Advances in Experimental Medicine and Biology 918,
DOI 10.1007/978-94-024-0890-4_7

rapidly, resulting in hot, swollen, tender, and hemorrhagic lymph nodes, which are called buboes. Within a few days after the initial fleabite, the infection spreads to the bloodstream and disseminates to the spleen, liver, and other organs. Victims of bubonic plague often die of septic shock that is caused by the large numbers of bacteria that are released into the blood stream. Primary septicemic plague patients develop systemic *Y. pestis* sepsis with no obvious buboes. Pneumonic plague is the most deadly form of plague, and it can be transmitted person-to-person via the inhalation of respiratory droplets that are produced by severe coughing of infected patients. After a short incubation period, generally 1–6 days after exposure to airborne *Y. pestis*, symptoms typically begin with a flu-like illness, such as severe headache, nausea, and malaise, which rapidly leads to coughing, difficulty breathing, and the production of bloody sputum. Primary pneumonic plague usually progresses rapidly, and victims often die from respiratory failure and the sequelae of severe sepsis, such as hemorrhage, coagulopathy, and circulatory collapse [1]. Pneumonic plague has a very high mortality rate that approaches 100% if not treated in a timely manner with effective antibiotics and supporting therapies.

7.1 Pathological Changes of Plague in Different Animal Models

Studies of plague have focused on its pathogenic mechanism, vaccines, and treatments. It is essential to develop effective animal models of plague, because these studies cannot be performed ethically in humans. For these purposes, various types of animal models of plague have been developed, including mice, rats, guinea pigs, and nonhuman primates [2]. Studies have indicated that rodent and nonhuman primate models of pneumonic plague closely resemble the human disease and that the pathologic changes that occur during bubonic plague are very similar in rodents, nonhuman primates, and humans. In this section, the pathological changes caused by *Y. pestis* in different animal models are described.

*7.1.1 Pathologic Changes of Plague in a Mouse (*Mus musculus*) Model [3–7]*

Mice have been the most commonly used model for molecular pathogenesis and vaccine studies of plague since the 1950s. The extent of pathological lesions depends on the challenge dose, the inoculation route, and the infection time. When mice were administered five median lethal doses (LD_{50}) (1×10^3colony-forming units (CFU)) of the virulent *Y. pestis* CO92 strain by the intranasal (*i.n.*) route, mild vacuolation with minute amounts of polymorphonuclear leukocytes (PMNs) was observed in focal areas of the bronchial epithelia at 12 h postinfection (hpi), whereas the liver and spleen did not show any histopathological changes.

At 24 hpi, a mild infiltration of alveolar septa by PMNs, as well as blood capillary dilation, was seen in the lungs after mice were administered 1×10^3 CFU of the *Y. pestis* CO92 by the *i.n.* route. The lymphoid follicles in the spleen showed a mild increase in the number of apoptotic bodies in the germinal centers, whereas no histopathological lesions were found in the liver. When mice were infected with 1.3×10^4 CFU of the *Y. pestis* CO92 strain via the aerosolized route, the mice had acute inflammation in the lungs. Small areas of minimal acute inflammation and necrosis in the liver and a mild inflammatory cell infiltrate in the red pulp of the spleen were observed, but there were no significant pathological changes in the heart.

At 36 hpi, when mice were administered 1×10^3 CFU of the *Y. pestis* CO92 by the *i.n.* route, the bronchi in the lungs showed more prominent infiltrates by PMNs, and there was patchy edema in the alveolar spaces. The alveolar septa became more congested and more heavily infiltrated by PMNs compared with earlier time points. The splenic changes were similar to those at 24 hpi, while the liver showed occasional PMNs and lymphocytes in the sinusoids.

At 48 hpi, when mice were administered 1×10^3 CFU of the *Y. pestis* CO92 by the *i.n.* route, the lungs exhibited diffuse, vascular congestion in alveolar septa, and more PMNs were found in the interstitium, while there was bronchopneumonia with microabscess formation and necrosis. The liver showed focal hepatocyte death and the formation of microabscesses in the parenchyma. The spleen revealed a remarkable increase in the number of apoptotic bodies in germinal centers and mildly increased numbers of PMNs in the red pulp. When mice were challenged with 1.3×10^4 CFU of the *Y. pestis* CO92 strain by the aerosolized route, inflammation in the lungs was similar to that observed at 24 hpi. The liver showed mild acute inflammation, necrosis, and occasional bacteria. Neutrophilic infiltrates and some necrotic cells were observed in the red pulp of the spleen, while there were no significant pathological lesions in the heart.

At 60 hpi, when mice were administered 1×10^3 CFU of the *Y. pestis* CO92 by the *i.n.* route, extensive pulmonary edema, bronchopneumonia, and bacterial clumps in areas of necrosis were observed in the lungs. Bronchi had luminal secretions that contained PMNs. The liver had microabscesses with necrosis of the hepatic parenchyma, and the spleen showed more prominent pathological changes than those observed at 48 hpi.

At 72 hpi, the mice infected with 1.3×10^4 CFU of aerosolized *Y. pestis* CO92 began to die. The lungs showed severe pneumonia with edema and the presence of small-to-large groups of bacteria. The liver showed acute inflammatory changes with necrosis, the presence of fibrin, and various numbers of bacteria. The mice had moderate lymphoid depletion in the white pulp of the spleen, whereas various numbers of bacteria, congestion, edema, fibrin, and cellular loss were observed in the red pulp of the spleen. Occasionally, bacteria were found in blood vessels in the heart.

When outbred CD1 and inbred C57BL/6 mice were infected by the *i.n.* route with 0.8×10^4 to 2.9×10^4 CFU of *Y. pestis* CO92, their histopathological changes were very similar, and no gross differences were observed between the two strains of mice. Following *i.n.* infection, at 24 hpi, pulmonary congestion and a remarkable lack of PMNs were observed in the lungs of both strains of mice. In some CD1 mice, there were a few small focal infiltrations of inflammatory cells in the lungs at 24 hpi, but no focal infiltrations were observed in the C57BL/6 mice.

At 48 hpi, PMN infiltration was observed in the interstitium and airspaces of the two strains of mice, along with lobular consolidation, areas of necrosis, fibrin deposition, and hemorrhaging. At this time point, pulmonary effusion was evident in both strains of mice, and there were large areas of bacterial colonization embedded within inflammatory cells and debris throughout the lungs. In the conducting airways, dense exudates of PMNs, bacteria, and red blood cells were observed, whereas inflammatory cells obliterated the smaller airways. The spleen had an obscure structure, with foci of bacteria present throughout the tissue. In the liver, small bacterial foci were dispersed throughout the tissue without appreciable inflammatory cell infiltration.

In a mouse model of bubonic plague, Swiss-Webster mice were inoculated by the intradermal (*i.d.*) route with the wild-type 195/P strain or a *caf1*-negative mutant, which fails to produce the fraction 1 (F1) capsule, of the fully virulent *Y. pestis* 195/P strain. The histopathological changes in the spleen, liver, and lymph nodes proximal to the fleabite site were similar following infection with either strain. The normal nodal architecture of the lymph nodes was severely destroyed or completely effaced by larges masses of bacteria that were mixed with abundant, necrotic cellular debris, which is typical of bubonic plague. The red pulp of the spleen was diffusely obscured by moderate hemorrhage, numerous bacteria, and moderate amounts of cellular debris. Moderate to severe lymphocytolysis was observed in the white pulp of the spleen, with a multifocal loss of periarteriolar lymphoid sheaths, and the affected architecture was replaced by cellular debris, fibrin, and numerous bacteria. Diffuse hepatocellular degeneration was observed in the liver, as were mildly swollen and vacuolated hepatocytes. Variably sized bacterial masses were also observed multifocally in the liver, which replaced lost hepatocytes. Bacteria were seen in medium-sized veins, and fibrin thrombi were found occasionally in blood vessels. Despite the similar pathologies that were observed in the lymph nodes, spleens, and livers from the Swiss-Webster mice infected with the *Y. pestis*195/P strain or the *caf1*-negative mutant, the localizations and morphologies of the bacterial masses differed. In the spleen, large discrete bacterial colonies formed by the *caf1* mutant were confined to the marginal zone of the white pulp, whereas those formed by the wild-type strain were more loosely and evenly distributed. In the liver, the colonies formed by the *caf1* mutant were larger and more numerous than those formed by the wild-type strain.

7.1.2 Pathologic Changes of Plague in a Rat Model [8–10]

Plague in the Brown Norway rat (*Rattus norvegicus*) closely resembles human plague, and the pathology in this model is also similar to that observed in human bubonic plague. In a Brown Norway rat model of bubonic plague, rats were infected with 500 CFU of the fully virulent *Y. pestis* strain 195/P by an *i.d.* inoculation. At the time of euthanasia, inguinal lymph nodes proximal to the inoculation site were surrounded by an edematous, hemorrhagic, and gelatinous capsule. The lymph nodes showed necrosis and were two to four times larger than normal, which exactly matches the descriptions of buboes in human plague patients at autopsy. At 24 hpi, a few extracellular bacteria appeared in the marginal sinus of the inguinal lymph node proximal to the inoculation site; their numbers increased and they continued to spread within the marginal sinus, with limited PMN recruitment. At 36 hpi, multifocal bacterial aggregates extended from the marginal sinus into the cortex, with an increased number of PMNs and cellular debris. The marginal and medullary sinuses were edematous, and late in the infection, the normal architecture of the lymph node was replaced by numerous bacteria that were mixed with abundant cellular debris, necrotic PMNs, and fibrin. Vascular fibrin thrombi and variable amounts of hemorrhage were also observed throughout the lymph node. Moderate lymphocyte hyperplasia expanded toward the center of the lymph node 24–36 hpi, increased from 36 to 48 hpi, and then decreased at 72 hpi, which correlated with the destruction of the germinal centers of the lymph node. In the perinodal tissues of the lymph node, many PMNs, lymphocytes, and macrophages were observed, but no bacteria were found.

Finally, numerous bacteria that were mixed with necrotic PMNs, hemorrhage, and cellular debris were observed in the perinodal region. Vascular luminal bacteria and perinodal vasculitis were also found. An acute, necrotizing, fibrinous, and septic lymphadenitis and periadenitis with large masses of bacteria were observed, which led to hemorrhage, septicemia, and necrotizing vasculitis. The architecture of the spleen was markedly changed. The marginal zone decreased in thickness or disappeared. The white pulp displayed multifocal lymphocytolysis, which led to the loss of periarteriolar lymphoid sheaths, in which there were numerous bacteria mixed with abundant fibrin and cellular debris. The red pulp sinuses also contained a large number of bacteria mixed with abundant fibrin and cellular debris.

In a rat aerosol model, rats were exposed to 8.6 LD_{50} doses of the *Y. pestis* CO92 strain in a whole-body Madison chamber. At 24 hpi, the rats showed minimal perivascular edema and acute inflammation in the lungs, and some animals exhibited additional subacute inflammation within the lung parenchyma. An influx of neutrophils was observed in the acute inflammation, whereas subacute inflammation was characterized by a mixture of neutrophils and mononuclear cells. However, no changes were observed in the liver or spleen.

At 48 hpi, the lungs showed a mild-to-moderate perivascular edema and acute inflammation, subacute inflammation in the lung parenchyma, mild leukocytosis in blood vessels, bacteria in many alveoli, and a moderate and diffuse bacteremia. In some animals, a mild, multifocal bacteremia was seen in the liver. In the spleen, all of the animals showed mild-to-moderate numbers of bacteria and minimal-to-moderate acute inflammation and fibrin.

At 72 hpi, the lungs showed mild-to-moderate perivascular and alveolar edema, mild congestion, subacute inflammation, leukocytosis in the blood vessels, and mild-to-moderate bacteremia. The livers of the rats showed multifocal bacteremia, acute inflammatory cells, necrotic hepatocytes, and clumps of bacteria within sinusoids. The spleens of rats showed edema, acute inflammatory cells, and moderate numbers of bacteria. At 72 hpi, histopathology indicated that bacteria had disseminated from the lungs to other peripheral organs, such as the spleen and liver.

*7.1.3 Pathological Changes of Plague in a Guinea Pig (*Cavia porcellus*) Model [11, 12]*

Guinea pigs have been traditionally used as an animal model for the evaluation of plague vaccines, but only a few studies have focused on the pathological lesions in this model. After guinea pigs were challenged with 10^4 CFU of the *Y. pestis* GB strain by the *s.c.* route, postmortems showed extensive abscesses in the spleen and liver, but no evidence of lesions was observed in the lungs. Remarkable swelling of the lymph nodes was observed in all animals, and upon dissection, they were found to contain pus.

In another study, adult guinea pigs were inoculated with different doses of the virulent *Y. pestis* 141 strain by the *s.c.* route, including low-dose (25 CFU), medium-dose (100 CFU), and high-dose (5000 CFU) groups. Except for the heart tissues, other tissues examined in the study exhibited significant pathological changes, including congestion, hemorrhage, necrosis, and fatty degeneration. Bleeding and tissue necrosis were more severe with increasing inoculation doses.

In the low-dose group, the livers showed extension and congestion in the central veins of hepatic lobules, blood clots in hepatic sinusoids, congestion in hepatocellular interstitia, lymphocyte infiltration in part of the hepatic lobules, and hepatocyte degeneration with mild edema. In the spleens, an altered white pulp architecture and lymphocyte hyperplasia were observed, but no necrotic foci were found. The red pulp region was widened, and it contained much congestion and occasional spotty necrosis. In the lungs, mild extension and congestion of capillaries in the alveolar wall, a mild widening of pulmonary alveoli, formation of pulmonary bullae by dissolution of part of the alveolar walls, extension and congestion of blood vessels, and inflammatory cell infiltration in the pulmonary interstitium were found. In the kidneys, mild extension and congestion in the capillaries of the renal glomerulus were observed, but no inflammatory cell infiltration was seen in the glomerulus. Mild edema in the epithelial cells of the proximal or distal convoluted tubules and mild congestion in the renal interstitium were observed.

In the medium-dose group, destruction of a part of the architecture of the hepatic lobules, lymphocyte infiltration, small spots of necrosis, lymphocyte infiltration, blood clots in the hepatic cord interstitium, extension and congestion in central veins, and hepatocyte degeneration with edema were found in the liver tissues. In the spleen, tissues were destroyed, and the architecture of the white pulp was altered. Additionally, there was lymphocyte hyperplasia, and the architecture of the red pulp disappeared; the pulp was covered with blood, and focal, necrotic areas were found. Eosinophil, granulocyte, and centrocyte aggregation increased in the spleen tissues. In the lungs, extension and congestion of capillaries in the alveolar wall were found, and the alveolar wall was widened. Additionally, many blood clots were found in the pulmonary interstitium, and large numbers of infiltrating lymphocytes were observed in the alveolar wall and alveolar spaces, while small spots of necrosis were found in the lungs. In the kidneys, clear extension and congestion in the capillaries of the renal glomerulus were seen, but no inflammatory cell infiltration was found in the glomerulus. Mild parenchymatous degeneration in the epithelial cells of the proximal or distal convoluted tubules and marked congestion in the renal interstitium were observed.

In the high-dose group, there was a disruption of the architecture of a part of the hepatic lobules, which was replaced by a large amount of blood clotting organization. Hepatocytes atrophied and disappeared, and a mass of adipose tissue appeared. Flaky, necrotic foci became larger and increased in number. In the spleen, tissue destruction, a regional distribution of necrotic foci, extensive congestion in the red pulp, splenocyte edema, and an increase in inflammatory cell aggregation were observed. In the lungs, the capillaries in the alveolar walls showed severe extension and congestion, and the widths of the alveolar walls increased. Additionally, many blood clots were observed in the pulmonary interstitium, and the numbers of inflammatory cells markedly increased, while patchy, necrotic foci were found in pulmonary parenchyma. In the spleen, the number of renal glomeruli showing extension and congestion remarkably increased, and the architecture of the renal glomerulus was partly destroyed. Proximal or distal convoluted tubules showed more obvious congestion, and the epithelial cells of the proximal or distal convoluted tubules displayed severe parenchymatous degeneration. Congestion in the renal interstitium increased.

7.1.4 Pathologic Changes of Plague in Nonhuman Primates [13–19]

Generally, plague vaccines are first evaluated in rodents, then in nonhuman primates, and, finally, in human patients. The disease in rhesus macaques (*Macaca mulatta*) differs from that in humans, as rhesus macaques frequently develop disseminated intravascular coagulation (DIC) and chronic pneumonia as a result of pneumonic plague, while humans usually develop acute pneumonia without

DIC. Several species of nonhuman primates, including Indonesian cynomolgus macaques (*Macaca fascicularis*), cynomolgus macaques, Chinese-origin rhesus macaques, and Indian-origin rhesus macaques, have been challenged with virulent *Y. pestis*. The pathological changes in these animal models are described below.

When Indonesian cynomolgus macaques were infected with 12–42,700 CFU of the *Y. pestis* CO92 strain by the aerosolized route, they exhibited acute suppurative bronchopneumonia, splenic leukocytosis, suppurative splenitis, tracheobronchial lymphadenitis, and acute fibrinous interstitial pneumonia with bacteria in major pulmonary vessels and septal capillaries. In another study, when cynomolgus macaques of Indonesian origin were infected with 3300–9900 CFU of the *Y. pestis* CO92 strain by the aerosol route, the gross appearance of the lungs was characterized by multifocal areas of pneumonia, along with enlarged tracheobronchial lymph nodes. Consolidated areas with pneumonia were apparent and palpably firm, and moderately well-demarcated regions with different amounts of dark-red discoloration and hemorrhage were observed. No detectable lesions were observed, although bacteria could be isolated from lung tissues at 24 hpi. At 48 hpi, an early focus of infection was occasionally found in the lungs, and alveolar thickening caused by mononuclear cell infiltration was observed. By 72 hpi, distinct focal lesions were found, with fibrinosuppurative pneumonia and alveolar edema, whereas in the non-lesion regions, there was septal and/or alveolar infiltration of neutrophils and mononuclear cells. By 96 hpi, lesions were extensive in the lungs. A wide distribution of necrosis was present throughout all areas of the lungs, and the areas of pneumonia were typically deep in the pulmonary parenchyma.

When cynomolgus macaques were infected with 200 ± 25 LD_{50} of the *Y. pestis* CO92 strain by the aerosolized route, tissue sections showed moderate-to-severe multifocal and acute bronchopneumonia with multiple areas of severe congestion and hemorrhage. In many areas, portions of alveolar lumina were filled with edematous fluid, fibrin filaments, numerous neutrophils, monocytes, and macrophages. Occasionally, alveolar macrophages contained bacteria or necrotic cellular debris. Large aggregates of bacteria were distributed in pulmonary interstitial spaces and alveoli. There were multifocal, coalescing areas of necrosis within the lung parenchyma, which obliterated the normal architecture and large abscesses. Abscesses were characterized by central foci of lytic necrosis surrounded by degenerate neutrophils that were interspersed with numerous bacteria entangled in fibrin and erythrocyte deposits.

When Chinese-origin rhesus macaques were infected by the *s.c.* route with 6×10^6 CFU of the *Y. pestis* 141 strain, hemorrhage, effusion, edema, inflammatory cell infiltration, and abscesses containing *Y. pestis* were observed in the lung tissues. The lymph nodes showed severe congestion, edema, a reduced number of lymphocytes, and the disappearance of recognizable architecture. The liver tissues showed hepatocyte swelling, vacuolar degeneration, dilatation, hyperemia of the central vein of the hepatic lobules, and slight congestion within the sinus hepaticus. The spleen tissues displayed reduced amounts of white pulp and acinus lienalis, and fewer lymphocytes, as well as splenic cord edema. Acinus renis analosis, renal capsule effusion, interstitial edema, and vascular engorgement were observed in the kidney tissues of the animals infected with *Y. pestis* strain 141. There were no evident changes in the heart tissues. However, when Chinese-origin rhesus macaques were infected with 1.74×10^9 CFU of *Y. pestis* strain 141 by the *s.c.* route, a disappearance of recognizable lung tissue architecture and the formulation of inflammatory cell foci, which were mainly composed of lymphocytes, were observed in the lung tissues. In addition, edema, congestion, and inflammatory cell infiltration were found in the lung tissues. Lymph node tissues showed architectural deterioration and low numbers of lymphocytes. Slight congestion within the sinus hepaticus and hepatic cell degeneration were observed. The spleen tissues showed a reduced amount of white pulp and acinus lienalis, as well as fewer lymphocytes. Interstitial lymphocyte infiltration and vascular engorgement were also observed. Segmentation of cardiac muscle fibers and vascular engorgement were found in heart tissues.

The *Y. pestis*201 strain was isolated from a Brandt's vole (*Microtus brandti*) in Inner Mongolia, China, and it has an LD_{50} of 3 CFU for BALB/c mice by the *s.c.* route. Strain 201 belongs to a newly established *Y. pestis* biovar, Microtus, which is supposed to have low virulence in guinea pigs, nonhuman primates, and humans. When Chinese-origin rhesus macaques were infected intravenously with 9×10^8 CFU of the *Y. pestis* Microtus 201 strain, no changes in histopathology were observed in the heart, liver, spleen, lungs, kidneys, and lymph nodes, compared with normal tissues. However, when the animals were intravenously injected with 1.28×10^{10} CFU of the *Y. pestis* Microtus 201 strain, their tissues were altered substantially. Vascular engorgement was observed in heart tissues. Liver tissues showed slight liver cell degeneration. Atrophic white pulp, splenic sinus congestion, and reduced numbers of lymphocytes were found in the spleen tissues. A disappearance of recognizable lung tissue architecture, edema, congestion, and inflammatory cell infiltration were observed in the lung tissues. Renal interstitium congestion and edema were observed in the kidney tissues, and the lymph node tissues displayed lymphocyte hyperplasia, edema, and neutrophil infiltration.

When rhesus macaques that originated from India were infected with 3×10^4 CFU of the KIM-10 strain of *Y. pestis* by the aerosolized route, the pathologic lesions in the organs were similar to classical descriptions of pneumonic plague. Renal fibrin thrombi formed, and they were located primarily in glomerular capillaries. A focus of lobular consolidation was observed in the peripheral portion of the pulmonary lobules. Upon sectioning of the pulmonary lobules, hepatization was found, occasionally with central liquefaction and hemorrhage. The surrounding tissue was edematous, and its area was greater than that of the consolidated portion. Microscopically advanced lesions consisted of a central abscess. In such areas, hemorrhages and inflamed vessels were frequently observed. Dilatation of lymphatics around bronchi and veins was found in areas of edematous alveoli that were filled with edema fluid teeming with *Y. pestis*, and the lymphatics were also heavily colonized by bacteria. Bronchial lymph nodes displayed extensive liquefactive necrosis, enormous numbers of bacilli, and hemorrhages in all of the infected animals. A reduction of the red pulp was observed in the spleen tissue, with hemorrhages and necrotic foci in some cases. Enormous numbers of bacilli and fibrin thrombi were observed in the sinusoids. Lesions in the liver generally included sinusoidal congestion, periportal or diffuse sinusoidal leukocytosis, and mild fatty change. Occasionally, bacteria and fibrin thrombi were present in the sinusoids. Occasionally, focal hemorrhages were observed in the adrenal tissues, but fibrin thrombi were rarely found, even in animals that exhibited extensive renal deposits. When animals were infected subcutaneously with 3×10^9 CFU of the Alexander strain, they showed local ulceration and liquefactive necrosis in the subcutaneous tissue at the site of inoculation, inguinal buboes, septicemia, and multiple hemorrhages. Fibrin thrombi were extensively found in the glomerular capillaries of the animals.

7.1.5 Pathological Changes in Human Plague [20]

There are three clinicopathological forms of human plague: bubonic, septicemic, and pneumonic. Primary pneumonic plague in humans begins as a bronchopneumonia characterized by numerous bacteria and proteinaceous effusion in the alveoli. The pneumonia spreads rapidly and is accompanied by pleuritis. Microscopic examination reveals hemorrhage, necrotic foci, scant suppuration, and numerous bacilli in most of the alveoli. Pathological changes of bubonic plague in humans mainly consist of congestion, edema, hemorrhage, and bacterial aggregates in swollen lymph nodes. Lymphoid tissues show degeneration, and necrosis, with infiltration and proliferation of lymphocytes, neutrophils, and macrophages. The pulmonary tissues often display congestion swelling, hemorrhagic lobular pneumonia, and inflammatory cell infiltration in the bronchi.

Pathologic changes in formalin-fixed, paraffin-embedded autopsy tissues were detected in six patients, three of whom were from Ecuador, two patients were associated with the 1998 outbreak, and one was a sporadic suspect case from the southwestern United States. Histopathologic abnormalities were observed in the lungs, lymph nodes, kidneys, and spleen, but the lesions in other tissues appeared unremarkable. The lung tissue from case 1 showed acute pneumonia with intra-alveolar infiltration by neutrophils and mononuclear cells, whereas intra-alveolar hemorrhage and edema were found in case 2. The pathologic changes in the lung tissues from cases 4 and 6 were not notable except for focal intra-alveolar edema. The lymph nodes from cases 4–6 were examined, and all three cases showed various degrees of lymphoid depletion, as well as one case of necrosis, and two cases of foamy macrophages in the sinusoids. Pathologic changes in the spleen tissues were detected in cases 4 and 6. Multiple abscesses were found in case 4, whereas a large number of macrophages in the sinusoids were observed in case 6. The kidney tissue from case 6 displayed fibrin thrombi in the glomeruli, which is indicative of DIC.

Using an immunohistochemical detection method based on a monoclonal anti-F1 antibody, formalin-fixed tissues were examined for the presence of bacteria. Intact bacteria were found inside macrophages and in the blood vessels in all of the tissues, and large numbers of extracellular bacteria were observed in the alveolar spaces in the lung tissues from cases 1 and 2. The lung tissues from cases 4 and 6 also contained intact bacteria, as well as granular staining in the interstitial vessels and connective tissues, but no bacteria were observed in the alveolar spaces or in the lumen of the bronchi. In cases 1, 2, 4, and 6, intact bacteria were present in Kupffer cells, whereas in the liver sinusoids, there were intact bacteria and granular staining. In the kidney tissues from cases 4 and 6, there were abundant granular antigen staining and bacteria in the blood vessels, but in case 2, the tissue showed intact bacterial foci inside mononuclear cells in the interstitium. An amputated finger specimen from case 3 displayed granular antigen staining and intact bacteria in the blood vessels of the dermis, subcutaneous tissues, and bone marrow. The spleen tissue from case 4 showed granular staining and intact bacteria in the abscesses. In the lymph nodes, the tissues showing lymphoid depletion contained bacteria in the blood vessels, but those from patients who had necrosis showed intact intracellular bacteria in macrophages and abundant granular staining in the subcapsular sinusoids. Intact bacteria and granular antigen staining were also observed in the brain from case 5. Bacteria were present predominantly in the meninges, but they were also found in the parenchymal blood vessels in smaller numbers. In the tissues of the gastrointestinal tract from case 6, bacteria were also found in the blood vessels and focally in the lamina propria of the mucosa.

7.2 Virulence Mechanisms That Differentiate *Y. pestis* from Its Enteropathogenic Progenitor

Y. pestis, the causative agent of plague, has recently evolved from the enteropathogenic *Y. pseudotuberculosis* [21]. Three pathogenic yersiniae share a number of common virulence mechanisms, such as a type III secretion system (T3SS) encoded by a 70-kb plasmid (pCD1 in *Y. pestis* and pYV in *Y. enterocolitica* and *Y. pseudotuberculosis*), and virulence factors that promote coordinated gene expression and iron acquisition. Although closely related in terms of phylogeny, these three pathogens are extremely diversified in terms of their clinical symptoms, ecological niches, typical infection routes, and host specificities. *Y. pestis* is the etiologic agent of plague that has caused more than 200 million deaths in three pandemics, whereas the others merely cause limited gastrointestinal manifestations. *Y. pestis* has unique nutrient requirements, routes of transmission, and virulence mechanisms that enable it to be a successful flea-borne and highly virulent pathogen. Massive gene losses played more important roles than gene acquisitions in the evolution of *Y. pestis* [22]. Several newly acquired features that contribute to the unique pathogenicity and transmission of *Y. pestis* were the result of gene inactivation, a type of reductive evolution.

7.2.1 Lipopolysaccharide Modification

Y. pestis produces a short-chain, rough lipopolysaccharide (LPS) that lacks the O antigen because of several mutations in LPS biosynthesis genes. The O antigen is an important virulence factor of many enterobacterial pathogens, including *Y. enterocolitica* and *Y. pseudotuberculosis*, and it confers resistance to complement-mediated phagocyte bacterial killing. The Pla protease is a plasminogen activator that is encoded by the pPCP1 plasmid that has been considered unique to *Y. pestis*; however, a *pla*-bearing fragment was found in *Citrobacter koseri* [23]. The major activity of Pla is to cleave plasminogen to plasmin, and plasmin can further degrade fibrin clots and extracellular matrix (ECM) proteins such as laminin and fibronectin, which promotes bacterial dissemination. The proteolytic activities of Pla require rough LPS, but are inhibited by the O antigen, thereby revealing the nature of the selective advantage of rough LPS [24]. The mutation of LPS biosynthesis genes was a critical event during the evolution of *Y. pestis* pathogenesis.

Lipid A modification in *Y. pestis* is different from those of its enteropathogenic relatives. As the temperature shifts from 26 °C (the flea temperature) to 37 °C (the mammalian temperature), lipid A of *Y. pestis* LPS switches from a hexa-acylated to tetra-acylated form that is poorly recognized by Toll-like receptor 4 (TLR4), rendering it 500-fold less toxic than those of other *Enterobacteriaceae* grown at 37 °C [25]. This is due to the loss of the *lpxL* gene in *Y. pestis*, which encodes an acyltransferase that adds secondary acyl chains to the tetra-acylated lipid A precursor in *Y. pseudotuberculosis* [26]. The less toxic lipid A benefits the bacteria, since an extreme bacteremia will not cause immediate death, which promotes high titers of *Y. pestis* bacilli in the blood stream, thereby making flea-borne transmission feasible [27].

7.2.2 Biofilm Formation

Biofilm formation by *Y. pestis* is required for blocking the flea proventriculus, which is a critical step for the flea-borne transmission of plague [28, 29]. Biofilm formation in *Yersinia* is positively regulated by the second messenger cyclic-di-GMP (c-di-GMP) [30] and negatively regulated by the two-component Rcs regulatory system. Rcs consists of two membrane proteins, RcsD and RcsC; a DNA-binding regulator, RcsB; and an auxiliary protein, RcsA. Upon stimulation, the RcsC sensor is autophosphorylated and transfers a phosphate group to RcsD and then to RcsB. Phosphorylated RcsB forms an RcsB homodimer or a heterodimer with RcsA. RcsB/RcsA represses the expression of HmsT [31], a diguanylate cyclase enzyme that synthesizes c-di-GMP, leading to the inhibition of biofilm production. Because of the inhibitory effect of the RcsB/RcsA regulator, *Y. pseudotuberculosis* cannot form biofilms in insect vectors. In *Y. pestis*, *rcsA* is inactivated and replacing it with the functional *Y. pseudotuberculosis rcsA* resulted in a strong repression of biofilm formation in the flea [32, 33]. The mutation of *rcsA* was required for the food-borne *Y. pseudotuberculosis* to evolve into the flea-borne *Y. pestis* [33].

7.2.3 Loss of Adhesins for Enteric Colonization

The adhesins Inv and YadA are indispensable for enteropathogenic yersiniae to colonize the Peyer's patches and mesenteric lymph nodes [34, 35]; however, both *inv* and *yadA* are pseudogenes in *Y. pestis*. The introduction of functional Inv and YadA to *Y. pestis* led to a significant attenuation of virulence [36, 37]. YadA of *Y. pseudotuberculosis* ($YadA_{pstb}$) promotes the efficient internalization of bacteria into host cells via binding to β-integrins, whereas YadA of *Y. enterocolitica* has no such activity because it lacks the unique amino-terminal amino acid sequence of $YadA_{pstb}$. In addition to β-integrins, YadA binds to various ECM molecules (such as laminin, collagen, and fibronectin) and confers complement-mediated serum resistance. Neutrophils release granule protein complexes called neutrophil extracellular traps (NETs) that can bind and kill invading bacteria. YadA promotes the binding of bacteria to NETs; thus, enteropathogenic *Yersinia* is sensitive to NET-mediated killing by neutrophils [37, 38]. The double-edged sword effect of YadA adhesion on bacterial pathogenesis explains the negative selection of the *yadA* gene during the evolution of *Y. pestis.*

7.2.4 Pla

As previously noted, the plasminogen activator Pla is encoded by pPCP1, a plasmid that is unique to *Y. pestis* [39]. The acquisition of pPCP1 was a critical event in the evolution of *Y. pestis* from its enteric progenitor *Y. pseudotuberculosis* [40]. Pla belongs to the omptin family of outer membrane aspartate proteases, and it contains a conserved β-barrel fold. The major activity of Pla is to cleave plasminogen to plasmin, which can further degrade fibrin clots and ECMs such as laminin and fibronectin [41]. Laminin is a major glycoprotein of mammalian basement membranes, and it is degraded by plasmin, but not by Pla. The protease activities of Pla toward a range of substrates significantly promote the dissemination of *Y. pestis* from subcutaneous infection sites. Inactivation of *pla* dramatically decreases the virulence of *Y. pestis* in subcutaneously infected mice, as the infection is restricted to the local inoculation sites. Pla is also an adhesion that is specific for laminin and heparan sulfate proteoglycan, and purified Pla-coated microparticles bound laminin in a reconstituted basement membrane that was immobilized on Permanox slides. Pla cleaves the C3 component of the complement system and inhibits complement activation.

7.2.5 F1 Capsular Protein

The F1 capsular protein, which is encoded by the pMT1 plasmid that is unique to *Y. pestis*, is one of the major virulence factors of *Y. pestis*. The gel-like F1 capsule is composed of a polymer protein subunit, Caf1. *Y. pestis* strains with spontaneous mutations in the *caf1* gene or those in which *caf1* is deleted are still virulent [42, 43], although the F1 antigen seems to contribute to bacterial virulence in bubonic plague, but is not essential for pneumonic plague in a mouse and guinea pig infection model [44, 45]. An anti-F1 antibody provides high protection against F1-positive, but not F1-negative, strains. F1 appears to have some role in antiphagocytosis, although the primary *Y. pestis* virulence factor that blocks uptake by phagocytosis is the T3SS [46].

7.3 Iron Acquisition from the Host

The survival of bacteria in a mammalian host requires an ability to obtain iron from mammalian iron-binding proteins. The yersiniabactin (Ybt) siderophore-dependent iron transport system, which is critical for iron utilization by *Y. pestis*, is encoded in the high-pathogenicity island that is common to the pathogenic *Yersinia* species and is contained in the unstable102-kb pigmentation (*pgm*) chromosomal locus. The *ybt* locus that encodes the Ybt siderophore-dependent iron transport system includes gene clusters for Ybt siderophore synthesis (high-molecular-weight protein 1 (HMWP1), HMWP2, YbtD, YbtE, YbtS, YbtT, and YbtU) and Ybt uptake (YbtQ, YbtP, and Psn). Ybt has high affinity for ferric iron, and it is necessary for the nutritional use of iron (obtained from transferrin and lactoferrin) by yersiniae [47]. Spontaneous *pgm*⁻ mutants of *Y. pestis* or specific Ybt synthesis or uptake mutants are almost completely incapable of causing bubonic plague, but can cause septicemic plague [47–50]. Mutations of different components of the Ybt system have various impacts on the virulence of aerosolized bacteria, which cause pneumonic plague [47]. *Y. pestis* mutants that lack the Ybt iron transport system are avirulent in subcutaneously challenged mice, but are fully virulent in intravenously challenged mice [51]. In addition to the Ybt system, other ion transport systems, including Yfe, Yfu, Yiu, and Hmu, have been functionally characterized, but play minor roles in bacterium iron utilization [52]. Yfe is an ATP-binding cassette transport system that accumulates both iron and manganese, and a *yfeAB* mutant of *Y. pestis* exhibited an approximately 100-fold increase in the LD_{50} in an *s.c.* infection [53]. Yfu, Yiu, and Hmu were shown to be nonessential for virulence in the mouse [52].

7.4 Adhesion to and Invasion of Host Cells by *Y. pestis*

Y. pestis is lymphophilic and prefers to attack certain type of cells, such as macrophages and alveolar epithelial cells of the lung. The pH6 antigen (PsaA) adhesin contributes to the cell-type preference of *Y. pestis*. PsaA is expressed under low-pH conditions (pH5–6.0) at mammalian temperatures [54]. Acidification of phagolysosomes induces the synthesis of PsaA by intracellular bacteria in macrophages [55]. The PsaA fimbria is made up of a surface homopolymer of PsaA subunits, and it is assembled via the chaperone-usher pathway. Two major receptors for Psa have been shown to be β1-linked galactosyl residues in glycosphingolipids and phosphocholine and phosphatidylcholine in phospholipids, which are present on alveolar epithelial cells [56, 57]. The crystal structure of PsaA in complex with galactose and phosphocholine reveals that the two receptor-binding sites share a common structural motif and that Tyr126 of PsaA is important for binding the two receptors [58]. PsaA can bind to the Fc portion of human, but not mouse, rabbit, or sheep, IgG on the surfaces of cells, and it interacts with low-density lipoproteins [59]. An in vitro cell infection assay showed that PsaA does not enhance adhesion to mouse macrophages, but promotes resistance to phagocytosis [60]. Additionally, PsaA facilitates *Yersinia* outer protein (Yop) delivery by the T3SS when bacteria are grown at 37 °C and pH6 by mediating intimate contacts between host cells and bacteria, as do other adhesions, such as the attachment invasion locus (Ail) and Pla [61]. Mutation of the *psaA* gene locus of attenuated strain results in a 100-fold increase in the $LD_{50}$when mice are infected intravenously, but there is no any attenuation when mice are infected subcutaneously with *psaA*$^-$ variants of wild-type strains [62, 63].

Ail is a chromosomally encoded, small-membrane protein that is common to all three pathogenic *Yersinia*. It belongs to the Ail/Lom outer membrane protein family that is found in enterobacteria, which includes Rck and PagC in *Salmonella enterica serovar typhimurium* and OmpX in *Escherichia coli*. While the major adhesins YadA and Inv in *Y. pseudotuberculosis* are inactivated in *Y. pestis*, Ail is the primary adhesin of *Y. pestis*. An *ail* mutant of the *Y. pestis* KIM5 strain (which is *pgm*$^-$) exhibited a greatly decreased ability to bind cultured cells, and it was attenuated over 10^3-fold in intravenously infected mice. Deletion of *ail* in the fully virulent CO92 strain greatly delayed the time to death in mice that were intranasally challenged with 10 or 100 LD_{50} of the bacteria, and the mutant was attenuated approximately10^5-fold in a rat pneumonic plague model [64, 65]. In *Y. pestis*, Ail is expressed at high levels at both 26 and 37 °C. However, its expression in *Y. enterocolitica* is temperature-regulated and is only detectable at 37 °C, but not at lower temperatures, in stationary phase cells [66, 67]. A pioneering study found that Ail is involved in the invasion of eukaryotic cells and serum resistance in *Y. enterocolitica*. *Y. pestis* strains contain four loci that have sequence homology to Ail. Analysis of the isogenic mutants lacking each gene in the KIM8-E background demonstrated that only Ail confers serum resistance [67]. The crystal structure of Ail reveals an eight-stranded, amphipathic, transmembrane β-barrel with four extracellular loops [68]. The active sites for the invasion and serum resistance functions reside in extracellular loops 2 and 3, which might be separate because of the presence of mutants that are defective in the invasion or serum resistance functions [69]. Ail binds to the ECM components fibronectin, laminin, and heparin [68, 70], which are critical for invasion and Yop delivery into host cells. Ail also binds to various substrates that are associated with the complement pathway, which include the complement inhibitor C4b-binding protein and vitro nectin, a host protein that is involved in cell attachment, fibrinolysis, and inhibition of the complement system [71, 72]. *Y. pestis* Ail shows higher binding affinity to ECM substrates than *Y. pseudotuberculosis* Ail because of two amino acid differences, and the rough LPS of *Y. pestis* increases the accessibility of Ail to eukaryotic cells. The influence of LPS length on the binding to host cells has also been observed in *Y. enterocolitica*, and *Y. enterocolitica* mutants that express LPS without the O side chain have an increased ability to enter mammalian cells [66].

Two novel, chromosomally encoded orthologs of *yadA*, *yadB* and *yadC*, have been shown to be thermally induced virulence factors that promote the invasion of *Y. pestis*, and mutation of *yadBC* led to decreased epithelial cell invasion and attenuated virulence in bubonic, but not pneumonic, plague in mice [73, 74].

7.5 Intracellular Replication of *Y. pestis*

All pathogenic *Yersinia* species predominately localize to the extracellular environment of the host. The initial stage of intracellular replication of *Y. pestis* is critical for its preadaptation to the mammalian host environment before establishing a successful systemic infection. In a typical natural transmission, *Y. pestis* bacilli enter the host's subcutaneous tissues via vomiting when infected fleas attempt to ingest blood meals, and they are ingested readily by macrophages, which actually become an intracellular niche where *Y. pestis* replicates and evades the host immune system [75]. Thereafter, *Y. pestis* rapidly adapts to the adverse environments of the mammalian host by arming itself with a variety of virulence factors, such as the F1 capsule and the T3SS, to avoid phagocytosis and initiate its extracellular life cycle [76].

Y. pestis can survive and replicate in naïve macrophages in vitro and in vivo [77], but is efficiently killed by neutrophils and activated macrophages [75, 78]. Professional phagocytes devour and digest bacteria, and antigen-presenting cells can present peptides from bacteria on their cell surface to stimulate host defense signaling to clear the invading bacteria. Phagosomes containing trapped bacteria generally fuse with lysosomes, and mature to phagolysosomes, which is accompanied by the acidification of vacuoles; this creates a harsh environment that destroys invading bacteria. However, *Y. pestis* has been shown to inhibit the acidification of phagosomes, as well as autophagy, thereby rendering immature phagosomes to be protective cocoons that permit bacterial replication [77].

Many studies have investigated the initial intracellular survival of *Y. pestis* in host phagocytes. Lukaszewski et al. showed that *Y. pestis* bacilli could be consistently detected in splenic macrophages until 5 days postinfection (dpi), but could only be detected in splenic neutrophils prior to 72 hpi [75] in a mouse bubonic plague model. Furthermore, they showed that *Y. pestis* bacilli in splenic neutrophils are nonviable when cultured in vitro, but those in splenic macrophages are viable, indicating that neutrophils restrict the growth of *Y. pestis*, whereas macrophages cannot. Spinner et al. showed that intracellular *Y. pestis* bacteria can survive and replicate in cultured human PMNs, and a high percentage of the infected PMNs underwent apoptosis within 12 h of infection. These PMNs can be recognized and internalized by autologous macrophages, and *Y. pestis* survives and replicates within the macrophages following efferocytosis [79]. After the early infection stage, *Y. pestis* largely replicates and lives extracellularly, and this can be supported by histopathology results that detected a large number of extracellular bacteria in animal tissues [8, 75]. These results from cell culture infections and a mouse infection model prove that macrophages are the primary intracellular niches that provide a permissible condition for *Y. pestis* replication.

Pujol et al. showed that *Y. pestis* was able to replicate in macrophages that were activated with interferon-γ and this ability required the *rip* (required for intracellular proliferation) operon (consisting of *ripA*, *ripB*, and *ripC*) within the *pgm* locus [78], which is conserved among several macrophage-residing pathogens, such as *Burkholderia* and *Salmonella* species [80]. The *rip* operon is responsible for the ability of *Y. pestis* to lower intracellular nitric oxide (NO) levels, which is required for the replication of *Y. pestis* in post-activated macrophages. RipA was subsequently characterized as a butyryl-coenzyme A (CoA) transferase that produces butyrate, which is an anti-inflammatory molecule shown to lower macrophage-produced NO levels [80]. Crystal structures of RipA have been determined [81]. RipC is a putative citrate lyase beta subunit, and a crystal structure analysis revealed that RipC can form homotrimers and that it might be a CoA or CoA derivative-binding protein [82].

PhoP/PhoQ comprises a two-component system. PhoQ is the sensor kinase and PhoP is the response regulator. The involvement of PhoP in bacterial survival and replication in macrophages has been proven in *Salmonella enterica serovar* Typhimurium. In *Yersinia*, PhoP has been shown to regulate many genes that are important for virulence. Grabenstein et al. showed that a *phoP* mutant of *Y. pseudotuberculosis* cannot replicate in low-magnesium medium or in macrophages and that it was attenuated over 100-fold compared with a wild-type strain in an intestinal infection mouse model [83]. Oyston et al. showed that PhoPQ is important for *Y. pestis* survival in J774A.1 cells, which is a mouse macrophage cell line, and under conditions of low pH and oxidative stress in vitro, and that the LD_{50} of the *phoP* mutant increased 75-fold [84]. A recent report investigated the impacts of PhoP on virulence, systemic organ colonization, and immune response modulation using an isogenic *phoP* mutant of the fully virulent *Y. pestis* CO92 strain, as well as *phoP* mutants of the *Y. pseudotuberculosis* IP32953 and YPIII strains. The results found that PhoP is not as essential for oral infections of *Y. pseudotuberculosis* [85] and that the *phoP* mutant showed no significant difference in virulence in bubonic and pneumonic plague infection models [86]. The intraspecies variations of PhoPQ function might reflect the unique lifestyle of *Y. pestis*. In the flea digestive tract, the *Y. pestis*

PhoPQ regulatory system is induced and modulates bacterial adaptation to environmental stress, which results in the modification of the bacterial outer surface and the promotion of biofilm formation.

Several studies have tried to identify novel factors that contribute to the virulence of *Y. pestis*. Using a transposon site hybridization-based approach, a library of *Y. pestis* mutants was screened in an in vitro infection system, and OmpA was found to enhance the intracellular survival of *Y. pestis* [87]. The library contains 31,500 KIM6+ transposon insertion mutants, and the screening was performed in primary murine macrophages. In addition to genes that were already known to be important for intracellular survival, novel genes were identified that are required for intracellular survival, including glucose-1-phosphate uridylyltransferase (*galU*), UDP-N-acetylglucosamine 2-epimerase (*wecB*), and UDP-N-acetyl-d-mannosamine dehydrogenase (*wecC*). A subsequent investigation demonstrated that a *galU* mutant of the KIM6 + strain is more susceptible to polymyxin B and cathelicidin-related antimicrobial peptide than the parent strain. Mutation of *galU* prevents the amino-arabinose modification of lipid A. *wecBC* encodes the enterobacterial common antigen [88].

7.6 Type III Secretion System (T3SS)

The three human pathogenic bacteria in the genus *Yersinia* share a common T3SS that is encoded by the 70-kb plasmid, named pCD1 in *Y. pestis* and pYV1 in *Y. enterocolitica* and *Y. pseudotuberculosis*, which is indispensable for all three pathogens to establish systematic infections. The T3SS assembles a macromolecular device called an injectisome that can directly deliver virulence effectors, called Yops, into the cytosol of mammalian host cells, leading to the disruption of the cytoskeleton and host immune response singling. Although *Y. pestis* delivers Yop effectors to almost all types of cells in tissue cultures, evidence showed that it preferentially targets host immune cells during infections [89, 90]. Dendritic cells, macrophages, and neutrophils are the most frequently injected cells, and T and B lymphocyte are less selected [90]. The T3SS is a temperature- and low-calcium-induced virulence mechanism, and the expression and secretion function of the T3SS are triggered at 37 °C when bacteria are grown in vitro in a calcium-depleted medium. Therefore, theT3SS was initially recognized as a low-calcium response (LCR) element. During infection of a mammalian host, the T3SS is activated and delivers Yops into the cytosol to modulate the host innate immune response, which enables bacterial survival and replication. Many studies of the biochemical functions of Yop effectors have been reported.

LcrV, the needle tip component of the T3SS injectisome, induces the production of interleukin (IL)-10 and suppresses the production of proinflammatory cytokines [91].

YopH is a potent tyrosine phosphatase that dephosphorylates a variety of functionally distinct substrates. It contains a multifunctional amino-terminal domain that mediates the translocation of YopH and targets it to its substrates and a carboxyl-terminal protein tyrosine phosphatase domain that resembles eukaryotic tyrosine phosphatases. Pioneering studies demonstrated that YopH dephosphorylates p130Cas and Fyb, leading to the loss of focal adhesions and the inhibition of phagocytosis [92–94]. Purified YopH inhibits T cell and B-lymphocyte activation in vitro, and YopH shows inhibitory effects when cell cultures were infected with *Y. pseudotuberculosis* [95]. Subsequent studies revealed that YopH dephosphorylates Lck and ZAP70, the major signal transducer for the T cell antigen receptor (TCR), which enables YopH to block the first step of TCR and suppress the development of an immune response against *Yersinia* [96]. *Y. pseudotuberculosis* infection of cultured cells showed an inhibitory effect on Ca^{2+} signaling in PMNs, which was dependent on YopH tyrosine phosphatase activity [97]. Using a β-lactamase activity-based reporter system, PMNs and inflammatory monocytes containing Yop effectors were isolated from mice infected with *Y. pseudotuberculosis*, and proteins that were significantly dephosphorylated in mice infected with a wild-type strain, but not those infected with a Δ*yopH* strain, were identified. The authors present an approach that can be applied to the study of effectors in animal model systems, which is still challenging, and the results revealed that YopH targets the PRAM-1/SKAP-HOM (PRAM-1, a promyelocytic leukemia (PML)-retinoic acid receptor α (RARα) target gene encoding an adaptor molecule-1; SKAP-HOM, SKAP55 homologue; SKAP55, the Src kinase-associated phosphoprotein) and the SLP-76/Vav/PLCg2 (SLP-76, SH2-containing leukocyte-specific protein of 76 kDa; Vav, proto-oncogene, critical for lymphocyte development and activation; PLCg2, phospholipase C, gamma 2) signaling axis to inhibit calcium flux in PMNs [98]. Mutation of *yopH* greatly attenuated virulence (3.4×10^6-fold increase in theLD_{50}) of *Y. pestis* in intravenously challenged mice. Notably, the virulence of *yopH* mutants of enteropathogenic *Yersinia* is attenuated as well, but to a much lesser degree (1000-fold increase in theLD_{50}), and it seems that YopH is more important for virulence in *Y. pestis* than in the enteropathogenic *Yersinia*. Because of the critical importance of the YopH tyrosine phosphatase in the inhibition of the host immune response, numerous groups are searching for small molecule inhibitors for YopH to develop new therapeutic reagents.

The T3SS injectisome delivers at least three effectors (YopE, YopT, and YpkA) that inhibit phagocytosis by professional phagocytes. All three effectors belong to a large family of bacterial toxins that target the Rho family of small GTP-binding proteins (Rho GTPases). YopE is a Rho GTPase-activating protein (GAP) that inactivates multiple Rho GTPases, including RhoA, Rac1, and Cdc42, by inducing their conversion from the active GTP-bound state to the inactive GDP-bound state [99, 100]. In the GDP-bound state, Rho GTPases are associated with the cellular membranes, where they carry out their cellular functions, via a thioether linkage bond between a prenyl group and a cysteine residue at their carboxyl terminus. Guanine nucleotide dissociation inhibitors (GDIs) negatively regulate the activity of GTPases by inhibiting the dissociation of GDP. YopE localizes to the plasma membrane, and the GAP activity of YopE promotes the hydrolysis of GTP to GDP, which prevents GTPases from responding to upstream signaling events. Although YopE is equally effective on Rac1, RhoA, and Cdc42 in vitro, it is preferably active on Rac1 and RhoA, but not Cdc42, in vivo. The virulence of a *Y. pestis yopE* mutant is attenuated greatly (1×10^4-fold increase in the LD_{50}) in an *i.v.* infection mouse model.

YpkA is a multidomain effector that contains an amino-terminal secretion signal peptide that is recognized by the T3SSapparatus, followed by a eukaryotic-like Ser/Thr kinase domain, a GDI domain, and a carboxyl-terminal actin-binding domain [101, 102]. YpkA is an inactive kinase in bacteria, and it is activated inside the host cytosol by binding to the coactivator actin that is abundant in eukaryotic cells [103]. Upon activation, YpkA undergoes autophosphorylation, and it phosphorylates a subset of intracellular substrates, including Gαq (the α subunit of heterotrimeric G proteins), vasodilator-stimulated phosphoprotein (VASP), otubain-1, and actin. The cytotoxic effects of YpkA predominately manifest as the disruption of the actin cytoskeleton, the disappearance of stress fibers, and the rounding up of cells, which were thought to be dependent on its kinase activity and GDI domain [101, 102, 104–106]. A membrane-localization domain located in its amino-terminal region has been shown to direct YpkA to the inner surface of the cell membrane where the GTPases carry out their functions [107]. The GDI domain of YpkA binds to GTP- or GDP-bound GTPases, and it mimics host GDI proteins to inhibit nucleotide exchange by RhoA and Rac1 [102]. YpkA has been shown to disrupt Gαq-mediated signaling pathways by phosphorylating Gαq at Ser47 in the GTP-binding pocket [101]. G proteins are involved in stimulating phospholipase-C-β, and some members activate GTPase-mediated pathways. VASP is phosphorylated by YpkA predominantly at Ser157, which leads to the inhibition of VASP-driven actin polymerization, the disruption of stress fiber formation, and macrophage phagocytosis. VASP belongs to the enabled/vasodilator-stimulated phosphoprotein family, and it plays pivotal roles in the regulation of the actin cytoskeleton, which has been shown to be targeted in infections caused by *Helicobacter pylori* and *Listeria monocytogenes* [108–110]. The kinase activity and the GDI domain of YpkA act synergistically to remodel the host cytoskeleton and disturb host cellular processes that are related to cytoskeleton rearrangements, such as phagocytosis, cell-cell junctions, and cell motility. In addition to its effects on the cytoskeleton, YpkA has been shown to trigger cell death, and amino-terminal amino acids 133–262 are responsible for inducing apoptosis [111]. The virulence of *Y. pseudotuberculosis* expressing a YpkA kinase-deficient mutant was greatly attenuated in a mouse infection model [112].

YopT acts as a papain-like cysteine protease that removes the prenyl group from RhoA, RhoG, Rac1, and Cdc42, which releases these GTPases from the membrane and inactivates them [113, 114]. YopT acts equally on GTP- and GDP-bound Rho GTPases, and a carboxyl-terminal sequence of basic amino acids of Rho GTPases is required for recognition by YopT. YopT has been shown to contribute to the antiphagocytic activity of bacteria, but it is not essential for the virulence of pathogenic *Yersinia*, because *yopT* mutants colonize tissues and kill mice as efficiently as wild-type strains. This may be because YopT is functionally redundant in the presence of YopE and YpkA.

YopM contains variable numbers of leucine-rich repeats (LRRs), which consist of approximately 20 amino acids, and the size of YopM proteins in different *Yersinia* strains is dependent on the number of LRRs. In *Y. pestis*, YopM contains 15 LRRs, except in biovar Microtus strains. A cryptographic analysis found that four monomers of YopM form a hollow cylindrical structure [115]. YopM has a nuclear localization signal at its carboxyl terminus, and it traffics to the nucleus of target cells by means of an endocytic pathway [116]. YopM can form a complex with ribosomal S6 protein kinase 1 (Rsk1) and protein kinase c-like 2 (Prk2) and activate them. Recent studies revealed that YopM is implicated in the downregulation of proinflammatory immune responses [117, 118]. *Y. pestis* infections cause a systemic depletion of natural killer cells in mice, which is YopM-dependent. In mice that were intravenously infected with *Y. pseudotuberculosis*, YopM was required for the induction of a high level of the immunosuppressive cytokine IL-10, and the IL-10 levels correlated with bacterial colonization in the spleen and liver of the infected mice [119]. LaRock et al. showed that YopM directly binds caspase-1 to inhibit caspase-1 activity and the maturation of the inflammasome. Nucleotide-binding oligomerization domain-like receptors (NLRs) are cytosolic receptors in eukaryotic cells, among which NLRC4, NLRP1, and NLRP3 form inflammasomes by recruiting an apoptosis-associated speck-like protein containing a CARD (carboxyl-terminal caspase recruitment domain) to activate caspase-1, which leads to the cleavage of pro-IL-1β to mature IL-β and inflammatory cell death via pyroptosis. YopM enables bacteria to avoid this host innate immunity strategy [117]. YopM is important for the virulence of pathogenic *Yersinia*. A *yopM* null mutant of *Y. pestis* was attenuated at least 10^5-fold in intravenously challenged mice, and it was rapidly eliminated from the spleen and liver.

YopJ functions as an acetyltransferase that inactivates NF-kB and MAPK pathways by acetylating MAPKK or IKK [99, 100]. Once translocated into host cells, Yop effectors act in a finely tuned and coordinated manner to hijack various host-signaling pathways to thwart the innate immune response [99, 100, 120]. YopJ is a cysteine protease of the CE clan. YopJ is thought to remove ubiquitin or a ubiquitin-like modification from target proteins in host cells. Mutation of *yopJ* in *Y. pestis* showed no obvious virulence attenuation, although greater virulence attenuations have been observed in enteropathogenic *Yersinia* [9].

YopK is identical among the three pathogenic *Yersinia* species, and the homologous *yopK* gene in *Y. pestis* and *Y. pseudotuberculosis* is called *yopQ* in *Y. enterocolitica*. YopK was initially proven to be essential for a systemic infection in mice by *Y. pestis* and *Y. pseudotuberculosis* [121–123]. Thereafter, YopK was shown to play important roles in controlling Yop translocation across the eukaryotic cell membrane, and a *yopK* mutant translocated more Yop effectors into host cells, thereby inducing more rapid cytotoxic responses than a wild-type strain [124]. Using a β-lactamase reporter assay, Dewoody et al. demonstrated that YopK controls the rate and fidelity of Yop injection into host cells [125, 126], and they further confirmed that YopE and YopK work at different steps in controlling the translocation of Yops, as YopK acts independently of YopE and controls Yop translocation from inside host cells [125]. Brodsky et al. proved that YopK interacts with the YopB-

YopD T3SS translocon and prevents the recognition of the *Yersinia* T3SS via an unknown mechanism, thereby leading to the inhibition of NLRP3 inflammasome activation [127]. Thorslund et al. found that YopK interacts with the receptor for activated C kinase (RACK1) and that this interaction promotes the phagocytosis resistance of *Y. pseudotuberculosis* [128]. Despite the invaluable information provided by these studies, the mechanism underlying YopK-dependent regulation of Yop translocation has yet to be clarified.

7.7 Interaction with the Host Immune Defense

A very high titer of bacteria in the bloodstream, which actually causes a grave terminal bacteremia (1×10^8 bacteria/mL), is required for *Y. pestis* to be transmitted by fleabites [129]. *Y. pestis* harbors a variety of anti-inflammatory mechanisms that allow bacteria to grow rapidly without being killed at the early stage of infection. To achieve this goal, *Y. pestis* has evolved a set of effectors to inhibit host inflammatory responses. LcrV of *Y. pestis* induces the production of the anti-inflammatory cytokine IL-10 via an interaction with TLR2 [130]. The Yop effectors that are translocated into the host cytoplasm by the T3SS can either disrupt the cytoskeleton (YopE, YopT, and YpkA), inhibit phagocytosis (YopH, YopE, YopT, and YpkA), or downregulate the inflammatory responses (YopJ) [100]. During the battle between *Y. pestis* and its host, the host responds strongly by upregulating immunological functions and the expression of many other immune response-associated genes.

After the onset of an infection, *Y. pestis* multiplies with extraordinary rapidity, and this feature is largely facilitated by its ability to suppress host innate immune responses. Studies of bubonic plague rodent models demonstrated that host infections with *Y. pestis* are typically biphasic [8, 131, 132]. *Y. pestis* multiplies rapidly in draining lymph nodes near the site of a fleabite, with no detectable inflammation at the early stage (6–36 hpi) of infection, and then the bacteria escape from the bubo and disseminate into the blood to colonize the liver, spleen, and lungs. The host immune response manifests itself by phagocyte infiltration, the production of inflammatory cytokines, and tissue necrosis during this stage of infection. A pathology investigation of a mouse intranasal model of primary pneumonic plague confirmed this impressive biphasic feature, in which the infection begins with an anti-inflammatory state in the first 24–36 hpi and rapidly progresses to a highly proinflammatory state by 48 hpi and death by 3 dpi [5]. Another study showed that robust neutrophil recruitment to the lungs was not observed until 48 hpi and proinflammatory chemokines in bronchoalveolar lavage fluids could not be readily detected as well, which coincided with an increase in PMN recruitment to the lungs [6]. A study compared disease progression and evolution in mice after *i.n.* inoculation of *Y. pestis* or *Y. pseudotuberculosis* [133]. The results demonstrated that both species reached draining lymph nodes in similar numbers at 24 hpi; however, bacterial loads were higher in *Y. pestis*-infected draining lymph nodes than in *Y. pseudotuberculosis*-infected draining lymph nodes at 2 dpi. Histopathology inspections illustrated that *Y. pseudotuberculosis* infection was accompanied by an abscess-type PMN cell infiltration, while *Y. pestis*-infected lymph nodes were typified by an invasion of the tissue by bacteria without PMN activation. This again shows that the ability of *Y. pestis* to inhibit the inflammatory response at the early stage of infection is vital for its extreme virulence following infection by the *s.c.* route.

Transcriptomic responses of animals or cultured cells infected with *Y. pestis* have been reported [131, 134–137], and the results revealed that *Y. pestis* inhibits the innate immune response in various aspects in different infection models. Among them, RNA sequencing technologies have been used to investigate the dynamic innate immune response in THP1 cells to *Y. pestis* [137]. It was found that the RIG-I- like receptor (RLR) signaling pathway that is thought to be responsible for antiviral defense is involved in the host response to *Y. pestis* infection, and a subsequent study showed that mice deficient in the mitochondrial antiviral signaling protein, an adaptor of the RLR signaling pathway, have higher resistance to *Y. pestis* in bubonic and pneumonic plague animal models [137].

References

1. Smiley ST. Immune defense against pneumonic plague. Immunol Rev. 2008;225:256–71.
2. Smiley ST. Current challenges in the development of vaccines for pneumonic plague. Expert Rev Vaccines. 2008;7(2):209–21.
3. Agar SL, Sha J, Foltz SM, Erova TE, Walberg KG, Parham TE, et al. Characterization of a mouse model of plague after aerosolization of Yersinia pestis CO92. Microbiology. 2008;154(Pt 7):1939–48.
4. Sha J, Agar SL, Baze WB, Olano JP, Fadl AA, Erova TE, et al. Braun lipoprotein (Lpp) contributes to virulence of yersiniae: potential role of Lpp in inducing bubonic and pneumonic plague. Infect Immun. 2008;76(4):1390–409.
5. Lathem WW, Crosby SD, Miller VL, Goldman WE. Progression of primary pneumonic plague: a mouse model of infection, pathology, and bacterial transcriptional activity. Proc Natl Acad Sci U S A. 2005;102(49):17786–91.
6. Bubeck SS, Cantwell AM, Dube PH. Delayed inflammatory response to primary pneumonic plague occurs in both outbred and inbred mice. Infect Immun. 2007;75(2):697–705.
7. Sebbane F, Jarrett C, Gardner D, Long D, Hinnebusch BJ. The Yersinia pestis caf1M1A1 fimbrial capsule operon promotes transmission by flea bite in a mouse model of bubonic plague. Infect Immun. 2009;77(3):1222–9.
8. Sebbane F, Gardner D, Long D, Gowen BB, Hinnebusch BJ. Kinetics of disease progression and host response in a rat model of bubonic plague. Am J Pathol. 2005;166(5):1427–39.
9. Lemaitre N, Sebbane F, Long D, Hinnebusch BJ. *Yersinia pestis* YopJ suppresses tumor necrosis factor alpha induction and contributes to apoptosis of immune cells in the lymph node but is not required for virulence in a rat model of bubonic plague. Infect Immun. 2006;74(9):5126–31.
10. Agar SL, Sha J, Foltz SM, Erova TE, Walberg KG, Baze WB, et al. Characterization of the rat pneumonic plague model: infection kinetics following aerosolization of *Yersinia pestis* CO92. Microbes Infect/Institut Pasteur. 2009;11(2):205–14.
11. Jones SM, Griffin KF, Hodgson I, Williamson ED. Protective efficacy of a fully recombinant plague vaccine in the guinea pig. Vaccine. 2003;21(25–26):3912–8.

12. Zhancui D, Yonghai Y, Shouhong Y, Yi Z, Zhimin Y, Aipin Z, et al. Pathology of in guinea pigs infected with plague bacillus. Mod Prev Med. 2015;42(5):899–901.
13. Cornelius CA, Quenee LE, Overheim KA, Koster F, Brasel TL, Elli D, et al. Immunization with recombinant V10 protects cynomolgus macaques from lethal pneumonic plague. Infect Immun. 2008;76(12):5588–97.
14. Andel RV, Sherwood R, Gennings C, Lyons CR, Hutt J, Gigliotti A, et al. Clinical and pathologic features of Cynomolgus Macaques (Macaca fascicularis) infected with aerosolized Yersinia pestis. Comp Med. 2008;58(1):68–75.
15. Koster F, Perlin DS, Park S, Brasel T, Gigliotti A, Barr E, et al. Milestones in progression of primary pneumonic plague in cynomolgus macaques. Infect Immun. 2010;78(7):2946–55.
16. Tian G, Qiu Y, Qi Z, Wu X, Zhang Q, Bi Y, et al. Histopathological observation of immunized rhesus macaques with plague vaccines after subcutaneous infection of Yersinia pestis. PLoS One. 2011;6(4):e19260.
17. Zhang Q, Wang Q, Tian G, Qi Z, Zhang X, Wu X, et al. Yersinia pestis biovar Microtus strain 201, an avirulent strain to humans, provides protection against bubonic plague in rhesus macaques. Hum Vacci Immunother. 2014;10(2):1–10.
18. Tian G, Qi Z, Qiu Y, Wu X, Zhang Q, Yang X, et al. Comparison of virulence between the Yersinia pestis Microtus 201, an avirulent strain to humans, and the vaccine strain EV in rhesus macaques, Macaca mulatta. Hum Vaccin Immunother. 2015;10(12):3552–60.
19. Finegold MJ, Petery JJ, Berendt RF, Adams HR. Studies on the pathogenesis of plague. Blood coagulation and tissue responses of Macaca mulatta following exposure to aerosols of Pasteurella pestis. Am J Pathol. 1968;53(1):99–114.
20. Guarner J, Shieh WJ, Greer PW, Gabastou JM, Chu M, Hayes E, et al. Immunohistochemical detection of Yersinia pestis in formalin-fixed, paraffin-embedded tissue. Am J Clin Pathol. 2002;117:205–9.
21. Achtman M, Zurth K, Morelli G, Torrea G, Guiyoule A, Carniel E. *Yersinia pestis*, the cause of plague, is a recently emerged clone of *Yersinia pseudotuberculosis*. Proc Natl Acad Sci U S A. 1999;96(24):14043–8.
22. Chain PS, Carniel E, Larimer FW, Lamerdin J, Stoutland PO, Regala WM, et al. Insights into the evolution of *Yersinia pestis* through whole-genome comparison with *Yersinia pseudotuberculosis*. Proc Natl Acad Sci U S A. 2004;101(38):13826–31.
23. Armougom F, Bitam I, Croce O, Merhej V, Barassi L, Nguyen TT, et al. Genomic insights into a new Citrobacter koseri strain revealed gene exchanges with the virulence-associated Yersinia pestis pPCP1 plasmid. Front Microbiol. 2016;7:340.
24. Kukkonen M, Suomalainen M, Kyllonen P, Lahteenmaki K, Lang H, Virkola R, et al. Lack of O-antigen is essential for plasminogen activation by *Yersinia pestis* and *Salmonella enterica*. Mol Microbiol. 2004;51(1):215–25.
25. Montminy SW, Khan N, McGrath S, Walkowicz MJ, Sharp F, Conlon JE, et al. Virulence factors of *Yersinia pestis* are overcome by a strong lipopolysaccharide response. Nat Immunol. 2006;7(10):1066–73.
26. Kawahara K, Tsukano H, Watanabe H, Lindner B, Matsuura M. Modification of the structure and activity of lipid A in *Yersinia pestis* lipopolysaccharide by growth temperature. Infect Immun. 2002;70(8):4092–8.
27. Prentice MB, Rahalison L. Plague. Lancet. 2007;369:1196–207.
28. Hinnebusch BJ. Biofilm-dependent and biofilm-independent mechanisms of transmission of Yersinia pestis by fleas. Adv Exp Med Biol. 2012;954:237–43.
29. Jarrett CO, Deak E, Isherwood KE, Oyston PC, Fischer ER, Whitney AR, et al. Transmission of *Yersinia pestis* from an infectious biofilm in the flea vector. J Infect Dis. 2004;190(4):783–92.
30. Bobrov AG, Kirillina O, Ryjenkov DA, Waters CM, Price PA, Fetherston JD, et al. Systematic analysis of cyclic di-GMP signalling enzymes and their role in biofilm formation and virulence in Yersinia pestis. Mol Microbiol. 2011;79(2):533–51.

31. Sun YC, Guo XP, Hinnebusch BJ, Darby C. The Yersinia pestis Rcs phosphorelay inhibits biofilm formation by repressing transcription of the diguanylate cyclase gene hmsT. J Bacteriol. 2012;194(8):2020–6.
32. Sun YC, Hinnebusch BJ, Darby C. Experimental evidence for negative selection in the evolution of a *Yersinia pestis* pseudogene. Proc Natl Acad Sci U S A. 2008;105(23):8097–101.
33. Sun YC, Jarrett CO, Bosio CF, Hinnebusch BJ. Retracing the evolutionary path that led to flea-borne transmission of Yersinia pestis. Cell Host Microbe. 2014;15(5):578–86.
34. Heise T, Dersch P. Identification of a domain in Yersinia virulence factor YadA that is crucial for extracellular matrix-specific cell adhesion and uptake. Proc Natl Acad Sci U S A. 2006;103(9):3375–80.
35. Marra A, Isberg RR. Invasin-dependent and invasin-independent pathways for translocation of Yersinia pseudotuberculosis across the Peyer's patch intestinal epithelium. Infect Immun. 1997;65(8):3412–21.
36. Rosqvist R, Skurnik M, Wolf-Watz H. Increased virulence of Yersinia pseudotuberculosis by two independent mutations. Nature. 1988;334(6182):522–4.
37. Casutt-Meyer S, Renzi F, Schmaler M, Jann NJ, Amstutz M, Cornelis GR. Oligomeric coiled-coil adhesin YadA is a double-edged sword. PLoS One. 2010;5(12):e15159.
38. Muhlenkamp M, Oberhettinger P, Leo JC, Linke D, Schutz MS. Yersinia adhesin A (YadA)–beauty & beast. Int J Med Microbiol. 2015;305(2):252–8.
39. Sodeinde OA, Goguen JD. Genetic analysis of the 9.5-kilobase virulence plasmid of Yersinia pestis. Infect Immun. 1988;56(10):2743–8.
40. Zimbler DL, Schroeder JA, Eddy JL, Lathem WW. Early emergence of Yersinia pestis as a severe respiratory pathogen. Nat Commun. 2015;6:7487.
41. Lahteenmaki K, Virkola R, Saren A, Emody L, Korhonen TK. Expression of plasminogen activator pla of *Yersinia pestis* enhances bacterial attachment to the mammalian extracellular matrix. Infect Immun. 1998;66(12):5755–62.
42. Friedlander AM, Welkos SL, Worsham PL, Andrews GP, Heath DG, Anderson Jr GW, et al. Relationship between virulence and immunity as revealed in recent studies of the F1 capsule of Yersinia pestis. Clin Infect Dis. 1995;21 Suppl 2:S178–81.
43. Drozdov IG, Anisimov AP, Samoilova SV, Yezhov IN, Yeremin SA, Karlyshev AV, et al. Virulent non-capsulate Yersinia pestis variants constructed by insertion mutagenesis. J Med Microbiol. 1995;42(4):264–8.
44. Sha J, Endsley JJ, Kirtley ML, Foltz SM, Huante MB, Erova TE, et al. Characterization of an F1 deletion mutant of Yersinia pestis CO92, pathogenic role of F1 antigen in bubonic and pneumonic plague, and evaluation of sensitivity and specificity of F1 antigen capture-based dipsticks. J Clin Microbiol. 2011;49(5):1708–15.
45. Samoilova SV, Samoilova LV, Yezhov IN, Drozdov IG, Anisimov AP. Virulence of pPst + and pPst- strains of Yersinia pestis for guinea-pigs. J Med Microbiol. 1996;45(6):440–4.
46. Du Y, Rosqvist R, Forsberg A. Role of fraction 1 antigen of *Yersinia pestis* in inhibition of phagocytosis. Infect Immun. 2002;70(3):1453–60.
47. Fetherston JD, Kirillina O, Bobrov AG, Paulley JT, Perry RD. The yersiniabactin transport system is critical for the pathogenesis of bubonic and pneumonic plague. Infect Immun. 2010;78(5):2045–52.
48. Perry RD, Fetherston JD. Yersiniabactin iron uptake: mechanisms and role in Yersinia pestis pathogenesis. Microbes Infect/Institut Pasteur. 2011;13(10):808–17.
49. Bearden SW, Fetherston JD, Perry RD. Genetic organization of the yersiniabactin biosynthetic region and construction of avirulent mutants in *Yersinia pestis*. Infect Immun. 1997;65(5):1659–68.
50. Fetherston JD, Bertolino VJ, Perry RD. YbtP and YbtQ: two ABC transporters required for iron uptake in *Yersinia pestis*. Mol Microbiol. 1999;32(2):289–99.
51. Sebbane F, Jarrett C, Gardner D, Long D, Hinnebusch BJ. Role of the *Yersinia pestis* yersiniabactin iron acquisition system in the incidence of flea-borne plague. PLoS One. 2010;5(12):e14379.

52. Pieper R, Huang ST, Parmar PP, Clark DJ, Alami H, Fleischmann RD, et al. Proteomic analysis of iron acquisition, metabolic and regulatory responses of Yersinia pestis to iron starvation. BMC Microbiol. 2010;10:30.
53. Bearden SW, Perry RD. The Yfe system of *Yersinia pestis* transports iron and manganese and is required for full virulence of plague. Mol Microbiol. 1999;32(2):403–14.
54. Perry RD, Fetherston JD. *Yersinia pestis*–etiologic agent of plague. Clin Microbiol Rev. 1997;10(1):35–66.
55. Lindler LE, Tall BD. Yersinia pestis pH 6 antigen forms fimbriae and is induced by intracellular association with macrophages. Mol Microbiol. 1993;8(2):311–24.
56. Payne D, Tatham D, Williamson ED, Titball RW. The pH 6 antigen of *Yersinia pestis* binds to beta1-linked galactosyl residues in glycosphingolipids. Infect Immun. 1998;66(9):4545–8.
57. Galvan EM, Chen H, Schifferli DM. The Psa fimbriae of Yersinia pestis interact with phosphatidylcholine on alveolar epithelial cells and pulmonary surfactant. Infect Immun. 2007;75(3):1272–9.
58. Bao R, Nair MK, Tang WK, Esser L, Sadhukhan A, Holland RL, et al. Structural basis for the specific recognition of dual receptors by the homopolymeric pH 6 antigen (Psa) fimbriae of Yersinia pestis. Proc Natl Acad Sci U S A. 2013;110(3):1065–70.
59. Zav'yalov VP, Abramov VM, Cherepanov PG, Spirina GV, Chernovskaya TV, Vasiliev AM, et al. pH6 antigen (PsaA protein) of *Yersinia pestis*, a novel bacterial Fc-receptor. FEMS Immunol Med Microbiol. 1996;14(1):53–7.
60. Huang XZ, Lindler LE. The pH 6 antigen is an antiphagocytic factor produced by *Yersinia pestis* independent of *Yersinia* outer proteins and capsule antigen. Infect Immun. 2004;72(12):7212–9.
61. Felek S, Tsang TM, Krukonis ES. Three Yersinia pestis adhesins facilitate Yop delivery to eukaryotic cells and contribute to plague virulence. Infect Immun. 2010;78(10):4134–50.
62. Lindler LE, Klempner MS, Straley SC. *Yersinia pestis* pH 6 antigen: genetic, biochemical, and virulence characterization of a protein involved in the pathogenesis of bubonic plague. Infect Immun. 1990;58(8):2569–77.
63. Anisimov AP, Bakhteeva IV, Panfertsev EA, Svetoch TE, Kravchenko TB, Platonov ME, et al. The subcutaneous inoculation of pH 6 antigen mutants of Yersinia pestis does not affect virulence and immune response in mice. J Med Microbiol. 2009;58(Pt 1):26–36.
64. Felek S, Krukonis ES. The Yersinia pestis Ail protein mediates binding and Yop delivery to host cells required for plague virulence. Infect Immun. 2009;77(2):825–36.
65. Kolodziejek AM, Schnider DR, Rohde HN, Wojtowicz AJ, Bohach GA, Minnich SA, et al. Outer membrane protein X (Ail) contributes to Yersinia pestis virulence in pneumonic plague and its activity is dependent on the lipopolysaccharide core length. Infect Immun. 2010;78(12):5233–43.
66. Pierson DE. Mutations affecting lipopolysaccharide enhance ail-mediated entry of Yersinia enterocolitica into mammalian cells. J Bacteriol. 1994;176(13):4043–51.
67. Bartra SS, Styer KL, O'Bryant DM, Nilles ML, Hinnebusch BJ, Aballay A, et al. Resistance of Yersinia pestis to complement-dependent killing is mediated by the Ail outer membrane protein. Infect Immun. 2008;76(2):612–22.
68. Yamashita S, Lukacik P, Barnard TJ, Noinaj N, Felek S, Tsang TM, et al. Structural insights into Ail-mediated adhesion in Yersinia pestis. Structure. 2011;19(11):1672–82.
69. Miller VL, Beer KB, Heusipp G, Young BM, Wachtel MR. Identification of regions of Ail required for the invasion and serum resistance phenotypes. Mol Microbiol. 2001;41(5):1053–62.
70. Tsang TM, Felek S, Krukonis ES. Ail binding to fibronectin facilitates Yersinia pestis binding to host cells and Yop delivery. Infect Immun. 2010;78(8):3358–68.
71. Ho DK, Riva R, Kirjavainen V, Jarva H, Ginstrom E, Blom AM, et al. Functional recruitment of the human complement inhibitor C4BP to Yersinia pseudotuberculosis outer membrane protein Ail. J Immunol. 2012;188(9):4450–9.

72. Schesser Bartra S, Ding Y, Fujimoto LM, Ring JG, Jain V, Ram S, et al. Yersinia pestis uses the Ail outer membrane protein to recruit vitronectin. Microbiology. 2015;116(11):2174–83.
73. Forman S, Wulff CR, Myers-Morales T, Cowan C, Perry RD. Straley SC: *yadBC* of Yersinia pestis, a new virulence determinant for bubonic plague. Infect Immun. 2008;76(2):578–87.
74. Uittenbogaard AM, Myers-Morales T, Gorman AA, Welsh E, Wulff C, Hinnebusch BJ, et al. Temperature-dependence of yadBC phenotypes in Yersinia pestis. Microbiology. 2014;160(Pt 2):396–405.
75. Lukaszewski RA, Kenny DJ, Taylor R, Rees DG, Hartley MG, Oyston PC. Pathogenesis of Yersinia pestis infection in BALB/c mice: effects on host macrophages and neutrophils. Infect Immun. 2005;73(11):7142–50.
76. Cornelis GR, Boland A, Boyd AP, Geuijen C, Iriarte M, Neyt C, et al. The virulence plasmid of *Yersinia*, an antihost genome. Microbiol Mol Biol Rev. 1998;62(4):1315–52.
77. Pujol C, Klein KA, Romanov GA, Palmer LE, Cirota C, Zhao Z, et al. *Yersinia pestis* can reside in autophagosomes and avoid xenophagy in murine macrophages by preventing vacuole acidification. Infect Immun. 2009;77(6):2251–61.
78. Pujol C, Grabenstein JP, Perry RD, Bliska JB. Replication of Yersinia pestis in interferon gamma-activated macrophages requires ripA, a gene encoded in the pigmentation locus. Proc Natl Acad Sci U S A. 2005;102(36):12909–14.
79. Spinner JL, Winfree S, Starr T, Shannon JG, Nair V, Steele-Mortimer O, et al. Yersinia pestis survival and replication within human neutrophil phagosomes and uptake of infected neutrophils by macrophages. J Leukoc Biol. 2014;95(3):389–98.
80. Torres R, Swift RV, Chim N, Wheatley N, Lan B, Atwood BR, et al. Biochemical, structural and molecular dynamics analyses of the potential virulence factor RipA from Yersinia pestis. PLoS One. 2011;6(9):e25084.
81. Torres R, Lan B, Latif Y, Chim N, Goulding CW. Structural snapshots along the reaction pathway of Yersinia pestis RipA, a putative butyryl-CoA transferase. Acta Crystallogr D Biol Crystallogr. 2014;70(Pt 4):1074–85.
82. Torres R, Chim N, Sankaran B, Pujol C, Bliska JB, Goulding CW. Structural insights into RipC, a putative citrate lyase beta subunit from a *Yersinia pestis* virulence operon. Acta Crystallogr Sect F: Struct Biol Cryst Commun. 2012;68(Pt 1):2–7.
83. Grabenstein JP, Marceau M, Pujol C, Simonet M, Bliska JB. The response regulator PhoP of *Yersinia pseudotuberculosis* is important for replication in macrophages and for virulence. Infect Immun. 2004;72(9):4973–84.
84. Oyston PC, Dorrell N, Williams K, Li SR, Green M, Titball RW, et al. The response regulator PhoP is important for survival under conditions of macrophage-induced stress and virulence in *Yersinia pestis*. Infect Immun. 2000;68(6):3419–25.
85. Pisano F, Heine W, Rosenheinrich M, Schweer J, Nuss AM, Dersch P. Influence of PhoP and intra-species variations on virulence of Yersinia pseudotuberculosis during the natural oral infection route. PLoS One. 2014;9(7):e103541.
86. Bozue J, Mou S, Moody KL, Cote CK, Trevino S, Fritz D, et al. The role of the phoPQ operon in the pathogenesis of the fully virulent CO92 strain of Yersinia pestis and the IP32953 strain of Yersinia pseudotuberculosis. Microb Pathog. 2011;50(6):314–21.
87. Bartra SS, Gong X, Lorica CD, Jain C, Nair MK, Schifferli D, et al. The outer membrane protein A (OmpA) of Yersinia pestis promotes intracellular survival and virulence in mice. Microb Pathog. 2012;52(1):41–6.
88. Klein KA, Fukuto HS, Pelletier M, Romanov G, Grabenstein JP, Palmer LE, et al. A transposon site hybridization screen identifies galU and wecBC as important for survival of Yersinia pestis in murine macrophages. J Bacteriol. 2012;194(3):653–62.
89. Maldonado-Arocho FJ, Green C, Fisher ML, Paczosa MK, Mecsas J. Adhesins and host serum factors drive Yop translocation by yersinia into professional phagocytes during animal infection. PLoS Pathog. 2013;9(6):e1003415.

90. Marketon MM, DePaolo RW, DeBord KL, Jabri B, Schneewind O. Plague bacteria target immune cells during infection. Science. 2005;309(5741):1739–41.
91. Sing A, Rost D, Tvardovskaia N, Roggenkamp A, Wiedemann A, Kirschning CJ, et al. Yersinia V-antigen exploits toll-like receptor 2 and CD14 for interleukin 10-mediated immunosuppression. J Exp Med. 2002;196(8):1017–24.
92. Black DS, Bliska JB. Identification of p130Cas as a substrate of Yersinia YopH (Yop51), a bacterial protein tyrosine phosphatase that translocates into mammalian cells and targets focal adhesions. EMBO J. 1997;16(10):2730–44.
93. Persson C, Carballeira N, Wolf-Watz H, Fallman M. The PTPase YopH inhibits uptake of Yersinia, tyrosine phosphorylation of p130Cas and FAK, and the associated accumulation of these proteins in peripheral focal adhesions. EMBO J. 1997;16(9):2307–18.
94. Black DS, Marie-Cardine A, Schraven B, Bliska JB. The Yersinia tyrosine phosphatase YopH targets a novel adhesion-regulated signalling complex in macrophages. Cell Microbiol. 2000;2(5):401–14.
95. Yao T, Mecsas J, Healy JI, Falkow S, Chien Y. Suppression of T and B lymphocyte activation by a Yersinia pseudotuberculosis virulence factor, yopH. J Exp Med. 1999;190(9):1343–50.
96. Alonso A, Bottini N, Bruckner S, Rahmouni S, Williams S, Schoenberger SP, et al. Lck dephosphorylation at Tyr-394 and inhibition of T cell antigen receptor signaling by Yersinia phosphatase YopH. J Biol Chem. 2004;279(6):4922–8.
97. Andersson K, Magnusson KE, Majeed M, Stendahl O, Fallman M. Yersinia pseudotuberculosis-induced calcium signaling in neutrophils is blocked by the virulence effector YopH. Infect Immun. 1999;67(5):2567–74.
98. Rolan HG, Durand EA, Mecsas J. Identifying Yersinia YopH-targeted signal transduction pathways that impair neutrophil responses during in vivo murine infection. Cell Host Microbe. 2013;14(3):306–17.
99. Shao F. Biochemical functions of Yersinia type III effectors. Curr Opin Microbiol. 2008;11(1):21–9.
100. Trosky JE, Liverman AD, Orth K. Yersinia outer proteins: Yops. Cell Microbiol. 2008;10(3):557–65.
101. Navarro L, Koller A, Nordfelth R, Wolf-Watz H, Taylor S, Dixon JE. Identification of a molecular target for the Yersinia protein kinase A. Mol Cell. 2007;26(4):465–77.
102. Prehna G, Ivanov MI, Bliska JB, Stebbins CE. Yersinia virulence depends on mimicry of host Rho-family nucleotide dissociation inhibitors. Cell. 2006;126(5):869–80.
103. Trasak C, Zenner G, Vogel A, Yuksekdag G, Rost R, Haase I, et al. Yersinia protein kinase YopO is activated by a novel G-actin binding process. J Biol Chem. 2007;282(4):2268–77.
104. Dukuzumuremyi JM, Rosqvist R, Hallberg B, Akerstrom B, Wolf-Watz H, Schesser K. The Yersinia protein kinase A is a host factor inducible RhoA/Rac-binding virulence factor. J Biol Chem. 2000;275(45):35281–90.
105. Juris SJ, Rudolph AE, Huddler D, Orth K, Dixon JE. A distinctive role for the Yersinia protein kinase: actin binding, kinase activation, and cytoskeleton disruption. Proc Natl Acad Sci U S A. 2000;97(17):9431–6.
106. Letzelter M, Sorg I, Mota LJ, Meyer S, Stalder J, Feldman M, et al. The discovery of SycO highlights a new function for type III secretion effector chaperones. EMBO J. 2006;25(13):3223–33.
107. Hakansson S, Galyov EE, Rosqvist R, Wolf-Watz H. The Yersinia YpkA Ser/Thr kinase is translocated and subsequently targeted to the inner surface of the HeLa cell plasma membrane. Mol Microbiol. 1996;20(3):593–603.
108. Bear JE, Gertler FB. Ena/VASP: towards resolving a pointed controversy at the barbed end. J Cell Sci. 2009;122(Pt 12):1947–53.
109. Bierne H, Miki H, Innocenti M, Scita G, Gertler FB, Takenawa T, et al. WASP-related proteins, Abi1 and Ena/VASP are required for Listeria invasion induced by the Met receptor. J Cell Sci. 2005;118(Pt 7):1537–47.

110. Auerbuch V, Loureiro JJ, Gertler FB, Theriot JA, Portnoy DA. Ena/VASP proteins contribute to Listeria monocytogenes pathogenesis by controlling temporal and spatial persistence of bacterial actin-based motility. Mol Microbiol. 2003;49(5):1361–75.
111. Park H, Teja K, O'Shea JJ, Siegel RM. The Yersinia effector protein YpkA induces apoptosis independently of actin depolymerization. J Immunol. 2007;178(10):6426–34.
112. Wiley DJ, Nordfeldth R, Rosenzweig J, DaFonseca CJ, Gustin R, Wolf-Watz H, et al. The Ser/Thr kinase activity of the Yersinia protein kinase A (YpkA) is necessary for full virulence in the mouse, mollifying phagocytes, and disrupting the eukaryotic cytoskeleton. Microb Pathog. 2006;40(5):234–43.
113. Shao F, Merritt PM, Bao Z, Innes RW, Dixon JE. A Yersinia effector and a Pseudomonas avirulence protein define a family of cysteine proteases functioning in bacterial pathogenesis. Cell. 2002;109(5):575–88.
114. Shao F, Vacratsis PO, Bao Z, Bowers KE, Fierke CA, Dixon JE. Biochemical characterization of the Yersinia YopT protease: cleavage site and recognition elements in Rho GTPases. Proc Natl Acad Sci U S A. 2003;100(3):904–9.
115. Evdokimov AG, Anderson DE, Routzahn KM, Waugh DS. Unusual molecular architecture of the *Yersinia pestis* cytotoxin YopM: a leucine-rich repeat protein with the shortest repeating unit. J Mol Biol. 2001;312(4):807–21.
116. Skrzypek E, Cowan C, Straley SC. Targeting of the *Yersinia pestis* YopM protein into HeLa cells and intracellular trafficking to the nucleus. Mol Microbiol. 1998;30(5):1051–65.
117. LaRock CN, Cookson BT. The Yersinia virulence effector YopM binds caspase-1 to arrest inflammasome assembly and processing. Cell Host Microbe. 2012;12(6):799–805.
118. Kerschen EJ, Cohen DA, Kaplan AM, Straley SC. The plague virulence protein YopM targets the innate immune response by causing a global depletion of NK cells. Infect Immun. 2004;72(8):4589–602.
119. McPhee JB, Mena P, Zhang Y, Bliska JB. Interleukin-10 induction is an important virulence function of the Yersinia pseudotuberculosis type III effector YopM. Infect Immun. 2012;80(7):2519–27.
120. Viboud GI, Bliska JB. *Yersinia* outer proteins: role in modulation of host cell signaling responses and pathogenesis. Annu Rev Microbiol. 2005;59:69–89.
121. Straley SC, Cibull ML. Differential clearance and host-pathogen interactions of YopE- and YopK- YopL- Yersinia pestis in BALB/c mice. Infect Immun. 1989;57(4):1200–10.
122. Holmstrom A, Rosqvist R, Wolf-Watz H, Forsberg A. Virulence plasmid-encoded YopK is essential for Yersinia pseudotuberculosis to cause systemic infection in mice. Infect Immun. 1995;63(6):2269–76.
123. Holmstrom A, Rosqvist R, Wolf-Watz H, Forsberg A. YopK, a novel virulence determinant of Yersinia pseudotuberculosis. Contrib Microbiol Immunol. 1995;13:239–43.
124. Holmstrom A, Petterson J, Rosqvist R, Hakansson S, Tafazoli F, Fallman M, et al. YopK of Yersinia pseudotuberculosis controls translocation of Yop effectors across the eukaryotic cell membrane. Mol Microbiol. 1997;24(1):73–91.
125. Dewoody R, Merritt PM, Houppert AS, Marketon MM. YopK regulates the Yersinia pestis type III secretion system from within host cells. Mol Microbiol. 2011;79(6):1445–61.
126. Dewoody R, Merritt PM, Marketon MM. YopK controls both rate and fidelity of Yop translocation. Mol Microbiol. 2013;87(2):301–17.
127. Brodsky IE, Palm NW, Sadanand S, Ryndak MB, Sutterwala FS, Flavell RA, et al. A Yersinia effector protein promotes virulence by preventing inflammasome recognition of the type III secretion system. Cell Host Microbe. 2010;7(5):376–87.
128. Thorslund SE, Edgren T, Pettersson J, Nordfelth R, Sellin ME, Ivanova E, et al. The RACK1 signaling scaffold protein selectively interacts with Yersinia pseudotuberculosis virulence function. PLoS One. 2011;6(2):e16784.
129. Butler T. Plague into the 21st century. Clin Infect Dis. 2009;49(5):736–42.

130. Sing A, Reithmeier-Rost D, Granfors K, Hill J, Roggenkamp A, Heesemann J. A hypervariable N-terminal region of Yersinia LcrV determines Toll-like receptor 2-mediated IL-10 induction and mouse virulence. Proc Natl Acad Sci U S A. 2005;102(44):16049–54.
131. Comer JE, Sturdevant DE, Carmody AB, Virtaneva K, Gardner D, Long D, et al. Transcriptomic and innate immune responses to Yersinia pestis in the lymph node during bubonic plague. Infect Immun. 2010;78(12):5086–98.
132. Bergsbaken T, Cookson BT. Innate immune response during Yersinia infection: critical modulation of cell death mechanisms through phagocyte activation. J Leukoc Biol. 2009;86(5):1153–8.
133. Guinet F, Ave P, Jones L, Huerre M, Carniel E. Defective innate cell response and lymph node infiltration specify Yersinia pestis infection. PLoS One. 2008;3(2):e1688.
134. Rogers JV, Choi YW, Giannunzio LF, Sabourin PJ, Bornman DM, Blosser EG, et al. Transcriptional responses in spleens from mice exposed to Yersinia pestis CO92. Microb Pathog. 2007;43(2–3):67–77.
135. Galindo CL, Moen ST, Kozlova EV, Sha J, Garner HR, Agar SL, et al. Comparative analyses of transcriptional profiles in mouse organs using a pneumonic plague model after infection with wild-type Yersinia pestis CO92 and its Braun lipoprotein mutant. Comp Funct Genomics. 2009;2009:914762.
136. Liu H, Wang H, Qiu J, Wang X, Guo Z, Qiu Y, et al. Transcriptional profiling of a mice plague model: insights into interaction between Yersinia pestis and its host. J Basic Microbiol. 2009;49(1):92–9.
137. Du Z, Yang H, Tan Y, Tian G, Zhang Q, Cui Y, et al. Transcriptomic response to Yersinia pestis: RIG-I like receptor signaling response is detrimental to the host against plague. J Genet Genomics Yi chuan xue bao. 2014;41(7):379–96.

Chapter 8
Genetic Regulation of *Yersinia pestis*

Yanping Han, Haihong Fang, Lei Liu, and Dongsheng Zhou

Abstract *Y. pestis* exhibits dramatically different traits of pathogenicity and transmission, albeit their close genetic relationship with its ancestor—*Y. pseudotuberculosis*, a self-limiting gastroenteric pathogen. *Y. pestis* is evolved into a deadly pathogen and transmitted to mammals and/or human beings by infected flea biting or directly contacting with the infected animals. Various kinds of environmental changes are implicated into its complex life cycle and pathogenesis. Dynamic regulation of gene expression is critical for environmental adaptation or survival, primarily reflected by genetic regulation mediated by transcriptional factors and small regulatory RNAs at the transcriptional and posttranscriptional level, respectively. The effects of genetic regulation have been shown to profoundly influence *Y. pestis* physiology and pathogenesis such as stress resistance, biofilm formation, intracellular survival, and replication. In this chapter, we mainly summarize the progresses on popular methods of genetic regulation and on regulatory patterns and consequences of many key transcriptional and posttranscriptional regulators, with a particular emphasis on how genetic regulation influences the biofilm and virulence of *Y. pestis*.

Keywords *Yersinia pestis* • Gene expression • Regulation • Biofilm • Virulence

Yersinia pestis experiences various kinds of environmental changes during transmission and infection. Environmental modulation of gene expression in *Y. pestis* is critical for its complex life cycle and pathogenesis. *Y. pestis* adapts appropriate responses, primarily reflected by changes in transcription of specific sets of genes or small regulatory sRNAs.

Y. Han (✉) • H. Fang • L. Liu • D. Zhou
Beijing Institute of Microbiology and Epidemiology,
No. Dongdajie, Fengtai, Beijing 100071, China
e-mail: hypiota@hotmail.com

R. Yang, A. Anisimov (eds.), *Yersinia pestis: Retrospective and Perspective*,
Advances in Experimental Medicine and Biology 918,
DOI 10.1007/978-94-024-0890-4_8

8.1 Transcriptional Regulation of Gene Expression

To adapt to various environmental stresses, including those caused by host environments, bacteria have developed complex regulatory networks that respond to multiple changes in growth conditions. Bacterial gene regulation, especially transcriptional regulation of gene expression, plays a crucial role in the adaptation of bacteria to their environments. Although other regulatory mechanisms exist, transcriptional regulation is the most fundamental and well studied in bacteria. The transcription factors (TFs) involved in this process are key components of adaptive responses, and their study is essential. The methods and patterns of gene expression regulation by TFs and transcriptional regulation of gene expression in *Y. pestis* will be discussed below.

8.1.1 Methods for Revealing Transcriptional Regulation by TFs

During transcriptional regulation in prokaryotes, gene expression is controlled during RNA synthesis by regulators that can interact with specific regulatory DNA elements. Transcriptional regulators are a type of *trans*-acting factor in bacteria. They also act as cofactors of RNA polymerase during initiation of transcription [1]. Prior to functional studies being performed, the type and target genes of a candidate TF must be identified. There are several methods to investigate TFs and their target genes in bacteria that will be summarized as follows.

8.1.1.1 Prediction of Transcription Factors Based on the Genome Information

The preliminary approaches of TF prediction are based on the complete nucleotide information and the annotation by aligning factors homologous and the special functional domains to known TFs [2]. The further illustration of TF is derived from the computational collection and evaluation of DNA-binding motifs. In prokaryote, TF prediction is usually referred to the information of *E. coli* as the model [3, 4]. It is also able to predict genome-wide TFs according to the information from the isogenic organisms, especially ones with verified functions [5, 6].

The online or offline results can be received through the professional software and database utilizing to compare the homology and identity of the domains, protein families, or three-dimensional structure with the large majority of *E. coli* TFs and the regulated genes. Clusters of orthologous groups (COGs) involved in transcriptional regulation were analyzed from the COG database [7, 8]. The profile of candidate TFs can be evaluated using hidden Markov models (HMMs) of domains from the SUPERFAMILY [9] and Pfam [10] databases. CATH-Gene3D (http://www.cathdb.info/)

database has collected 26 million protein domains classified into 2738 superfamilies, and it can be used to predict protein functions based on the structures [11]. Candidated TFs can be predicted and verified primarily in this website.

More public databases for prokaryotic transcriptional regulation prediction were collected and reviewed in reference [12].

8.1.1.2 DNA Pull-down Based on Affinity Chromatography Strategies to Determine TF Target Specificity

Once a candidate TF is identified using the prediction methods described above, its regulatory target genes need to be identified. To isolate and identify sequence-specific DNA-binding proteins from cellular extracts, DNA pull-down strategies, including DNA-affinity chromatography and gel mobility shift assays, can be used.

DNA probes containing a TF-binding site are either adsorbed or linked covalently to a chromatographic support, and whole-cell extract can be flowed through them. Proteins that bind to the immobilized probes are retained, while unbound material is washed away. TFs specifically binding to the DNA probes are then eluted, isolated, and characterized. Eluted proteins are separated using -SDS-polyacrylamide gel electrophoresis (SDS-PAGE) and identified using mass spectrometry (MS) [13].

DNA-affinity chromatography has been improved with many modifications and has been widely applied in the isolation and identification of TFs [14, 15]. DNA-affinity capture assay (DACA) coupled with liquid chromatography-tandem mass spectrometry analysis was applied to identify the transcriptional regulators involved in the biosynthesis of actinorhodin (Act) and undecylprodigiosin (Red) in *Streptomyces coelicolor* [16].

8.1.1.3 cDNA Microarray Expression Profiling and RNA Sequencing Methods Used to Screen Target Gene Expression

Complementary DNA (cDNA) microarray is widely used to evaluate mRNA abundance of all the genes from the whole cell on a pretreated chip. After the interesting TF is selected and the corresponding gene is deleted, a typical two-sample experiment based on the cDNA microarray can be designed. RNAs are extracted under the instructions of PureLink ™ RNA Mini Kit (Ambion, USA) from the wild-type and deletion strains, respectively, labeled with different fluorescein dyes. The special kinds of cDNA originated from the information of genome are synthetized and then dotted on a chip of microarray using an instrument automatically. The labeled RNAs are co-hybridized with the pretreated cDNA microarray, and the hybridized signals are scanned, and data are analyzed with the corresponding procedures. In general, it takes two times change of fluorescence intensity as the difference judgment standard. The differentially expressed genes can be recognized in the target genes modulated by the interesting TF for further research [17, 18].

With the advent of low-cost, high-throughput sequencing technology, RNA sequencing has been widely used as an effective tool to investigate genes modulated by specific TFs [19, 20]. A number of innovations for this method that permit the investigation of gene expression in single cells have also been developed. Techniques such as duplex-specific nuclease (DSN) treatment and total transcript amplification (TTA) will be preferred to conventional hybridization-based experiments in prokaryotes [21, 22].

8.1.1.4 Chromatin Immunoprecipitation to Study Direct Interactions Between TFs and DNA

Chromatin immunoprecipitation (ChIP) is a powerful tool for studying the interaction of specific proteins and DNA in vivo. ChIP can be used to determine whether a transcription factor interacts with a candidate target gene and the special binding sites. Combining ChIP with next-generation sequencing technology allows for ChIP-seq experiments, which can efficiently identify both TF- and histone-binding sites across the entire genome in parallel [23].

Genomic DNA is sheared under conditions that preserve protein-DNA interactions, and an antibody targeting the TF of interest is used to enrich for TF-DNA complexes by ChIP. The library of isolated DNA fragments is prepared for deep sequencing to identify sequence tags associated with the TF of interest. Following assembly of millions of isolated sequence tags, precise locations where the TF binds can be identified genome wide.

A statistical framework named CSSP (ChIP-seq Statistical Power) provides an analytical approach for power calculations of ChIP-seq experiments. It enables the researchers to take advantage of their own or public data to determine required sequencing depths of their ChIP-seq experiments while maintaining the false discovery rate at a user-specified level. This web page (www.stat.wisc.edu/~zuo/CSSP) provides download links to the R package for powerful computations of ChIP-seq experiments [24].

ChIP sequencing (ChIP-seq) and RNA sequencing (RNA-seq) were used in tandem to map the AraC regulon in *Escherichia coli* and *Salmonella enterica* [25]. Merrikh group has reported the *Bacillus subtilis* accessory helicase PcrA as an essential factor involved in mitigating transcription-replication conflicts and identifies chromosomal regions with which it routinely acts using ChIP-seq experiments [26].

8.1.1.5 Methods to Identify Target Genes Modulated by the Candidate TFs

Several methods need to be used to identify target genes up- or down-regulated by a candidate TF. In general, experiments that probe in vivo levels of gene regulation, such as primer extension, quantitative RT-PCR, Western blot and LacZ fusion, and β-galactosidase assay can be used to investigate TF-dependent changes in gene expression. Specific binding sites can be confirmed using in vitro experiments such as electrophoretic mobility shift assays (EMSAs) and DNase I footprinting.

8.1.1.5.1 RNA Isolation and Primer Extension Assay

Determining the transcriptional start site by primer extension is a fast and accurate method. Additionally, it permits a parallel semiquantitative investigation of the extent of regulation between wild-type (WT) and TF gene deletion strains.

For a primer extension assay following the AMV Reverse Transcriptase (Promega, USA), a 5′-^{32}P-labeled oligonucleotide primer complementary to a portion of the RNA transcript of a target gene is employed to synthesize cDNAs from RNA templates isolated from both WT and TF gene deletion strains, respectively. The same labeled primer is also used for sequencing with the Top DNA sequencing kit (Bioneer, Korea). The primer extension and sequencing products are analyzed in a 6 % polyacrylamide eight-molar urea gel, and the result is detected by autoradiography (Fuji Medical X-ray Film, Tokyo, Japan).

8.1.1.5.2 Quantitative RT-PCR (qRT-PCR)

Compared with a primer extension assay, qRT-PCR can quantitatively analyze the RNA expression differences of the target genes between the WT and TF gene deletion strains, which are normalized to the 16S rRNA gene.

Gene-specific primers are designed to produce a 100–200 bp amplicon for each gene. Then, cDNAs are generated using five micrograms of purified RNA and three micrograms of random hexamer primers. qRT-PCR is performed with the SYBR Green Master Mix (TaKaRa, Japan) through the LightCycler System (Roche, Switzerland) [27, 28]. The expression of each gene is normalized to the abundance of the 16S rRNA gene to account for experimental variations. The relative mRNA level of each probed gene is determined by standard curves generated from 16S rRNA expression.

8.1.1.5.3 LacZ Fusion and β-Galactosidase Assay

A PCR product of a promoter-proximal DNA region of each gene of interest is cloned into a low-copy-number transcriptional fusion vector upstream of a promoter-less LacZ reporter gene. The recombinant LacZ plasmid is transformed into both the WT strain and the TF deletion strain. Then, β-galactosidase activity in cellular extracts is measured using the β-Galactosidase Enzyme Assay System (Promega, USA) to evaluate the extent to which transcription of the gene of interest is modulated by the chosen TF.

8.1.1.5.4 Western Blot

Western blots can be used to measure the extent to which translation is modulated by the selected TF between the WT and TF deletion strains. The protein encoded by a candidate target gene is expressed in *E. coli* and used to immunize rabbits to produce and harvest polyclonal antibodies (pAb) that recognize the protein of interest. Total protein extracts isolated from the WT and TF deletion strains are separated by SDS-PAGE gel electrophoresis and transferred onto PVDF membrane, then successively hybridized with pAb and goat anti-rabbit IRDYeR 800CW antibody, respectively. Translational differences between the WT and TF deletion strains can be determined from the signal observed following blotting.

8.1.1.5.5 Electrophoretic Mobility Shift Assay (EMSA)

Radiolabeled target DNA is incubated with in vitro purified TF protein at room temperature for 20–30 min, and the mixture is then loaded onto a native 4 % (w/v) polyacrylamide gel. If the TF binds the labeled DNA, it retards the mobility of the probe during PAGE, which can easily be detected by autoradiography. The specificity of interaction can be established using an antibody specific to the TF being investigated. When this antibody is included in the mobility shift assay, it will further decrease the mobility of the DNA-TF complex, causing a supershift.

8.1.1.5.6 DNase I Footprinting

Based on EMSA results, the TF-binding site can be investigated by DNase I footprinting. Multiple DNA-binding experiments using radiolabeled probe and increasing concentrations of TF are set up in parallel with a no protein control. Each of these reactions is partially digested with RQ1 RNase-Free DNase I (Promega, USA), producing a ladder of nucleotide resolution products. The digested DNA samples are purified and analyzed in an eight-molar urea 6 % polyacrylamide gel along with the Sanger sequencing products as a reference. The TF-binding site can be determined by comparing the digested samples with the sequencing ladder after autoradiography.

8.1.2 Patterns of Gene Expression Regulation by TFs

Bacteria have complex regulatory networks that respond and adapt to environmental changes. Both gene regulations by TFs at the transcriptional level and by sRNAs at the posttranscriptional level play an important role in bacterial regulatory networks. Gene regulation in bacteria predominantly occurs at the transcriptional level. In addition, transcriptional initiation is the most highly regulated step of gene

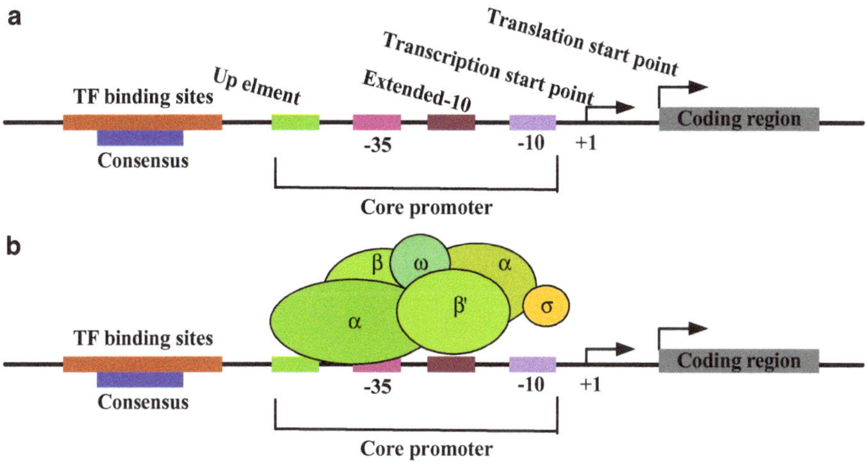

Fig. 8.1 Schematic of a prokaryotic promoter and RNA polymerase-binding region

expression. Bacteria possess a variety of TFs that are required for different functions in physiological pathways, which regulate the expression of target genes in a precise and orderly fashion.

8.1.2.1 Mechanisms of Transcriptional Regulation

In bacteria, genes with related functions are arranged in operons that are complete units for gene expression and regulation. An operon is composed of structural genes, regulatory genes, and DNA elements recognized by TFs. Consecutive genes in an operon can be transcribed as a single primary RNA. During RNA synthesis, gene transcription is regulated by TFs that specifically interact with a DNA element (Fig. 8.1). RNA polymerase (RNAP) synthesizes RNA using bound DNA as a template. Bacterial RNAP consists of a core enzyme that is responsible for catalyzing RNA synthesis and a specific (sigma, σ) subunit that is required for recognition of DNA sequences contained within the promoter region [1, 29]. Sigma factor is necessary for transcription initiation, which contributes considerably to gene regulation.

8.1.2.2 Transcriptional Activation or Repression

Some bacterial TFs only bind to a *cis*-acting DNA sequence, while others may bind to both *cis*-acting DNA sequences and other TFs or themselves. DNA-binding TFs are a crucial component of gene regulation. They can activate or repress transcription of target genes by binding to DNA of their respective promoter regions. Some TFs act solely as activators or repressors, while others can both act as an activator

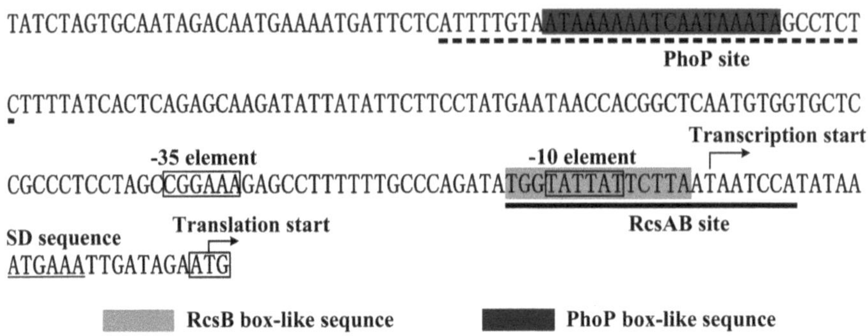

Fig. 8.2 Structure of promoter-proximal DNA region of *waaA*. Shown are the translation/transcription starts, predicted core promoter −10 and −35 elements, predicted Shine-Dalgarno (SD) sequences for ribosomal binding, PhoP and RcsAB sites, and PhoP and RcsAB boxlike sequences

and a repressor at the same time. When an activator promotes the transcription of a gene, its effect on the promoter appears to facilitate the function of RNAP. In contrast, a repressor inhibits the expression of its target genes by binding to DNA to prevent RNAP from initiating transcription [12]. Taking PhoP- and RcsB-dependent regulation of *waaA* transcription in *Yersinia pestis* as an example, binding of PhoP to PhoP sites located upstream of the core region of the *waaA* promoter may change the DNA structure and thereby promote RNAP initiation of transcription. However, RcsB sites cover the −10 element of the *waaA* promoter core region and thereby prevent transcription of *waaA* (Fig. 8.2) [30]. Some regulators are able to promote or repress the transcription of a gene depending on environmental conditions. The global regulator RovA in *Yersinia pestis* can reciprocally control the transcription of itself in a fashion dependent on the nutritional environment and the concentration of RovA in vivo [31].

8.1.2.3 Characterization of Transcription Factor-Target Gene Interactions

Characterization of the interactions between transcription factors and their target genes will help with understanding and predicting the transcriptional relationship. TFs can bind to multiple sequences with varying affinity. However, the binding sites of most TFs contain a short (5–20 bp) conserved region. If a collection of binding sites of a TF has been determined from its target genes by DNase I footprinting, sequence alignments of these TF-binding sites will generate a consensus motif. A consensus motif can be represented as either a consensus string or a position-specific scoring matrix (PSSM). A PSSM contains more information and is better to evaluate new potential sites.

8.1.3 Transcriptional Regulation of Gene Expression in **Yersinia pestis**

Yersinia pestis is the causative agent of a plague that has caused at least three pandemics through human history, which has, in total, claimed hundreds of thousands of lives [32]. Plague has a global distribution of endemic foci, and it is considered to be a reemerging infectious disease today [33]. *Y. pestis* is able to cause severe invasive infections including bubonic plague in lymph nodes, septicemic plague in blood vessels, and pneumonic plague in lungs. In natural environments, *Y. pestis* circulates between rodent reservoir hosts and flea vectors. Plague infections mostly occur in animals, while humans are accidental hosts of this pathogen. When *Y. pestis* shifts its lifestyle from flea vectors to mammalian reservoir hosts, it will encounter a series of intense and even life-threatening environmental changes. Accommodative gene expression in response to such situations appears to be essential for the survival of *Y. pestis*, and thus the regulation of gene expression involved in these processes are correspondingly important.

8.1.3.1 Transcriptional Regulation of Gene Expression in *Y. pestis* Biofilm Formation

Y. pestis evolved from its close progenitor *Yersinia pseudotuberculosis*, which is a mild foodborne pathogen. However, flea-borne transmission of *Y. pestis* is an essential mode that distinguishes it from *Y. pseudotuberculosis* [34]. *Y. pestis* forms a biofilm that can attach to and physically block the proventriculus of fleas. In turn, this causes fleas to feel hungry and bite repeatedly, promoting the spread of *Y. pestis* into new individuals of mammalian reservoirs [35]. Therefore, biofilm production of *Y. pestis* plays a crucial role in its transmission. Simultaneously, regulation of gene expression associated with biofilm formation is equally important, and thus the regulators involved in these processes should be treated as being as important as the effectors themselves.

8.1.3.1.1 Major Determining Factors of Biofilm Production in *Y. pestis*

The *Y. pestis hmsHFRS* operon is responsible for the synthesis and transport of exopolysaccharide (a primary dry component of the biofilm matrix) [36]. Cyclic-di-GMP (c-di-GMP) is a second messenger molecule promoting biofilm exopolysaccharide production in bacteria. In *Y. pestis*, HmsT and HmsD (*hmsD* is located in the three-gene operon *hmsCDE*) are the only diguanylate cyclases catalyzing c-di-GMP synthesis [37], while HmsP is the sole phosphodiesterase catalyzing c-di-GMP degradation [38].

The gene *gmhA* encodes a phosphoheptose isomerase that controls heptose synthesis, a conserved component of lipopolysaccharide (LPS). Deletion of *gmhA* induces severely hampered biofilm formation that is inadequate for flea blockage [39]. YrbH is an arabinose 5-phosphate isomerase catalyzing the conversion of ribulose 5-phosphate into arabinose 5-phosphate, the first committed step in the 3-deoxy-D-manno-oct-2-ulosonic acid (Kdo) biosynthesis pathway. WaaA is a Kdo transferase that is involved in the synthesis of lipopolysaccharide [40]. In *Y. pestis*, the *yrbH* mutant produces almost no biofilm, while the *waaA* mutant produces defective but detectable biofilms [40].

SpeA (arginine decarboxylase) and SpeC (ornithine decarboxylase) are necessary for polyamine synthesis. Polyamines are required for the normal expression of the key Hms proteins including HmsR, HmsS, and HmsT via a posttranscriptional regulatory mechanism. Deletion of *speA* or *speC* leads to a dramatic loss of *Y. pestis* biofilm formation in vitro [41].

8.1.3.1.2 Rcs Phosphorelay Regulatory System Represses Biofilm Production

The Rcs phosphorelay system is an atypical two-component regulatory system that is composed of three proteins, RcsB, RcsC, and RcsD [42]. RcsC and RcsD are membrane-bound proteins, and RcsC acts as a sensor kinase that catalyzes autophosphorylation. The resulting phosphate group is then transferred to RcsD and finally to RcsB (a cytoplasmic protein). Phosphorylated RcsB acts as a transcriptional regulator independently or upon binding of an auxiliary protein RcsA. RcsA is functional in *Y. pseudotuberculosis*, and RcsAB tightly represses *Yersinia* biofilm formation, leading to a biofilm-negative phenotype in *Y. pseudotuberculosis* [43, 44]. The *rcsA* gene is inactivated in *Y. pestis* because of a 30-bp duplication insertion in its open reading frame (ORF), and replacing the *rcsA* pseudogene with a functional *Y. pseudotuberculosis rcsA* allele strongly represses *Y. pestis* biofilm formation and essentially abolishes flea blockage [44]. The pseudogenization of *rcsA* is a prerequisite of potent *Y. pestis* biofilm formation, enabling the efficient flea-borne transmission of this pathogen [44, 45]. The RcsAB complex recognizes the conserved core sequence TAAGAAT-ATTCCTA (a 7–7 inverted repeat). The RcsAB complex is a strong repressor of *Y. pestis* biofilm formation through directly repressing transcription of *hmsCDE*, *hmsT*, *hmsHFRS*, and *waaAE-coaD* while positively regulating *hmsP* in an indirect manner. Thus, RcsA inactivation leads to a dramatically elevated expression of RcsAB-dependent biofilm-activated genes in *Y. pestis* relative to *Y. pseudotuberculosis*, thereby promoting biofilm production. This is a representative example of favorable evolution to promote *Y. pestis* transmission via fleas.

8.1.3.1.3 The Ferric Uptake Regulator Fur Is a Repressor of Biofilm Formation

The ferric uptake regulator Fur predominantly controls iron metabolism in bacteria [46]. Fur regulates not only almost all iron assimilation functions in *Y. pestis* but also multiple genes involved in various non-iron functions. Fur governs a complex regulatory cascade in *Y. pestis* [47, 48]. A 19-bp box and a position-specific scoring matrix (PSSM) have been built to represent the conserved *cis*-acting signals recognized by Fur [47]. The 9-1-9 inverted repeat sequence *AATGATAATNATTATCATT* represents the Fur-binding motif or Fur box. The deletion of *fur* results in abundant in vitro biofilm production in *Y. pestis*, and Fur can directly bind to the promoter-proximal region of *hmsT* and repress its transcription [49].

8.1.3.1.4 RovM and YfbA are Involved in Colonization and Metabolic Adaption to the Flea Gut

The LysR family of transcriptional regulators represents the most abundant type of transcriptional regulator in bacteria. Members of the LysR family appear to possess a conserved structure with an N-terminal DNA-binding helix-turn-helix motif and a C-terminal co-inducer-binding domain. LysR-type regulators are responsible for regulating genes that are involved in a variety of functions, including biofilm production, virulence, motility, and metabolism [50]. RovM is a LysR-type transcriptional regulator (LTTR) that directly inhibits the transcription of *rovA* in *Y. pseudotuberculosis*. It is the first LTTR found to be involved in virulence regulation in a mammalian pathogen [51]. In addition, RovM is able to sense specific nutritional cues in the flea gut environment and affects metabolic adaptation to enhance *Y. pestis* colonization of the flea gut [52]. Our work has demonstrated that RovM is a fundamental activator of *Y. pestis* biofilm production. RovM accomplishes this by directly stimulating the transcription of *hmsT*, *hmsCDE*, and indirectly enhancing *hmsHFRS* transcription while simultaneously indirectly repressing *hmsP* transcription (unpublished data). YfbA is also a regulator belonging to the LysR family that is essential for *Y. pestis* colonization and biofilm formation [53]. However, it is still unknown what biofilm factors are regulated by YfbA.

8.1.3.2 Transcriptional Regulation of Gene Expression in *Y. pestis* Virulence

Three *Yersinia* species are known to be pathogenic to humans: *Y. enterocolitica*, *Y. pseudotuberculosis*, and *Y. pestis*, of which *Y. pestis* causes the most deadly infectious disease in humans. The three pathogens share a subset of virulence determinative factors, while each having distinct components. *Y. pestis* has multiple virulence-associated factors and regulators, and the expression of most virulence factors is regulated. Most *Y. pestis* virulence factors are thermally regulated and are active at either 26 °C (the temperature in flea vectors) or 37 °C (the host temperature). Nevertheless, temperature-dependent gene regulation always accompanies with other environmental cues such as pH, ion concentration, nutrient availability, osmolarity, oxygen tension, and DNA damage [54]. Thus, the question of how *Y. pestis* orchestrates this process requires further study.

8.1.3.2.1 Major Virulence Determinative Factors in *Y. pestis*

Y. pestis harbors a 102-kb chromosomal region (the *pgm* locus) encoding elements critical for virulence. The *pgm* locus is composed of two distinct parts: a 35-kb high-pathogenicity island (HPI) which carries virulence determinative genes [55, 56] and an additional 68-kb segment that contains genetic elements of the *hms* locus responsible for conferring the *Y. pestis* pigmentation phenotype on Congo red agar at 28 °C [57]. Two parts of this region carry crucial genes that are required either for virulence (Ybt system) or for pathogen transmission.

Y. pestis psa loci are composed of two operons: *psaABC* and *psaEF* [58]. The expression of *psaABC* is strongly stimulated following a temperature upshift from 26 to 37 °C and in acidic environments. The operon *psaEF* encodes the transcriptional activators of *psaABC* [59]. The pH6 antigen plays important roles in bubonic and pneumonic plague in mice [60, 61]. It acts as an adhesin to facilitate the bacterial colonization of host cells [62–64]. The pH6 antigen-mediated binding of *Y. pestis* with eukaryotic cells facilitates the delivery of virulence determinants Yops (effectors of plasmid pCD1-encoding type III secretion system) to target cells [65]. In addition, the pH6 antigen does not only function as an antiphagocytic factor to block bacterial uptake but also efficiently binds to apolipoprotein B-containing lipoproteins in human plasma [66, 67]. It has been speculated that once *Y. pestis* cells are released from infected macrophages to adopt an extracellular lifestyle, pH6 antigen-mediated binding to the lipoprotein will prevent the pathogen from being phagocytosed by phagocytes [67].

Typical strains of *Y. pestis* harbor three virulence plasmids (pCD1, pMT1, and pPCP1) encoding a variety of virulence determinants. The 70-kb plasmid pCD1 controls the synthesis of two significant virulence factors: first, Yops (*Yersinia* outer membrane proteins) are necessary for *Yersinia* host infection, and the second virulence determinant YadA functions as a major adhesin involved in bacterial adherence to eukaryotic extracellular matrix elements while simultaneously conferring resistance to nonspecific host defenses [54]. F1 antigen is an important virulence antigen responsible for *Y. pestis* antiphagocytosis, the expression of which requires four genes (*caf1*, *caf1A*, *caf1M*, and *caf1R*) located on the 110-kb plasmid pMT1. Expression of the *caf* operon is thermally regulated, which is demonstrated by a dramatic increase in Caf1A levels following a rise in temperature from 28 to 37 °C [68]. The AraC-family regulator Caf1R takes part in this procedure. Another crucial virulence determinant on pMT1 is *Yersinia* murine toxin (Ymt) that plays crucial roles in flea-borne transmission [69]. The plasminogen activator/coagulase (*pla*) gene is harbored by the pPCP1 plasmid, which is required for the full virulence of *Y. pestis* during both bubonic and pneumonic plague [70, 71].

8.1.3.2.2 RovA Is a Global Virulence-Required Regulator

The MarR (multiple antibiotic resistance regulator) family of bacterial transcriptional regulators plays an important role in virulence factor production and responses to antibiotic and oxidative stresses [72]. As a member of the MarR family of transcriptional regulators, RovA is required for the full virulence of all three pathogenic *yersiniae* through regulation of various virulence loci [61, 73–76]. RovA stimulates the transcription of *inv* that encodes an invasin that mediates translocation of *Y. pseudotuberculosis* and *Y. enterocolitica* across the intestinal epithelium [74–77]. However, the *inv* gene is naturally inactivated in *Y. pestis* [78]. In *Y. pestis*, RovA stimulates the transcription of the *psaEF*, *psaABC*, and CUS-2 prophage loci [61]. The CUS-2 prophage is stably integrated in *Orientalis*, but unstable in *Antiqua*, *Medievalis*, and *Microtus* [79, 80]. The acquisition of this prophage is not associated with flea transmission, but contributes to virulence in mice [79]. RovA still plays critical roles in the construction and functioning of the bacterial membrane, which manifests the regulatory functions of RovA in antibiotic resistance and environmental adaptation [81].

8.1.3.2.3 CRP Acts as a Global Transcriptional Factor to Control Pathogenicity

Cyclic AMP receptor protein (CRP) is a virulence-required regulator that has been characterized in many pathogens [82–86]. It controls more than 6 % of the genes in *Y. pestis* [86]. CRP is active only in the presence of cyclic AMP (cAMP), a classic small-molecule inducer. The cAMP-CRP complex specifically recognizes the symmetric consensus sequence TGTGA-N6-TCACA (known as the CRP box sequence) [87]. CRP serves as a transcriptional regulator of the T3SS/Yop machinery and the plasminogen activator protease Pla [88, 89]. This most likely explains why CRP is critical for the development of bubonic and pneumonic plague. Similarly, CRP is necessary for colonization and persistence of *Y. pseudotuberculosis* in the MLNs and organs during infection and is required for the expression of the flagellar Ysc/Yop and Ysa T3SS in *Y. enterocolitica* [83, 90]. In addition, CRP also influences *Y. pestis* virulence through a posttranscriptional mechanism [91].

8.1.3.2.4 The PhoP-PhoQ Regulatory System Is Crucial for *Y. pestis* Survival in Macrophages

PhoP and PhoQ constitute a two-component regulatory system [92]. The sensor PhoQ is activated by phosphorylation of one of its histidine (His) residues upon specific environmental stimuli. It then phosphorylates PhoP. Phosphorylated PhoP functions as an active transcriptional regulator and recognizes an 18-bp motif (TGTTTAWN$_4$TGTTTAW; W=A or T) in *Y. pestis* [93]. The *phoP* null mutant shows significantly reduced virulence in macrophages and under in vitro conditions of low pH and oxidative stress [94]. In addition, PhoP-regulated genes are involved in promoting resistance to antimicrobial peptides or low-Mg^{2+} conditions found in phagosomes [95]. This indicates that PhoP plays important roles in *Y. pestis* survival in macrophages through regulation of multiple target genes during the early infection stage.

8.2 Posttranscriptional Regulation of Gene Expression

Regulation of gene expression takes place at different regulatory levels in bacteria, of which transcriptional regulation is perhaps the most studied. Nevertheless, the crucial role of posttranscriptional regulation is beginning to be appreciated in gene regulatory networks. It mainly functions through the modulation of RNA decay, transcript elongation, and translation initiation. Small regulatory RNAs (sRNAs) and RNA-binding proteins (RBPs) have been identified as two primary types of posttranscriptional regulator. The posttranscriptional regulation mediated by sRNAs and RBPs will be discussed below.

8.2.1 Methods for Revealing sRNA Regulation in Bacteria

Known sRNAs generally exert their regulatory functions by binding to specific mRNAs or proteins [96]. Consequently, to understand the genetic regulation of a given sRNA in bacteria, discovery of the molecules it interacts within the cell is essential. It is currently accepted that *trans*-acting sRNAs typically interact with the 5′ UTR of their mRNA targets via short imperfect complementarity and thus influence translation and mRNA stability [97]. In many cases, the sRNA-mRNA interactions are assisted by the RNA chaperone Hfq [97]. Prediction and validation of mRNA targets have been performed by various computational and experimental methods [98–100]. Mutants that result in overexpression or deletion of sRNAs were used to screen for regulated targets in an individual or systematic fashion. Changes of individual mRNA abundance can be detected by a variety of methods including Northern blot, primer extension, and RT-PCR [101]. The consequences of translational activation or repression can be evaluated at the protein level by SDS-PAGE or Western blot [101]. In addition, reporter gene fusion systems are commonly used to detect translational changes of a given target gene [101]. However, an increasing number of sRNAs are thought to have one or more targets owing to limited complementarity with their targets. Therefore, global methods are frequently used in the initial phase of sRNA characterization [102–104]. For example, potential target mRNAs have been discovered by combining microarray- or RNA-seq-based transcriptomics with transient sRNA overexpression or with immunoprecipitation with Hfq protein. In addition, in vitro experiments including gel shift assays, RNA footprinting, and ribosome toeprinting have also been used to confirm sRNA-mRNA interactions [101]. RNA pull-down assays coupled with mass spectrometry or Western blot can be used to identify the ensemble of RNA-binding proteins interacting with a specific RNA [105].

8.2.1.1 Global Discovery of Expressed sRNAs by RNomics and RNA Sequencing

Most of the well-characterized sRNAs are between 50 and 400 nt in length. They are usually encoded in the intergenic regions (IGRs) of bacterial chromosomes or plasmids and expressed under certain growth, stress, or virulence conditions [106]. Given that sRNAs are expressed in sufficient abundance, they can be detected individually or globally by a variety of methods. Here, we will mainly focus on the global screening method based on RNomics and RNA sequencing.

A variety of new sRNAs and antisense transcripts have been recently identified in bacteria via sequencing of cDNA prepared from size-fractionated or total RNA [102, 107, 108]. Generally, total RNAs are extracted from bacteria grown under different growth conditions or from samples co-immunoprecipitated with proteins. RNA species within a size range of 50–500 nt are enriched and purified by gel extraction. For the RNomic approach, the RNA is reverse transcribed and cloned into a plasmid vector. The resulting inserts are analyzed using conventional Sanger sequencing. For RNA sequencing, the RNA is ligated with 5′ and 3′ adaptors and reverse transcribed. The cDNA is then subject to pyrosequencing at the genome-wide level.

8.2.1.2 Global Screening of Multiple mRNA Targets of sRNAs

Save for the fraction of sRNAs that act to modulate the activity of proteins, the majority of sRNAs are thought to function via direct base pairing with mRNAs and subsequently altering messenger stability [109]. Pulse expression of sRNA combined with transcriptome profiling has been used to identify mRNA targets [101]. The entire sequence of the sRNA of interest is cloned into a transcriptional fusion plasmid in which sRNA expression is under the control of an inducible promoter. After induction for 10–15 min, microarray or RNA-seq is used to monitor mRNA decay on a genome-wide scale. mRNAs with significantly altered abundance are regarded as direct targets of sRNA for further characterization.

8.2.1.3 Experiment Validation of sRNA-mRNA Interactions In Vivo

After the initial screening of multiple sRNA targets, direct evidence for the individual sRNA-mRNA base pairing should be collected in vivo. Reporter gene fusion assays have been used to validate mRNA targets in a two-compatible-plasmid reporter system [110]. In this system, sRNA is placed under the control of a constitutive promoter in a high-copy plasmid, while a translational fusion with the 5′ region of a putative mRNA target is carried on a low-copy plasmid. The reporter gene abundance is measured to confirm posttranscriptional regulation in vivo. The standard procedure of compensatory base pair exchanges is often performed to determine the precise interacting sites of a *bona fide* sRNA-mRNA pair [111]. Regulation of the translational fusion is impaired when a mutation is introduced into the interaction site of either the sRNA or at the cognate position in the target mRNA, but it is restored by introducing compensatory mutations.

8.2.2 Patterns of Gene Expression Regulation by sRNAs

Bacteria flexibly regulate specific sets of genes to survive and adapt to stressful environments. In bacteria, DNA-binding proteins, noncoding RNAs, and their target genes are coordinated in hierarchical regulatory networks. Broadly speaking, all of the untranslated RNAs in bacteria possessing regulatory functions, except for rRNAs and tRNAs, can be categorized as sRNAs. Similar to protein-based regulators, a single sRNA may regulate one or more genes and thus might have effects on multiple cellular pathways [109]. Compared with DNA-binding regulators, sRNA regulation at the posttranscriptional level seems to be more efficient and advantageous for quick responses to environmental stresses. Therefore, many bacterial sRNAs are frequently implicated in stress responses or virulence processes that are required for immediate responses to changing environments [112, 113]. According to their genomic location relative to their mRNA targets and their modes of action, regulatory RNAs can be broadly classified into four major classes: *trans*-encoded sRNAs, *cis*-encoded sRNAs, RNA thermometers, and riboswitches [113]. These sRNAs influence mRNA translation or protein activity by base pairing with target mRNAs or binding with proteins (Fig. 8.3).

8.2.2.1 Key Components of Posttranscriptional Regulation

8.2.2.1.1 *Trans*-encoded sRNAs

A *trans*-encoded sRNA is defined as an sRNA that acts upon their targets at distinct genomic loci. They interact through limited complementarity with target mRNAs and usually bind to the Shine-Dalgarno region or the coding region of the mRNA, which often results in translation inhibition and mRNA cleavage and degradation. A large portion of well-characterized sRNAs in bacteria is *trans*-encoded and range in size between 50 and 300 nts. These mainly influence mRNA translation and/or stability through limited and imperfect complementarity with target transcripts [113]. The RNA chaperone Hfq is often required for loose binding of most *trans*-encoded sRNAs with their targets. Hfq binding facilitates the formation of sRNA-mRNA duplexes and thereby protects them from the attack of RNases [114].

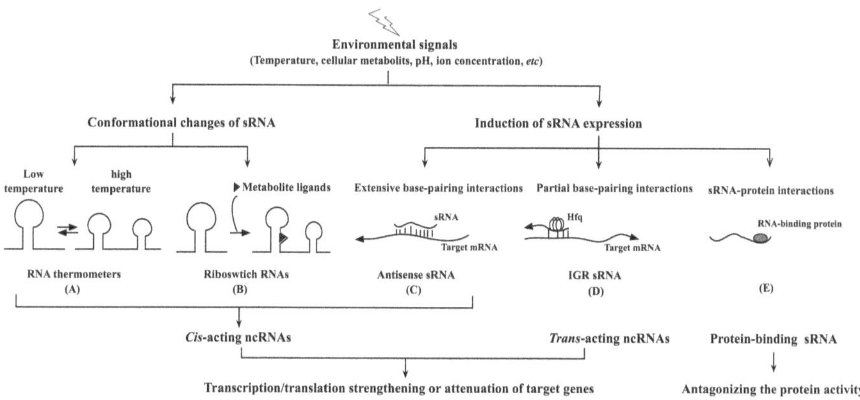

Fig. 8.3 Action modes of *cis*- and *trans*-acting and protein-binding sRNAs in bacteria. *Cis*-acting sRNAs include RNA thermometers that experience conformational changes in response to temperature shift (**a**), riboswitches capable of binding metabolite ligands (**b**), antisense sRNAs that extensively base pair with sense RNA (**c**), and IGR sRNA and some antisense RNAs belonging to the family of *trans*-acting sRNAs partially base pair with their mRNA targets (**d**), while protein-binding sRNAs bind with specific proteins and sequester their activity (**e**)

8.2.2.1.2 *Cis*-encoded sRNAs

A *cis*-encoded sRNA is located directly opposite from one or more target ORFs. It interacts with its target transcript in a fully antisense manner. In the majority of cases, the base pairing between a *cis*-encoded ncRNA and its target transcript forms a duplex and mediates the posttranscriptional downregulation of one or more target genes, but activating mechanisms have also been found in a few cases [113].

A particular type of sRNA, riboswitch RNA, is contained in some mRNA noncoding regions and includes a conserved ligand-binding domain and an expression platform. A riboswitch is typically located at the 5′-UTR of an RNA transcript and acts in cis to regulate the gene expression of the downstream coding sequence [115]. Regulation occurs as a shift in the secondary or tertiary structures in the aptamer domain upon binding of a small molecule cofactor.

8.2.2.1.3 Protein-Binding sRNAs

Except for mRNA-interacting sRNAs, the protein-binding sRNA modulates the expression of target genes by binding the corresponding regulatory protein or RNA polymerase instead of base pairing with RNA targets. The CsrB/C RNAs are involved in the carbon storage regulatory system of *E. coli* by directly binding to the global regulator CsrA and thus inhibiting its activity [116]. The 6S RNA regulates RNA polymerase activity by forming a structure that mimics an open promoter complex that can be recognized by the polymerase holoenzyme containing σ^{70} [117].

8.2.2.1.4 RNA-Binding Proteins

RNA-binding proteins are involved in the posttranscriptional regulatory processes of RNA stability and mRNA translation efficiency. The RNA-protein interactions modulate the accessibility of the ribosome-binding site (RBS) for ribosome binding or of RNases for RNA degradation or function as chaperones for interaction with other effector molecules [118]. CsrA and Hfq protein are the two best-studied RNA-binding proteins in bacteria. CsrA works predominantly by competing with the ribosome for binding to the RBS of its mRNA targets [119]. The sRNA molecules CsrB and CsrC bind to CsrA and antagonize its activity [116]. Hfq is a small, highly abundant protein presently involved in physiological fitness and pathogenesis in many bacteria [120]. The pleiotropic effects upon *hfq* inactivation have been observed in a variety of bacteria. Hfq forms a doughnut-like homohexameric structure and belongs to the Sm/Lsm superfamily of RNA-binding proteins [121]. It typically binds diverse single-stranded RNAs and is thus implicated in the control of gene expression [114, 121]. It is generally thought that Hfq is a key player in controlling gene expression posttranscriptionally mediated by *trans*-acting sRNA [97]. They can imperfectly base pair with target mRNAs in the canonical SD/AUG region or other sites of the 5′ UTR or coding sequence [122–124]. The primary role of Hfq in sRNA-mediated regulation is to modulate translation by facilitating interactions between sRNAs and target mRNAs and/or to alter the stability of sRNA-mRNA hybrids by recruiting RNase E to target mRNAs [125]. Increasing numbers of sRNAs have been shown to simultaneously act on multiple mRNAs, and likewise, many mRNA transcripts are emerging as shared targets of multiple cognate sRNAs.

8.2.2.2 Mechanisms of sRNA-Mediated Regulation

8.2.2.2.1 Translational Repression

Translation Repression by Blocking Ribosome Access to Target RNAs

Translation initiation is the most highly regulated step of protein biosynthesis in bacteria. The crucial rate-limiting step is the assembly of the 30S small ribosomal subunit with mRNAs. Initial binding of the 30S ribosome to specific mRNAs is primarily mediated by RNA-RNA interactions. The anti-SD sequence of the 16S rRNA recognizes and base pairs with the SD sequence of mRNA. Correspondingly, the interaction between 16S rRNA and mRNA directs the initiation codon to the P site of the small ribosomal subunit, where it interacts with the initiator tRNA molecule. When the 50S ribosomal subunit enters along with initiation factors, the 70S initiation complex is well prepared to initiate the elongation phase of translation. The mRNA region occupied by the ribosome upon translation initiation includes the RBS and an extensive 30-nt region.

Canonical Mechanism: sRNAs Competing with Ribosome-Binding Sites

Bacterial sRNAs typically repress translation of target mRNAs by pairing directly to the SD sequence or its flanking sites encompassing sequences extended to the fifth codon of the ORF (−20 to +15 relative to the translational start site AUG) and blocking ribosome access, leading to translation inhibition and the concomitant decay of mRNA [106].

Noncanonical Mechanisms: Targeting Elements Outside of RBS

Several noncanonical mechanisms of sRNA-induced translation repression have been recently described, including transcriptional blocking through occlusion of ribosome standby sites, upstream leader ORFs, and translational enhancers [126]. Additionally, Hfq recruited by sRNAs can bind at the RBS of an mRNA and thus inhibit ribosome loading, in which sRNAs act as indirect translational repressors [127].

Translation Repression by Degrading Target mRNA

It is thought that sRNAs act primarily by repressing translation and that many sRNAs only block translation without directly influencing mRNA decay. However, some sRNA-mRNA base pairing leads to mRNA decay instead of translational inhibition [128]. Moreover, translational repression by sRNAs is often helped by rapid mRNA decay to remove target mRNA and make repression irreversible. In *E. coli*, this degradation is usually achieved by the action of RNase III or the RNase E-containing RNA degradosome [129]. Those mRNAs base pairing with sRNAs lose the protection from translating ribosomes and become more sensitive to RNase attacks. Interestingly, it has shown that Hfq is able to recruit the RNA degradosome by interacting with the C-terminal region of RNase E, which functions as an assembly scaffold [125]. RNase E recruitment could actively stimulate nucleolytic attacks on the target mRNA due to an indirect effect of mRNA translation inhibition, which contrasts with the simple and passive model of RNase E recruitment.

In contrast with the view that sRNA-induced mRNA degradation is regarded as a consequence of translation repression, an alternative theory proposed that mRNA degradation mediated by certain sRNAs is independent of translation repression. This model is exemplified by the sRNA MicC, which promotes RNase E-dependent cleavage in the coding sequence of *ompD* mRNA but does not affect its translation [122]. The long-distance downstream of the start codon excludes the possibility of competition with initiating ribosomes.

8.2.2.2.2 Target Activation

As mentioned above, sRNAs commonly sequester the 5′ UTR of target mRNAs and thus often repress their targets. However, more and more examples of sRNA-mediated translation activation and mRNA stabilization have now been documented.

The mechanisms underlying target activation are distinct from target repression mentioned above. The sRNAs can interact with the 5′ UTR, the coding sequence, or the 3′ UTR of the target mRNAs and promote translation or stabilize target mRNAs [130]. RNases and Hfq are frequently required for translational activation. Interestingly, a given sRNA can function as a dual-function regulator that causes activation and repression of different mRNA targets.

Translational Activation by Improving mRNA Accessibility to Ribosomes

Most commonly, sRNAs activate target mRNA translation by an anti-antisense mechanism in which an intrinsic mRNA secondary structure covering the RBS is disrupted by sRNA-mRNA base pairing. This base pairing interaction exposes the mRNA RBS improving mRNA accessibility to ribosomes and thus prevents the formation of a translation-inhibitory structure and enhances mRNA translational initiation. In addition, sRNA-mediated mRNA activation is also enhanced via an mRNA-processing event. For example, a stem-loop structure is formed with the long 5′ UTR of the *colA* mRNA covering the SD sequence and inhibiting translation in *Clostridium perfringens*. The sRNA VR-RNA base pairing with *colA* mRNA elicits mRNA processing by removing the upstream portion of the stem-loop. The processed *colA* mRNA is consequently more accessible to ribosomes [131].

Translational Activation by Interfering with Ribonucleolytic Decay

In contrast to translational activation that occurs through increasing accessibility of an mRNA's RBS, certain sRNAs base pair with their target mRNAs and induce translational activation by directly adjusting mRNA susceptibility to ribonucleolytic degradation [130]. For instance, a protein-coding mRNA is susceptible to nuclease-mediated decay. Base pairing between antisense sRNAs and target mRNA alters their secondary structures and protects them from RNase E-mediated decay.

8.2.3 *Posttranscriptional Gene Regulation in* Y. pestis

Y. pestis must acclimatize itself to diverse environmental conditions during its complex life cycle by modulating gene expression of metabolic genes, cell surface genes, and virulence factors. Small RNAs allow bacteria to rapidly adjust to the changing conditions [32]. More than 300 sRNAs identified by RNomics and deep sequencing facilitate the study of posttranscriptional mechanisms of gene regulation in *Y. pestis* [132–136]. Posttranscriptional regulation and its underlying role in virulence and host adaptation are beginning to be addressed in this pathogen over the past few years [137, 138].

8.2.3.1 Global Identification of Candidate sRNAs in *Y. pestis*

Initially, 1478 potential sRNAs were computationally predicted in *Y. pestis* based on gene homology, secondary structure, genomic position, and conservation [139]. Many efforts have been performed to globally discover expressed sRNA candidates in *Y. pestis* grown in vitro and in vivo using RNomics or RNA sequencing [132–136]. Five reports by three research groups have identified 354 sRNAs expressed in *Y. pestis* and/or *Y. pseudotuberculosis* with approximately one third of sRNAs validated by Northern blot or quantitative PCR [137]. The expression and Hfq dependence of sRNAs were similarly reported to differ in *Y. pestis* grown under different growth conditions [132–136]. Koo et al. first identified 150 Ysrs (*Yersinia* small RNAs) in *Y. pseudotuberculosis* grown in vitro at 26 or 37 °C, of which 144 sequences are present in the *Y. pestis* CO92 genome, and 32 sRNAs are homologous with the previously annotated sRNAs in *E. coli* and *Salmonella* [132]. Further investigation by the Lathem group revealed another 63 sRNA candidates expressed in *Y. pestis* strain CO92 grown in rich medium to different growth phases [135] and 31 sRNAs in *Y. pestis* strain KIM6+ [134]. The Yang research group identified 43 and 104 sRNAs in enzootic *Y. pestis* strain 201 grown under various growth conditions in vitro and/or within mice lungs and spleens by using RNomics and RNA sequencing, respectively [133]. An unexpectedly low overlap was found between these studies. The discrepancy is likely explained by the use of different strains, growth conditions, library sequencing, and analysis pipelines. For example, five sRNAs (sR028, sR041, sR050, sR066, and sR070) have been recently re-annotated as 5′ UTRs [140].

8.2.3.2 Posttranscriptional Regulatory Factors with Assigned Functions in *Y. pestis*

Here, we will summarize the progress on dozens of key posttranscriptional regulatory factors with assignable functions in *Y. pestis* (Table 8.1). The posttranscriptional regulators of *Y. pestis* include RNA-binding proteins, RNases, *trans*-acting, and thermosensing sRNAs.

8.2.3.2.1 Hfq

The RNA chaperone Hfq is a central hub of sRNA-mediated regulation and has been shown to be essential for bacterial physiology and pathogenesis [114, 120]. Mutation of *hfq* typically confers pleiotropic phenotypes presumably by affecting outer membrane biogenesis, protein secretion, virulence gene expression, and general stress response pathways [120]. Hfq is a pleiotropic regulator of gene expression involved in *Y. pestis* stress resistance, intracellular survival, and pathogenesis [141]. In addition, Hfq is also required for biofilm formation by modulating intracellular levels of c-di-GMP [142, 143]. Deletion of *hfq* significantly impairs the

Table 8.1 Posttranscriptional regulators with assigned physiological roles in *Y. pestis*

Name	Known definition	Direct targets	Regulatory roles in *Y. pestis*	References
Hfq	RNA chaperone, pleiotropic	Multiple mRNA and sRNA targets	Participates in environmental stress, intracellular survival, and pathogenesis	[136]
			Functionally coordinated with CRP protein	[65]
CsrA	Global carbon storage regulator	CsrB and CsrC	Enhances biofilm production perhaps via cyclic diguanylate regulation	[164]
SsrA/tmRNA	*Trans*-translation regulator	SmpB	Modulates pathogenesis partially by regulating the expression of YopB, YopD, and LcrV proteins	[150]
RyhB	Regulator of iron homeostasis	Dozens of mRNA transcript n*sodB*, *sdhCDAB*, etc.	Two RyhB homologues tightly regulated by iron, Fur, and Hfq	[153]
GcvB	Regulator of the peptide transport systems	*dppA* and *oppA* transcript	Activated by GcvA and repressed by GcvR; involving in bipeptide transport	[154]
FourU	Thermosensor	*lcrF* transcript	Thermosensing conformational changes of the fourU thermosensor alter the translational initiation of *lcrF* and thus elicits *Y. pestis* virulence	[157]
Ysr141	Unknown	*yopJ* transcript	An intergenic sRNA encoded by pCD1	[135]
HmsB	Biofilm regulator	*hmsP* transcript?	Modulates biofilm production by regulating the expression of DGC and PDE enzyme	[35]

ability of *Y. pestis* to resist phagocytosis and survive within macrophages at the initial stage of infection. Furthermore, the *hfq* deletion mutant is highly attenuated in a mouse infection model. Approximately half of characterized sRNAs require Hfq to facilitate the formation of sRNA-target mRNA duplexes. Therefore, the phenotypes observed in the absence of *hfq* might be attributable to many pairs of Hfq-dependent sRNAs that regulate critical factors required for physiology and pathogenesis (Fig. 8.4). Transcriptomic analysis showed that 243 genes belonging to 13 functional classes were regulated upon *hfq* mutation, in which approximately 23 % are known or hypothesized to be involved in stress resistance and virulence.

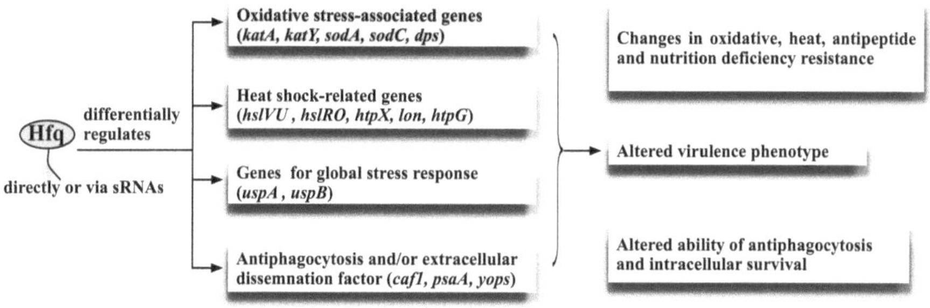

Fig. 8.4 Hfq-regulated stress- and virulence-associated genes and their potential functional consequences

These include the plasminogen activator protease gene *pla*, the F1 antigen gene *caf1*, the diguanylate cyclase gene *hmsT*, and half of the genes of the T3SS [141]. Schiano et al. also present detailed information demonstrating that Hfq is involved in posttranscriptional regulation of the T3SS in *Y. pestis* [135]. Dozens of known or novel sRNAs have been demonstrated to be Hfq dependent or not in *Y. pestis* or *Y. pseudotuberculosis* [132, 134, 135].

8.2.3.2.2 RNases

RNA degradation is indispensable for normal intracellular metabolism in bacteria. In *E. coli*, a multienzyme assembly termed the RNA degradosome is important for mRNA decay and processing. Two of the principal components are ribonuclease E (RNase E) and polynucleotide phosphorylase (PNPase) [144]. Targeted mRNA decay determined by sRNAs plays a crucial role in the posttranscriptional regulation of gene expression. Accessibility of target mRNAs to sRNA results in target degradation of the sRNA-mRNA duplex, usually with the help of Hfq. Deletion of the *rne* gene-encoding RNase E is lethal in *Y. pestis* [145]. RNase E and PNPase have also been shown to be involved in T3SS regulation [146, 147].

8.2.3.2.3 sRNA

Less than 20 % of sRNA candidates were identified by more than one study in a group of several studies on the identification of *Y. pestis* sRNAs [138]. Expectedly, most of these overlapping species were previously annotated sRNAs. While the functions of many known sRNAs conserved in *E. coli* and/or other pathogens have been generally inferred and validated during physiological and pathogenic processes, the majority of novel sRNAs remain a mystery in *Y. pestis*. Here, we will focus on those sRNAs, which regulate *Y. pestis* physiology and/or pathogenesis.

Maintaining Cellular Homeostasis

Trans-translation is a unique process found in a bacterial quality control pathway that contributes to the release of stalled ribosomes [148]. SsrA RNA functions as both a tRNA and mRNA and is therefore also known as tmRNA, which is a key molecule in *trans*-translation. SsrA is a unique component of the bacterial translational control system that resolves ribosomes stalled on defective mRNAs that lack stop codons. SmpB protein may recognize the tRNA-like domain of SsrA and deliver it to target stalled ribosomes [149]. The deletion of *smpB*-*ssrA* significantly attenuates the deadly pathogen upon mouse infection via both the intranasal and intravenous routes partially due to the inhibition of T3SS-related proteins (YopB, YopD, and LcrV) [150]. The *ssrA*-deficient strain is a potential vaccine candidate since a strong antibody response was induced upon intranasal inoculation with the *ssrA* mutant, and the vaccinated animals were well protected against pneumonic plague [150].

RyhB, an sRNA conserved in *Enterobacteriaceae*, accounts for the posttranscriptional regulation of several genes mainly responsible for iron homeostasis [151]. It is tightly regulated by the ferric uptake regulator (Fur) and interacts with the RNA chaperone Hfq [151, 152]. It is also involved in other cellular processes such as bacterial growth, biofilm formation, chemotaxis, and acid resistance. *Y. pestis* encodes two RyhB homologues (RyhB1 and RyhB2) that are located on different chromosomal loci [153]. Both RyhBs are detected in the lungs of mice infected with *Y. pestis*, suggesting that RyhB is likely to play a role in iron acquisition during the progression of pneumonic plague in mice. However, the *ryhB* mutant shows no discernable effects on virulence within the host following subcutaneous or intranasal infection, perhaps due to a redundancy in iron uptake systems [153].

Yersinia pestis gcvB encodes two sRNAs that repress expression of *dppA*, which encodes a periplasm-binding protein component of the dipeptide transport system [154]. The *gcvB* sRNA is activated by GcvA protein and repressed by GcvR protein. Deletion of *gcvB* altered the growth rate and colony morphology of *Y. pestis*, suggesting that GcvB might function as a global regulator [154]. The pleiotropic phenotypes of the *Y. pestis gcvB* mutant might be attributed to differential expression of specific genes regulated by GcvB RNA. More direct targets of GcvB in addition to *dppA* remain to be found in *Y. pestis*.

Targeting the Virulence Factor T3SS

The plasmid pCD1-borne type III secretion system (T3SS) in *Yersinia pestis* is committed to produce effector proteins and translocate across membranes of cells of a eukaryotic host and is essential for mammalian infection [155]. Transcriptional regulation of T3SS has been widely studied, while the roles of posttranscriptional regulation have recently begun to be recognized as well.

RNA thermosensors regulate transcriptional activators to promote optimal infection programs in bacterial pathogens. LcrF is a transcriptional activator of T3SS genes in *Y. pestis* [156]. A fourU RNA thermometer located in the 5′ UTR of *lcrF* allows ribosomes to access or sequesters the RBS at different temperatures.

Specifically, a fourU thermosensor in the 5′ UTR of *lcrF* forms a double stem-loop structure, which represses mRNA translation at 26 °C. Upon hairpin melting at 37 °C, the SD is liberated, allowing translation initiation to occur [157]. At the body temperature of warm-blooded hosts (37 °C), translational initiation of *lcrF* results in differential synthesis of YscW and LcrF protein and further elicits *Y. pestis* virulence [157].

An unstable sRNA, Ysr141, is encoded on the opposite strand within the intergenic region between *yopH* and the gene YPCD1.68c derived from the pCD1 plasmid. It is shown to interact with a 34-nt site of the *yopJ* 5′ UTR and thus activate synthesis of the YopJ protein [135].

Regulating the Transcriptional Regulator CRP

CRP regulates a large number of genes including many crucial virulence genes (such as *pla* and T3SS) as a global transcriptional regulator upon catabolite repression in *Y. pestis* [86]. Hfq contributes to positive regulation of *crp* at the posttranscriptional level at normal physiological temperature. Hfq might regulate CRP expression by recruiting sRNA to the 5′ UTR of the *crp* mRNA [91]. Posttranscriptional regulation is necessary for the expression of *pla* and thus for primary pneumonic plague infection. Evidence that the growth defect due to *hfq* deletion can be partially restored by CRP synthesis hints that Hfq and CRP are functionall y coordinated during *Y. pestis* infection [91].

Modulating Cellular Skeletons

Many other sRNAs (GlmY/GlmZ, MicF/MicC, and SgrS) have also been identified in *Y. pestis* by RNA sequencing and/or bioinformatic analysis. They have been characterized to modulate the cytoskeleton in *E. coli* and many other bacteria. Further efforts should be made to identify their mRNA targets and evaluate their precise functions in *Y. pestis* pathogenesis and physiology.

Regulating Biofilm Formation

Biofilms are known as complex communities of bacterial cells covered with a self-produced exopolysaccharide matrix (EPS). Biofilm formation is helpful for blocking the foregut of fleas during *Y. pestis* transmission [158, 159]. Cyclic diguanylate (c-di-GMP), an important bacterial second messenger, is central to the mechanism of posttranscriptional regulation of biofilm formation [159]. Cellular c-di-GMP is synthesized by diguanylate cyclase (DGC) enzyme and degraded by phosphodiesterase (PDE) enzyme, respectively. *Y. pestis* contains 10 genes encoding DGCs and PDEs in its genome [159, 160]. In addition, adhesive and aggregative factors such as pili, flagella, fimbriae, cell surface, and exopolysaccharides are produced to constitute the biofilm extracellular matrix [161].

A recent study showed that Hfq has an impact on intracellular c-di-GMP levels in *Y. pestis* by regulating the abundance of both HmsP (PDE) and HmsT (DGC), which further elicits the biofilm formation process [142]. Hfq contributes to the regulation of HmsP at the transcriptional level, while HmsT is directly regulated via Hfq at the posttranscriptional level [142]. Many sRNAs participate in biofilm regulatory networks at the posttranscriptional level by base pairing with target mRNA or binding to protein and thus influencing the expression or activity of transcriptional regulators essential for biofilm developmental processes [162].

HmsB RNA (previously called sRNA035) has been recently identified as a biofilm regulator [35]. The *hmsB* mutant has a reduced ability to form biofilm, perhaps resulting from decreased c-di-GMP production. Fang et al. also demonstrated that HmsB positively regulates the expression of DGC enzyme but negatively regulates the expression of PDE enzyme [35]. However, the exact molecular interactions of HmsB with its mRNA targets involved in biofilm formation remain to be elucidated.

CsrA is well known to inhibit biofilm formation in *E. coli* by regulating glycogen synthesis [163]. In contrast, it has recently been shown that CsrA stimulates *Y. pestis* biofilm production independent of glycogen formation [164]. CsrA did not affect the relative abundance of *hmsH*, *hmsP*, *hmsT*, or cyclic diguanylate production. But interestingly, the excessive biofilm production of the *csrA* mutant was restored upon the deletion of *hmsP*, suggesting CsrA might enhance biofilm formation through c-di-GMP regulation [164].

8.3 Perspective

The abundance of 50% of all identified sRNAs has been recently shown to be affected by the transcriptional regulator CRP in *Y. pseudotuberculosis* [140]. Pathogenicity mechanisms have been extensively studied in *Y. pestis*, which deepen our understanding of virulence and fitness-related factors essential for establishing an infection. Information on the great amount of sRNAs is available in *Yersinia*, which promotes opportunities to reveal the mechanisms of sRNA-mediated posttranscriptional regulation. However, we are far from completely elucidating the concerted networks of transcriptional and posttranscriptional regulation implicated in *Y. pestis* pathogenesis. Further efforts should be made to illuminate the posttranscriptional regulatory mechanisms in vivo and to reveal the relationships between transcriptional regulators and regulatory sRNA-Hfq protein complexes in detail.

References

1. Borukhov S, Nudler E. RNA polymerase holoenzyme: structure, function and biological implications. Curr Opin Microbiol. 2003;6(2):93–100.
2. Stein L. Genome annotation: from sequence to biology. Nat Rev Genet. 2001;2(7):493–503.
3. Perez-Rueda E, Collado-Vides J. The repertoire of DNA-binding transcriptional regulators in *Escherichia coli* K-12. Nucleic Acids Res. 2000;28(8):1838–47.

4. Madan Babu M, Teichmann SA. Evolution of transcription factors and the gene regulatory network in *Escherichia coli*. Nucleic Acids Res. 2003;31(4):1234–44.
5. Doerks T, Andrade MA, Lathe 3rd W, von Mering C, Bork P. Global analysis of bacterial transcription factors to predict cellular target processes. Trends Genet. 2004;20(3):126–31.
6. Kummerfeld SK, Teichmann SA. DBD: a transcription factor prediction database. Nucleic Acids Res. 2006;34(Database issue):D74–81.
7. Tatusov RL, Fedorova ND, Jackson JD, Jacobs AR, Kiryutin B, Koonin EV, Krylov DM, Mazumder R, Mekhedov SL, Nikolskaya AN, et al. The COG database: an updated version includes eukaryotes. BMC Bioinf. 2003;4:41.
8. Tatusov RL, Galperin MY, Natale DA, Koonin EV. The COG database: a tool for genome-scale analysis of protein functions and evolution. Nucleic Acids Res. 2000;28(1):33–6.
9. Madera M, Vogel C, Kummerfeld SK, Chothia C, Gough J. The SUPERFAMILY database in 2004: additions and improvements. Nucleic Acids Res. 2004;32(Database issue):D235–9.
10. Bateman A, Coin L, Durbin R, Finn RD, Hollich V, Griffiths-Jones S, Khanna A, Marshall M, Moxon S, Sonnhammer EL, et al. The Pfam protein families database. Nucleic Acids Res. 2004;32(Database issue):D138–41.
11. Sillitoe I, Lewis TE, Cuff A, Das S, Ashford P, Dawson NL, Furnham N, Laskowski RA, Lee D, Lees JG, et al. CATH: comprehensive structural and functional annotations for genome sequences. Nucleic Acids Res. 2015;43(Database issue):D376–81.
12. Zhou D, Yang R. Global analysis of gene transcription regulation in prokaryotes. Cell Mol Life Sci CMLS. 2006;63(19–20):2260–90.
13. Gadgil H, Jurado LA, Jarrett HW. DNA affinity chromatography of transcription factors. Anal Biochem. 2001;290(2):147–78.
14. Park SS, Ko BJ, Kim BG. Mass spectrometric screening of transcriptional regulators using DNA affinity capture assay. Anal Biochem. 2005;344(1):152–4.
15. Yaneva M, Tempst P. Affinity capture of specific DNA-binding proteins for mass spectrometric identification. Anal Chem. 2003;75(23):6437–48.
16. Park SS, Yang YH, Song E, Kim EJ, Kim WS, Sohng JK, Lee HC, Liou KK, Kim BG. Mass spectrometric screening of transcriptional regulators involved in antibiotic biosynthesis in *Streptomyces coelicolor* A3(2). J Ind Microbiol Biotechnol. 2009;36(8):1073–83.
17. Yang YH, Buckley MJ, Speed TP. Analysis of cDNA microarray images. Brief Bioinform. 2001;2(4):341–9.
18. Quackenbush J. Microarray data normalization and transformation. Nat Genet. 2002;32(Suppl):496–501.
19. Vandernoot VA, Langevin SA, Solberg OD, Lane PD, Curtis DJ, Bent ZW, Williams KP, Patel KD, Schoeniger JS, Branda SS, et al. cDNA normalization by hydroxyapatite chromatography to enrich transcriptome diversity in RNA-seq applications. Biotechniques. 2012;53(6):373–80.
20. Mentz A, Neshat A, Pfeifer-Sancar K, Puhler A, Ruckert C, Kalinowski J. Comprehensive discovery and characterization of small RNAs in *Corynebacterium glutamicum* ATCC 13032. BMC Genomics. 2013;14:714.
21. Kang Y, Norris MH, Zarzycki-Siek J, Nierman WC, Donachie SP, Hoang TT. Transcript amplification from single bacterium for transcriptome analysis. Genome Res. 2011;21(6):925–35.
22. Yi H, Cho YJ, Won S, Lee JE, Jin Yu H, Kim S, Schroth GP, Luo S, Chun J. Duplex-specific nuclease efficiently removes rRNA for prokaryotic RNA-seq. Nucleic Acids Res. 2011;39(20):e140.
23. Schmid CD, Bucher P. ChIP-Seq data reveal nucleosome architecture of human promoters. Cell. 2007;131(5):831–2; author reply 832–3.
24. Zuo C, Keles S. A statistical framework for power calculations in ChIP-seq experiments. Bioinformatics. 2014;30(6):753–60.
25. Stringer AM, Currenti S, Bonocora RP, Baranowski C, Petrone BL, Palumbo MJ, Reilly AA, Zhang Z, Erill I, Wade JT. Genome-scale analyses of *Escherichia coli* and *Salmonella*

enterica AraC reveal noncanonical targets and an expanded core regulon. J Bacteriol. 2014;196(3):660–71.
26. Merrikh CN, Brewer BJ, Merrikh H. The *B. subtilis* accessory helicase PcrA facilitates DNA replication through transcription units. PLoS Genet. 2015;11(6):e1005289.
27. Gao H, Zhang Y, Yang L, Liu X, Guo Z, Tan Y, Han Y, Huang X, Zhou D, Yang R. Regulatory effects of cAMP receptor protein (CRP) on porin genes and its own gene in *Yersinia pestis*. BMC Microbiol. 2011;11:40.
28. Gao H, Zhang Y, Han Y, Yang L, Liu X, Guo Z, Tan Y, Huang X, Zhou D, Yang R. Phenotypic and transcriptional analysis of the osmotic regulator OmpR in *Yersinia pestis*. BMC Microbiol. 2011;11:39.
29. Wosten MM. Eubacterial sigma-factors. FEMS Microbiol Rev. 1998;22(3):127–50.
30. Liu L, Fang N, Sun Y, Yang H, Zhang Y, Han Y, Zhou D, Yang R. Transcriptional regulation of the *waaAE-coaD* operon by PhoP and RcsAB in *Yersinia pestis* biovar *Microtus*. Protein Cell. 2014;5(12):940–4.
31. Zhang Y, Gao H, Wang L, Xiao X, Tan Y, Guo Z, Zhou D, Yang R. Molecular characterization of transcriptional regulation of *rovA* by PhoP and RovA in *Yersinia pestis*. PLoS One. 2011;6(9):e25484.
32. Perry RD, Fetherston JD. *Yersinia pestis*-etiologic agent of plague. Clin Microbiol Rev. 1997;10(1):35–66.
33. Human plague: review of regional morbidity and mortality, 2004–2009. Releve epidemiologique hebdomadaire/Section d'hygiene du Secretariat de la Societe des Nations = Weekly epidemiological record/Health Section of the Secretariat of the League of Nations. 2009;85(6):40–5.
34. Zhou D, Yang R. Formation and regulation of *Yersinia* biofilms. Protein Cell. 2011;2(3):173–9.
35. Fang N, Qu S, Yang H, Fang H, Liu L, Zhang Y, Wang L, Han Y, Zhou D, Yang R. HmsB enhances biofilm formation in *Yersinia pestis*. Front Microbiol. 2014;5:685.
36. Bobrov AG, Kirillina O, Forman S, Mack D, Perry RD. Insights into *Yersinia pestis* biofilm development: topology and co-interaction of Hms inner membrane proteins involved in exopolysaccharide production. Environ Microbiol. 2008;10(6):1419–32.
37. Bobrov AG, Kirillina O, Ryjenkov DA, Waters CM, Price PA, Fetherston JD, Mack D, Goldman WE, Gomelsky M, Perry RD. Systematic analysis of cyclic di-GMP signalling enzymes and their role in biofilm formation and virulence in *Yersinia pestis*. Mol Microbiol. 2011;79(2):533–51.
38. Kirillina O, Fetherston JD, Bobrov AG, Abney J, Perry RD. HmsP, a putative phosphodiesterase, and HmsT, a putative diguanylate cyclase, control Hms-dependent biofilm formation in *Yersinia pestis*. Mol Microbiol. 2004;54(1):75–88.
39. Darby C, Ananth SL, Tan L, Hinnebusch BJ. Identification of gmhA, a *Yersinia pestis* gene required for flea blockage, by using a *Caenorhabditis elegans* biofilm system. Infect Immun. 2005;73(11):7236–42.
40. Tan L, Darby C. *Yersinia pestis* is viable with endotoxin composed of only lipid A. J Bacteriol. 2005;187(18):6599–600.
41. Patel CN, Wortham BW, Lines JL, Fetherston JD, Perry RD, Oliveira MA. Polyamines are essential for the formation of plague biofilm. J Bacteriol. 2006;188(7):2355–63.
42. Majdalani N, Gottesman S. The Rcs phosphorelay: a complex signal transduction system. Annu Rev Microbiol. 2005;59:379–405.
43. Sun YC, Guo XP, Hinnebusch BJ, Darby C. The *Yersinia pestis* Rcs phosphorelay inhibits biofilm formation by repressing transcription of the diguanylate cyclase gene hmsT. J Bacteriol. 2012;194(8):2020–6.
44. Sun YC, Hinnebusch BJ, Darby C. Experimental evidence for negative selection in the evolution of a *Yersinia pestis* pseudogene. Proc Natl Acad Sci U S A. 2008;105(23):8097–101.
45. Zhang J. Positive selection, not negative selection, in the pseudogenization of *rcsA* in *Yersinia pestis*. Proc Natl Acad Sci U S A. 2008;105(42):E69; author reply E70.

46. Escolar L, Perez-Martin J, de Lorenzo V. Opening the iron box: transcriptional metalloregulation by the Fur protein. J Bacteriol. 1999;181(20):6223–9.
47. Zhou D, Qin L, Han Y, Qiu J, Chen Z, Li B, Song Y, Wang J, Guo Z, Zhai J, et al. Global analysis of iron assimilation and fur regulation in *Yersinia pestis*. FEMS Microbiol Lett. 2006;258(1):9–17.
48. Gao H, Zhou D, Li Y, Guo Z, Han Y, Song Y, Zhai J, Du Z, Wang X, Lu J, et al. The iron-responsive Fur regulon in *Yersinia pestis*. J Bacteriol. 2008;190(8):3063–75.
49. Sun F, Gao H, Zhang Y, Wang L, Fang N, Tan Y, Guo Z, Xia P, Zhou D, Yang R. Fur is a repressor of biofilm formation in *Yersinia pestis*. PLoS One. 2012;7(12):e52392.
50. Maddocks SE, Oyston PC. Structure and function of the LysR-type transcriptional regulator (LTTR) family proteins. Microbiology. 2008;154(Pt 12):3609–23.
51. Heroven AK, Dersch P. RovM, a novel LysR-type regulator of the virulence activator gene *rovA*, controls cell invasion, virulence and motility of *Yersinia pseudotuberculosis*. Mol Microbiol. 2006;62(5):1469–83.
52. Vadyvaloo V, Hinz AK. A LysR-type transcriptional regulator, RovM, senses nutritional cues suggesting that it is involved in metabolic adaptation of *Yersinia pestis* to the flea gut. PLoS One. 2015;10(9):e0137508.
53. Tam C, Demke O, Hermanas T, Mitchell A, Hendrickx AP, Schneewind O. YfbA, a *yersinia pestis* regulator required for colonization and biofilm formation in the gut of cat fleas. J Bacteriol. 2014;196(6):1165–73.
54. Marceau M. Transcriptional regulation in *Yersinia*: an update. Curr Issues Mol Biol. 2005;7(2):151–77.
55. Buchrieser C, Prentice M, Carniel E. The 102-kilobase unstable region of *Yersinia pestis* comprises a high-pathogenicity island linked to a pigmentation segment which undergoes internal rearrangement. J Bacteriol. 1998;180(9):2321–9.
56. Hare JM, Wagner AK, McDonough KA. Independent acquisition and insertion into different chromosomal locations of the same pathogenicity island in *Yersinia pestis* and *Yersinia pseudotuberculosis*. Mol Microbiol. 1999;31(1):291–303.
57. Lillard Jr JW, Fetherston JD, Pedersen L, Pendrak ML, Perry RD. Sequence and genetic analysis of the hemin storage (*hms*) system of *Yersinia pestis*. Gene. 1997;193(1):13–21.
58. Price SB, Freeman MD, Yeh KS. Transcriptional analysis of the *Yersinia pestis* pH 6 antigen gene. J Bacteriol. 1995;177(20):5997–6000.
59. Yang Y, Isberg RR. Transcriptional regulation of the *Yersinia pseudotuberculosis* pH6 antigen adhesin by two envelope-associated components. Mol Microbiol. 1997;24(3):499–510.
60. Weening EH, Cathelyn JS, Kaufman G, Lawrenz MB, Price P, Goldman WE, Miller VL. The dependence of the *Yersinia pestis* capsule on pathogenesis is influenced by the mouse background. Infect Immun. 2011;79(2):644–52.
61. Cathelyn JS, Crosby SD, Lathem WW, Goldman WE, Miller VL. RovA, a global regulator of *Yersinia pestis*, specifically required for bubonic plague. Proc Natl Acad Sci U S A. 2006;103(36):13514–9.
62. Lindler LE, Tall BD. *Yersinia pestis* pH 6 antigen forms fimbriae and is induced by intracellular association with macrophages. Mol Microbiol. 1993;8(2):311–24.
63. Galvan EM, Chen H, Schifferli DM. The Psa fimbriae of *Yersinia pestis* interact with phosphatidylcholine on alveolar epithelial cells and pulmonary surfactant. Infect Immun. 2007;75(3):1272–9.
64. Liu F, Chen H, Galvan EM, Lasaro MA, Schifferli DM. Effects of Psa and F1 on the adhesive and invasive interactions of *Yersinia pestis* with human respiratory tract epithelial cells. Infect Immun. 2006;74(10):5636–44.
65. Felek S, Tsang TM, Krukonis ES. Three *Yersinia pestis* adhesins facilitate Yop delivery to eukaryotic cells and contribute to plague virulence. Infect Immun. 2010;78(10):4134–50.
66. Huang XZ, Lindler LE. The pH 6 antigen is an antiphagocytic factor produced by Yersinia pestis independent of *Yersinia* outer proteins and capsule antigen. Infect Immun. 2004;72(12):7212–9.

67. Makoveichuk E, Cherepanov P, Lundberg S, Forsberg A, Olivecrona G. pH6 antigen of *Yersinia pestis* interacts with plasma lipoproteins and cell membranes. J Lipid Res. 2003;44(2):320–30.
68. Karlyshev AV, Galyov EE, Abramov VM, Zav'yalov VP. Caf1R gene and its role in the regulation of capsule formation of *Y. pestis*. FEBS Lett. 1992;305(1):37–40.
69. Hinnebusch BJ, Rudolph AE, Cherepanov P, Dixon JE, Schwan TG, Forsberg A. Role of *Yersinia* murine toxin in survival of *Yersinia pestis* in the midgut of the flea vector. Science. 2002;296(5568):733–5.
70. Lathem WW, Price PA, Miller VL, Goldman WE. A plasminogen-activating protease specifically controls the development of primary pneumonic plague. Science. 2007;315(5811):509–13.
71. Sodeinde OA, Subrahmanyam YV, Stark K, Quan T, Bao Y, Goguen JD. A surface protease and the invasive character of plague. Science. 1992;258(5084):1004–7.
72. Wilkinson SP, Grove A. Ligand-responsive transcriptional regulation by members of the MarR family of winged helix proteins. Curr Issues Mol Biol. 2006;8(1):51–62.
73. Dube PH, Handley SA, Revell PA, Miller VL. The *rovA* mutant of *Yersinia enterocolitica* displays differential degrees of virulence depending on the route of infection. Infect Immun. 2003;71(6):3512–20.
74. Nagel G, Lahrz A, Dersch P. Environmental control of invasin expression in *Yersinia pseudotuberculosis* is mediated by regulation of RovA, a transcriptional activator of the SlyA/Hor family. Mol Microbiol. 2001;41(6):1249–69.
75. Revell PA, Miller VL. A chromosomally encoded regulator is required for expression of the *Yersinia enterocolitica inv* gene and for virulence. Mol Microbiol. 2000;35(3):677–85.
76. Ellison DW, Lawrenz MB, Miller VL. Invasin and beyond: regulation of *Yersinia* virulence by RovA. Trends Microbiol. 2004;12(6):296–300.
77. Heroven AK, Nagel G, Tran HJ, Parr S, Dersch P. RovA is autoregulated and antagonizes H-NS-mediated silencing of invasin and *rovA* expression in *Yersinia pseudotuberculosis*. Mol Microbiol. 2004;53(3):871–88.
78. Simonet M, Riot B, Fortineau N, Berche P. Invasin production by *Yersinia pestis* is abolished by insertion of an IS200-like element within the *inv* gene. Infect Immun. 1996;64(1):375–9.
79. Derbise A, Chenal-Francisque V, Pouillot F, Fayolle C, Prevost MC, Medigue C, Hinnebusch BJ, Carniel E. A horizontally acquired filamentous phage contributes to the pathogenicity of the plague bacillus. Mol Microbiol. 2007;63(4):1145–57.
80. Li Y, Dai E, Cui Y, Li M, Zhang Y, Wu M, Zhou D, Guo Z, Dai X, Cui B, et al. Different region analysis for genotyping *Yersinia pestis* isolates from China. PLoS One. 2008;3(5):e2166.
81. Yang F, Ke Y, Tan Y, Bi Y, Shi Q, Yang H, Qiu J, Wang X, Guo Z, Ling H, et al. Cell membrane is impaired, accompanied by enhanced type III secretion system expression in *Yersinia pestis* deficient in RovA regulator. PLoS One. 2010;5:e12840.
82. Oh MH, Lee SM, Lee DH, Choi SH. Regulation of the *Vibrio vulnificus hupA* gene by temperature alteration and cyclic AMP receptor protein and evaluation of its role in virulence. Infect Immun. 2009;77(3):1208–15.
83. Petersen S, Young GM. Essential role for cyclic AMP and its receptor protein in *Yersinia enterocolitica* virulence. Infect Immun. 2002;70(7):3665–72.
84. Rickman L, Scott C, Hunt DM, Hutchinson T, Menendez MC, Whalan R, Hinds J, Colston MJ, Green J, Buxton RS. A member of the cAMP receptor protein family of transcription regulators in *Mycobacterium tuberculosis* is required for virulence in mice and controls transcription of the *rpfA* gene coding for a resuscitation promoting factor. Mol Microbiol. 2005;56(5):1274–86.
85. Skorupski K, Taylor RK. Cyclic AMP and its receptor protein negatively regulate the coordinate expression of cholera toxin and toxin-coregulated pilus in *Vibrio cholerae*. Proc Natl Acad Sci U S A. 1997;94(1):265–70.

86. Zhan L, Han Y, Yang L, Geng J, Li Y, Gao H, Guo Z, Fan W, Li G, Zhang L, et al. The cyclic AMP receptor protein, CRP, is required for both virulence and expression of the minimal CRP regulon in *Yersinia pestis* biovar *microtus*. Infect Immun. 2008;76(11):5028–37.
87. Gunasekera A, Ebright YW, Ebright RH. DNA sequence determinants for binding of the *Escherichia coli* catabolite gene activator protein. J Biol Chem. 1992;267(21):14713–20.
88. Kim TJ, Chauhan S, Motin VL, Goh EB, Igo MM, Young GM. Direct transcriptional control of the plasminogen activator gene of *Yersinia pestis* by the cyclic AMP receptor protein. J Bacteriol. 2007;189(24):8890–900.
89. Liu H, Wang H, Qiu J, Wang X, Guo Z, Qiu Y, Zhou D, Han Y, Du Z, Li C, et al. Transcriptional profiling of a mice plague model: insights into interaction between *Yersinia pestis* and its host. J Basic Microbiol. 2009;49(1):92–9.
90. Heroven AK, Sest M, Pisano F, Scheb-Wetzel M, Steinmann R, Bohme K, Klein J, Munch R, Schomburg D, Dersch P. Crp induces switching of the CsrB and CsrC RNAs in *Yersinia pseudotuberculosis* and links nutritional status to virulence. Front Cell Infect Microbiol. 2012;2:158.
91. Lathem WW, Schroeder JA, Bellows LE, Ritzert JT, Koo JT, Price PA, Goldman WE. Posttranscriptional regulation of the *Yersinia pestis* cyclic AMP receptor protein Crp and impact on virulence. mBio. 2014;5(1):e01038–13.
92. Groisman EA. The pleiotropic two-component regulatory system PhoP-PhoQ. J Bacteriol. 2001;183(6):1835–42.
93. Li YL, Gao H, Qin L, Li B, Han YP, Guo ZB, Song YJ, Zhai JH, Du ZM, Wang XY, et al. Identification and characterization of PhoP regulon members in *Yersinia pestis* biovar *Microtus*. BMC Genomics. 2008;9(1):143.
94. Oyston PC, Dorrell N, Williams K, Li SR, Green M, Titball RW, Wren BW. The response regulator PhoP is important for survival under conditions of macrophage-induced stress and virulence in *Yersinia pestis*. Infect Immun. 2000;68(6):3419–25.
95. Grabenstein JP, Fukuto HS, Palmer LE, Bliska JB. Characterization of phagosome trafficking and identification of PhoP-regulated genes important for survival of *Yersinia pestis* in macrophages. Infect Immun. 2006;74(7):3727–41.
96. Masse E, Majdalani N, Gottesman S. Regulatory roles for small RNAs in bacteria. Curr Opin Microbiol. 2003;6(2):120–4.
97. De Lay N, Schu DJ, Gottesman S. Bacterial small RNA-based negative regulation: Hfq and its accomplices. J Biol Chem. 2013;288(12):7996–8003.
98. Vogel J, Wagner EG. Target identification of small noncoding RNAs in bacteria. Curr Opin Microbiol. 2007;10(3):262–70.
99. Tjaden B, Goodwin SS, Opdyke JA, Guillier M, Fu DX, Gottesman S, Storz G. Target prediction for small, noncoding RNAs in bacteria. Nucleic Acids Res. 2006;34(9):2791–802.
100. Chen S, Lesnik EA, Hall TA, Sampath R, Griffey RH, Ecker DJ, Blyn LB. A bioinformatics based approach to discover small RNA genes in the *Escherichia coli* genome. Biosystems. 2002;65(2–3):157–77.
101. Sharma CM, Vogel J. Experimental approaches for the discovery and characterization of regulatory small RNA. Curr Opin Microbiol. 2009;12(5):536–46.
102. Vogel J, Bartels V, Tang TH, Churakov G, Slagter-Jager JG, Huttenhofer A, Wagner EG. RNomics in *Escherichia coli* detects new sRNA species and indicates parallel transcriptional output in bacteria. Nucleic Acids Res. 2003;31(22):6435–43.
103. Zhang A, Wassarman KM, Rosenow C, Tjaden BC, Storz G, Gottesman S. Global analysis of small RNA and mRNA targets of Hfq. Mol Microbiol. 2003;50(4):1111–24.
104. Wassarman KM, Repoila F, Rosenow C, Storz G, Gottesman S. Identification of novel small RNAs using comparative genomics and microarrays. Genes Dev. 2001;15(13):1637–51.
105. Windbichler N, von Pelchrzim F, Mayer O, Csaszar E, Schroeder R. Isolation of small RNA-binding proteins from *E. coli*: evidence for frequent interaction of RNAs with RNA polymerase. RNA Biol. 2008;5(1):30–40.

106. Waters LS, Storz G. Regulatory RNAs in bacteria. Cell. 2009;136(4):615–28.
107. Sharma CM, Hoffmann S, Darfeuille F, Reignier J, Findeiss S, Sittka A, Chabas S, Reiche K, Hackermuller J, Reinhardt R, et al. The primary transcriptome of the major human pathogen *Helicobacter pylori*. Nature. 2010;464(7286):250–5.
108. Liu JM, Livny J, Lawrence MS, Kimball MD, Waldor MK, Camilli A. Experimental discovery of sRNAs in *Vibrio cholerae* by direct cloning, 5S/tRNA depletion and parallel sequencing. Nucleic Acids Res. 2009;37(6):e46.
109. Storz G, Vogel J, Wassarman KM. Regulation by small RNAs in bacteria: expanding frontiers. Mol Cell. 2011;43(6):880–91.
110. Urban JH, Vogel J. Translational control and target recognition by *Escherichia coli* small RNAs in vivo. Nucleic Acids Res. 2007;35(3):1018–37.
111. Corcoran CP, Podkaminski D, Papenfort K, Urban JH, Hinton JC, Vogel J. Superfolder GFP reporters validate diverse new mRNA targets of the classic porin regulator, MicF RNA. Mol Microbiol. 2012;84(3):428–45.
112. Papenfort K, Vogel J. Regulatory RNA in bacterial pathogens. Cell Host Microbe. 2010;8(1):116–27.
113. Bobrovskyy M, Vanderpool CK. Regulation of bacterial metabolism by small RNAs using diverse mechanisms. Annu Rev Genet. 2013;47:209–32.
114. Vogel J, Luisi BF. Hfq and its constellation of RNA. Nat Rev Microbiol. 2011;9(8):578–89.
115. Garst AD, Edwards AL, Batey RT. Riboswitches: structures and mechanisms. Cold Spring Harb Perspect Biol. 2010;3:1–13.
116. Babitzke P, Romeo T. CsrB sRNA family: sequestration of RNA-binding regulatory proteins. Curr Opin Microbiol. 2007;10(2):156–63.
117. Wassarman KM. 6S RNA: a regulator of transcription. Mol Microbiol. 2007;65(6):1425–31.
118. Van Assche E, Van Puyvelde S, Vanderleyden J, Steenackers HP. RNA-binding proteins involved in post-transcriptional regulation in bacteria. Front Microbiol. 2015;6:141.
119. Timmermans J, Van Melderen L. Post-transcriptional global regulation by CsrA in bacteria. Cell Mol Life Sci. 2010;67(17):2897–908.
120. Chao Y, Vogel J. The role of Hfq in bacterial pathogens. Curr Opin Microbiol. 2010;13(1):24–33.
121. Sauer E. Structure and RNA-binding properties of the bacterial LSm protein Hfq. RNA Biol. 2013;10(4):610–8.
122. Pfeiffer V, Papenfort K, Lucchini S, Hinton JC, Vogel J. Coding sequence targeting by MicC RNA reveals bacterial mRNA silencing downstream of translational initiation. Nat Struct Mol Biol. 2009;16(8):840–6.
123. Sharma CM, Darfeuille F, Plantinga TH, Vogel J. A small RNA regulates multiple ABC transporter mRNAs by targeting C/A-rich elements inside and upstream of ribosome-binding sites. Genes Dev. 2007;21(21):2804–17.
124. Bouvier M, Sharma CM, Mika F, Nierhaus KH, Vogel J. Small RNA binding to 5′ mRNA coding region inhibits translational initiation. Mol Cell. 2008;32(6):827–37.
125. Morita T, Aiba H. RNase E action at a distance: degradation of target mRNAs mediated by an Hfq-binding small RNA in bacteria. Genes Dev. 2011;25(4):294–8.
126. Desnoyers G, Bouchard MP, Masse E. New insights into small RNA-dependent translational regulation in prokaryotes. Trends Genet TIG. 2013;29(2):92–8.
127. Desnoyers G, Masse E. Noncanonical repression of translation initiation through small RNA recruitment of the RNA chaperone Hfq. Genes Dev. 2012;26(7):726–39.
128. Prevost K, Desnoyers G, Jacques JF, Lavoie F, Masse E. Small RNA-induced mRNA degradation achieved through both translation block and activated cleavage. Genes Dev. 2011;25(4):385–96.
129. Saramago M, Barria C, Dos Santos RF, Silva IJ, Pobre V, Domingues S, Andrade JM, Viegas SC, Arraiano CM. The role of RNases in the regulation of small RNAs. Curr Opin Microbiol. 2014;18:105–15.

130. Papenfort K, Vanderpool CK. Target activation by regulatory RNAs in bacteria. FEMS Microbiol Rev. 2015;39(3):362–78.
131. Obana N, Nomura N, Nakamura K. Structural requirement in *Clostridium perfringens* collagenase mRNA 5′ leader sequence for translational induction through small RNA-mRNA base pairing. J Bacteriol. 2013;195(12):2937–46.
132. Koo JT, Alleyne TM, Schiano CA, Jafari N, Lathem WW. Global discovery of small RNAs in *Yersinia pseudotuberculosis* identifies *Yersinia*-specific small, noncoding RNAs required for virulence. Proc Natl Acad Sci U S A. 2011;108(37):E709–17.
133. Qu Y, Bi L, Ji X, Deng Z, Zhang H, Yan Y, Wang M, Li A, Huang X, Yang R, et al. Identification by cDNA cloning of abundant sRNAs in a human-avirulent *Yersinia pestis* strain grown under five different growth conditions. Future Microbiol. 2012;7(4):535–47.
134. Beauregard A, Smith EA, Petrone BL, Singh N, Karch C, McDonough KA, Wade JT. Identification and characterization of small RNAs in *Yersinia pestis*. RNA Biol. 2013;10(3):397–405.
135. Schiano CA, Koo JT, Schipma MJ, Caulfield AJ, Jafari N, Lathem WW. Genome-wide analysis of small RNAs expressed by *Yersinia pestis* identifies a regulator of the Yop-Ysc type III secretion system. J Bacteriol. 2014;196(9):1659–70.
136. Yan Y, Su S, Meng X, Ji X, Qu Y, Liu Z, Wang X, Cui Y, Deng Z, Zhou D, et al. Determination of sRNA expressions by RNA-seq in *Yersinia pestis* grown *in vitro* and during infection. PLoS One. 2013;8(9):e74495.
137. Martinez-Chavarria LC, Vadyvaloo V. *Yersinia pestis* and *Yersinia pseudotuberculosis* infection: a regulatory RNA perspective. Front Microbiol. 2015;6:956.
138. Schiano CA, Lathem WW. Post-transcriptional regulation of gene expression in *Yersinia* species. Front Cellular Infect Microbiol. 2012;2:129.
139. Livny J, Brencic A, Lory S, Waldor MK. Identification of 17 *Pseudomonas aeruginosa* sRNAs and prediction of sRNA-encoding genes in 10 diverse pathogens using the bioinformatic tool sRNAPredict2. Nucleic Acids Res. 2006;34(12):3484–93.
140. Nuss AM, Heroven AK, Waldmann B, Reinkensmeier J, Jarek M, Beckstette M, Dersch P. Transcriptomic profiling of *Yersinia pseudotuberculosis* reveals reprogramming of the Crp regulon by temperature and uncovers Crp as a master regulator of small RNAs. PLoS Genet. 2015;11(3):e1005087.
141. Geng J, Song Y, Yang L, Feng Y, Qiu Y, Li G, Guo J, Bi Y, Qu Y, Wang W, et al. Involvement of the post-transcriptional regulator Hfq in *Yersinia pestis* virulence. PLoS One. 2009;4(7):e6213.
142. Bellows LE, Koestler BJ, Karaba SM, Waters CM, Lathem WW. Hfq-dependent, co-ordinate control of cyclic diguanylate synthesis and catabolism in the plague pathogen *Yersinia pestis*. Mol Microbiol. 2012;86:661–74.
143. Rempe KA, Hinz AK, Vadyvaloo V. Hfq regulates biofilm gut blockage that facilitates flea-borne transmission of *Yersinia pestis*. J Bacteriol. 2012;194(8):2036–40.
144. Carpousis AJ. The *Escherichia coli* RNA degradosome: structure, function and relationship in other ribonucleolytic multienzyme complexes. Biochem Soc Trans. 2002;30(2):150–5.
145. Deng Z, Liu Z, Bi Y, Wang X, Zhou D, Yang R, Han Y. Rapid degradation of Hfq-free RyhB in *Yersinia pestis* by PNPase independent of putative ribonucleolytic complexes. BioMed Res Int. 2014;2014:798918.
146. Rosenzweig JA, Chromy B, Echeverry A, Yang J, Adkins B, Plano GV, McCutchen-Maloney S, Schesser K. Polynucleotide phosphorylase independently controls virulence factor expression levels and export in *Yersinia spp.* FEMS Microbiol Lett. 2007;270(2):255–64.
147. Yang J, Jain C, Schesser K. RNase E regulates the *Yersinia* type 3 secretion system. J Bacteriol. 2008;190(10):3774–8.
148. Keiler KC. Biology of trans-translation. Annu Rev Microbiol. 2008;62:133–51.
149. Karzai AW, Susskind MM, Sauer RT. SmpB, a unique RNA-binding protein essential for the peptide-tagging activity of SsrA (tmRNA). EMBO J. 1999;18(13):3793–9.

150. Okan NA, Mena P, Benach JL, Bliska JB, Karzai AW. The *smpB-ssrA* mutant of *Yersinia pestis* functions as a live attenuated vaccine to protect mice against pulmonary plague Infection. Infect Immun. 2010;78(3):1284–93.
151. Masse E, Gottesman S. A small RNA regulates the expression of genes involved in iron metabolism in *Escherichia coli*. Proc Natl Acad Sci U S A. 2002;99(7):4620–5.
152. Geissmann TA, Touati D. Hfq, a new chaperoning role: binding to messenger RNA determines access for small RNA regulator. EMBO J. 2004;23(2):396–405.
153. Deng Z, Meng X, Su S, Liu Z, Ji X, Zhang Y, Zhao X, Wang X, Yang R, Han Y. Two sRNA RyhB homologs from *Yersinia pestis* biovar *microtus* expressed in vivo have differential Hfq-dependent stability. Res Microbiol. 2012;163(6–7):413–8.
154. McArthur SD, Pulvermacher SC, Stauffer GV. The *Yersinia pestis gcvB* gene encodes two small regulatory RNA molecules. BMC Microbiol. 2006;6:52.
155. Pan NJ, Brady MJ, Leong JM, Goguen JD. Targeting type III secretion in *Yersinia pestis*. Antimicrob Agents Chemother. 2009;53(2):385–92.
156. Hoe NP, Goguen JD. Temperature sensing in *Yersinia pestis*: translation of the LcrF activator protein is thermally regulated. J Bacteriol. 1993;175(24):7901–9.
157. Bohme K, Steinmann R, Kortmann J, Seekircher S, Heroven AK, Berger E, Pisano F, Thiermann T, Wolf-Watz H, Narberhaus F, et al. Concerted actions of a thermo-labile regulator and a unique intergenic RNA thermosensor control *Yersinia* virulence. PLoS Pathog. 2012;8(2):e1002518.
158. Jarrett CO, Deak E, Isherwood KE, Oyston PC, Fischer ER, Whitney AR, Kobayashi SD, DeLeo FR, Hinnebusch BJ. Transmission of *Yersinia pestis* from an infectious biofilm in the flea vector. J Infect Dis. 2004;190(4):783–92.
159. Hinnebusch BJ, Erickson DL. *Yersinia pestis* biofilm in the flea vector and its role in the transmission of plague. Curr Top Microbiol Immunol. 2008;322:229–48.
160. Bobrov AG, Kirillina O, Perry RD. Regulation of biofilm formation in *Yersinia pestis*. Adv Exp Med Biol. 2007;603:201–10.
161. Pratt LA, Kolter R. Genetic analysis of *Escherichia coli* biofilm formation: roles of flagella, motility, chemotaxis and type I pili. Mol Microbiol. 1998;30(2):285–93.
162. Martinez LC, Vadyvaloo V. Mechanisms of post-transcriptional gene regulation in bacterial biofilms. Front Cellular Infect Microbiol. 2014;4:38.
163. Baker CS, Morozov I, Suzuki K, Romeo T, Babitzke P. CsrA regulates glycogen biosynthesis by preventing translation of *glgC* in *Escherichia coli*. Mol Microbiol. 2002;44(6):1599–610.
164. Willias SP, Chauhan S, Lo CC, Chain PS, Motin VL. CRP-mediated carbon catabolite regulation of *Yersinia pestis* biofilm formation is enhanced by the carbon storage regulator protein, CsrA. PLoS One. 2015;10(8):e0135481.

Chapter 9
Yersinia pestis in the Age of Big Data

Ruifu Yang and Vladimir L. Motin

Abstract As omics-driven technologies developed rapidly, genomics, transcriptomics, proteomics, metabolomics and other omics-based data have been accumulated in unprecedented speed. Omics-driven big data in biology have changed our way of research. "Big science" has promoted our understanding of biology in a holistic overview that is impossibly achieved by traditional hypothesis-driven research. In this chapter, we gave an overview of omics-driven research on *Y. pestis*, provided a way of thinking on *Yersinia pestis* research in the age of big data, and made some suggestions to integrate omics-based data for systems understanding of *Y. pestis*.

Keywords *Yersinia pestis* • Big data • Omics-driven research

9.1 The Age of Big Data

The well-known company, McKinsey, was the first to propose the term "big data" because data has been widely employed when describing all walks of life and has become an important factor of production. In facing the sea of data, new methods to glean the most important pieces of information to guide our decision-making and activities in daily life will promote a new wave of data-based objectives. In the era of big data, it is sometimes easier to make all decisions based on objective analysis of massive data, rather than using intuitive insight.

There is no commonly accepted definition of big data. At the beginning of the proposal, we understood this term to mean a dataset. However, we soon recognized

R. Yang (✉)
Beijing Institute of Microbiology and Epidemiology,
No. Dongdajie, Fengtai, Beijing 100071, China
e-mail: ruifuyang@gmail.com

V.L. Motin
Departments of Pathology and Microbiology & Immunology, University of Texas Medical Branch, Galveston, TX 77555, USA
e-mail: vlmotin@utmb.edu

R. Yang, A. Anisimov (eds.), *Yersinia pestis: Retrospective and Perspective*, Advances in Experimental Medicine and Biology 918,
DOI 10.1007/978-94-024-0890-4_9

that this term indeed changed our way of doing things. We gradually expanded this term from "dataset" to "process" because we do not only need to analyze and understand the dataset itself but also the whole process utilized to generate the dataset. This ensures better decision making and comprehension of the whole picture. Below are two definitions of big data in order to give the readers a better idea of term from different angles.

According to the McKinsey Global Institute's definition [1], *"Big Data" refers to datasets whose size is beyond the ability of typical data-base software tools to capture, store, manage, and analyze. This definition is intentionally subjective and incorporates a moving definition of how big a dataset needs to be in order to be considered Big Data – i.e., we don't define Big Data in terms of being larger than a certain number of terabytes (thousands of gigabytes). We assume that, as technology advances over time, the size of datasets that qualify as Big Data will also increase. Also note that the definition can vary by sector, depending on what kinds of software tools are commonly available and what sizes of datasets are common in a particular industry. With those caveats, Big Data in many sectors today will range from a few dozen terabytes to multiple petabytes (thousands of terabytes).*

According to Gartner's IT glossary [2], *Big Data is high-volume, high-velocity and high-variety information assets that demand cost-effective, innovative forms of information processing for enhanced insight and decision making.*

According to Kalyvas and Overly's book [3], *Big Data is a process to deliver decision-making insights. The process uses people and technology to quickly analyze large amounts of data of different types (traditional table structured data and unstructured data, such as pictures, video, email, transaction data, and social media interactions) from a variety of sources to produce a stream of actionable knowledge.*

The distinguishing characteristics of big data are "three Vs," volume (amount of data), velocity (the speed of processing and the pace of change to data), and variety (sources of data and types of data), which are the most notable features distinguishing big data from traditional data [3].

Massive data has already profoundly revolutionized the academic world. In the past, most of research was conducted via hypothesis-driven thinking, and problems were solved step by step. In contrast, big data-based research is based on objective data, and the relationship between different datasets is discovered by bioinformatics analysis. Big data-based research is a kind of holistic investigation, intending to reveal the complex biological activities as a whole picture. In the field of medicine, never before has the role of informatics been so crucial for clinical diagnosis and treatment. Effective application of big data in clinics has promoted the development of personal medicine or precision medicine [4–8] and also P4 (predictive, preventive, personalized, and participatory) medicine [9].

In the field of infectious diseases, big data provided unprecedented opportunities for global infectious disease surveillance [10]. For instance, the Google team has successfully employed search engine data to forecast seasonal outbreaks of influenza [11–15], which was a great step in infectious disease forecasting without using any disease surveillance data. If the surveillance data and genome data were integrated in their predicting model, the forecasting accuracy would increase.

In the age of omics, it is easy for us to acquire massive data of genomics, proteomics, transcriptomics, metabolomics, interactomics, phenomics, and other omics-based research. For zoonosis, we can also correlate the relevant environmental, ecological, surveillance, and other data together for finding more interesting clues to understand life in nature [16].

Big data or massive data is a relative term without the criterion of data volume. For investigators involved in biological research, we have different origins of data and face challenges during data analysis [17]. Sometimes too much data is not the ideal choice for solving an issue because it instead results in a complex network that is difficult to decipher [18]. The challenges of the big data era span generation of the data, novel analysis strategies, data sharing and use, etc., which explains the urgent need for development of new approaches to cope with the big data revolution [19]. To take full advantage of what big data can provide, interdisciplinary scientists are critical and must work together to overcome the hurdles associated with massive data. Big data building requires computer scientists, while statisticians and mathematicians play critical roles in data modeling. Biologists, epidemiologists, ecologists, geologists, etc., are also indispensible to understand the mechanisms of life in nature. Therefore, big data can help to solve complex problems only with the participation of large teams of experts from different disciplines.

9.2 Technologies for Big Data Production

High-throughput omics technologies (e.g., genomics, transcriptomics, proteomics, metabolomics, interactomics, phenomics, lipidomics, and beyond) produce massive data in a speed unimaginable previously [17, 20].

The advent of next generation sequencing (NGS) technology is a kind of high-throughput DNA sequencing technology using strategies different from the traditional Sanger-based sequencing method [21]. The 454 Genome Sequencing System (Roche) and the SOLiD (Applied Biosystems, ABI), Solexa (Illumina), and Ion Torrent (Life Technologies) sequencing methods belong to the second-generation sequencing technology. These technologies can perform thousands to billions of sequencing reactions in parallel without the need of a fragment cloning library and can produce numerous reads covering whole genome at an unprecedented speed. The third-generation single-molecule sequencing technology (the PacBio RS system from Pacific Biosciences) employs the single-molecule, real-time (SMRT) DNA sequencing technology, which can achieve long reads up to 8500 bp and can also directly identify epigenetic modifications of DNA methylation [22, 23]. Nanopore-based sequencer (Oxford MinION) was described as the fourth-generation DNA sequencing technology in literature, which showed promise in lowering sequencer's size and sequencing price [24, 25].

Transcriptome refers to a subset of genes transcribed in any given organism under a specific condition. It is the first step to execute a gene's function, linking between the genome, the proteome, and the cellular phenotype [20]. Transcriptome-wide RNA sequencing (RNA-seq) is a powerful technique to obtain data for all expressed RNAs in a bacterial population or in a single cell, including noncoding RNAs, which is transcriptomics [26].

Proteome refers to all proteins of a given organism at a specific condition, which could be identified by using mass spectrometry either qualitatively or quantitatively. Proteomics identifies and quantifies the entire proteome of an organism and interprets the proteins' biological functions [27].

Metabolomics aims to achieve high-throughput identification, quantification, and characterization of endogenous and exogenous metabolites for a holistic interpretation of biological systems [28, 29], usually using ^{1}H-NMR, UHPLC/MS/MS^2 (ultrahigh-performance liquid chromatography-tandem mass spectrometry), UPLC/ESI-SYNAPT-HDMS (ultra-performance liquid chromatography/electrospray ionization – SYNAPT – high-definition mass spectrometry), and UPLC-Q-TOF-HDMS analyses [29].

All cellular processes require protein-protein interactions, which is fundamental for many areas of biochemical and biomedical research [30]. During bacterial pathogenesis, pathogenic bacteria invade the host, and the host will eliminate them by innate or acquired immunity. This pathogen-host interaction was studied using traditional techniques in the past, which revealed dynamic interactions between a single protein or multiple proteins from a bacterium and host protein(s). This approach can only help us understand the bacterial pathogenesis ex parte. We now can use high-throughput technologies to study the interactome between a bacterial pathogen and the host.

Pathogen-host interactomes attempt to systematically identify interaction networks among pathogen proteins and host proteins [31–33] and can be studied by mass spectrometry and systematic yeast two-hybrid (Y2H) screening [32]. For instance, a *Yersinia pestis*-human protein interaction network was identified by Y2H system in combination with the previously published interactions [34].

Lipidomics investigates the structural and functional complexity of lipids in biological systems to answer biological questions associated with lipid metabolism. Liquid chromatography and mass spectrometry are the two most commonly used analytical approaches [35, 36].

Phenomics globally and systematically reveals cellular phenotypes, which closely associates with the genotypes and environmental context [37]. Phenotype MicroArray (PM) technology has been applied to characterize different groups of microorganisms in a high-throughput manner [38]. This platform can analyze nearly 2000 phenotypes together in one experiment if all the preformulated 96-well microplates are used. It could be used for comparing a wild-type strain and mutant of a bacterium. However, depending on the bacteria tested, one must carefully select the suitable carbon source in metabolic experiments, the dye mix, and the buffered medium. The proper statistical software must also be used to analyze the massive data [37].

All the omics technologies mentioned above have been widely used and have produced massive data for our understanding of certain key scientific questions, such as bacterial evolution and pathogenicity. Unfortunately, until now this type of analysis has not been reported for a bacterium using all available omics data to systematical uncover a key issue. For *Y. pestis*, genomics, transcriptomics, proteomics, and interactomics have all been applied to answer different questions. For example, our group applied these omics strategies to a *Microtus* strain 91001 for systematically understanding its microevolution and virulence. However, there still lacks parallel systematic research on other strains for comparison. Although we have done some work to try to holistically describe the unique features of *Y. pestis*, we still face the major challenge of how to integrate these omics data in order to systematically extract fascinating details for further study. In summary, application of omics technologies to *Y. pestis* is a good start for understanding this deadly pathogen's microevolution, unique lifestyle, and virulence.

9.3 Omics Research for *Yersinia pestis*

As mentioned above, through various omics-based research, we can acquire massive data for dynamically comprehensive and quantitative analysis of the interactions between all of the components in a bacterium, which is the goal of systems microbiology. Although we have performed genomic, transcriptomic, proteomic, and metabolomic studies on *Y. pestis*, we still cannot obtain a holistic view on its physiological, pathogenic, or evolutionary phenomena. Currently, there are several strains (CO92, KIM, and 91001) of *Y. pestis* used as a model to study various features of this pathogen. Due to its high-risk nature as a bioterrorism agent, most laboratories use attenuated strains to study *Y. pestis* in the laboratory. The current published data generated via omics-based research involve different *Y. pestis* strains subjected to a wide variety of conditions, such as different culture media and conditions, diverse environmental cues, multiple animal models with different infectious routes, and even various types of analyzing technologies. Our group made the first attempt to apply a multi-omics system approach to cross investigate temporal transcriptomic, proteomic, and metabolomic data of *Y. pestis* epidemic strain CO92 with *Y. pestis* nonepidemic strain Pestoides F and *Yersinia pseudotuberculosis* PB1 in parallel sample-matched experiments [39, 40]. This use of multiple pathogenic *Yersinia* strains in a single study coupled with a computational network modeling, as well as comparative analysis of similar data on *Salmonella*, allowed for the prediction of genes putatively involved in core pathogenic processes important for virulence mechanisms of *Yersinia* species [41]. Nevertheless, the massive volume of data obtained from this study still requires additional analysis and verification of major key points to gain further insight into *Y. pestis* physiology, virulence, and evolution.

9.3.1 Genomics

Genome sequencing became significantly easier with the advance of next generation sequencing technologies. Identification of complete genome sequences will unravel the minor differences of *Y. pestis* isolates from different plague foci by analyzing the single nucleotide polymorphisms (SNPs) and other genetic variations, such as variable number tandem repeats (VNTRs), clustered regularly interspaced short palindromic repeats (CRISPRs), small insertions and deletions (INDELs), gene acquisition/loss, etc. In Chap. 6, *Y. pestis* genome sequence variations and evolution are described in detail. Here, we will summarize the trends of *Y. pestis* genome sequencing for understanding its evolution and virulence.

First, more whole genome sequences are needed from different foci around the world for full understanding of this pathogen's evolution. The genome sequences of *Y. pestis* from Eurasia region, South Asia, and Africa are needed, in particular, for constructing a comprehensive phylogenetic tree of *Y. pestis* [42, 43], which can be used to reveal the path of *Y. pestis* evolution. Second, the sequencing of ancient DNA from the buried remains of individuals suspected to have died from plague has unraveled historical events surrounding plague outbreaks, including the three pandemics. Sequencing of additional ancient DNA samples from Asia, Eurasia, Africa, and America would be ideal so that we could reconstruct the evolutionary history of *Y. pestis*. Third, the third generation of sequencing technology, such as the PacBio sequencing, has been successfully applied to obtain complete bacterial genomes. This will allow the in-depth study of bacterial populations. Moreover, we need to obtain more finished sequences of *Y. pestis* isolates for analysis of genetic variations other than SNPs. This way, we can systematically map all kinds of polymorphisms onto the SNPs-based phylogenetic tree for fully understanding the role of different variations in *Y. pestis* microevolution. In addition, we can also correlate the phenotypic and ecological data with the tree for understanding of its functional evolution. Fourth, epigenetic modifications of the *Y. pestis* genome should be investigated with correlates to phenotypes. This will illustrate the fine adjustments that the *Y. pestis* microbe utilizes to adapt to varying environmental conditions. Finally, a genomic polymorphism database is needed for both basic research and microbial forensics. After obtaining genome sequences of *Y. pestis* isolates from around the globe, we will be in a position to develop an internationally shared database for the purposes mentioned above. Moreover, due to the potential bioterrorism/bioweapon potential of this pathogen, the development of such a database for international sharing is not only a scientific issue but is also a political decision. Nevertheless, it is the responsibility of scientists to facilitate this database-building process.

9.3.2 Transcriptomics

Transcriptomic research pertaining to the virulence network of *Y. pestis* has been reviewed comprehensively [17]. Here, we will provide a brief summary of the key points of *Y. pestis* transcriptomic studies.

The microarray-based and RNA-seq-based transcriptomics techniques are the main tools for investigating *Y. pestis* virulence, metabolism, and gene regulation. For understanding *Y. pestis* lifestyle and environmental modulation of global gene expression, microarray-based expression data were used to dissect bacterial adaptation to various environment conditions [17], including temperature alteration, increased osmolarity, ion deficiency, antibiotic treatment, oxidative and acidic stresses, antibacterial peptide treatment, as well as nutrition limitation. In addition, the expression profiles were analyzed using a microarray-based method for *Y. pestis* in cell line infection and animal challenge models.

RNA-seq technology can provide a better signal-to-noise ratio compared to microarray-based approaches. Using this technology, novel sRNAs have been identified in *Y. pseudotuberculosis* and the most of them were closely related to *Y. pestis*. One of the sRNA has been preliminarily confirmed to play a role in pneumonic plague. In our laboratory, we applied RNomics to *Y. pestis* and found 43 highly abundant sRNAs under multiple environmental cues; some of them might play critical roles in pathogenesis and adaptability [17].

Host responses to *Y. pestis* infection have been studied using the transcriptomic strategy as well [44]. Das et al. first applied cDNA microarray to study the transcriptomic responses of cultured macrophage-like cells to *Y. pestis* infection [45], providing mechanistic insights to understanding some of the symptoms induced by this pathogen. Several reports subsequently described rodent responses to *Y. pestis* infection using gene chip technologies [46–48], including Affymetrix GeneChip mouse and rat genome arrays. The analysis of patterns of up- or downregulated genes, including genes relating to immune functions, cytoskeletal structure, apoptosis, cell cycle, and protein degradation helped in understanding the host responses to *Y. pestis* bubonic infection. The in vivo transcriptional responses to pneumonic plague were also revealed using mouse PancChip [49]. Another study also compared in vivo transcriptional responses in mouse organs using pneumonic plague model after infection with wild-type CO92 and its Braun lipoprotein mutant [47]. The most abundant responses were found in liver and lungs; however, the spleen, which was traditionally thought as main invading organ by *Y. pestis*, demonstrated relatively lower transcriptional responses at different times postinfection [49].

From the studies mentioned above and in the previous review [17], we can conclude that transcriptomics can provide a lot of information about responses of *Y. pestis* to different environmental cues at the RNA expression level, including conditions found during animal infection. Currently, the transcriptomic approach offers the vast majority of information among the omic techniques to unravel the function of the gene or sRNA. For linking the transcriptomic data with function, we still need to employ traditional research tools, including gel shift, DNase I footprinting, primer extension, and new high-throughput technologies, such as RNA immunoprecipitation (RIP) and cross-linking immunoprecipitation coupled with deep sequencing (CLIP-seq), to verify the functions of the selected target genes or sRNAs. The transcriptomic data also needs to be integrated with other omics data, including genomic, proteomic, metabolomic, and interactomic data for an overall understanding of functions from genotypes to phenotypes.

9.3.3 Proteomics

Proteomics strategies have been successfully applied to *Y. pestis* studies, such as proteome mapping [50–52], secretome [53], membrane proteome [54], comparative proteomes [55], subcellular proteomes [56], periplasmic proteome [57], *Y. pestis* responses to iron starvation [58], and host responses to *Y. pestis* infection [59]. Most of the reports have been reviewed comprehensively, and here we will add new advances that were not mentioned in the review [17].

In 2015, Nozadze et al. compared proteomes of four representative *Y. pestis* strains isolated from natural plague foci in the Republic of Georgia using two-dimensional gel electrophoresis and mass spectrometry [55]. Four culture conditions, 28 °C with or without calcium and 37 °C with or without calcium, were compared for simulating type three secretion system (T3SS) in secretion or nonsecretion environments. They found that the protein expression patterns and cytotoxic activities were significantly differed among these strains, possibly reflecting their physiological functions.

Different proteomic reports on *Y. pestis* have provided us with rich protein expression data under different conditions, mostly at the descriptive level. Quantitative studies of *Y. pestis* proteomics for secretome and different parts of the cell (such as membrane and cytoplasmic proteome) under different growth conditions in vitro, or even in vivo, will be required to finely dissect the proteomic characteristics exhibited during *Y. pestis* physiology or pathogenesis. Particularly, the quantitative temporal analysis of both the secretome and proteome might be helpful for understanding the series of events that occur during the assembly process of T3SS and secretion of *Yersinia* outer proteins (YOPs).

Proteomics techniques can also help define specific pathways in the bacterium. This was observed during investigation of *Y. pestis* growth under iron-starved conditions for understanding the Fur-dependent iron metabolic pathway [58]. The iron acquisition systems, such as Ybt, Yfe, Yfu, Yiu, and Hmu were confirmed to be important in iron metabolism. The authors also found that an uncharacterized TonB-dependent OM receptor encoded by *y0850* might be regulated by Fur, suggesting its involvement in iron metabolism.

As mentioned for transcriptomic data, the results of proteomic studies have to be verified by conventional experiments, e.g., Western blot, and other advanced techniques, such as protein microarrays [60, 61]. By using protein microarrays, it is possible to conduct antibody profiling of the host exposed to *Y. pestis* infection, as well as changes in protein expression in *Y. pestis* under different cultivation conditions. We applied protein microarray to analyze the pooled sera obtained from mice infected with *Y. pestis* to observe the antibody responses to different proteins. By absorbing the sera with *Y. pestis* bacteria cultivated at 26 °C (mimicking the flea cycle), 37 °C (mimicking the host), or both cultures, it was possible to obtain the antibody profile specific to *Y. pestis* proteins expressed at each specific temperature, or expressed only in vivo. For instance, we identified 12 proteins that were expressed at 37 °C but not at 26 °C, or expressed at significantly higher levels at 37 °C versus 26 °C. Seven proteins appeared to only be expressed in vivo following infection in mice [60].

The growing number of genomes in GenBank has resulted in an almost exclusive automated annotation of deposited sequences. Therefore, a correct and complete annotation of the reference genomes is crucial in preventing a multiplication of annotation errors and improvement of the annotation quality. Annotation refinement should be applied to previously sequenced genomes, and the omics approach can significantly help in identification of the ORFs missed or miscalled by the original annotation. We used the proteogenomic approach based on analysis of the transcriptome and proteome of *Y. pestis* CO92, Pestoides F, and *Y. pseudotuberculosis* PB1 to refine the genome annotation of these strains. The genome annotation refinements included discovery of novel OFRs, extended start sites, identification of frameshifts, and translated "pseudogenes"[39].

Comparative proteomics should help to understand the pathogenic differences between *Y. pestis* and its ancestor, *Y. pseudotuberculosis*. These two pathogens share about 97 % homology of chromosomal DNA; however, the pathogenicity is vastly different: *Y. pestis* causes systemic infection with severe and lethal outcomes, while *Y. pseudotuberculosis* only causes mild gastrointestinal symptoms with favorable outcomes. When these two organisms infected human monocyte U937 cells, differentially expressed proteins were observed by MALDI TOF MS analysis [56]. A total of 16 and 13 proteins were identified as differentially expressed in response to *Y. pestis* or *Y. pseudotuberculosis* exposure, respectively. Only two of them (γ-actin, ACTG and monocyte/neutrophil elastase inhibitor, PI2) were shared between the two sets of exposures. Although only a few of proteins were identified, they contribute to a wide range of functions, including DNA replication and transcription, apoptosis, protein synthesis and degradation, cytoskeletal rearrangement, cell signaling, metabolism, and protein folding. These preliminary results warrant a continued search for additional targets to reveal the pathogenic differences between two *Yersinia* species.

9.3.4 Metabolomics

In contrast to transcriptomic and proteomic studies, the metabolics investigation of *Y. pestis* is quite limited. A few studies have contributed to the understanding of the *Y. pestis* metabolome [40, 41, 58, 62–64]. The first genome-scale metabolomic model of *Y. pestis* 91001, a strain of newly designated biovar microtus [65–67], was reconstructed using available genomic annotation, physiological, and biochemical data [62]. This theoretical model demonstrated excellent agreement with the known metabolic performance of *Y. pestis*. The authors also identified the known cryptic genes for the urease, methionine, and phenylalanine pathways and found the *betB* as a cryptic gene responsible for the rhamnose fermentation pathway. In the genome of *Y. pestis* 91001 [65], there are 4037 genes annotated and about 28.4 % of them (1146/4037) are related to cellular metabolism. In the reconstructed metabolism model, Navid et al. revealed 1020 reactions and 825 metabolites [62]. The reactions include different pathways, 200 amino acid, 201 carbon, 159 cofactor, 142 transport, 134 membrane, 123 nucleotide, 27 energy, and 34 others. This theoretical model allows us to design effective drugs that target the identified cell's metabolic weaknesses and strengths.

The work of Navid et al. provided a good example of the benefit of integrating gene expression data with metabolomic pathways [64]. They employed gene expression data of *Y. pestis* strain 201, and Kim reported previously for deducing metabolic changes for temperature shift under different calcium concentrations and for antibiotic stresses.

The metabolic and mathematical model of *Y. pestis* CO92 was constructed at a genome-wide scale with supporting experimental data [63]. The constructed metabolic model, iPC815, contains 815 genes, 678 proteins, 963 unique metabolites, and 1678 reactions. Compared with the model of 91001, this model bridges some gaps in several key biosynthetic pathways, explaining certain known nutritional requirements characteristic of CO92. This work was a good example of how genomic, proteomic, and metabolic data can provide insight into *Y. pestis* metabolisms and physiological characteristics.

Importantly, understanding of the *Y. pestis* metabolome will shed light on mechanisms of metabolomic adaptations to in vivo microenvironments that the pathogen can encounter during different stages of infection, which promotes cross talk between metabolism and "nutritional virulence" [68].

9.3.5 Interactomics

Interaction between a pathogen and host involves constant communication at the protein and RNA levels. Typically, the protein-protein interactions (PPIs) between pathogen and host are investigated. For *Y. pestis*, the interaction network between secreted or membrane proteins of this pathogen and human proteins have been revealed by yeast two-hybrid technology [34, 69]. Dyer et al. employed a random yeast two-hybrid approach for human biodefense pathogens, such as *Bacillus anthracis*, *Francisella tularensis*, and *Y. pestis*, to identify interactions at protein level, unraveling 3073, 1383, and 4059 pairs of PPIs, respectively [69]. Moreover, the authors also computed the conserved modules among the three pathogen-human protein interaction networks, yielding 39–64 conserved modules between different pairs of pathogens. Another project performed by our laboratory used the yeast two-hybrid strategy to identify PPIs between *Y. pestis* and human proteins [34]. A total of 153 potential virulence-associated proteins from *Y. pestis* were chosen as bait proteins according to the following criteria: (1) all known virulence factors, including type III YOP effectors, Ail, Pla protease, pH 6 antigen, and F1 antigen; (2) genes that are disrupted in nonepidemic strain 91001, which is avirulent for humans, but intact in fully virulent strain CO92; (3) genes that have been reported to be highly regulated in the life cycle of *Y. pestis*; (4) proteins that have been found to trigger immune responses in infected rodents or humans; (5) the annotated membrane proteins or exported proteins; (6) genes that are annotated to be related to mobility, detoxification, and pathogenicity; and (7) all the plasmid-encoding genes from the three plasmids, pCD1, pMT1 and pPCP1. An interaction network including 204 interactions between 66 *Y. pestis* proteins and 109 human proteins was identified, followed by their integration with the previously published 23 interactions. *Y. pestis* was highly prone to interact with hub and bottleneck proteins essential for normal cellular functions in the human PPI network. Important signaling pathways of host immune responses were preferentially targeted by *Y. pestis*, including the pathways involved in Toll-like receptors (TLR), focal adhesion, cytoskeletal regulation, leukocyte transendoepithelial migration, and mitogen-activated protein kinase (MAPK) signaling. These results provided important clues for future functional characterization of some important interactions to understand the molecular mechanism of *Y. pestis* pathogenesis in humans. For example, a subsequent study on the interaction between vasodilator-stimulated phosphoprotein (VASP) and YpkA demonstrated that YpkA-mediated VASP phosphorylation-inhibited actin polymerization and promoted disruption of the actin cytoskeleton, which resulted in the inhibition of phagocytosis of *Y. pestis* by the host [70].

An interactomic strategy could also be applied to reveal complex bacterial structures. The type three secretion system (T3SS) is a complex syringe-like structure that plays a key role in bacteria-host interaction. The T3SS injects virulence factors, such as YOPs, into the cytosol of host cells, resulting in hijacking of the host cells by interrupting the immune response signaling pathways. In our laboratory, we cloned all 57 functional genes encoded by pCD1 for yeast two-hybrid analysis, trying to reveal all possible protein interactions important for T3SS structure [71]. A total of 21 pairs of interaction proteins were identified, including the previously reported 9 pairs of interactions and 12 novel interaction pairs. Of the novel ones, interaction pairs of YscG-SycD, YscG-TyeA, YscI-YscF, and YopN-YpCD1.09c were successfully validated by GST-tag pull-down assay, indicating their potential roles in assembly of the T3SS structure. This study demonstrated that YscI-YscF interaction plays a critical role in formation of T3SS (unpublished data).

9.4 Integration of Big Data for Systems Biology

Currently, a massive volume of data resulting from complex omics studies of *Y. pestis* life cycle, and pathogenesis is available from comparative genomics, gene expression, proteins, and metabolites investigations. However, the information obtained is mostly fragmented and provides us with only a partial explanation of some characteristics of this pathogen. The situation resembles the story about the blind men who touch an elephant to learn what it is like. Nevertheless, the fast growing accumulation of experimental data will undoubtedly provide enough information to make a whole story. Therefore, integration of the various types of big data for systematical understanding of the lifestyle and pathogenesis of *Y. pestis* is urgently required. However, we face several practical obstacles, including the following: (1) different laboratories use different strains of *Y. pestis* and often employ different techniques that make data integration difficult; (2) current omics research tools do not ensure that we acquire complete information at the genomic, RNA, protein, and metabolic levels, raising the potential for losing important results. Nevertheless, it is still worth the effort to use partial data from omics-driven studies to explain the story; (3) like other organisms, bacterial life is also a dynamic process. Current research efforts are often focused on static moments, causing the data to reflect only a transient process; (4) even with dynamic data, we still lack a mathematic model and effective computation to integrate all possible data to create a comprehensive story.

Overall, data integration requires data criterion from different laboratories. Two to three representative strains (currently, CO92, KIM, and 91001 are the mostly used in different research) should be selected for data production using identical and internationally accepted protocols. The data presentation formats also needed standardization. Finally, an integrating computational strategy is necessary for effective integration of big data. Importantly, any integration of omics-based data could not be complete without studying the phenotype of the organism, including its virulent properties.

References

1. McKinsey Global Institute. Big data: the next frontier for innovation, competition, and productivity. McKinsey & Company; 2011.
2. Gartner: IT glossary. http://www.gartner.com/it-glossary/big-data/. 2013.
3. Kalyvas JR, Overly MR: Big data: a business and legal guide. Taylor & Francis Group, LLC 2015.
4. Schouten P. Big data in health care. Healthc Financ Manage. 2013;67(2):40–2.
5. Pennisi E. How will big pictures emerge from a sea of biological data? Science. 2005;309(5731):94.
6. Howe D, Costanzo M, Fey P, Gojobori T, Hannick L, Hide W, Hill DP, Kania R, Schaeffer M, St Pierre S, et al. Big data: the future of biocuration. Nature. 2008;455(7209):47–50.
7. Jeong K, Jung E, Park DK. Trend of wireless u-health. IEEE, ISCIT. 2009;2009:829–33.

8. Steinbrook R. Personally controlled online health data–the next big thing in medical care? N Engl J Med. 2008;358(16):1653–6.
9. Hood L. Systems biology and p4 medicine: past, present, and future. Rambam Maimonides Med J. 2013;4(2):e0012.
10. Hay SI, George DB, Moyes CL, Brownstein JS. Big data opportunities for global infectious disease surveillance. PLoS Med. 2013;10(4):e1001413.
11. de la Barrera CA, Reyes-Teran G. Influenza: forecast for a pandemic. Arch Med Res. 2005;36(6):628–36.
12. Shaman J, Karspeck A, Yang W, Tamerius J, Lipsitch M. Real-time influenza forecasts during the 2012–2013 season. Nat Commun. 2013;4:2837.
13. Ginsberg J, Mohebbi MH, Patel RS, Brammer L, Smolinski MS, Brilliant L. Detecting influenza epidemics using search engine query data. Nature. 2009;457(7232):1012–4.
14. Nsoesie EO, Brownstein JS, Ramakrishnan N, Marathe MV. A systematic review of studies on forecasting the dynamics of influenza outbreaks. Influenza Other Respir Viruses. 2014;8(3):309–16.
15. Shaman J, Karspeck A. Forecasting seasonal outbreaks of influenza. Proc Natl Acad Sci U S A. 2012;109(50):20425–30.
16. Ben-Ari T, Neerinckx S, Gage KL, Kreppel K, Laudisoit A, Leirs H, Stenseth NC. Plague and climate: scales matter. PLoS Pathog. 2011;7(9):e1002160.
17. Yang R, Du Z, Han Y, Zhou L, Song Y, Zhou D, Cui Y. Omics strategies for revealing Yersinia pestis virulence. Front Cell Infect Microbiol. 2012;2:157.
18. Committee for Science and Technology Challenges to U.S. National Security Interests: Report of a Workshop on Big Data. National Academies Press; 2012.
19. National Research Council. Frontiers in massive data analysis. Washington, DC: The National Academies Press. This PDF is available from The National Academies Press at http://www.napedu/catalogphp?record_id=18374. 2013.
20. Singh OV, Nagaraj NS. Transcriptomics, proteomics and interactomics: unique approaches to track the insights of bioremediation. Brief Funct Genomic Proteomic. 2006;4(4):355–62.
21. Yang Y, Xie B, Yan J. Application of next-generation sequencing technology in forensic science. Genomic Proteomic Bioinforma. 2014;12(5):190–7.
22. Roberts RJ, Carneiro MO, Schatz MC. The advantages of SMRT sequencing. Genome Biol. 2013;14(7):405.
23. Chin CS, Alexander DH, Marks P, Klammer AA, Drake J, Heiner C, Clum A, Copeland A, Huddleston J, Eichler EE, et al. Nonhybrid, finished microbial genome assemblies from long-read SMRT sequencing data. Nat Methods. 2013;10(6):563–9.
24. Feng Y, Zhang Y, Ying C, Wang D, Du C. Nanopore-based fourth-generation DNA sequencing technology. Genomic Proteomic Bioinforma. 2015;13(1):4–16.
25. Eisenstein M. Oxford Nanopore announcement sets sequencing sector abuzz. Nat Biotechnol. 2012;30(4):295–6.
26. Burgess DJ. Technology: bead capture for single-cell transcriptomics. Nat Rev Genet. 2015;16(4):195.
27. Li N, Xu Z, Zhai L, Li Y, Fan F, Zheng J, Xu P, He F. Rapid development of proteomics in China: from the perspective of the Human Liver Proteome Project and technology development. Sci China Life Sci. 2014;57(12):1162–71.
28. Castro CC, Martins RC, Teixeira JA, Silva Ferreira AC. Application of a high-throughput process analytical technology metabolomics pipeline to Port wine forced ageing process. Food Chem. 2014;143:384–91.
29. Noto A, Dessi A, Puddu M, Mussap M, Fanos V. Metabolomics technology and their application to the study of the viral infection. J Matern Fetal Neonatal Med. 2014;27 Suppl 2:53–7.
30. Feng S, Zhou L, Huang C, Xie K, Nice EC. Interactomics: toward protein function and regulation. Expert Rev Proteomic. 2015;12(1):37–60.

31. Uetz P, Dong YA, Zeretzke C, Atzler C, Baiker A, Berger B, Rajagopala SV, Roupelieva M, Rose D, Fossum E, et al. Herpesviral protein networks and their interaction with the human proteome. Science. 2006;311(5758):239–42.
32. Mendez-Rios J, Uetz P. Global approaches to study protein-protein interactions among viruses and hosts. Future Microbiol. 2010;5(2):289–301.
33. Simonis N, Rual JF, Lemmens I, Boxus M, Hirozane-Kishikawa T, Gatot JS, Dricot A, Hao T, Vertommen D, Legros S, et al. Host-pathogen interactome mapping for HTLV-1 and –2 retroviruses. Retrovirology. 2012;9:26.
34. Yang H, Ke Y, Wang J, Tan Y, Myeni SK, Li D, Shi Q, Yan Y, Chen H, Guo Z, et al. Insight into bacterial virulence mechanisms against host immune response via the *Yersinia pestis*-human protein-protein interaction network. Infect Immun. 2011;79(11):4413–24.
35. Yang L, Li M, Shan Y, Shen S, Bai Y, Liu H. Recent advances in lipidomics for disease research. J Sep Sci. 2015.
36. Vaz FM, Pras-Raves M, Bootsma AH, van Kampen AH. Principles and practice of lipidomics. J Inherit Metab Dis. 2015;38(1):41–52.
37. Sturino J, Zorych I, Mallick B, Pokusaeva K, Chang YY, Carroll RJ, Bliznuyk N. Statistical methods for comparative phenomics using high-throughput phenotype microarrays. J Biostat. 2010;6(1):Article 29.
38. Viti C, Decorosi F, Marchi E, Galardini M, Giovannetti L. High-throughput phenomics. Methods Mol Biol. 2015;1231:99–123.
39. Schrimpe-Rutledge AC, Jones MB, Chauhan S, Purvine SO, Sanford JA, Monroe ME, Brewer HM, Payne SH, Ansong C, Frank BC, et al. Comparative omics-driven genome annotation refinement: application across Yersiniae. PLoS ONE. 2012;7(3):e33903.
40. Ansong C, Schrimpe-Rutledge AC, Mitchell HD, Chauhan S, Jones MB, Kim YM, McAteer K, Deatherage Kaiser BL, Dubois JL, Brewer HM, et al. A multi-omic systems approach to elucidating Yersinia virulence mechanisms. Mol BioSyst. 2013;9(1):44–54.
41. Ansong C, Deatherage BL, Hyduke D, Schmidt B, McDermott JE, Jones MB, Chauhan S, Charusanti P, Kim YM, Nakayasu ES, et al. Studying Salmonellae and Yersiniae host-pathogen interactions using integrated 'omics and modeling. Curr Top Microbiol Immunol. 2013;363:21–41.
42. Johnson SL, Daligault HE, Davenport KW, Jaissle J, Frey KG, Ladner JT, Broomall SM, Bishop-Lilly KA, Bruce DC, Coyne SR et al. Thirty-two complete genome assemblies of nine Yersinia species, including Y. pestis, Y. pseudotuberculosis, and Y. enterocolitica. Genome announcements. 2015;3(2).
43. Zhgenti E, Johnson SL, Davenport KW, Chanturia G, Daligault HE, Chain PS, Nikolich MP. Genome assemblies for 11 Yersinia pestis strains isolated in the caucasus region. Genome announcements. 2015;3(5).
44. Du Z, Yang H, Tan Y, Tian G, Zhang Q, Cui Y, Yanfeng Y, Wu X, Chen Z, Cao S, et al. Transcriptomic response to Yersinia pestis: RIG-I like receptor signaling response is detrimental to the host against plague. J Genet Genomics. 2014;41(7):379–96.
45. Das R, Dhokalia A, Huang XZ, Hammamieh R, Chakraborty N, Lindler LE, Jett M. Study of proinflammatory responses induced by Yersinia pestis in human monocytes using cDNA arrays. Genes Immun. 2007;8(4):308–19.
46. Rogers JV, Choi YW, Giannunzio LF, Sabourin PJ, Bornman DM, Blosser EG, Sabourin CL. Transcriptional responses in spleens from mice exposed to Yersinia pestis CO92. Microb Pathog. 2007;43(2–3):67–77.
47. Galindo CL, Moen ST, Kozlova EV, Sha J, Garner HR, Agar SL, Chopra AK. Comparative analyses of transcriptional profiles in mouse organs using a pneumonic plague model after infection with wild-type Yersinia pestis CO92 and its Braun lipoprotein mutant. Comp Funct Genomics. 2009;2009:914762.
48. Comer JE, Sturdevant DE, Carmody AB, Virtaneva K, Gardner D, Long D, Rosenke R, Porcella SF, Hinnebusch BJ. Transcriptomic and innate immune responses to Yersinia pestis in the lymph node during bubonic plague. Infect Immun. 2010;78(12):5086–98.

49. Liu H, Wang H, Qiu J, Wang X, Guo Z, Qiu Y, Zhou D, Han Y, Du Z, Li C, et al. Transcriptional profiling of a mice plague model: insights into interaction between Yersinia pestis and its host. J Basic Microbiol. 2009;49(1):92–9.
50. Chromy BA, Choi MW, Murphy GA, Gonzales AD, Corzett CH, Chang BC, Fitch JP, McCutchen-Maloney SL. Proteomic characterization of Yersinia pestis virulence. J Bacteriol. 2005;187(23):8172–80.
51. Zhou L, Ying W, Han Y, Chen M, Yan Y, Li L, Zhu Z, Zheng Z, Jia W, Yang R, et al. A proteome reference map and virulence factors analysis of Yersinia pestis 91001. J Proteome. 2012;75(3):894–907.
52. Zhou S, Deng W, Anantharaman TS, Lim A, Dimalanta ET, Wang J, Wu T, Chunhong T, Creighton R, Kile A, et al. A whole-genome shotgun optical map of Yersinia pestis strain KIM. Appl Environ Microbiol. 2002;68(12):6321–31.
53. Yen YT, Bhattacharya M, Stathopoulos C. Genome-wide in silico mapping of the secretome in pathogenic Yersinia pestis KIM. FEMS Microbiol Lett. 2008;279(1):56–63.
54. Pieper R, Huang ST, Robinson JM, Clark DJ, Alami H, Parmar PP, Perry RD, Fleischmann RD, Peterson SN. Temperature and growth phase influence the outer-membrane proteome and the expression of a type VI secretion system in Yersinia pestis. Microbiology. 2009;155(Pt 2):498–512.
55. Nozadze M, Zhgenti E, Meparishvili M, Tsverava L, Kiguradze T, Chanturia G, Babuadze G, Kekelidze M, Bakanidze L, Shutkova T, et al. Comparative proteomic studies of Yersinia pestis strains isolated from natural foci in the republic of Georgia. Front Public Health. 2015;3:239.
56. Zhang CG, Gonzales AD, Choi MW, Chromy BA, Fitch JP, McCutchen-Maloney SL. Subcellular proteomic analysis of host-pathogen interactions using human monocytes exposed to Yersinia pestis and Yersinia pseudotuberculosis. Proteomics. 2005;5(7):1877–88.
57. Pieper R, Huang ST, Clark DJ, Robinson JM, Parmar PP, Alami H, Bunai CL, Perry RD, Fleischmann RD, Peterson SN. Characterizing the dynamic nature of the Yersinia pestis periplasmic proteome in response to nutrient exhaustion and temperature change. Proteomics. 2008;8(7):1442–58.
58. Pieper R, Huang ST, Parmar PP, Clark DJ, Alami H, Fleischmann RD, Perry RD, Peterson SN. Proteomic analysis of iron acquisition, metabolic and regulatory responses of Yersinia pestis to iron starvation. BMC Microbiol. 2010;10:30.
59. Chromy BA, Perkins J, Heidbrink JL, Gonzales AD, Murphy GA, Fitch JP, McCutchen-Maloney SL. Proteomic characterization of host response to *Yersinia pestis* and near neighbors. Biochem Biophys Res Commun. 2004;320(2):474–9.
60. Li B, Tan Y, Guo J, Cui B, Wang Z, Wang H, Zhou L, Guo Z, Zhu Z, Du Z, et al. Use of protein microarray to identify gene expression changes of Yersinia pestis at different temperatures. Can J Microbiol. 2011;57(4):287–94.
61. Li B, Jiang L, Song Q, Yang J, Chen Z, Guo Z, Zhou D, Du Z, Song Y, Wang J, et al. Protein microarray for profiling antibody responses to Yersinia pestis live vaccine. Infect Immun. 2005;73(6):3734–9.
62. Navid A, Almaas E. Genome-scale reconstruction of the metabolic network in Yersinia pestis, strain 91001. Mol BioSyst. 2009;5(4):368–75.
63. Charusanti P, Chauhan S, McAteer K, Lerman JA, Hyduke DR, Motin VL, Ansong C, Adkins JN, Palsson BO. An experimentally-supported genome-scale metabolic network reconstruction for Yersinia pestis CO92. BMC Syst Biol. 2011;5:163.
64. Navid A, Almaas E. Genome-level transcription data of Yersinia pestis analyzed with a new metabolic constraint-based approach. BMC Syst Biol. 2012;6:150.
65. Song Y, Tong Z, Wang J, Wang L, Guo Z, Han Y, Zhang J, Pei D, Zhou D, Qin H, et al. Complete genome sequence of Yersinia pestis strain 91001, an isolate avirulent to humans. DNA Res. 2004;11(3):179–97.
66. Zhou D, Han Y, Song Y, Tong Z, Wang J, Guo Z, Pei D, Pang X, Zhai J, Li M, et al. DNA microarray analysis of genome dynamics in Yersinia pestis: insights into bacterial genome microevolution and niche adaptation. J Bacteriol. 2004;186(15):5138–46.

67. Zhou D, Tong Z, Song Y, Han Y, Pei D, Pang X, Zhai J, Li M, Cui B, Qi Z, et al. Genetics of metabolic variations between Yersinia pestis biovars and the proposal of a new biovar, microtus. J Bacteriol. 2004;186(15):5147–52.
68. Abu Kwaik Y, Bumann D. Microbial quest for food in vivo: 'nutritional virulence' as an emerging paradigm. Cell Microbiol. 2013;15(6):882–90.
69. Dyer MD, Neff C, Dufford M, Rivera CG, Shattuck D, Bassaganya-Riera J, Murali TM, Sobral BW. The human-bacterial pathogen protein interaction networks of Bacillus anthracis, Francisella tularensis, and Yersinia pestis. PLoS ONE. 2010;5(8):e12089.
70. Ke Y, Tan Y, Wei N, Yang F, Yang H, Cao S, Wang X, Wang J, Han Y, Bi Y, et al. Yersinia protein kinase A phosphorylates vasodilator-stimulated phosphoprotein to modify the host cytoskeleton. Cell Microbiol. 2015;17(4):473–85.
71. Yang H, Tan Y, Zhang T, Tang L, Wang J, Ke Y, Guo Z, Yang X, Yang R, Du Z. Identification of novel protein-protein interactions of Yersinia pestis type III secretion system by yeast two hybrid system. PLoS ONE. 2013;8(1):e54121.

Chapter 10
Immunology of *Yersinia pestis* Infection

Yujing Bi

Abstract As a pathogen of plague, *Yersinia pestis* caused three massive pandemics in history that killed hundreds of millions of people. *Yersinia pestis* is highly invasive, causing severe septicemia which, if untreated, is usually fatal to its host. To survive in the host and maintain a persistent infection, *Yersinia pestis* uses several stratagems to evade the innate and the adaptive immune responses. For example, infections with this organism are biphasic, involving an initial "noninflammatory" phase where bacterial replication occurs initially with little inflammation and following by extensive phagocyte influx, inflammatory cytokine production, and considerable tissue destruction, which is called "proinflammatory" phase. In contrast, the host also utilizes its immune system to eliminate the invading bacteria. Neutrophil and macrophage are the first defense against *Yersinia pestis* invading through phagocytosis and killing. Other innate immune cells also play different roles, such as dendritic cells which help to generate more T helper cells. After several days post infection, the adaptive immune response begins to provide organism-specific protection and has a long-lasting immunological memory. Thus, with the cooperation and collaboration of innate and acquired immunity, the bacterium may be eliminated from the host. The research of *Yersinia pestis* and host immune systems provides an important topic to understand pathogen-host interaction and consequently develop effective countermeasures.

Keywords *Yersinia pestis* • Immunology • Inflammatory • Innate immunity • Acquired immunity

Y. Bi (✉)
Beijing Institute of Microbiology and Epidemiology,
No. Dongdajie, Fengtai, Beijing 100071, China
e-mail: byj7801@sina.com

R. Yang, A. Anisimov (eds.), *Yersinia pestis: Retrospective and Perspective*,
Advances in Experimental Medicine and Biology 918,
DOI 10.1007/978-94-024-0890-4_10

10.1 Interactions of *Yersinia pestis* and Immune Cells

Most human plague cases present clinically as one of three primary forms: bubonic, septicemic, or pneumonic plague. The most common form is bubonic plague, which is transmitted from rodent reservoirs to humans via the biting of infected fleas. Patients with primary bubonic plague can develop secondary septic or pneumonic infections, the latter of which can then be spread from person to person via respiratory droplets generated from severe sneezing and coughing of patients. Pneumonic plague is nearly always fatal unless treated with effective antibiotics within 20 h post symptom onset. All types of infection will cause a response of the host immune system against the invading *Yersinia pestis*.

During the early stages of infection, *Y. pestis* can enter both macrophages and neutrophils through either active or passive entry mechanisms [1]. Although early researches showed *Y. pestis* is typically killed in neutrophils, whereas in macrophages, it can survive and acquire antiphagocytic capabilities, which enables its extracellular survival in vivo. Recent studies indicated that a small percent of *Y. pestis* can also survive and replicate in neutrophils and send a PS (a marker of early apoptosis) to macrophages [2, 3]. This progress may provide *Y. pestis* with a less inflammatory route of entry into macrophages, resulting in decreased proinflammatory signaling. To survive inside the host and maintain a persistent infection, *Y. pestis* uses a variety of different mechanisms to evade or overcome the host immune system, especially the innate immune system, as shown in Fig. 10.1.

10.2 *Yersinia pestis* Overcomes the Immune Response

Y. pestis uses many different virulence factors to resist host immune responses, facilitate cellular attachment or invasion, subvert endocytic trafficking, block phagocytosis, modulate apoptotic pathways, and manipulate innate immunity and host responses as part of the initial infection process.

10.2.1 Dampening of the Inflammatory Response

Many studies have demonstrated that infection by *Y. pestis* elicits a notably delayed inflammatory response [4–6]. Type III secretion system (T3SS), which is one of the pathogens' major virulence factors, plays a crucial role in dampening host inflammatory responses. Multiple signaling pathways are repressed when the host is infected by *Y. pestis* by various *Yersinia* outer proteins (Yops). YopH inactivates the phosphatidylinositol-3 kinase (PI3K)-AKT cascade in macrophages, which correlates with the downregulation of mRNA coding for monocyte chemoattractant protein 1 (MCP-1) [7]. YopJ/P suppresses the mitogen-activated protein kinase (MAPK)

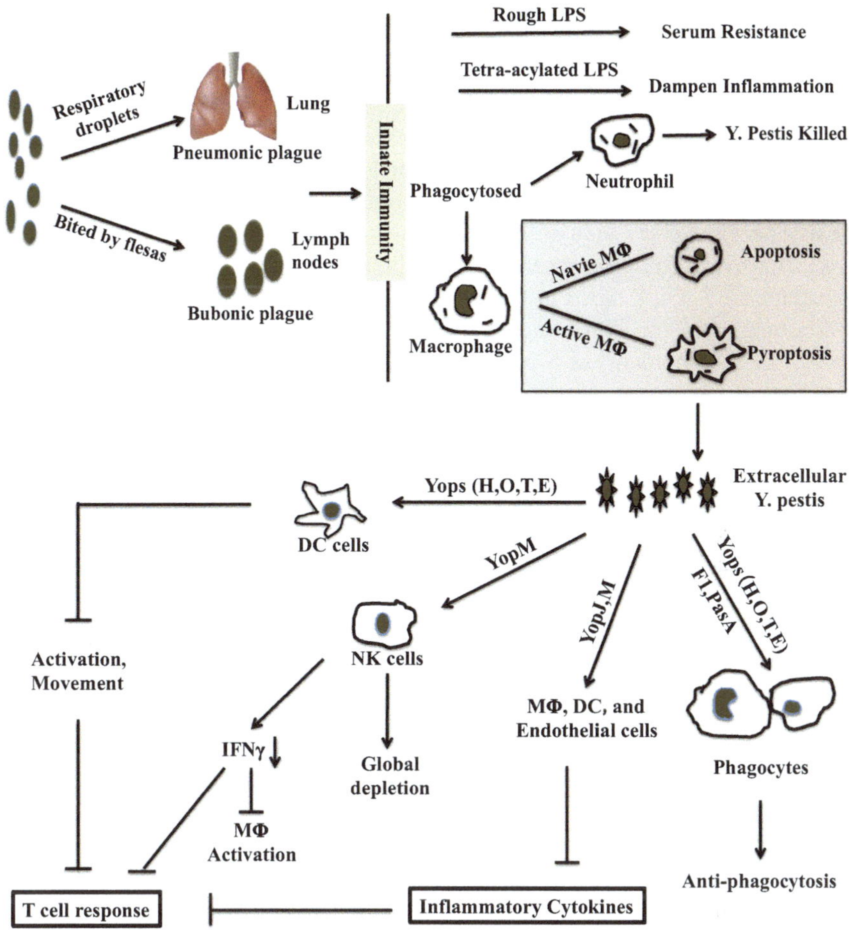

Fig. 10.1 Interactions between *Yersinia pestis* and the host immune system

signaling pathway by an unknown mechanism [8]. In vitro experiments indicated that YopE strongly inhibits nuclear factor-kappa B (NF-kB) activation and JNK and ERK activation, whereas YopT only moderately inhibits these responses [9]. YopK and YopM both inhibit inflammasomes [10, 11]. In addition, *Y. pestis* can suppress the production of cytokines and chemokines. LcrV, which is the needle tip protein that assembles into the needle tip complex, is associated with suppression of tumor necrosis factor alpha (TNFα) and interferon gamma (IFNγ) and the induction of immunosuppressive interleukin (IL)-10 in mice [12]. YopE can also prevent the production of IL-8 [9]. YopM interacts with protein kinase C-like 2 and ribosomal protein S6 kinase, which are also involved in proinflammatory signaling [13]. Recently, the N terminus of YscF was also found to decrease cytokine induction [14]. Inhibition of signaling pathways and suppression of cytokines not only reduces the activation of natural killer (NK) cells and phagocytes but also destroys the inflammatory environment necessary for adaptive immunity.

Lipopolysaccharide (LPS) is a major component of the outer membrane in Gram-negative bacteria and a ligand for toll-like receptor 4 (TLR4). In different host-specific environments, the expression and formation of LPS in *Y. pestis* change accordingly [15, 16]. When the bacteria grow in the flea gut (21–26 °C), they produce a typical hexa-acylated LPS, which activates TLR4-mediated immune signaling to induce the expression of proinflammatory cytokines (TNFα, IL-1, IL-6, and IL-8). However, after the temperature transition from the flea (26 °C) to the mammalian host (37 °C), *Y. pestis* instantly begins to produce tetra-acylated LPS, which is nonstimulatory for TLR4 and an antagonist for the stimulatory hexa-acylated form of LPS. Importantly, a *Y. pestis* strain that produces hexa-acylated lipid A at 37 °C was found to be more than 100,000-fold attenuated in a mouse model of bubonic plague [17]. The attenuated phenotype is accompanied by increased production of TNFα and depends upon host expression of TLR4. It is thus clear that *Y. pestis* evades innate immunity, at least in part, by avoiding TLR-mediated activation of innate immunity.

10.2.2 Resistance to Phagocytosis

Because *Y. pestis* proliferates extracellularly, it is essential to block phagocytosis after contact with host cells. At least four Yop effectors (YopH, YopE, YopT, and YopO) are involved in inhibiting the phagocytosis of yersiniae; however, their mechanisms are all different. YopH is a tyrosine phosphatase, which acts on several cytoskeletal proteins including p130Cas [18–20]. Dephosphorylation of p130Cas disrupts its interaction with Crk and subsequent Rac activation required for *Yersinia* internalization [21]. YopE, which acts as a GTPase-activating protein for the RhoA family of GTPases (RhoA, Rac and Cdc42) [22], inhibits Rac-dependent actin polymerization either directly or through inactivation of the upstream RhoG [23, 24]. YopT is a protease that specifically cleaves RhoA, Rac, and Cdc42 near their carboxyl termini [25], which irreversibly inactivates host cell proteins. YopO interferes with the host cell regulation of Rho GTPases by actin [26–30]. Using various models, several different cellular activities have been attributed to YopJ, including deubiquination of Ikβ and TNF receptor-associated factors (TRAFs), as well as acetylation of MAPK kinases. However, precisely how YopJ exerts its activity during infection is presently unclear. In summary, the translocated Yop effectors are exotoxins that disable the phagocytic machinery by (1) disrupting the host cell cytoskeleton, (2) suppressing cytokine production, and (3) interfering with cell signal pathways.

Another crucial antiphagocytic factor is the F1 capsule protein (Caf1), which is solely produced by *Y. pestis*. However, the mechanism of resistance to phagocytosis induced by this capsule protein is different from that of T3SS and presumably by preventing interactions with receptors that could potentially recognize and result in the uptake of pathogens [31]. PsaA was also shown to inhibit phagocytosis, most likely by binding lipoproteins that prevent recognition by host cells [32, 33].

10.2.3 Resistance to Complement-Dependent Bacteriolysis

Complement-dependent bacterial killing is one of the first lines of innate immunity against pathogens. *Y. pestis* must survive in blood to cause disease and be transmitted from host to host by fleas. Thus, resistance to serum complement is an essential phenotype for bacterial survival in blood. Unlike the enteropathogenic yersiniae (*Y. enterocolitica* and *Y. pseudotuberculosis*), which are fully resistant to complement when they are grown at 37 °C, but not when they are grown at 26 °C [5, 30], *Y. pestis* is constitutively resistant to complement at both 26 and 37 °C. In *Y. enterocolitica,* resistance to complement has been shown to involve YadA, Ail, OmpR, and LPS O antigen [34]. By binding to the complement regulatory proteins factor H and the C4b-binding protein, Ail and YadA provide resistance against complement-mediated killing [35, 36]. OmpR might alter the susceptibility of *Y. enterocolitica* to complement-mediated killing through remodeling of the outer membrane [37]. However, *Y. pestis* does not express YadA; therefore, *Y. pestis* LPS may mediate serum resistance [38], although there is no evidence of a direct role for LPS in serum resistance. Thus, Ail was found to play the crucial role in the inhibition of the bactericidal properties of complement by *Y. pestis*. Notably, there are four Ail/OmpX homologues encoded by the *Y. pestis* genome, but only *y1324* (in KIM strain, corresponding to YPO2506 in CO92, *ompX*) confers resistance to human serum. Deletion of this gene results in a rapid, essentially 100 % loss of serum resistance [39, 40]. This loss is attributed to the action of complement because heat-inactivated serum does not have lethal properties. Although currently the exact mechanisms of resistance to complement-mediated killing conferred by *Y. pestis* Ail are still unknown, based on the studies from two other yersiniae (*Y. enterocolitica* and *Y. pseudotuberculosis),* it is likely that the protein binds to negative regulators of alternative (factor H) and classical and lectin [C4b-binding protein (C4BP)] complement pathways, thus preventing complement-mediated attack of the pathogen [36, 41–43].

10.2.4 Other Impacts on Host Immunity

Many bacteria have evolved means to convert host plasminogen to plasmin, a protease that degrades extracellular matrix. In *Y. pestis* plasminogen activator (Pla), encoded by the *Y. pestis*-specific plasmid pPCP1, plays an important role. In comparison with the wild type, *Y. pestis* lacking Pla has been reported to have greatly reduced virulence when introduced via the intradermal and subcutaneous routes but produced equivalent or nearly equivalent virulence when introduced by the intravenous route [5, 44, 45].These results partly different from early studies, which indicated even lack of plasmid pPst (also call pPCP1), did not lead to an increase in LD50 with either by subcutaneously or by respiratory route challenge [46, 47]. In the pneumonic plague model, the dissemination of Pla-deficient *Y. pestis* to the circulatory system was found to be unaffected, whereas bacterial growth in the lung was greatly reduced [44, 48]. Further studies are required to discern the specific mechanisms by which Pla impacts *Y. pestis* virulence. In addition, Pla has been reported to be a ligand for a macrophage and DC surface receptor, DEC-205, which is related to antigen presentation [49].

The 2-component regulatory system (2CS) OmpR-EnvZ is required to resist innate immunity during the early and late stages of the plague [50]. Different from *Y. pestis*' 23 other 2CSs, OmpR-EnvZ is the only one required for production of bubonic, septicemic, and pneumonic plague. In in vitro studies, OmpR-EnvZ was required to counter serum complement and leukocytes but was not required for the secretion of antiphagocytic exotoxins. In contrast, in in vivo studies, *Y. pestis* lacking OmpR-EnvZ did not induce an early immune response in the skin and was fully virulent in neutropenic mice.

10.3 Innate Immunity in Plague

Normally, once a bacterium infects the host, the innate immune response provides immediate protection and 4–5 days post infection, the T or B cell-mediated adaptive immune responses begins to provide organism-specific protection. In 2005, using Yop beta-lactamase hybrids and fluorescent staining of live cells from plague-infected animals; Marketon et al. found that *Y. pestis* selected immune cells for injection. Further research in vivo showed that macrophages, dendritic cells, and granulocytes/neutrophils are early targets for injection, whereas B and T lymphocytes were rarely selected. Thus, it appears that *Y. pestis* appears to destroy cells with innate immune functions that represent the first line of defense [51]. A similar study identified the pulmonary cells targeted by *Y. pestis* during primary pneumonic plague using a FRET-based probe to quantitate injection of effector proteins by the *Y. pestis* T3SS [52]. They found that these bacteria target alveolar macrophages early during infection of mice, followed by a switch in host cell preference to neutrophils. In addition to mouse models, the Gunnison's prairie dog, an important natural host of plague, was investigated [53]. This study highlights the importance of innate immunity against plague in wild Gunnison's prairie dogs.

10.3.1 Macrophages

Y. pestis has long been considered to be a facultative intracellular pathogen [54]. Macrophages are often regarded as permissive sites for survival and replication of *Y. pestis* at early stages of infection. Bubonic plague is typically initiated as an intradermal infection following the bite of an infected flea. Then, *Y. pestis* may invade the host directly through the skin and encounter phagocytes such as neutrophils and macrophages at the site of invasion. Most *Y. pestis* bacilli initially present are likely killed by neutrophils; however, the bacteria phagocytosed by macrophages can survive. *Y. pestis* preferentially infects host macrophages, probably via recognition of specific surface-associated CCR5 molecules [55]. This intracellular growth is essential for the pathogenesis of *Y. pestis* in three ways. First, macrophages provide a niche, allowing bacteria to proliferate and acquire the ability to evade phagocytosis. Second, intracellular growth in macrophages provides a protected environment for the bacteria to avoid contact with other components of the host immune system such as complement. Third, *Y. pestis* in macrophages can express various virulence determinants [56, 57]. Furthermore, macrophages may provide the bacterium with a vehicle for transport from the initial site of infection to deeper lymph tissues [58, 59]. The late stage of *Y. pestis* infection is characterized by a rapid increase in the number of bacteria within the spleen and escape of bacteria from macrophages into the extracellular compartment of the spleen. The cause of this escape is likely to be related to the macrophage necrosis or apoptosis observed during in vitro studies [57, 60]. Once *Y. pestis* replicates and expresses various virulence determinants in macrophages, they can be released into the extracellular compartment and spread systemically with the acquisition of phagocytosis resistance. Although *Y. pestis* can survive and replicate in macrophages during the early stage of infection, this macrophage compliance can be overcome in vitro by stimulation with a combination of IFNγ and TNFα [1].

Although macrophages provide a niche for *Y. pestis* survival and replication in the early stages of infection, *Y. pestis* can also cause macrophage death. Two distinct processes, corresponding to the inflammatory crescendo, are observed in vivo. *Y. pestis* causes apoptosis in naïve macrophages, but cell death in activated macrophages by inflammatory pyroptosis [61]. During macrophage apoptosis, *Y. pestis* YopJ is necessary. When the bacterial LPS contacts host macrophages, proapoptotic signaling is activated [62, 63]. LPS also upregulates cell survival genes and inflammatory cytokine production controlled by MAPK and NF-kB [62, 64, 65]; however, YopJ inhibits their activation [66, 67], and, therefore, apoptotic signaling predominates [62, 68]. Apoptotic cells are cleared by phagocytes, and this encounter triggers production of the anti-inflammatory cytokines TGFα and IL-10 [69, 70], making the process noninflammatory. Pyroptosis results from the activation of caspase-1, which is functionally distinct from the structurally related apoptotic caspases [36]. *Y. pestis*-induced pyroptosis requires plasmid-encoded T3SS, but not YopJ or any of the other known effector molecules [61]. Caspase-1 stimulates maturation and secretion of multiple inflammatory cytokines such as IL-1 and IL-18

[71]. Thus, pyroptosis causes inflammation. Why are two kinds of macrophage death necessary in *Y. pestis* infection? In bubonic plague, the infections are obviously biphasic: bacteria initially replicate without a measurable host response for periods up to 36–48 h, with a noticeable lack of inflammation, but, eventually, phagocyte influx into infected tissues and lymph nodes results in inflammation, cytokine production, and tissue necrosis [6, 72–74]. The change in mode of macrophage death is partly explained by observation of the host responses to *Y. pestis* infection. Apoptosis (noninflammatory death) [75–77] of naïve macrophages is consistent with initial bacterial growth in the relative absence of inflammation. Pyroptosis (inflammatory death) [61, 78] in activated macrophages corresponds to later stages of infection, accompanied by enhanced cytokine production and tissue damage. Induction of both apoptosis and pyroptosis in macrophages may be a mechanism by which pathogens preferentially trigger immune cell death, resulting in bacterial dissemination and disruption of host innate immune signaling.

The available data strongly suggest that *Y. pestis* growth within macrophages plays an important, perhaps critical, pathogenic role during plague [58]. During bubonic plague, macrophages may provide a protected intracellular niche that allows time for flea-transmitted *Y. pestis* bacilli to adjust to growth within mammals, in part by upregulating expression of capsular F1 protein, LcrV, and Yops, thus enabling subsequent growth as extracellular, phagocyte-resistant bacteria. However, in pneumonic plague the role of macrophages seems different from that in bubonic plague. Dr. Goldman's group found that depletion of roughly 92 % of alveolar macrophages had little or no effect on bacterial burden in the lungs [52]. They gave two possible reasons: macrophages are not involved in limiting yersiniae survival in the lungs or that *Y. pestis* is able to neutralize the antibacterial effects of sentinel alveolar macrophages, presumably as a result of expressing antiphagocytic/anti-inflammatory factors including the Yops, F1 capsular protein, and pH 6 antigen.

10.3.2 Neutrophils

Histological, flow cytometric, and laser confocal microscopy evidence indicates that *Y. pestis* phagocytosed by neutrophils are killed [1, 54]. However, *Y. pestis* primarily targets neutrophils early after inoculation in the lung, presumably to limit host innate immune mechanisms aimed at bacterial killing and clearance. Thus, it seems that the interaction between *Y. pestis* and host neutrophils may have a strong bearing on the outcome of infection [79].

Researchers often assess bubonic plague through needle-inoculated *Y. pestis*. However, Shannon et al. found that the innate cellular host responses to flea-transmitted *Y. pestis* differed from and were more variable than responses to needle-inoculated bacteria [80]. They therefore developed new tools allowing for intravital microscopy of *Y. pestis* in the dermis of an infected mouse after transmission by its natural route of infection, the bite of an infected flea. They found that uninfected flea bites typically induced minimal neutrophil recruitment. The magnitude of neutrophil response to flea-transmitted *Y. pestis* varied considerably and appeared to correspond to the number of bacteria deposited at the bite site.

Dr. Goldman's group first identified the initial host cell targets of fully virulent *Y. pestis* during pulmonary infection. Using flow cytometry to monitor injection of a YopE-TEM fusion protein by the T3SS, they showed that *Y. pestis* initially targets CD11c high alveolar macrophages and neutrophils in the lungs. However, in contrast to the bubonic plague, during the first 24 h after pulmonary infection with a fully virulent *Y. pestis* strain, no significant changes were observed in the lungs in the levels of neutrophil infiltrate, the expression of adhesion molecules, or the expression of the major neutrophil chemoattractant CXCL1 (also known as keratinocyte cell-derived chemokine, KC) [4]. These results indicate that *Y. pestis* can slow the rate of neutrophil influx to the lungs by delaying the onset of chemokine and cytokine release. Moreover, neutrophils were observed to be "tightly packed" within spaces 72 h post infection during pneumonic plague [6]. However, in mice infected with an avirulent *Y. pestis* strain, early induction of chemokines, rapid neutrophil infiltration, and reduced bacterial burden were observed in the lungs of mice [81]. These results indicate that strain virulence may determine the host immune responses. Moreover, it seems that prevention of the early influx of neutrophils to the lungs is of major importance for *Y. pestis* virulence.

A growing volume of literature highlights interactions between *Y. pestis* and neutrophils, but the deletion of neutrophils during pneumonic plague shows significantly different results in various studies. Dr. Goldman's group showed that depletion of neutrophils had little to no effect on bacterial burden in the lungs [52]. This is in contrast to others studies, such as by Laws et al. who reported a modest increase in bacterial titers in lungs of infected mice early after neutrophil depletion [79]. Additionally, our group found neutrophil deletion with anti-Ly6G antibodies decreased survival in an intranasal *Y. pestis* mouse model [82]. Notably, a fully virulent *Yersinia pestis* strain, CO92, was used in Dr. Goldman's research; however, our study used a *Y. pestis* strain (strain 201) virulent to mice but nonvirulent to humans. Thus, the disparity in findings may come from differences in mouse lines and *Y. pestis* strains.

The mechanisms used by *Y. pestis* for targeting neutrophils have also been described. The *Y. pestis* adhesin Ail is required for efficient targeting of neutrophils in vivo [83]. The Ail protein was previously reported to inhibit the innate immune response, in particular the recruitment of a protective polymorphonuclear leukocyte (PMN) responses to the infected lymph node [84]. To identify factors conferring specificity to neutrophil targeting, a study investigated the role of serum [83]. They found that neutrophil targeting is mediated by complement receptor 3 (CR3) and, to a lesser extent, CD14. However, the exact nature of the receptor-ligand interactions and their contributions toward target cell selection remain unknown.

10.3.3 *Other Innate Immune Cells*

10.3.3.1 Dendritic Cells

Dendritic cells (DCs) are potent and specialized antigen-presenting cells, which help to generate more effective T helper cells. Therefore, an effective immune evasion strategy by a pathogen would be to target DCs and induce them to become tolerogenic, thus priming a regulatory IL-10 response that blocks inflammation and allows the pathogen to multiply without restraint. Furthermore, impairing DC maturation and promoting apoptosis of DCs can also help pathogens to disarm host defenses. During *Y. pestis* infection, DCs are one of the early targets of T3SS effectors [51]. Shannon et al. observed minimal interaction between *Y. pestis* and DCs; however, DCs consistently migrate toward flea-bitten sites containing *Y. pestis* [80]. The most pronounced effect of *Y. pestis* on DCs appears to be the paralysis of DC movement by impairing the cytoskeleton rearrangement function, attenuating the presentation of *Y. pestis* antigens by DCs [85]. Richard et al. found that the ability of *Y. pestis* to initiate DC activation is determined by its lipid A structure and depends on the pattern recognition receptor TLR4 [86]. In the bubonic plague model, IL-10 and TLR6 deficient mice are protected from plague infection, the mechanism of which was that TLR6 drove differentiation of tolerogenic DC and contributed to LcrV-mediated plague pathogenesis [87]. In plague vaccine studies, the interaction between vaccine and DCs also plays an important role. The protective mechanisms induced by rF1 + rV probably involve the activation of DCs, which initiate a primary immune response in naïve T cells [88]. Further study proved that LcrV targeting of DCs elicits combined humoral and cellular immunity and induces protection in a mouse model of pneumonic plague [89].

10.3.3.2 Natural Killer Cell

NK cells are a subset of lymphocytes that arrive at inflammatory sites and directly kill pathogen-infected cells without the recognition of antigenic peptides. Notably, *Y. pestis* can cause a global depletion of NK cells and decrease the secretion of IFNγ, resulting in reduced production of reactive nitrogen intermediates by macrophages. The cause of these anti-NK effects is the effector YopM, possibly by affecting the expression of IL-15 and its receptor IL-15Rα [90]. To elucidate whether $NK1.1^+$ cells were critical for the virulence effect of YopM, Ye et al. continued to test the effects of NK cell depletion on bacterial growth and found no effect of ablation on viable bacterial numbers of either the wild type or the YopM mutant strain in either the liver or spleen [91]. Thus, they concluded that NK cells are redundant for YopM's pathogenic mechanism.

10.4 Adaptive Immunity in Plague

In addition to innate immunity, *Y. pestis* infection can induce adaptive immunity. Adaptive immunity is characterized by the expansion, differentiation, and persistence of antigen-specific B and T cells. The primary function of B cells is to produce antibodies, thereby facilitating humoral defense, while the primary function of T cells is to produce phagocyte-activating cytokines, thereby facilitating cellular defense. However, during *Y. pestis* infection, humoral immunity and cellular immunity are not separated, and indeed they cooperate and collaborate with each other against the plague. For example, antibodies can protect T cell-deficient mice [92], and conversely T cells can protect antibody-deficient mice [93]. Therefore, characterization of *Y. pestis*-specific adaptive immune response in the host will provide a wealth of information for illustrating bacterial virulence and promoting the development of specific vaccines.

10.4.1 T Cell-Mediated Immune Responses to Yersinia pestis

A growing body of evidence demonstrates the crucial roles of T cell-mediated immunity against *Y. pestis*. A function of expanded pathogen-specific T cells is to secrete IFNγ and TNFα, and depletion of these proinflammatory cytokines prior to the passive transfer of the LcrV antibody into mice completely abrogated the protective effect observed in undepleted mice [94]. This indicates that a cellular proinflammatory response provides critical protective functions during humoral defense against lethal pulmonary *Y. pestis* infection. Direct evidence for the importance of the cellular response in protection against the plague came from studies in μMT mice, which are functional B cell-deficient mice and cannot produce antibodies. Transfer of *Y. pestis*-primed T cells into naïve μMT mice protected them against lethal intranasal *Y. pestis* challenge [93]. In another study, anti-F1 IgG transferred passively into T cell receptor knockout mice ($TCR^{-/-}$) at 24 h after infection was unable to rescue them, whereas it fully protected μMT mice, indicating that T cells play a critical role in the protection mediated by antibody to F1-antigen[95]. Thus, identification of *Y. pestis* antigens that stimulate a protective T cell response is one of the major goals for vaccine development. Using silico computer analysis and an in vitro IFNγ assay, we identified potential T cell antigens. In all, 34 individual proteins that stimulated a strong IFNγ response from splenocytes of mice immunized with *Y. pestis* live attenuated vaccine EV76 have been identified. Furthermore, in addition to LcrV, nine proteins may provide partial protection against challenge with a low dose of *Y. pestis* [96].

$CD4^+$ T helper (Th) cells include different subtypes based on their cytokine and transcription factor signatures [97]: Th1 cells produce IFNγ as their "signature" cytokine; Th2 cells produce IL-4, IL-5, and IL-13; Th17 cells produce IL-17; and regulatory T cells (Tregs) express the transcription factor FoxP3. Different T cell lineages play different roles in *Y. pestis* infection. Signal transducer and activator of transcription (STAT)-4 knockout mice, which have a reduced Th1 response when immunized with F1/V, were poorly protected against *Y. pestis* challenge. However, STAT-6 knockout mice with intact Th1 responses and diminished Th2 responses were fully protected [98]. These results indicate that Th1-mediated immune mechanisms, activated following Stat 4 phosphorylation, are essential in protection against the plague. Therefore, one potential way to improve the efficacy of plague vaccines would be to increase the Th1 response. Dinc et al. assessed the efficacy of the novel SA-4-1BBL costimulatory molecule as a Th1 adjuvant to improve cellular responses generated by the rF1-V vaccine [99]. They proved that addition of SA-4-1BBL improves the efficacy of the subunit vaccine by generating a strong Th1 cellular immune response without significant impact on the generation of Ab responses. Additionally, Smiley's group demonstrated that vaccination with live attenuated *Y. pestis* induces Th17 cells [100]. IL-17 was found to contribute to defense against pulmonary *Y. pestis* challenge [82], although this IL-17-mediated protection does not appear to result entirely from enhanced bacterial clearance. These results indicate that plague vaccines aiming to induce mixed Th1 and Th17 cellular responses would provide more powerful and comprehensive protection.

In addition to $CD4^+$ T cells, specific $CD8^+$ T cell responses also contribute to defense against pulmonary *Y. pestis* infection, although alone they may be insufficient to combat fully virulent *Y. pestis* strains. YopE of *Y. pestis* was found to contain a dominant CD8 T cell epitope, which can be recognized by nearly 20 % of pulmonary CD8 T cells. Moreover, immunizing mice with a single peptide, $YopE_{69-77}$, suffices to confer protection from lethal pulmonary challenge [101]. A further study investigated the effector functions of $YopE_{69-77}$-specific CD8 T cells during pulmonary *Y. pestis* infection [102]. They concluded that specific CD8 T cell-mediated protection against pneumonic plague is dependent on TNFα and IFNγ, but not on perforin. In addition, $CD4^+$ and $CD8^+$ T cells were found to synergistically protect against pneumonic plague in this mouse model [103].

10.4.2 Antibody-Mediated Defense Against Yersinia pestis

Because of the complex antigen structure of *Y. pestis*, a number of antibodies are found in *Y. pestis*-infected patients and model animals. Serum samples collected from plague convalescent patients can transfer passive protection to naïve mice, indicating that antibody-mediated defense plays an important role against *Y. pestis* challenge. In previous studies, F1 and LcrV were proven to provide a high degree of protection, and the corresponding vaccines also showed efficacy in small animal models [104–107]. Using an antigen microarray, which contained more than 140 *Y. pestis* virulence-associated proteins, we screened the antibody responses of plague patients [108]. Apart from F1, YopD, YopE, and pH6 antigens, which have been described previously as immunogens, ten other novel immunogenic proteins were found. Antibody titers and persistence are correlated with vaccine efficacy. Previous studies confirmed that the F1 antibody could persist for 1–4 years in humans [109]. Further studies in our lab explored the antibody profile from 65 plague patients who were in remission for more than 10 years using a protein microarray [110]. Results showed that antibody to F1 can persist in recovered patients for more than 10 years, while antibodies to LcrV and YopD were present for an even longer period of time. As artificial passive immunization has been demonstrated to be effective against *Y. pestis* infection in animals, how about maternal antibodies? We carried out a study evaluating the kinetics, protective efficacy, and transmission modes of maternal antibodies, using mice immunized with plague subunit vaccine [111]. The results indicated that maternal antibodies induced by the plague subunit vaccine in mother mice can be transferred to newborn mice via both the placenta and lactation were sustained for more than 10 weeks and provide early protection against plague for newborn mice.

Long-lived plasma cells and memory B cells are responsible for the long-term humoral immunity elicited by vaccination. Memory B cells are in charge of driving the rapid anamnestic antibody response that occurs after re-exposure to antigen. The serum antibody level is maintained by long-lived plasma cells. Thus, studies in our lab investigated the kinetics of memory B cell and plasma cell response in mice immunized with plague subunit vaccine F1 or the live attenuated vaccine EV76 [112]. The number of memory B cells in the spleens was significantly higher than that in the bone marrow, which is consistent with a previous study that found that the majority of memory B cells are present in the spleen [113]. We also found that the boost of antibody titer after revaccination may be dependent on the existence of memory B cells and an excess of antigen.

Although passive transfer of specific antibody can provide protection in rodents against pneumonic plague, in nonhuman primates vaccinated with F1/LcrV, high-titer specific antibody at the time of challenge cannot protect primates against pneumonic plague [114, 115]. These observations strongly suggest that antibody titer alone, at least as measured by standard ELISA, do not suffice in predicting the efficacy of pneumonic plague vaccines. Therefore, which factors are involved in antibody-mediated protection? First, neutrophils were found to contribute to antibody-mediated defense against pneumonic plague because neutrophil depletion abrogates serotherapy-mediated protection from pneumonic plague in a mouse model. Similar results were reported in a mouse model of septicemic plague, where neutrophil depletion abrogates protection mediated by polyclonal anti-LcrV [116]. Second, cytokines from phagocytes enhances antibody-mediated protection. Whether genetic deficiency or antibody neutralization of IFNγ and TNFα, cytokine depletion significantly impairs serotherapy-mediated protection [98, 117]. In addition, passive protection with antibody to F1 in infected mice was reduced in $C3^{-/-}$ mice that lack a functional complement system, indicating the contribution of Fc-effector mechanisms to antibody-mediated protection [95].

References

1. Lukaszewski RA, Kenny DJ, Taylor R, Rees DG, Hartley MG, Oyston PC. Pathogenesis of Yersinia pestis infection in BALB/c mice: effects on host macrophages and neutrophils. Infect Immun. 2005;73(11):7142–50.
2. Fukuto HS, Bliska JB. Editorial: Yersinia pestis survives in neutrophils and sends a PS to macrophages: bon appetit! J Leukoc Biol. 2014;95(3):383–5.
3. Spinner JL, Winfree S, Starr T, Shannon JG, Nair V, Steele-Mortimer O, Hinnebusch BJ. Yersinia pestis survival and replication within human neutrophil phagosomes and uptake of infected neutrophils by macrophages. J Leukoc Biol. 2014;95(3):389–98.
4. Bubeck SS, Cantwell AM, Dube PH. Delayed inflammatory response to primary pneumonic plague occurs in both outbred and inbred mice. Infect Immun. 2007;75(2):697–705.
5. Lathem WW, Price PA, Miller VL, Goldman WE. A plasminogen-activating protease specifically controls the development of primary pneumonic plague. Science. 2007;315(5811):509–13.
6. Lathem WW, Crosby SD, Miller VL, Goldman WE. Progression of primary pneumonic plague: a mouse model of infection, pathology, and bacterial transcriptional activity. Proc Natl Acad Sci U S A. 2005;102(49):17786–91.
7. Sauvonnet N, Lambermont I, van der Bruggen P, Cornelis GR. YopH prevents monocyte chemoattractant protein 1 expression in macrophages and T-cell proliferation through inactivation of the phosphatidylinositol 3-kinase pathway. Mol Microbiol. 2002;45(3):805–15.
8. Orth K, Palmer LE, Bao ZQ, Stewart S, Rudolph AE, Bliska JB, Dixon JE. Inhibition of the mitogen-activated protein kinase kinase superfamily by a Yersinia effector. Science. 1999;285(5435):1920–3.
9. Viboud GI, Mejia E, Bliska JB. Comparison of YopE and YopT activities in counteracting host signalling responses to Yersinia pseudotuberculosis infection. Cell Microbiol. 2006;8(9):1504–15.
10. Brodsky IE, Palm NW, Sadanand S, Ryndak MB, Sutterwala FS, Flavell RA, Bliska JB, Medzhitov R. A Yersinia effector protein promotes virulence by preventing inflammasome recognition of the type III secretion system. Cell Host Microbe. 2010;7(5):376–87.
11. LaRock CN, Cookson BT. The Yersinia virulence effector YopM binds caspase-1 to arrest inflammasome assembly and processing. Cell Host Microbe. 2012;12(6):799–805.
12. Brubaker RR. Interleukin-10 and inhibition of innate immunity to Yersiniae: roles of Yops and LcrV (V antigen). Infect Immun. 2003;71(7):3673–81.
13. McDonald C, Vacratsis PO, Bliska JB, Dixon JE. The yersinia virulence factor YopM forms a novel protein complex with two cellular kinases. J Biol Chem. 2003;278:18514–23.
14. Osei-Owusu P, Jessen Condry DL, Toosky M, Roughead W, Bradley DS, Nilles ML. The N terminus of type III secretion needle protein YscF from Yersinia pestis functions to modulate innate immune responses. Infect Immun. 2015;83(4):1507–22.
15. Rebeil R, Ernst RK, Jarrett CO, Adams KN, Miller SI, Hinnebusch BJ. Characterization of late acyltransferase genes of *Yersinia pestis* and their role in temperature-dependent lipid A variation. J Bacteriol. 2006;188(4):1381–8.
16. Tan L, Darby C. *Yersinia pestis* is viable with endotoxin composed of only lipid A. J Bacteriol. 2005;187(18):6599–600.
17. Montminy SW, Khan N, McGrath S, Walkowicz MJ, Sharp F, Conlon JE, Fukase K, Kusumoto S, Sweet C, Miyake K, et al. Virulence factors of *Yersinia pestis* are overcome by a strong lipopolysaccharide response. Nat Immunol. 2006;7(10):1066–73.
18. de la Puerta ML, Trinidad AG, del Carmen Rodriguez M, Bogetz J, Sanchez Crespo M, Mustelin T, Alonso A, Bayon Y. Characterization of new substrates targeted by Yersinia tyrosine phosphatase YopH. PLoS One. 2009;4(2):e4431.
19. Hamid N, Gustavsson A, Andersson K, McGee K, Persson C, Rudd CE, Fallman M. YopH dephosphorylates Cas and Fyn-binding protein in macrophages. Microb Pathog. 1999;27(4):231–42.

20. Persson C, Carballeira N, Wolf-Watz H, Fallman M. The PTPase YopH inhibits uptake of Yersinia, tyrosine phosphorylation of p130Cas and FAK, and the associated accumulation of these proteins in peripheral focal adhesions. EMBO J. 1997;16(9):2307–18.
21. Weidow CL, Black DS, Bliska JB, Bouton AH. CAS/Crk signalling mediates uptake of Yersinia into human epithelial cells. Cell Microbiol. 2000;2(6):549–60.
22. Von Pawel-Rammingen U, Telepnev MV, Schmidt G, Aktories K, Wolf-Watz H, Rosqvist R. GAP activity of the Yersinia YopE cytotoxin specifically targets the Rho pathway: a mechanism for disruption of actin microfilament structure. Mol Microbiol. 2000;36(3):737–48.
23. Roppenser B, Roder A, Hentschke M, Ruckdeschel K, Aepfelbacher M. Yersinia enterocolitica differentially modulates RhoG activity in host cells. J Cell Sci. 2009;122(Pt 5):696–705.
24. Andor A, Trulzsch K, Essler M, Roggenkamp A, Wiedemann A, Heesemann J, Aepfelbacher M. YopE of Yersinia, a GAP for Rho GTPases, selectively modulates Rac-dependent actin structures in endothelial cells. Cell Microbiol. 2001;3(5):301–10.
25. Shao F, Dixon JE. YopT is a cysteine protease cleaving Rho family GTPases. Adv Exp Med Biol. 2003;529:79–84.
26. Rosqvist R, Forsberg A, Rimpilainen M, Bergman T, Wolf-Watz H. The cytotoxic protein YopE of Yersinia obstructs the primary host defence. Mol Microbiol. 1990;4(4):657–67.
27. Andersson K, Carballeira N, Magnusson KE, Persson C, Stendahl O, Wolf-Watz H, Fallman M. YopH of Yersinia pseudotuberculosis interrupts early phosphotyrosine signalling associated with phagocytosis. Mol Microbiol. 1996;20(5):1057–69.
28. Iriarte M, Cornelis GR. YopT, a new Yersinia Yop effector protein, affects the cytoskeleton of host cells. Mol Microbiol. 1998;29(3):915–29.
29. Aepfelbacher M, Heesemann J. Modulation of Rho GTPases and the actin cytoskeleton by Yersinia outer proteins (Yops). Int J Med Microbiol. 2001;291(4):269–76.
30. Aepfelbacher M, Zumbihl R, Heesemann J. Modulation of Rho GTPases and the actin cytoskeleton by YopT of Yersinia. Curr Top Microbiol Immunol. 2005;291:167–75.
31. Du Y, Rosqvist R, Forsberg A. Role of fraction 1 antigen of *Yersinia pestis* in inhibition of phagocytosis. Infect Immun. 2002;70(3):1453–60.
32. Payne D, Tatham D, Williamson ED, Titball RW. The pH 6 antigen of *Yersinia pestis* binds to beta1-linked galactosyl residues in glycosphingolipids. Infect Immun. 1998;66(9):4545–8.
33. Makoveichuk E, Cherepanov P, Lundberg S, Forsberg A, Olivecrona G. pH6 antigen of *Yersinia pestis* interacts with plasma lipoproteins and cell membranes. J Lipid Res. 2003;44(2):320–30.
34. Anisimov AP, Dentovskaya SV, Titareva GM, Bakhteeva IV, Shaikhutdinova RZ, Balakhonov SV, Lindner B, Kocharova NA, Senchenkova SN, Holst O, et al. Intraspecies and temperature-dependent variations in susceptibility of *Yersinia pestis* to the bactericidal action of serum and to polymyxin B. Infect Immun. 2005;73(11):7324–31.
35. Bliska JB, Falkow S. Bacterial resistance to complement killing mediated by the Ail protein of Yersinia enterocolitica. Proc Natl Acad Sci U S A. 1992;89(8):3561–5.
36. Kirjavainen V, Jarva H, Biedzka-Sarek M, Blom AM, Skurnik M, Meri S. Yersinia enterocolitica serum resistance proteins YadA and ail bind the complement regulator C4b-binding protein. PLoS Pathog. 2008;4(8):e1000140.
37. Skorek K, Raczkowska A, Dudek B, Mietka K, Guz-Regner K, Pawlak A, Klausa E, Bugla-Ploskonska G, Brzostek K. Regulatory protein OmpR influences the serum resistance of Yersinia enterocolitica O:9 by modifying the structure of the outer membrane. PLoS One. 2013;8(11):e79525.
38. Porat R, McCabe WR, Brubaker RR. Lipopolysaccharide-associated resistance to killing of yersiniae by complement. J Endotoxin Res. 1995;2:91–7.
39. Kolodziejek AM, Sinclair DJ, Seo KS, Schnider DR, Deobald CF, Rohde HN, Viall AK, Minnich SS, Hovde CJ, Minnich SA, et al. Phenotypic characterization of OmpX, an Ail homologue of Yersinia pestis KIM. Microbiology. 2007;153(Pt 9):2941–51.

40. Bartra SS, Styer KL, O'Bryant DM, Nilles ML, Hinnebusch BJ, Aballay A, Plano GV. Resistance of Yersinia pestis to complement-dependent killing is mediated by the Ail outer membrane protein. Infect Immun. 2008;76(2):612–22.
41. Biedzka-Sarek M, Jarva H, Hyytiainen H, Meri S, Skurnik M. Characterization of complement factor H binding to Yersinia enterocolitica serotype O:3. Infect Immun. 2008;76(9):4100–9.
42. Biedzka-Sarek M, Salmenlinna S, Gruber M, Lupas AN, Meri S, Skurnik M. Functional mapping of YadA- and Ail-mediated binding of human factor H to Yersinia enterocolitica serotype O:3. Infect Immun. 2008;76(11):5016–27.
43. Ho DK, Riva R, Kirjavainen V, Jarva H, Ginstrom E, Blom AM, Skurnik M, Meri S. Functional recruitment of the human complement inhibitor C4BP to Yersinia pseudotuberculosis outer membrane protein Ail. J Immunol. 2012;188(9):4450–9.
44. Sebbane F, Jarrett CO, Gardner D, Long D, Hinnebusch BJ. Role of the *Yersinia pestis* plasminogen activator in the incidence of distinct septicemic and bubonic forms of flea-borne plague. Proc Natl Acad Sci U S A. 2006;103(14):5526–30.
45. Sodeinde OA, Subrahmanyam YV, Stark K, Quan T, Bao Y, Goguen JD. A surface protease and the invasive character of plague. Science. 1992;258(5084):1004–7.
46. Drozdov IG, Anisimov AP, Samoilova SV, Yezhov IN, Yeremin SA, Karlyshev AV, Krasilnikova VM, Kravchenko VI. Virulent non-capsulate Yersinia pestis variants constructed by insertion mutagenesis. J Med Microbiol. 1995;42(4):264–8.
47. Samoilova SV, Samoilova LV, Yezhov IN, Drozdov IG, Anisimov AP. Virulence of pPst+and pPst- strains of Yersinia pestis for guinea-pigs. J Med Microbiol. 1996;45(6):440–4.
48. Laudisoit A, Leirs H, Makundi RH, Van Dongen S, Davis S, Neerinckx S, Deckers J, Libois R. Plague and the human flea Tanzania. Emerg Infect Dis. 2007;13(5):687–93.
49. Zhang SS, Park CG, Zhang P, Bartra SS, Plano GV, Klena JD, Skurnik M, Hinnebusch BJ, Chen T. Plasminogen activator Pla of Yersinia pestis utilizes murine DEC-205 (CD205) as a receptor to promote dissemination. J Biol Chem. 2008;283(46):31511–21.
50. Reboul A, Lemaitre N, Titecat M, Merchez M, Deloison G, Ricard I, Pradel E, Marceau M, Sebbane F. Yersinia pestis requires the 2-component regulatory system OmpR-EnvZ to resist innate immunity during the early and late stages of plague. J Infect Dis. 2014;210(9):1367–75.
51. Marketon MM, DePaolo RW, DeBord KL, Jabri B, Schneewind O. Plague bacteria target immune cells during infection. Science. 2005;309(5741):1739–41.
52. Pechous RD, Sivaraman V, Price PA, Stasulli NM, Goldman WE. Early host cell targets of Yersinia pestis during primary pneumonic plague. PLoS Pathog. 2013;9(10):e1003679.
53. Busch JD, Van Andel R, Stone NE, Cobble KR, Nottingham R, Lee J, Versteeg M, Corcoran J, Cordova J, Van Pelt W, et al. The innate immune response may be important for surviving plague in wild gunnison's prairie dogs. J Wildl Dis. 2013;49(4):920–31.
54. Cavanaugh DC, Randall R. The role of multiplication of Pasteurella pestis in mononuclear phagocytes in the pathogenesis of flea-borne plague. J Immunol. 1959;83:348–63.
55. Elvin SJ, Williamson ED, Scott JC, Smith JN, De Lema Perez G, Chilla S, Clapham P, Pfeffer K, Schlondorff D, Luckow B. Evolutionary genetics: ambiguous role of CCR5 in Y. pestis infection. Nature. 2004;430(6998):417.
56. Pujol C, Bliska JB. The ability to replicate in macrophages is conserved between *Yersinia pestis* and *Yersinia pseudotuberculosis*. Infect Immun. 2003;71(10):5892–9.
57. Straley SC, Harmon PA. *Yersinia pestis* grows within phagolysosomes in mouse peritoneal macrophages. Infect Immun. 1984;45(3):655–9.
58. Pujol C, Bliska JB. Turning *Yersinia* pathogenesis outside in: subversion of macrophage function by intracellular yersiniae. Clin Immunol. 2005;114(3):216–26.
59. Zhou D, Han Y, Yang R. Molecular and physiological insights into plague transmission, virulence and etiology. Microbes Infect. 2006;8(1):273–84.

60. Weeks S, Hill J, Friedlander A, Welkos S. Anti-V antigen antibody protects macrophages from Yersinia pestis-induced cell death and promotes phagocytosis. Microb Pathog. 2002;32(5):227–37.
61. Bergsbaken T, Cookson BT. Macrophage activation redirects yersinia-infected host cell death from apoptosis to caspase-1-dependent pyroptosis. PLoS Pathog. 2007;3(11):e161.
62. Ruckdeschel K, Pfaffinger G, Haase R, Sing A, Weighardt H, Hacker G, Holzmann B, Heesemann J. Signaling of apoptosis through TLRs critically involves toll/IL-1 receptor domain-containing adapter inducing IFN-beta, but not MyD88, in bacteria-infected murine macrophages. J Immunol. 2004;173(5):3320–8.
63. Hoebe K, Du X, Georgel P, Janssen E, Tabeta K, Kim SO, Goode J, Lin P, Mann N, Mudd S, et al. Identification of Lps2 as a key transducer of MyD88-independent TIR signalling. Nature. 2003;424(6950):743–8.
64. Zhang Y, Ting AT, Marcu KB, Bliska JB. Inhibition of MAPK and NF-kappa B pathways is necessary for rapid apoptosis in macrophages infected with Yersinia. J Immunol. 2005;174(12):7939–49.
65. Karin M, Lin A. NF-kappaB at the crossroads of life and death. Nat Immunol. 2002;3(3):221–7.
66. Palmer LE, Hobbie S, Galan JE, Bliska JB. YopJ of Yersinia pseudotuberculosis is required for the inhibition of macrophage TNF-alpha production and downregulation of the MAP kinases p38 and JNK. Mol Microbiol. 1998;27(5):953–65.
67. Schesser K, Spiik AK, Dukuzumuremyi JM, Neurath MF, Pettersson S, Wolf-Watz H. The yopJ locus is required for Yersinia-mediated inhibition of NF-kappaB activation and cytokine expression: yopJ contains a eukaryotic SH2-like domain that is essential for its repressive activity. Mol Microbiol. 1998;28(6):1067–79.
68. Denecker G, Declercq W, Geuijen CA, Boland A, Benabdillah R, van Gurp M, Sory MP, Vandenabeele P, Cornelis GR. Yersinia enterocolitica YopP-induced apoptosis of macrophages involves the apoptotic signaling cascade upstream of bid. J Biol Chem. 2001;276(23):19706–14.
69. Savill J, Dransfield I, Gregory C, Haslett C. A blast from the past: clearance of apoptotic cells regulates immune responses. Nat Rev Immunol. 2002;2(12):965–75.
70. Voll RE, Herrmann M, Roth EA, Stach C, Kalden JR, Girkontaite I. Immunosuppressive effects of apoptotic cells. Nature. 1997;390(6658):350–1.
71. Fantuzzi G, Dinarello CA. Interleukin-18 and interleukin-1 beta: two cytokine substrates for ICE (caspase-1). J Clin Immunol. 1999;19(1):1–11.
72. Sebbane F, Gardner D, Long D, Gowen BB, Hinnebusch BJ. Kinetics of disease progression and host response in a rat model of bubonic plague. Am J Pathol. 2005;166(5):1427–39.
73. Balada-Llasat JM, Mecsas J. Yersinia has a tropism for B and T cell zones of lymph nodes that is independent of the type III secretion system. PLoS Pathog. 2006;2(9), e86.
74. Guinet F, Ave P, Jones L, Huerre M, Carniel E. Defective innate cell response and lymph node infiltration specify Yersinia pestis infection. PLoS One. 2008;3(2):e1688.
75. Mills SD, Boland A, Sory MP, van der Smissen P, Kerbourch C, Finlay BB, Cornelis GR. Yersinia enterocolitica induces apoptosis in macrophages by a process requiring functional type III secretion and translocation mechanisms and involving YopP, presumably acting as an effector protein. Proc Natl Acad Sci U S A. 1997;94(23):12638–43.
76. Monack DM, Mecsas J, Ghori N, Falkow S. Yersinia signals macrophages to undergo apoptosis and YopJ is necessary for this cell death. Proc Natl Acad Sci U S A. 1997;94(19):10385–90.
77. Ruckdeschel K, Roggenkamp A, Lafont V, Mangeat P, Heesemann J, Rouot B. Interaction of Yersinia enterocolitica with macrophages leads to macrophage cell death through apoptosis. Infect Immun. 1997;65(11):4813–21.
78. Shin H, Cornelis GR. Type III secretion translocation pores of Yersinia enterocolitica trigger maturation and release of pro-inflammatory IL-1beta. Cell Microbiol. 2007;9(12):2893–902.

79. Laws TR, Davey MS, Titball RW, Lukaszewski R. Neutrophils are important in early control of lung infection by Yersinia pestis. Microbes Infect. 2010;12(4):331–5.
80. Shannon JG, Bosio CF, Hinnebusch BJ. Dermal neutrophil, macrophage and dendritic cell responses to Yersinia pestis transmitted by fleas. PLoS Pathog. 2015;11(3), e1004734.
81. Vagima Y, Zauberman A, Levy Y, Gur D, Tidhar A, Aftalion M, Shafferman A, Mamroud E. Circumventing Y. pestis Virulence by Early Recruitment of Neutrophils to the Lungs during Pneumonic Plague. PLoS Pathog. 2015;11(5):e1004893.
82. Bi Y, Zhou J, Yang H, Wang X, Zhang X, Wang Q, Wu X, Han Y, Song Y, Tan Y, et al. IL-17A produced by neutrophils protects against pneumonic plague through orchestrating IFN-gamma-activated macrophage programming. J Immunol. 2014;192(2):704–13.
83. Merritt PM, Nero T, Bohman L, Felek S, Krukonis ES, Marketon MM. Yersinia pestis targets neutrophils via complement receptor 3. Cell Microbiol. 2015;17(5):666–87.
84. Hinnebusch BJ, Jarrett CO, Callison JA, Gardner D, Buchanan SK, Plano GV. Role of the Yersinia pestis Ail protein in preventing a protective polymorphonuclear leukocyte response during bubonic plague. Infect Immun. 2011;79(12):4984–9.
85. Velan B, Bar-Haim E, Zauberman A, Mamroud E, Shafferman A, Cohen S. Discordance in the effects of Yersinia pestis on the dendritic cell functions manifested by induction of maturation and paralysis of migration. Infect Immun. 2006;74(11):6365–76.
86. Robinson RT, Khader SA, Locksley RM, Lien E, Smiley ST, Cooper AM. Yersinia pestis evades TLR4-dependent induction of IL-12(p40)2 by dendritic cells and subsequent cell migration. J Immunol. 2008;181(8):5560–7.
87. Depaolo RW, Tang F, Kim I, Han M, Levin N, Ciletti N, Lin A, Anderson D, Schneewind O, Jabri B. Toll-like receptor 6 drives differentiation of tolerogenic dendritic cells and contributes to LcrV-mediated plague pathogenesis. Cell Host Microbe. 2008;4(4):350–61.
88. Kingston R, Burke F, Robinson JH, Bedford PA, Jones SM, Knight SC, Williamson ED. The fraction 1 and V protein antigens of Yersinia pestis activate dendritic cells to induce primary T cell responses. Clin Exp Immunol. 2007;149(3):561–9.
89. Do Y, Koh H, Park CG, Dudziak D, Seo P, Mehandru S, Choi JH, Cheong C, Park S, Perlin DS, et al. Targeting of LcrV virulence protein from Yersinia pestis to dendritic cells protects mice against pneumonic plague. Eur J Immunol. 2010;40(10):2791–6.
90. Kerschen EJ, Cohen DA, Kaplan AM, Straley SC. The plague virulence protein YopM targets the innate immune response by causing a global depletion of NK cells. Infect Immun. 2004;72(8):4589–602.
91. Ye Z, Kerschen EJ, Cohen DA, Kaplan AM, van Rooijen N, Straley SC. Gr1+ cells control growth of YopM-negative *Yersinia pestis* during systemic plague. Infect Immun. 2009;77(9):3791–806.
92. Green M, Rogers D, Russell P, Stagg AJ, Bell DL, Eley SM, Titball RW, Williamson ED. The SCID/Beige mouse as a model to investigate protection against *Yersinia pestis*. FEMS Immunol Med Microbiol. 1999;23(2):107–13.
93. Parent MA, Berggren KN, Kummer LW, Wilhelm LB, Szaba FM, Mullarky IK, Smiley ST. Cell-mediated protection against pulmonary *Yersinia pestis* infection. Infect Immun. 2005;73(11):7304–10.
94. Lin JS, Park S, Adamovicz JJ, Hill J, Bliska JB, Cote CK, Perlin DS, Amemiya K, Smiley ST. TNFalpha and IFNgamma contribute to F1/LcrV-targeted immune defense in mouse models of fully virulent pneumonic plague. Vaccine. 2010;29(2):357–62.
95. Levy Y, Flashner Y, Tidhar A, Zauberman A, Aftalion M, Lazar S, Gur D, Shafferman A, Mamroud E. T cells play an essential role in anti-F1 mediated rapid protection against bubonic plague. Vaccine. 2011;29(40):6866–73.
96. Li B, Zhou L, Guo J, Wang X, Ni B, Ke Y, Zhu Z, Guo Z, Yang R. High-throughput identification of new protective antigens from a Yersinia pestis live vaccine by enzyme-linked immunospot assay. Infect Immun. 2009;77(10):4356–61.
97. Bi Y, Liu G, Yang R. Reciprocal modulation between TH17 and other helper T cell lineages. J Cell Physiol. 2011;226(1):8–13.

98. Elvin SJ, Williamson ED. Stat 4 but not Stat 6 mediated immune mechanisms are essential in protection against plague. Microb Pathog. 2004;37(4):177–84.
99. Dinc G, Pennington JM, Yolcu ES, Lawrenz MB, Shirwan H. Improving the Th1 cellular efficacy of the lead Yersinia pestis rF1-V subunit vaccine using SA-4-1BBL as a novel adjuvant. Vaccine. 2014;32(39):5035–40.
100. Lin JS, Kummer LW, Szaba FM, Smiley ST. IL-17 contributes to cell-mediated defense against pulmonary Yersinia pestis infection. J Immunol. 2011;186(3):1675–84.
101. Lin JS, Szaba FM, Kummer LW, Chromy BA, Smiley ST. Yersinia pestis YopE contains a dominant CD8 T cell epitope that confers protection in a mouse model of pneumonic plague. J Immunol. 2011;187(2):897–904.
102. Szaba FM, Kummer LW, Duso DK, Koroleva EP, Tumanov AV, Cooper AM, Bliska JB, Smiley ST, Lin JS. TNFalpha and IFNgamma but not perforin are critical for CD8 T cell-mediated protection against pulmonary Yersinia pestis infection. PLoS Pathog. 2014;10(5):e1004142.
103. Philipovskiy AV, Smiley ST. Vaccination with live Yersinia pestis primes CD4 and CD8 T cells that synergistically protect against lethal pulmonary Y. pestis infection. Infect Immun. 2007;75(2):878–85.
104. Anderson Jr GW, Leary SE, Williamson ED, Titball RW, Welkos SL, Worsham PL, Friedlander AM. Recombinant V antigen protects mice against pneumonic and bubonic plague caused by F1-capsule-positive and -negative strains of Yersinia pestis. Infect Immun. 1996;64(11):4580–5.
105. Andrews GP, Heath DG, Anderson Jr GW, Welkos SL, Friedlander AM. Fraction 1 capsular antigen (F1) purification from *Yersinia pestis* CO92 and from an *Escherichia coli* recombinant strain and efficacy against lethal plague challenge. Infect Immun. 1996;64(6):2180–7.
106. DeBord KL, Anderson DM, Marketon MM, Overheim KA, DePaolo RW, Ciletti NA, Jabri B, Schneewind O. Immunogenicity and protective immunity against bubonic plague and pneumonic plague by immunization of mice with the recombinant V10 antigen, a variant of LcrV. Infect Immun. 2006;74(8):4910–4.
107. Leary SE, Griffin KF, Garmory HS, Williamson ED, Titball RW. Expression of an F1/V fusion protein in attenuated Salmonella typhimurium and protection of mice against plague. Microb Pathog. 1997;23(3):167–79.
108. Li B, Zhou D, Wang Z, Song Z, Wang H, Li M, Dong X, Wu M, Guo Z, Yang R. Antibody profiling in plague patients by protein microarray. Microbes Infect. 2008;10(1):45–51.
109. Rasoamanana B, Leroy F, Boisier P, Rasolomaharo M, Buchy P, Carniel E, Chanteau S. Field evaluation of an immunoglobulin G anti-F1 enzyme-linked immunosorbent assay for serodiagnosis of human plague in Madagascar. Clin Diagn Lab Immunol. 1997;4(5):587–91.
110. Li B, Du C, Zhou L, Bi Y, Wang X, Wen L, Guo Z, Song Z, Yang R. Humoral and cellular immune responses to Yersinia pestis infection in long-term recovered plague patients. Clin Vaccine Immunol. 2012;19(2):228–34.
111. Qi Z, Zhao H, Zhang Q, Bi Y, Ren L, Zhang X, Yang H, Yang X, Wang Q, Li C, et al. Acquisition of maternal antibodies both from the placenta and by lactation protects mouse offspring from Yersinia pestis challenge. Clin Vaccine Immunol. 2012;19(11):1746–50.
112. Zhang X, Wang Q, Bi Y, Kou Z, Zhou J, Cui Y, Yan Y, Zhou L, Tan Y, Yang H, et al. Kinetics of memory B cell and plasma cell responses in the mice immunized with plague vaccines. Scand J Immunol. 2014;79(3):157–62.
113. Shenoy GN, Chatterjee P, Kaw S, Mukherjee S, Rathore DK, Bal V, Rath S, George A. Recruitment of memory B cells to lymph nodes remote from the site of immunization requires an inflammatory stimulus. J Immunol. 2012;189(2):521–8.
114. Williamson ED, Flick-Smith HC, Waters E, Miller J, Hodgson I, Le Butt CS, Hill J. Immunogenicity of the rF1+rV vaccine for plague with identification of potential immune correlates. Microb Pathog. 2007;42(1):11–21.

115. Bashaw J, Norris S, Weeks S, Trevino S, Adamovicz JJ, Welkos S. Development of in vitro correlate assays of immunity to infection with Yersinia pestis. Clin Vaccine Immunol. 2007;14(5):605–16.
116. Cowan C, Philipovskiy AV, Wulff-Strobel CR, Ye Z, Straley SC. Anti-LcrV antibody inhibits delivery of Yops by Yersinia pestis KIM5 by directly promoting phagocytosis. Infect Immun. 2005;73(9):6127–37.
117. Parent MA, Wilhelm LB, Kummer LW, Szaba FM, Mullarky IK, Smiley ST. Gamma interferon, tumor necrosis factor alpha, and nitric oxide synthase 2, key elements of cellular immunity, perform critical protective functions during humoral defense against lethal pulmonary *Yersinia pestis* infection. Infect Immun. 2006;74(6):3381–6.

Chapter 11
Plague: Clinics, Diagnosis and Treatment

Vladimir V. Nikiforov, He Gao, Lei Zhou, and Andrey Anisimov

Abstract Plague still poses a significant threat to human health and as a reemerging infection is unfamiliar to the majority of the modern medical doctors. In this chapter, the plague is described according to Dr. Nikiforov's experiences in the diagnosis and treatment of patients, and also a review of the relevant literature on this subject is provided. The main modern methods and criteria for laboratory diagnosis of plague are briefly described. The clinical presentations include the bubonic and pneumonic form, septicemia, rarely pharyngitis, and meningitis. Early diagnosis and the prompt initiation of treatment reduce the mortality rate associated with bubonic plague and septicemic plague to 5–50 %; although a delay of more than 24 h in the administration of antibiotics and antishock treatment can be fatal for plague patients. Most human cases can successfully be treated with antibiotics.

Keywords Bubonic plague • Septicemic plague • Pneumonic plague • Plague symptoms • Plague treatment

V.V. Nikiforov (✉)
Institute of Advanced Training, Federal Medical-Biological Agency of Russia, Moscow, Russia
e-mail: v.v.nikiforov@gmail.com

H. Gao
National Institute for Communicable Disease Control and Prevention, China CDC, Beijing 102206, China
e-mail: gaohe@icdc.cn

L. Zhou
Beijing Institute of Microbiology and Epidemiology, Beijing 100071, China
e-mail: ammszhoulei@aliyun.com

A. Anisimov
State Research Center for Applied Microbiology, Obolensk, Moscow Region, Russia
e-mail: anisimov@obolensk.org

R. Yang, A. Anisimov (eds.), *Yersinia pestis: Retrospective and Perspective*, Advances in Experimental Medicine and Biology 918,
DOI 10.1007/978-94-024-0890-4_11

11.1 Clinical Forms of Plague and Their Manifestations

It is difficult to describe precisely the clinical manifestations of plague. This is because a definite near-mythical concept of this disease, based on the works of previous clinicians in the pre-antibiotic age, had acquired some canonical features in the academic literature. The descriptions from the end of the nineteenth century and the beginning of the twentieth century may be more reliable than earlier literature on this subject. In earlier publications, plague patients and those suffering from other severe infections were in many cases indistinguishable, indicating that in ancient times the plague was a term encompassing a collective of different infectious diseases.

It is written in Galen's *Commentaries* on Hippocratic *Epidemics* that if a disease affects many people in a region, it can be referred to as an epidemic disease, and if this disease causes many human deaths, it is then a "plague." The majority of modern authors have limited clinical experience of the plague because it is currently a relatively rare infectious disease. In addition, the clinical manifestations of the disease in endemic areas vary considerably owing to "natural immunization" of the local population [1] and the various host specificities of *Yersinia pestis* strains from different natural plague foci [2]. It is clear that the data on plague manifestations typical for the residents of Vietnam or India, for example, cannot unconditionally be extrapolated to Europeans. We can only speculate as to how different *Y. pestis* clones would interact with nonimmune European populations. In this chapter, we will describe the plague according to our experiences in the diagnosis and treatment of patients, and we will also provide a comprehensive review of the relevant literature on this subject.

In 1940, G. P. Rudnev [3] classified plague according to its epidemiological features, and this classification is still used to date in Russia:

(A) Mainly local forms (typically peripheral with relatively rare external dissemination):

  (i) Cutaneous
  (ii) Bubonic
  (iii) Cutaneous-bubonic

(B) Internally disseminated or generalized forms:

  (i) Primary septicemic
  (ii) Secondary septicemic

(C) Externally disseminated forms (often with abundant external dissemination):

  (i) Primary pneumonic
  (ii) Secondary pneumonic
  (iii) Intestinal

This classification, in general, meets the practical public health requirements; however, the "pure" cutaneous and intestinal forms of the plague have never been observed in practice.

From a clinician perspective, primary and secondary forms of pneumonic plague may be considered as primary and secondary septicemic (generalized) forms, mainly affecting the lungs. Accordingly, the intestinal form could be considered a septicemic form with a primary lesion of the intestine.

The incubation period for the plague is usually 2–3 days, occasionally 1–5 days, and, rarely, up to 6 days. In the case of insufficient prophylactic antibiotic therapy, the delitescence may increase up to 10 days [4]. However, if the infection is caused by a laboratory accident (a high-dose aerosol infection), the incubation period is only several hours. It is widely believed that a longer incubation period is associated with milder clinical symptoms and better outcomes. Hence, as early as 1926, the international sanitary convention proposed that persons who had come into contact with plague patients should be isolated for medical observation for 6 days [5].

In most cases, the onset of plague is sudden. In rare cases, patients display some preliminary nonspecific symptoms, such as anorexia, pain in the sacral area, weakness, and exhaustion. In some cases, patients appear healthy the day before disease onset but experience feelings of nervousness, anxiety, or depression, falling seriously ill the following day.

The real onset of plague is marked with fever, chills, and headaches. Headaches so closely accompany plague that it became commonly referred to as "head disease" [3]. Patients become anxious and their gait is unsteady, with characteristic hand waving. The skin becomes dry and hot. The face appears red, swollen, and masklike, losing the expression of emotions. There is conjunctival injection, making sparkling eyes appear lustrous but still, and the pupils become slightly dilated. Patients often refuse to be physically evaluated. Speech becomes slurred. Hearing is impaired. In general, the patient gives the impression of being intoxicated. Thus, one patient in the Central Asian region of the former Soviet Union was refused admission by a physician-therapist because of strange behavior thought to be linked to intoxication with alcohol or drugs. That patient was found dead on a bench near the hospital later the same day, and the diagnosis at postmortem was the plague.

Other symptoms include a dry, swollen tongue, covered with chalklike, white limescale, and dry lips. *Herpes labialis* is not characteristic of plague patients. Patients sometimes experience burning pains in the upper and lower abdomen that cannot be relieved by drinking large quantities of cold water. The dissonance between the patient's thirst and shaking chills is also noteworthy. There may be repeated vomiting and/or tachycardia and a decrease in blood pressure. At this stage, plague patients look similar in appearance to typhoid fever patients with *status typhosus*.

The initial stages of the disease do not exhibit any unique symptoms. Clinicians can therefore often only diagnose a severe general intoxication syndrome. Plague may be suspected due to epidemiological anamnesis, but no definitive diagnosis can be made. However, as the disease progresses, the skin lesion caused by *Y. pestis* may develop into a carbuncle (plague ulcer), and this primary carbuncle (at the site of the bite), secondary carbuncles (due to hematogenous dissemination) may also develop. At first, an intensive crimson fleck that looks like an insect bite emerges on the skin. The skin in this area becomes very painful. Over the following 24 h, the skin hardens (first only detectable by touch), and then the affected area increases in size and becomes visible, rising above the surface (forming a papule). The pain intensifies, and a so-called plague flictena, a small pea-sized bubble filled with fluid that is yellowish or darker in appearance if blood is present, develops at the center of the papule. In most patients, the envelope of the bubble becomes torn, and a skin ulcer is formed, the edges of which begin to swell and increasingly protrude above the skin surface, forming a pinkish-red inflammatory torus that soon becomes cyanotic. The ulcer becomes deeply embedded in the skin and at the surface becomes encased by a dark scab. The skin lesion is now referred to as a plague carbuncle (Fig. 11.1a). New vesicles may be formed along the edge of this ulcer, making plague carbuncles appear similar to the ulcers associated with anthrax. The inflammation expands to the periphery over a period of less than 2 days, causing eccentric growth of the ulcer, with carbuncles considerably varying in size from 0.5 to >5.0 cm in diameter. There may be only one or several primary and/or secondary carbuncles. Two or more secondary skin lesions may form due to hematogenous dissemination of the pathogen, whereas the presence of several primary carbuncles may result from multiple simultaneous infections caused by repeated bites by one or several infected fleas.

The healing of carbuncles occurs extremely slowly. It begins with the appearance of a demarcation zone characterized by the release of moderate amounts of pus at the edges of the scab. The subsequent casting off of the scab and the outcrop of a granulating surface is followed by the formation of rough scars.

Although the existence of "pure" skin plague has been proposed, whereby the pathogen does not spread to the nearby lymph nodes, we have no evidence for this in clinical practice, and there is no consistent data on this proposed phenomenon in the literature.

As a rule, cutaneous lesions in plague patients develop at the same time as, or a little earlier than, a typical bubo, which is located close to the ulcer. For example, crus ulcers are accompanied by inguinal buboes, while hand ulcers are accompanied by axillary lymphadenitis. So, the illness initially presents in cutaneous-bubonic form (Fig. 11.1b), which clinically is identical in terms of its main manifestations to the solely bubonic form that will be discussed below.

Bubonic plague is the most frequent and well-described form of the disease. The early signs of an emerging bubo are painful sensitivity of the inguinal, axillary, cervical or submandibular areas that can be detected by careful medical examination within the first few hours of the disease. Usually there are no obvious insect bite marks at the site of infection or signs of lymphangitis. In the regional lymph nodes, however, one or two swollen and extremely painful lymph nodes may be noted. As the disease progresses, a "tumor" (typically, singular) grows and lymph nodes cluster into a dense conglomerate, which appears linked to the skin, making it impossible to palpate individual nodes. The bubo varies in size from a hazelnut to a hen's egg or even larger (according to historical descriptions, it may reach the size of a

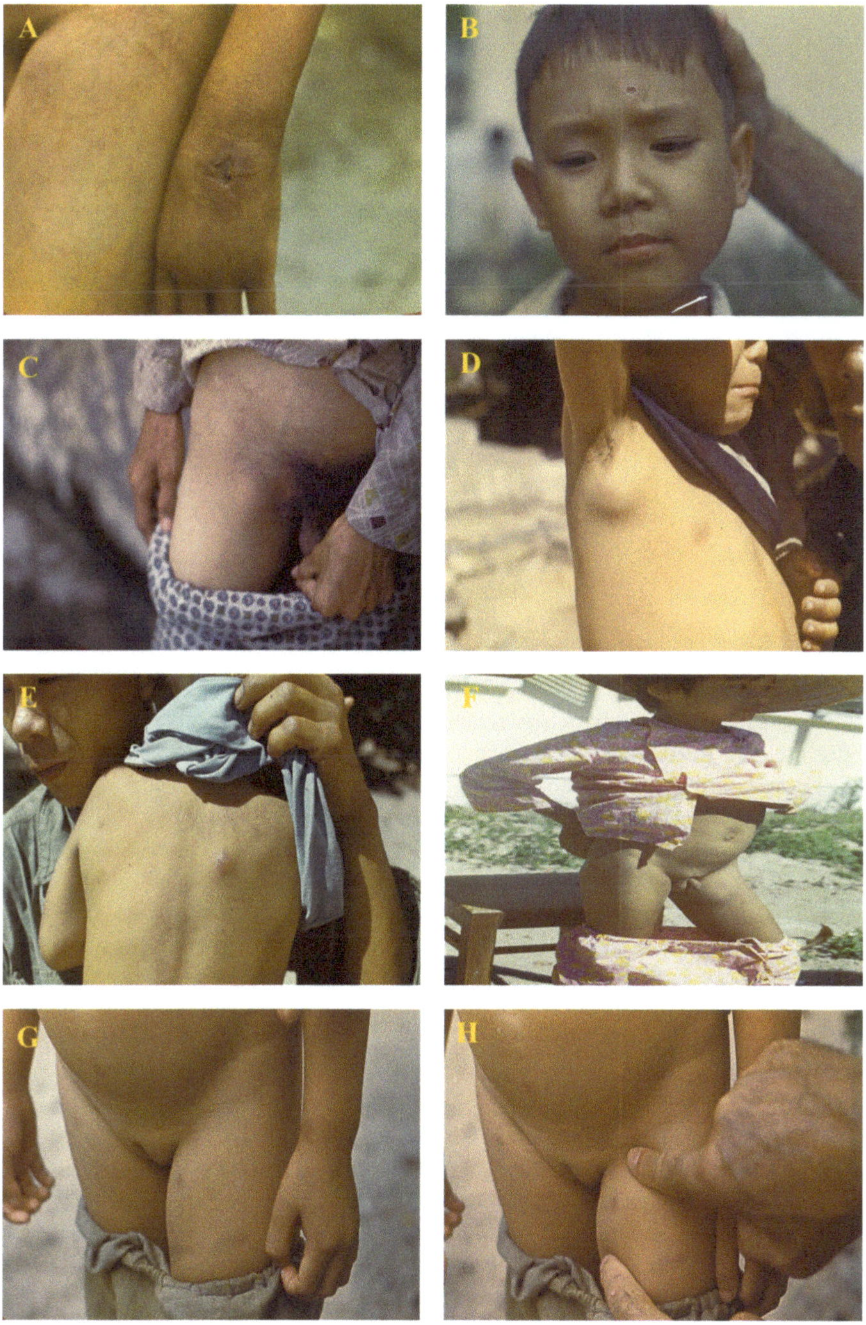

Fig. 11.1 Clinical manifestation of bubonic plague

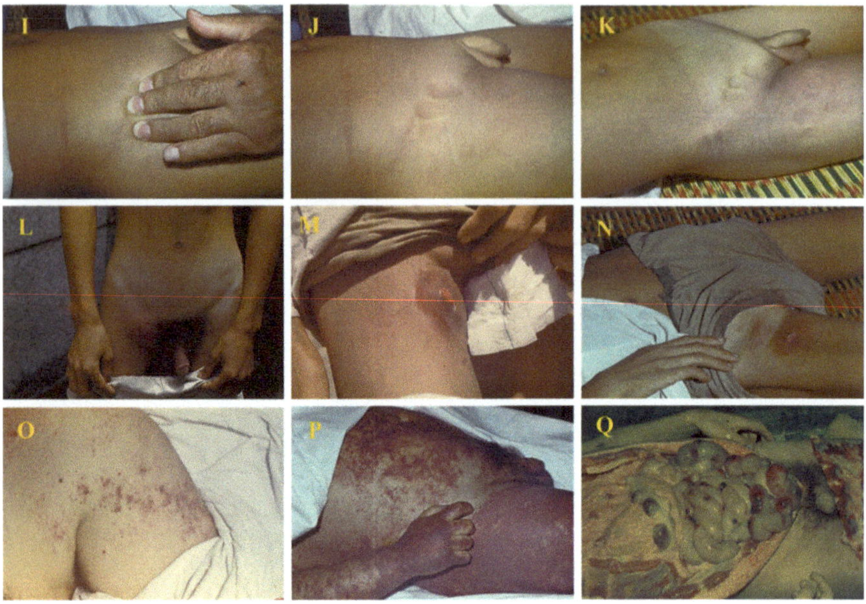

Fig. 11.2 Clinical manifestation of bubonic plague

newborn's head). It has sharp borders and is fixed and very painful. The severe pain often compels the patient to remain in a forced position. During the first days of the disease, the skin over the bubo remains unchanged (Fig. 11.1c). The most frequent locations of the buboes are the inguinal region (>50 % of all observations), the axillary area (c. 20 %) (Fig. 11.1d), and the neck (c. 5 %), with buboes on the elbow and popliteal region occurring rarely and even less frequently elsewhere on the body (Fig. 11.1e). Usually a patient displays only one bubo; however, there are exceptions as in the case of plague carbuncles (Fig. 11.1f). The simultaneous development of buboes at the groin, axilla, and neck is very rare. Most often, the affected area in the groin involves the lymph nodes lying 2–3 fingers below the femoral ring in the triangular space between the *sartorius* and *adductor longus* muscles. The higher frequency of femoral-inguinal buboes can be explained by the large quantity of lymph collected here due to the considerable skin surface area encompassed compared with other draining nodes. The general condition of patients progressively worsens as a bubo approaches the head. During the development of periadenitis, the skin over the bubo becomes denser making it impossible to fold, and inflammatory swelling of the skin gradually develops. The true size and boundaries of the bubo become uncertain because of compulsory periadenitis (Fig. 11.1g, h). A doughlike edema may spread a considerable distance from the bubo (e.g., above the inguinal ligament if the bubo is femoral). If one presses the swollen region, an indented impression is clearly observed (Fig. 11.2a–c). For axillary lymphadenitis that easily develops into secondary pneumonic plague, the edema can expand to the breast area, sometimes producing trembling from palpation of the "gelatinous tumor" [4]. Buboes located in the neck area generally correlate with severe symptoms and poor patient outcomes. The lymph glands enlarge rapidly and the hypodermis swells, forming a doughlike tumor, which becomes infiltrated and indurated during the course of periadenitis. Swelling can spread at the front down to the mammary glands and from behind down to the scapular area. The disease peaks 4–6 days after onset. Fever remains at 39–40 °C, with symptoms of intoxication reaching their maximum. The face becomes contorted, sometimes with an expression of anxiety and horror. Pronounced tachycardia, dull heart sounds, and low blood pressure are all characteristics of this stage of the disease. The liver and spleen become enlarged. If the patient survives this period, the buboes may follow a different clinical course. The buboes rarely disappear completely, but rather become sclerotized or suppurate. In the latter case, which often occurs around days 6–8, the skin over the bubo reddens and then becomes cyanotic (Fig. 11.2d). Bubo palpation, which in this period becomes less painful, begins to detect some movement within it. On days 8–10, the bubo bursts with the release of green-yellow homogeneous pus (Fig. 11.2e), sometimes mixed with blood, but with no strong odor. The epidemiological risk from suppurative masses is minimal, as bacteria were difficult to isolate after the first day. The mature purulent cavity becomes gradually filled with granulations and the process of cavity healing can take up to 3–4 weeks or even more. The purulent fistula continues to discharge during this period (Fig. 11.2f). The bubo eventually turns into a rough deep scar. In general, suppuration of a bubo is followed by a significant improvement in the general health of the patient. However, suppuration does not contribute to the favorable course of the disease, as the time necessary for a bubo to reach the stage of purulent melting is also sufficient for the patient to overcome the most difficult period of the illness. Furthermore, the purulent fistula is a potential site of entry for a secondary infection that can cause the development of phlegmonous adenitis.

All of the above applies to the so-called primary or early buboes, i.e., developing at the lymph nodes nearest to the site of infection. The occurrence of secondary or late buboes is possible during the course of any form of plague and at almost any time point owing to the hematogenous dissemination of *Y. pestis* to a distant lymph gland. Secondary buboes are smaller in size than primary buboes, are less painful, and generally do not suppurate, being dissolved in the case of recovery.

Cutaneous-bubonic and bubonic forms of plague, unless treated quickly with effective antibiotics, can cause systemic infection. Secondary pneumonia, secondary buboes, and meningitis, along with other manifestations, can develop when *Y. pestis* enters and multiplies in any secondary organs or target tissues (e.g., the lungs, lymph nodes irrelevant to the primary bubo, meninges). In such cases, the so-called secondary septicemic (secondary generalized) form of the plague develops. However, infectious-toxic shock syndrome (ITSS), which is manifested by a sharp increase in the signs of general intoxication and the emergence of new and specific septic shock symptoms, in particular skin and mucous bleeding, often develops in patients instead of the formation of bacterial metastasis.

In the case of systemic infection without the formation of a detectable primary lesion (e.g., bubo), the patient is usually diagnosed with primary septicemic plague.

However, if the development of secondary septicemic plague is suspected from the pathogenetic profile, the reason for the development of primary generalized forms of plague is unclear. The term "primary septicemic plague" implies the absence of a preliminary stage of *Y. pestis* accumulation in a regional bubo prior to its breakthrough into the blood. Whereas, real primary systemic plague may result from (i) a large infecting dose, (ii) unusually high virulence of the pathogen (both may be due to a laboratory-acquired infection or a bioterrorist attack), as well as (iii) the immunocompromised status of the patient. Other cases of primary generalized forms that do not adhere to these criteria remain to be clarified.

The issue of appearance of *Y. pestis* in the blood at different stages of the disease warrants further examination. In Vietnam during the 1980s, *Y. pestis* was often isolated from the blood of patients with the bubonic form of plague (or observed by microscopy of blood smears), whose general condition could be considered neither critical nor even severe. The presence of *Y. pestis* in the blood, regardless of the severity of the disease, was also noted earlier by other clinicians [6]. Thus, the detection of *Y. pestis* in the blood of a patient is not reason alone for diagnosing the septic form of the disease. In this regard, it remains a mystery where the distinction lies between a "simple" bacteremia and a systemic infectious bacteremia as the pathogenic basis for the generalized forms of plague (septicemia). This issue requires further clarification.

In the past, different terminology has been used to describe the cause of death in plague patients, especially during the first week of the disease with either the bubonic or pneumonic form; such terms have included "progressive heart failure," "sudden loss of strength," and "acute collapse." Such terms are still cited in some modern manuals [7]. However, plague and all that it involves clearly need demystification. So from the point of view of a modern clinician, all clinical manifestations of the terminal phase of plague are consistent with an ITSS caused by Gram-negative bacteria with some plague-specific features (see below).

Meningitis can appear at any stage of any form of plague infection. However, meningitis usually occurs either during severe forms of the disease, aggravating the main pathogenic process, or during the period of reversal of the symptoms. In the pre-antibiotic era, meningitis was frequently the cause of death, whereas today, the prognosis is somewhat more optimistic [8, 9]. The development of meningitis is associated with worsening of the patient's condition, increasing headaches, loss of consciousness, appearance of meningeal symptoms, and the characteristic positioning of patients (the classic "meningeal" posture), which may persist even after death. However, we have observed cases in which meningitis developed during the convalescent period of bubonic plague and was considered to be a relapse, without being deadly or particularly severe. Cerebrospinal fluid analysis indicates the purulent nature of inflammation, and *Y. pestis* can often be isolated from cerebrospinal fluid.

Secondary pneumonic plague is more common, developing in 8–10 % of bubonic plague patients. In addition to a worsening prognosis, the development of pneumonia increases the risk of pathogen transmission via coughing and breathing. As in the case of plague meningitis, the development of secondary plague pneumonia is possible at any stage of the infectious process and leads to a significant worsening of the patient's condition. Even when the fever subsides, the body temperature may return to 39 °C or higher. The involvement of the lungs in the infectious process is accompanied by the appearance of a cough and chest pain while breathing due to the development of pleurisy. It should be emphasized that pain (including headaches) is a major component of the plague infectious process and the absence of pain makes a diagnosis of plague unlikely, or even completely rejected.

Dry coughing is rarely seen in pneumonic plague. Scant and glassy sputum begins to excrete almost immediately. Then after about a day, the sputum quantity increases and it becomes contaminated with blood. On examination, lobar pneumonia of any etiology may be detected with localization of the lesions mostly in the middle lobe of the right lung or the upper lobes of both lungs. Such "atypical" localization of inflammatory foci allows the clinician to suspect that the pneumonia is caused by hematogenous dissemination of the pathogen, as more trivial secondary infections of the lungs caused by adhering microorganisms have the tendency to be localized mainly in the lower lobes.

More or less similar descriptions of the clinical profile of primary pneumonic plague can be found in many textbooks, manuals, and monographs. In Russian literature, the original detailed description of primary pneumonic plague was provided by G. P. Rudnev in 1940 [3]. Because there are few observations of this type of plague in the literature, this text is important, and we insert below practically word-for-word translation of its most recent adapted version [4].

G. P. Rudnev divided the disease into three stages: an initial stage, a peak stage, and a terminal stage (with progressive dyspnea, cyanosis, and sometimes coma). The second stage was considered the most dangerous in terms of transmission because patients develop a cough, expelling the pathogens as a respiratory spray.

The clinical picture, especially during the first hours of the disease, may be quite different. Usually, the disease has a rapid onset involving an abrupt and repeated fever, rapidly rising body temperature, an extremely severe headache, and frequent repeated vomiting. This may be followed by "knife-cutting" chest pain, palpitation, rapid pulse, severe dyspnea, and delirium. Later, the patient may collapse and fall into a coma, and this may subsequently lead to death. Coughing is associated with all stages of the disease, with variations in the quantity of sputum, from several spits to tens of liters. For a few patients, sputum production does not occur. Between the two types of pneumonic plague, wet (with sputum production in large quantities) and dry (without sputum production), there is a lot of variability in terms of sputum amount and appearance (sputum with odor, vitreous, liquid, or blood stained). In the final stage, the sputum is usually pure blood. In general, sputum viscosity is used as one of the diagnostic markers. In some nontypical cases, sputum may be rusty in appearance, for example, in croupous pneumonia or tuberculosis-like pneumonia. In extremely severe cases, palpation and auscultation are not particularly informative.

At the peak of primary pneumonic plague, patients usually manifest with depression followed by excitement, delirium, high fever, associated mild pneumonia, frequent coughing with bloody sputum, muffled heart sounds, excessive tachycardia, arrhythmia, and often vomiting of blood. Finally, sopor develops, shortness of breath increases, and the patient's faces become cyanotic as if suffocating, exhausted by a back-breaking struggle. At this stage, the physical strength of the patient is fading and the pulse becomes faster and weaker. Some patients fall into a coma; some die during repeated attempts to stand up and run away. These attempts to stand and run are very characteristic of plague delirium. The disease lasts only 3–5 days and without treatment is lethal. Just before death, the body temperature of some patients drops sharply to normal.

Analysis of the clinical profile described above from a contemporary point of view indicates that the final stages of the development of primary pneumonic plague are consistent with ITSS, along with all of the ensuing negative consequences for the patient. Recognition of this fact is extremely important because it emphasizes that any reduction in plague mortality is unlikely without significant progress in the treatment of ITSS [10].

As for primary septicemic plague, it differs clinically from primary pneumonic plague only by the absence of changes in the lungs and by a shorter clinical course (usually, <3 days from onset to death). We fully support the assertion of N. N. Zhukov-Verezhnikov [11] that this diagnosis is based only on laboratory data, because this form of plague has no specific clinical manifestations. We can speculate that both of these disease forms have identical pathogenic mechanisms, the only difference being that death in the case of primary septicemic plague occurs prior to the development of inflammatory lesions in the lungs.

Functional gastrointestinal disorders such as vomiting and diarrhea are frequently associated with severe and generalized forms of plague. It is often possible to isolate *Y. pestis* cultures from the feces of such patients. In some cases, the symptoms (e.g., nausea, vomiting, abdominal pain, diarrhea with mucus and blood) may, at some stage of the disease, become the dominant characteristic, tempting the classification of an "intestinal" form of plague. However, the generalized signs and symptoms associated with the infectious process in the end prevail over intestinal symptoms. So, we should speak not about a separate form of the disease, but about a systemic plague, mainly affecting the intestine.

It is well established that ITSS plays a crucial role in the mortality associated with the plague, and it displays features common to all types of septic shock caused by Gram-negative bacteria as well as some specific features. As is the case in other similar types of shock, tachycardia and tachypnea increase, body temperature and blood pressure decrease, and urination ceases. Other possible symptoms are nasal and gastrointestinal bleeding, nephrorrhagia, and metrorrhagia, but these are not common in uncomplicated cases of the plague. The peculiarities of the clinical manifestations of the plague are the unusually early and abundant multiple bluish-black hemorrhages (bruises) with a leaden hue that appear on the skin and mucosa. Their size may vary from a dot (Fig. 11.2g) or line to a much wider area, and they are commonly referred to as "spots of death" or *maculae mortis* (Fig. 11.2h). The similar signs can be seen on mucosal membranes and in the internal organs [12] (Fig. 11.2i). This particular characteristic feature together with the high mortality rate was the reason why the disease became known as the "Black Death." In addition,

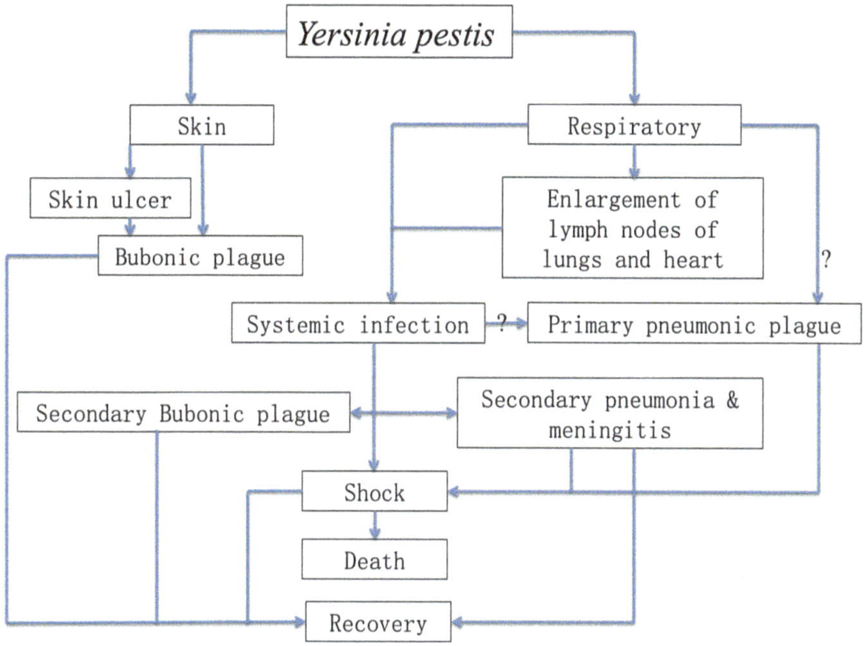

Fig. 11.3 Sketch map for the main clinical and pathogenic stages of the plague

the facial expression of a person dying from the plague is distinctive. G. P. Rudnev (1940) described the *facies pestica* as suffering and cyanotic, with an expression of horror and with sharp features and abundant sweat drops on the forehead ("dew of death").

Another characteristic feature of patients in the final stages of the plague is their behavior. As mentioned earlier, if patients are able to move, they always try to escape their house and die curled up at the gate of their own home [3], just as animals always die at the entrance of their burrows [13]. It is therefore important for medical staff in emergency care units to restrict patients from leaving.

In the absence of effective treatment, the human plague generally always leads to severe symptoms. However, some plague patients display relatively mild clinical symptoms, and such a variant of bubonic plague is named "pestis minor" or "pestis ambulans." Also, the possibility of asymptomatic (subclinical) plague should be considered and even the existence of a carrier state [3]. It may be that the plague actually occurs more frequently than it is diagnosed.

Figure 11.3 is a sketch map for the main clinical and pathogenic stages of the plague.

Recovery from the plague is a long process, often interspersed with periods of improvement and deterioration in the patient's general state of health. Complications of a nonspecific nature, such as nonspecific pneumonia, phlebitis, stomatitis, gingivitis, otitis, and erysipelas, as well as exacerbation of preexisting chronic diseases are possible.

It is not difficult to diagnose clinically cutaneous-bubonic or bubonic plague, but diagnosis of primary pneumonic plague is more challenging and diagnosis of primary septicemic plague is impossible. We do not see the benefit in discussing the differential diagnosis of the plague because the list of possible nosological forms of the disease would be considerable (from tularemia, anthrax, and glanders to meningococcemia). If a clinician is familiar with infectious pathology in general and in particular the clinical manifestations of plague, differentiating the specific form of the disease is not necessary. By contrast, if a clinician is not familiar with such rare diseases, there is also no benefit of a differential diagnosis.

For plague patients, blood and urine examinations are rarely performed. In routine blood examinations, the only change usually detected is an increase in white blood cells (25,000/mm^3), which is not normally seen in other infectious diseases. In routine urine examinations, elevated protein levels may be observed ("fever urine"). Microscopic examination of a whole blood smear could confirm the presence of *Y. pestis* unequivocally.

11.2 Rapid Detection of *Y. pestis* on Site

Alexandre Yersin was the first to discover the plague bacillus and paved the way for modern laboratory diagnosis of the disease. In 1894, he performed isolation of *Y. pestis* pure culture and identified it as the causative agent of plague using (i) examination of smears prepared from autopsy material (buboes) and pure cultures, (ii) bacterial culture on solid media, and (iii) infection of laboratory animals with autopsy material and pure cultures [14].

Laboratory diagnosis is problematic in many plague-affected countries, especially in the remote areas of developing countries, where the logistics, infrastructure, and resources are limited. Rapid and reliable detection of *Y. pestis* on site is essential for timely initiation of medical treatment and postexposure prophylactic measures when plague cases are suspected. However, the reference standard for confirmation of plague remains the isolation of *Y. pestis*, which is time-consuming and not possible in the field. Simple tools that require limited resources and allow for the rapid detection of *Y. pestis* on site are needed for the diagnosis of plague cases.

11.2.1 Immunological Diagnosis: Development of an Immunochromatographic Assay

Immunochromatographic assays (ICA) have become popular bedside or point-of-care (POC) diagnostic tools because they are sensitive, simple to perform, inexpensive to manufacture, and well suited for rapid on-site detection and can be performed in a variety of settings by nontechnical personnel. The Institute Pasteur of

Madagascar has developed an ICA that uses monoclonal antibodies to the F1 antigen of *Y. pestis* for the rapid diagnosis of pneumonic and bubonic plague [15]. Samples from patients suspected of plague were tested using the assay both in the laboratory and by health workers at 26 pilot sites in Madagascar. The assay detected concentrations of F1 antigen as low as 0.5 ng/mL within 15 min. With the combination of bacteriological methods and F1 ELISA as a reference standard, the positive and negative predictive values of the test were 90.6 % and 86.7 %, respectively. The agreement rate between assays performed at remote centers and those carried out in the laboratory was 89.8 %. Like other ICA assays using colloidal gold particles as the reporter, it is a semiquantitative test involving manual reading and a subjective threshold. The test must therefore be performed by professional staff.

Up-converting phosphor (UCP) technology-based ICA (UPT-ICA) has attracted considerable attention because of its advantages of quantification, reliability, and robustness [16]. UCPs are a type of rare-earth-containing crystal particles with the unique property of up-converting infrared excitation light (980 nm) to emit visible light. These unique optical properties render UCPs ideal labels and are more sensitive than other conventional reporters. UPT-ICAs for the rapid diagnosis of *Y. pestis* have been developed with high specificity and a low limit of detection of 10^4 CFU/mL (100 CFU/test) [17, 18]. This type of assay is applicable in the field with excellent tolerance to various complex samples, such as blood, viscera, (fresh and decomposed) and powders. Samples with a wide pH range (2–12), high ion strength, high viscosity, or the presence of bio-macromolecules could also be directly detected through simple dissolution or homogenization. The detection and quantification of *Y. pestis* with UPT-ICA can be performed using a biosensor reader in 15 min. Furthermore, it can be operated by nonprofessional staff.

11.2.2 Molecular Diagnosis by Real-Time PCR

Real-time PCR tests that target the structural gene for F1 antigen (*caf1*), the plasminogen activator gene (*pla*), and the gene encoding the specific chromosomal fragment (3a) have been developed and may prove useful as specific and rapid tests for plague diagnosis [19, 20]. The time required for the assay and the detection limit are less than those of conventional PCR amplification, and the risk of cross-contamination between amplification products is also reduced because post-amplification procedures are not required. Moreover, the assay could be adapted for use in the field by employing portable real-time PCR instruments and reagents able to be stored at room temperature. More recently, it was shown that a *pla*- and 3a-based PCR method could not reliably identify *Y. pestis* [2, 21–23] and that multilocus detection methods should be developed for this pathogen.

A novel portable real-time PCR thermocycler, PikoReal™ (Thermo Fisher Scientific Inc., USA), has been developed for detecting *Y. pestis* in the field [24]. The PikoReal system weighs less than the common PCR instruments (about 10 kg) and is equipped with five LED-illuminated optical channels, which increase durability when compared with standard light technologies, facilitating multiplexing. Analyses of *Y. pestis* with the PikoReal system proved to be simple and time-efficient compared with assays using the Applied Biosystems® 7300 equipment (Life Technologies Ltd., USA). In field trials, reliable results were achieved in approximately 90 min, from the beginning of sample preparation (45 min) to the completion of the diagnosis. All assay-specific primers and probes, in both the PikoReal and the ABI 7300 systems, correctly identified all tested isolates of *Y. pestis*, while no cross-reactivity was observed.

Real-time PCR reagents able to be stored at room temperature have also been developed using primers and probes targeting the 3a sequence and *caf1* in *Y. pestis* [20]. Carbohydrate mixtures were added to the PCR reagents, which were later vacuum-dried and evaluated for stability. The vacuum-dried reagents were stable at 37 °C for at least 49 days at a lower concentration of template DNA (10 copies/μl) and up to 79 days at a higher concentration ($\geq 10^2$ copies/μl). Soil samples spiked with *Y. pestis* (5×10^4 CFU/g) could be detected with the dried reagents, indicating they may be used at room temperature for field application.

In the absence of a national strategy for plague diagnosis [25–27], *it is necessary to use the WHO recommendations* [28].

11.3 Identification and Source Tracing of *Y. pestis* in the Laboratory

When the plague is suspected, clinical specimens should be collected immediately and sent to a professional laboratory for confirmation. Diagnosis can be made from a variety of specimens, including blood, aspirates from suspected buboes, pharyngeal swabs, cerebrospinal fluid, and sputum samples. All specimens sent from suspected cases should be labeled as high risk and handled in a biosafety cabinet.

11.3.1 Bacteriologic Tests (The Reference Standard Method)

The reference standard method for laboratory diagnosis of *Y. pestis* infection is based on the isolation and identification of the organism from clinical specimens [29]. The organism can be cultured and grows on most routine solid and liquid culture media (e.g., brain-heart infusion broth, sheep blood agar, or MacConkey agar). The optimal growth temperature of *Y. pestis* is 28 °C. However, standard 35–37 °C incubation is necessary for development of the F1 antigen of *Y. pestis* and for recovery of other pathogens that may be present. *Y. pestis* grow as gray-white, translucent colonies on solid media that are visible after incubation at 25–37 °C for 48 h. Colonies are about 1–2 mm in diameter and may have a raised center with a flat periphery (referred to as a "fried egg" or "hammered copper" appearance). *Y. pestis* is a pleomorphic Gram-negative rod. On staining of a smear with Giemsa or Wayson stain, this organism appears as bipolar coccobacilli. However, bipolar staining is not unique for *Y. pestis* and therefore is considered only suggestive of a diagnosis. Cultures can be conclusively identified as *Y. pestis* by specific phage lysis. Unlike F1 antigen expression, *Y. pestis* is susceptible to phage lysis at both 25 °C and 37 °C. Conventional biochemical identification systems can be used to assist in identification, but misidentification with other organisms, such as *Y. pseudotuberculosis* or other Enterobacteriaceae, is possible [30].

11.3.2 F1 Antigen Detection and Serological Diagnostic Techniques

In the absence of a definite *Y. pestis* isolate, a diagnosis can also be made using serological diagnostic techniques. A fourfold or greater change in the titer of antibodies to *Y. pestis* F1 antigen in a passive hemagglutination test of paired serum specimens is confirmatory for *Y. pestis*. The specificity of a positive passive hemagglutination test requires confirmation with the F1 antigen hemagglutination-inhibition test. Hemagglutinating antibodies directed against the F1 antigen of *Y. pestis* can appear as early as 5 days after the onset of symptoms but more commonly appear between 1 and 2 weeks after onset [30].

Detection of the F1 antigen in tissues or fluids by direct fluorescent antibody testing is considered an indicator of the presence of *Y. pestis*. Similarly, detection of an elevated F1 antibody titer (more than 1:10) in a single serum sample from a patient displaying plague-like symptoms who has not been vaccinated previously and has no history of *Y. pestis* infection is a positive diagnostic indicator. PCR analysis using primers specific for the gene encoding F1 antigen has also been reported as a method of *Y. pestis* identification. Furthermore, enzyme-linked immunosorbent assays (ELISAs) can be employed to detect *Y. pestis* and measure F1 antigen levels or levels of serum antibodies to F1 antigen [31]. An ELISA is available to measure antibodies against F1 antigen that involves both immunoglobulin M (IgM), indicative of a recent or current infection, and immunoglobulin G (IgG), indicative of a past infection.

11.4 Diagnostic Criteria for Plague

The standard case definition of plague (according to the WHO, weekly Epidemiological Record, no 28, July 2006) is as follows:

1. *Suspected case*: compatible clinical presentation and consistent epidemiological features, including exposure to infected animals or humans, and/or evidence of flea bites, and/or residence in or travel to a known endemic focus area within the previous 10 days

2. *Presumptive case*: meeting the definition of a suspected case *plus*:

(A) *Putative new or reemerging focus: at least two of the following tests proved positive:*

- Microscopy: material from bubo, blood, or sputum contains Gram-negative coccobacilli, bipolar after Wayson, or Giemsa staining
- F1 antigen detected in bubo aspirate, blood, or sputum
- A single anti-F1 serology without evidence of previous *Y. pestis* infection or immunization
- PCR detection of *Y. pestis* in bubo aspirate, blood, or sputum

(B) *Known endemic focus: at least one of the following tests proved positive:*

- Microscopic evidence from bubo, blood, or sputum sample of Gram-negative coccobacilli or bipolar coccobacilli observed after Wayson or Giemsa
- A single anti-F1 serology without evidence of previous plague infection or immunization
- F1 antigen detected in bubo aspirate, blood, or sputum
- PCR detection of *Y. pestis* in bubo aspirate, blood, or sputum

3. *Confirmed case*: meeting the definition of suspected case *plus*:

- An isolate from a clinical sample identified as *Y. pestis* by colony morphology and a positive result for at least two of the following four tests: phage lysis of cultures at 20–25 °C and 37 °C, F1 antigen detection, PCR, *Y. pestis* biochemical profile
- A fourfold rise in the anti-F1 antibody titer in paired serum samples
- In endemic areas when no other confirmatory test can be performed, a positive rapid diagnostic test with immunochromatography to detect F1 antigen

11.5 Treatment of Plague

In the absence of a national strategy for plague treatment [27], *it is necessary to use the WHO recommendations* [32].

Plague treatment involves two components: specific antibiotic treatment and nonspecific supporting therapy, such as antishock therapy. As mentioned previously, before the age of antibiotics, the mortality rate from bubonic plague was more than 50 %, with almost 100 % lethality reported for untreated cases of primary plague meningitis, pneumonia, or septicemia. Early diagnosis and the prompt initiation of treatment reduce the mortality rate associated with bubonic plague and septicemic plague to 5–50 % [33, 34], although a delay of more than 24 h in the administration of antibiotics and antishock treatment can be fatal for plague patients.

Antimicrobial therapy is the simplest component of the complex therapy required for plague patients. Streptomycin at a dose of 1.0 g twice a day (30 mg/kg/day) for

Table 11.1 Plague treatment [40] [recommended by the CDC]

Preferred for adults (including pregnant women and immunocompromised adults)	
Streptomycin	1 g, i.m., b.i.d.
Gentamicin	5 mg/kg, i.m. or i.v. daily, or 2 mg/kg loading with 1.7 mg/kg, i.m. or i.v., t.i.d.
Preferred for children (including immunocompromised children)	
Streptomycin	15 mg/kg, i.m., b.i.d. (maximum daily dose 2 g)
Gentamicin	2.5 mg/kg, i.m., or i.v., t.i.d.
Alternatives for adults	
Doxycycline	100 mg, i.v., b.i.d., or 200 mg, i.v.
Ciprofloxacin	400 mg, i.v., b.i.d.
Chloramphenicol	25 mg/kg, i.v., q.i.d.
Alternatives for children (including immunocompromised children)	
Doxycycline	If ≥ 45 kg, adult dosage; If < 45 kg, 2.2 mg/kg, i.v., b.i.d. (maximum 200 mg/dl)
Ciprofloxacin[a]	15 mg/kg, i.v., b.i.d.
Chloramphenicol	25 mg/kg, i.v., q.i.d.

[a]In Russia, this antibiotic is not permitted for use in children under the age of 15 years

10 days continues to be considered the most effective anti-plague remedy. There is also reason to believe that gentamicin will be highly effective at standard dosage for the management of patients with severe sepsis. In randomized comparative studies in Tanzania, a 7-day course of intramuscular therapy with gentamicin was equally effective against bubonic, septicemic, and pneumonic plague, as the oral use of doxycycline in both adults and children [35, 36]. Chloramphenicol is recommended for the treatment of plague meningitis. In all cases, antibiotic therapy should be continued for 10 days.

In the absence of clinical trials of various antimicrobials for the treatment of primary pneumonic plague in humans, the existing recommendations are based solely on the results of in vitro and animal experiments. The remedies of choice both in adults and children are the aminoglycosides, streptomycin, or gentamicin, at age-appropriate doses [35, 37] (Table 11.1).

Doxycycline, chloramphenicol, and ciprofloxacin can also be used as alternatives. When treating a limited number of plague patients, all of these drugs are administered parenterally, whereas during a plague outbreak involving a large number of patients and people who may have come into contact with the plague, oral administration of antibiotics is employed. In this situation, preference is given to doxycycline, which is administered at an oral dose of 100 mg twice a day or ciprofloxacin at an oral dose of 500 mg twice a day (Table 11.2). Chloramphenicol is administered orally at a dose of 25 mg/kg four times a day; however, this drug is not permitted for use in pregnant women owing to the high risk of toxic effects on the fetus [35, 37].

An adequate intensive pathogenetic treatment of plague [38, 39], and especially antishock therapy, requires much more knowledge and skills. ITSS is initially trig-

Table 11.2 Plague treatment in a mass casualty setting [40] [recommended by the CDC]

Preferred for adults (including pregnant women and immunocompromised adults)	
Doxycycline	100 mg, orally, b.i.d.
Ciprofloxacin	500 mg, orally, b.i.d.
Preferred for children (including immunocompromised children)	
Doxycycline	If ≥ 45 kg, adult dosage; If < 45 kg, 2.2 mg/kg orally, b.i.d.
Ciprofloxacin	20 mg/kg, orally, b.i.d.
Alternatives for adults	
Chloramphenicol	25 mg/kg, orally, q.i.d.
Alternatives for children (including immunocompromised children)	
Chloramphenicol	25 mg/kg, orally, q.i.d.
Ciprofloxacin[a]	15 mg/kg, i.v., b.i.d.

[a]In Russia, this antibiotic is not permitted for use in children under the age of 15 years

gered by bacterial toxic substances but is then a self-sustaining, progressive process. With simplification, the relationship between *Y. pestis* and ITSS can be explained using the analogy of a match and a fire. The match can easily light the fire, but quenching the match alone will not extinguish the fire. Removal of the causative agent is necessary for effective treatment of the plague but does not quell ITSS. The aim of intensive therapy is the maintenance of a patient's core indicators at the following levels:

- Central venous pressure: 8–12 mm Hg
- Mean arterial pressure: ≥65 mm Hg
- Urination: ≥0.5 ml/kg/h
- Degree of blood oxygen saturation: ≥70 %

Comprehensive antishock therapy for plague patients should include antibiotic treatment, infusion therapy, administration of vasopressors and corticosteroids, inotropy, and respiratory support (if necessary). Infusion therapy is carried out using a combination of crystalloids (e.g., physiological saline solution or plasma-Lyte 148) and colloids (modified fluid gelatin: gelofusine 4 %, polygeline, hydroxyethyl starch 200–500 ml of a 6 % or 10 % solution) at a ratio of 1:1.

Vasopressors, norepinephrine and dopamine, are the first choice drugs. The initial dosage for dopamine is 5 ml of a 4 % solution added to 200 ml of physiological saline, which is infused until the mean arterial pressure reaches 65 mm Hg. The dosage for norepinephrine is 2–5 μg/min, and this can be used in combination with dopamine.

Inotropic dobutaminum could be used for patients with reduced cardiac output at a dosage of 250 mg in 500 ml of physiological saline, potentially in combination with vasopressors at a calculated dosage is 2.5–40.0 μg/kg/min. Steroids (hydrocortisone, 240–300 mg/day for 5–7 days) can hasten hemodynamics to a normal level. We speculate that the use of extracorporeal purification methods, such as plasmapheresis or hemo-/plasma filtration, as used in cases of extreme hemorrhagic fever, or transfusion of erythrocytes en masse, would be an appropriate therapeutic strategy in the case of plague.

During intensive care, patients should be provided 25–30 kilocalories/kg/day. The level of blood glucose should be maintained at 4.5–6.1 mM/L. Phlebothrombosis and stress-induced gastrointestinal ulcers should be prevented during the entire treatment process. For all types of plague, only fully recovered patients with negative bacteriological results can be discharged. Patients who have suffered from bubonic plague can be discharged after repeated lymph node examinations at a 2-day interval, and pneumonic plague patients should undergo triplicate sputum examinations.

References

1. Williams JE, Harrison DN, Quan TJ, Mullins JL, Barnes AM, Cavanaugh DC. Atypical plague bacilli isolated from rodents, fleas, and man. Am J Public Health. 1978;68(3):262–4.
2. Anisimov AP, Lindler LE, Pier GB. Intraspecific diversity of Yersinia pestis. Clin Microbiol Rev. 2004;17(2):434–64.
3. Rudnev GP. Clinical picture of plague. Moscow: Medgiz; 1940.
4. Rudnev GP. Clinical picture of particularly dangerous infections. Moscow: Meditsina; 1966.
5. Text in League of Nations Treaty Series. 1926;78:230–349.
6. Nikanorov SM. Plague. In: Guidance on acute infectious diseases. Moscow: Leningrad; 1931.
7. Luchshev VI, Zharov SN. Selected lectures on infectious diseases and epidemiology. Moscow: Meditsina; 2004.
8. Butler T. Plague and other Yersinia infections. New York: Plenum Press; 1983.
9. Becker T, Polan JD, Quan T. Plague meningitis – a retrospective analysis of cases reported in the United States 1970–1979. WestJMed. 1987;147:554–7.
10. Kumar A. An alternate pathophysiologic paradigm of sepsis and septic shock: implications for optimizing antimicrobial therapy. Virulence. 2014;5(1):80–97.
11. Zhukov-Verezhnikov NN. The diagnosis of plague and cholera. Moscow: Medgiz; 1943.
12. Chalisov IA, Khazanov AT. Guide to pathologoanatomic diagnosis of the most important infectious diseases of man. Medicine: Leningrad; 1980.
13. Cully Jr JF, Barnes AM, Quan TJ, Maupin G. Dynamics of plague in a Gunnison's prairie dog colony complex from New Mexico. J Wildl Dis. 1997;33(4):706–19.
14. Hawgood BJ. Alexandre Yersin: discoverer of the plague bacillus, explorer and agronomist. J Med Biogr. 2008;16(3):167–72.
15. Chanteau S, Rahalison L, Ralafiarisoa L, Foulon J, Ratsitorahina M, Ratsifasoamanana L, Carniel E, Nato F. Development and testing of a rapid diagnostic test for bubonic and pneumonic plague. Lancet. 2003;361(9353):211–6.
16. Hampl J, Hall M, Mufti NA, Yao YM, MacQueen DB, Wright WH, Cooper DE. Upconverting phosphor reporters in immunochromatographic assays. Anal Biochem. 2001;288(2):176–87.
17. Yan ZQ, Zhou L, Zhao YK, Wang J, Huang LH, Hu KX, Liu HH, Wang H, Guo ZB, Song YJ, et al. Rapid quantitative detection of Yersinia pestis by lateral-flow immunoassay and up-converting phosphor technology-based biosensor. Sensor Actuat B-Chem. 2006;119(2):656–63.

18. Zhang PP, Liu X, Wang CB, Zhao Y, Hua F, Li CF, Yang RF, Zhou L. Evaluation of up-converting phosphor technology-based lateral flow strips for rapid detection of Bacillus anthracis Spore, Brucella spp., and Yersinia pestis. Plos One. 2014;9(8):e105305.
19. Butler T. Plague into the 21st century. Clin Infect Dis. 2009;49(5):736–42.
20. Qu S, Shi QH, Zhou L, Guo ZB, Zhou DS, Zhai JH, Yang RF. Ambient stable quantitative PCR reagents for the detection of Yersinia pestis. Plos Neglect Trop D. 2010;4(3):e629.
21. Qi Z, Wu Y, Li Y, Li C, Yang X, Zhang Q, Xin Y, Jin Y, Wei R, Cui Y. 3a-negative Yersinia pestis, China. Infect Dis Transl Med. 2015;1(2):61–2. doi:10.11979/idtm.201502004.
22. Hänsch S, Cilli E, Catalano G, Gruppioni G, Bianucci R, Stenseth NC, Bramanti B, Pallen MJ. The pla gene, encoding plasminogen activator, is not specific to Yersinia pestis. BMC Res Notes. 2004;8:535. doi:10.1186/s13104-13015-11525-x.
23. Armougom F, Bitam I, Croce O, Merhej V, Barassi L, Nguyen TT, La Scola B, Raoult D. Genomic insights into a new Citrobacter koseri strain revealed gene exchanges with the virulence-associated Yersinia pestis pPCP1 plasmid. Front Microbiol. 2016;7:340.
24. Molsa M, Hemmila H, Katz A, Niemimaa J, Forbes KM, Huitu O, Stuart P, Henttonen H, Nikkari S. Monitoring biothreat agents (Francisella tularensis, Bacillus anthracis and Yersinia pestis) with a portable real-time PCR instrument. J Microbiol Meth. 2015;115:89–93.
25. National Department of Health. National plague control guidelines. Plague control guidelines for South Africa. Pretoria: NDoH; 2008.
26. Kutyrev VV, Sharova IN, Osina NA, Kazakova ES, Eroshenko GA, Plotnikova EA, Piontkovskiy SA, Kulichenko AN, Taran TV, Maletskaya OV, et al. The order of organization and realization of laboratory diagnostics of plague for laboratories of local, regional and federal levels: methodological guidelines MUK. 4.2.2940-11. 2. Moscow: Federal service on customers' rights protection and human well-being surveillance; 2011. p. 2940–1.
27. Lesson 4. Laboratory issues. http://www.Bt.cdc.gov/agent/plague/trainingmodule/4/.
28. World Health Organization: Operational guidelines on plague surveillance, diagnosis, prevention and control. http://www.searowhoint/entity/emerging_diseases/documents/ISBN_9789_92_9022_376_4/en/. 2009.
29. Prevention CfDCa: Basic laboratory protocols for the presumptive identification of Yersinia pestis. Atlanta: Centers for Disease Control and Prevention. http://www.bt.cdc.gov/Agent/Plague/ype_la_cp_121301.pdf. 2001.
30. Prentice MB, Rahalison L. Plague. Lancet. 2007;369(9568):1196–207.
31. Rollins SE, Rollins SM, Ryan ET. Yersinia pestis and the plague. Am J Clin Pathol. 2003;119(Suppl):S78–85.
32. Poland JD, Dennis DT. Treatment of plague. Plague manual: epidemiology, distribution, surveillance and control. Geneva: World Health Organization; 1999.
33. Perry RD, Fetherston JD. Yersinia pestis—etiologic agent of plague. Clin Microbiol Rev. 1997;10:35–66.
34. Prentice MB, Rahalison L. Plague. Lancet. 2007;369:1196–207.
35. Inglesby TV, Dennis DT, Henderson DA, Bartlett JG, Ascher MS, Eitzen E, Fine AD, Friedlander AM, Hauer J, Koerner JF, et al. Plague as a biological weapon: medical and public health management. Working group on civilian biodefense. JAMA. 2000;283(17):2218–90.
36. Mwengee W, Butler T, Mgema S, Mhina G, Almasi Y, Bradley C, Formanik JB, Rochester CG. Treatment of plague with gentamicin or doxycycline in a randomized clinical trial in Tanzania. Clin Infect Dis. 2006;42(5):614–21.
37. Rubinstein E. Bioterrorism: role of antimicrobial agents (Lecture). In: IV IACMAC International conference "Antimicrobial therapy"; Moscow. 2001.
38. Anisimov AP, Amoako KK. Treatment of plague: promising alternatives to antibiotics. J Med Microbiol. 2006;55:1461–75.
39. Dennis DT, Gage KL, Gratz N, Poland JD, Tikhomirov E. Plague manual: epidemiology, distribution, surveillance and control. Geneva: World Health Organization; 1999.
40. Lesson 5. Medical management. http://www.Bt.cdc.gov/agent/plague/trainingmodule/5/.

Chapter 12
Plague Vaccines: Status and Future

Wei Sun

Abstract Three major plague pandemics caused by the gram-negative bacterium *Yersinia pestis* have killed nearly 200 million people in human history. Due to its extreme virulence and the ease of its transmission, *Y. pestis* has been used purposefully for biowarfare in the past. Currently, plague epidemics are still breaking out sporadically in most of parts of the world, including the United States. Approximately 2000 cases of plague are reported each year to the World Health Organization. However, the potential use of the bacteria in modern times as an agent of bioterrorism and the emergence of a *Y. pestis* strain resistant to eight antibiotics bring out severe public health concerns. Therefore, prophylactic vaccination against this disease holds the brightest prospect for its long-term prevention. Here, we summarize the progress of the current vaccine development for counteracting plague.

Keywords *Yersinia pestis* • Plague • Vaccines

12.1 Brief History of Plague Vaccines

Y*ersinia pestis*, the causative agent of plague, is an aerobic, nonmotile, gram-negative bacillus. Plague is a deadly disease that has impacted humans for at least 1500 years [1] and it continues to be a disease of significant concern. The destructive potential of plague is evident from three major pandemics: the Justinian plague of the sixth and seventh centuries that affected North Africa, Europe, Central and Southern Asia, and Arabia; the second pandemic in Europe, which killed one third of the Western European population (including the Black Death of 1347–1351 A.D.); and the third pandemic, which originated in China in 1855 and spread around the world via shipborne rats [2]. Overall, *Y. pestis* is estimated to have killed 100–200 million individuals throughout history, making it one of the worst human infectious diseases. It is also considered a reemerging disease [3, 4], and most of the

W. Sun (✉)
Department of Infectious Diseases and Pathology, College of Veterinary Medicine,
University of Florida, P.O. Box 110880, Gainesville, FL 32611-0880, USA
e-mail: wsun8@ufl.edu

R. Yang, A. Anisimov (eds.), *Yersinia pestis: Retrospective and Perspective*,
Advances in Experimental Medicine and Biology 918,
DOI 10.1007/978-94-024-0890-4_12

several thousands of human cases each year are now reported from Madagascar and other countries in Africa.

Although it does not match the "big three" (malaria, HIV/AIDS, and tuberculosis) in the number of people annually affected in the contemporary era, it is far more pathogenic and has the potential to spread much more rapidly than these other diseases [5]. Plague remains as one of the top five bioterrorism threats [6] and a CDC Tier 1 select agent pathogen. Therefore, there is an urgent need for effective means of preexposure and postexposure prophylaxis. Owing to the short incubation period, treatment with antibiotics, and possibly monoclonal antibodies and drugs inhibiting mediators of pathogenicity, offers the best prospect for postexposure prevention of disease. However, *Y. pestis* strains resistant to multiple drugs have been isolated from plague patients in Madagascar, which may spread multiple antibiotic resistance-encoding genes to plague reservoirs [7–9]. For longer-term protection and to counter drug resistance, vaccination is believed to be crucial [10, 11]. The development of vaccines got an early start in 1897, when Waldemar Haffkine (1860–1930) showed that a heat-killed culture of plague bacteria protected rabbits against experimental infection. This preparation was tested in humans in India, with >20 million doses being given, resulting in observations of reduced incidence and mortality in immunized persons [12]. In an effort led by Meyer, starting in 1939 [13, 14], the US Army developed a formalin-killed *Y. pestis* vaccine that was given to more than a million American servicemen deployed to Vietnam [14]. Plague vaccine (USP), a formalin preparation of the fully virulent strain *Y. pestis* 195/P, was the first FDA-licensed plague vaccine for human use in the United States and the United Kingdom [13, 14]. Controlled clinical trials have not been reported, but studies of US military personnel during the Vietnam War strongly suggest that formalin-killed, whole-cell vaccines protect against bubonic plague [14, 15]. However, these vaccines cause significant adverse reactions, particularly after booster injections, which are needed to maintain protection [16]. Moreover, they generally fail to protect mice and nonhuman primates against pulmonary *Y. pestis* challenge, and several humans contracted pneumonic plague despite immunization with this vaccine [17]. In the United States since 1999, a lack of effective protection against pneumonic plague; adverse reactions such as fever, headache, and lymphadenopathy; and the need for booster injections eventually resulted in diminished interest in the USP vaccine [6, 18, 19]. Currently, USP vaccine is still used for research only in the United Kingdom [20, 21]. Thus, killed whole-cell vaccines are probably not suitable for defense against weaponized pneumonic plague.

In 1931, Georges Girard and Jean Robic developed a live attenuated nonpigmented strain of the plague bacillus in Madagascar called EV [22]. This vaccine or similar live attenuated bacteria with designations including EV76, EV NIIEG, and Tjiwide were administered to millions of people in Madagascar, Indonesia, Vietnam, and the Soviet Union [22, 23]. *Vaccinum pestosum vivum siccum* on the base of the strain EV line NIIEG is still used and commercially available in Russia (http://www.epidemiolog.ru/catalog_vac/?SECTION_ID=&ELEMENT_ID=476) and Kazakhstan (http://pharmprice.kz/annotations/vakcina-chumnaya-zhivaya-suhaya/) [24]. By the end of the twentieth century, these vaccines were rarely used outside of

Russia due to their strong adverse reactions. Although considerable progresses have been made for developing safe effective vaccines against plague for human use, a licensed plague vaccine has not been released into commercial market yet. Here, we summarized current progresses in the development of plague vaccines.

12.2 Subunit Vaccines

Searching for new antigens from *Y. pestis* is a continuous endeavor for developing plague vaccines. Table 12.1 listed *Y. pestis* antigens that were evaluated for vaccine purpose.

Lipopolysaccharide (LPS) is an integral component of the outer membrane of gram-negative bacteria and can be used as an immunogenic molecule [25–32]. Prior et al. showed that the LPS extracted from *Y. pestis* strain GB stimulated the production of TNF-α and IL-6 from mouse macrophages, but was less active in these assays than the LPS isolated from *Escherichia coli* strain 0111. They also indicated that an antibody response to LPS in mice was primed by LPS immunization, but this response did not provide any protection against 100 MLD of *Y. pestis* strain GB [33]. The pH 6 antigen (PsaA) was initially described in 1961 as an antigen synthesized by *Y. pestis* and formed fimbria-like structures on the bacterial surface at the temperature close to body temperature of mammals (35–41 °C) and acidic pH values (5.8–6.0) close to the pH of abscesses or phagolysosomes in macrophages [34]. PsaA serves as an important adhesin in the establishment of *Y. pestis* infections [35–37]. The lack of PsaA synthesis in the *Y. pestis* KIM5 strain caused virulence reduction and an increase in the LD_{50} of at least 100-fold in mice after retro-orbital injection [38]. However, the loss of synthesis or constitutive production of pH 6 antigen in the fully virulent wild-type strains 231 and I-1996 did not influence their virulences or the average survival time of subcutaneously inoculated BALB/c mice [39]. Rabbits immunized with a live EV76 vaccine strain primed high levels of anti-PsaA (IgG) at 42 days after initial immunization [40], and also mice immunized with the EV76 produced a strong T-cell response to PsaA [41]. Schifferli's group showed that mice immunized with 40 μg of PsaA adjuvanted with Alhydrogel primed a strong humoral immune response and provided a significant protection (70 %) against an intranasal infection with *Y. pestis* KIM5 (Pgm-) in the iron dextran-treated mouse model [42]. However, no protection was shown in the case of immunization with PsaA protein against subcutaneous infection with fully virulent *Y. pestis* strains 231 and I-1996 [39].

Benner et al. determined the humoral immune response to *Y. pestis* antigens in mice that survived lethal *Y. pestis* aerosol challenge after antibiotic treatment, such as F1, V antigen, YpkA, YopH, YopM, YopB, YopD, YopN, YopE, YopK, plasminogen activator protease (Pla), and pH 6 antigen as well as purified lipopolysaccharide [43]. Their results indicated that the major antigens recognized by murine convalescent sera were F1, LcrV, YopH, YopM, YopD, and Pla [43]. Andrews et al. purified the recombinant proteins (YpkA, YopD, YopE, YopH, YopK, and YopN)

Table 12.1 Live attenuated *Y. pestis* strains as vaccines against plague

Y. pestis mutant	LD_{50}	Immunization	Protective efficacy	References
Y. pestis Kimberley53 Δ*pcm*	>10^7 CFU in female OF1 outbred mice by s.c.	s.c. vaccination with 10^6 CFU	Complete protection against s.c. challenge with 10^5 CFU of *Y. pestis* Kimberley53	[130]
Y. pestis Kimberley53 Δ*nlpD*	>10^7 CFU for s.c and airway routes of infection in Female OF1 mice	s.c. immunization with 10^7 CFU of mutant strain	Provides complete protection against s.c. challenge with 10^5 LD_{50} of *Y. pestis* Kimberley53 and 82 % protection against i.n. challenge with 5500 CFU of *Y. pestis* Kimberley53	[131]
Y. pestis 231 Δ*nlpD*, *Y. pestis* I-3455 Δ*nlpD*	>10^7 CFU in BALB/c mice by s.c. and >1.5×10^{10} CFU in guinea pigs	s.c. vaccination with 10^4 CFU for mice and 5×10^3 CFU for guinea pigs	Δ*nlpD* mutants induce immunity 10^5 times more potent than the one induced by the administration of the EV vaccine strain. At the same time, $NlpD^-$ bacteria failed to protect guinea pigs in the case of a subcutaneous challenge with *Y. pestis*, inducing a 10^6 times less potent protection compared with that conferred by immunization with the EV vaccine strain	[132]
Y. pestis GB Δ*dam*	2.3×10^3 CFU in female BALB/c mice by s.c.	s.c. vaccination with 1.5×10^3 CFU	Complete protection against s.c. challenge with 7500 CFU of *Y. pestis* GB	[279]
Y. pestis CO92 Δ*yopH*	>10^7 CFU in female outbred CD1mice by i.n. and s.c., respectively	s.c. and i.n. vaccination	s.c. vaccination with ~10^7 provides 70 % protection against s.c. challenge with ~10^5 CFU of *Y. pestis* CO92. i.n. vaccination with ~10^7 CFU of CO92 Δ*yopH* provides 100 % protection against s.c. challenge and 61.5 % protection against i.n. challenge with ~10^5 CFU CO92, respectively	[129]
Y. pestis EV Δ*lpxM*	>2.5×10^4 CFU in female BALB/c mice by s.c.	s.c. vaccination with 2.5×10^4 CFU of mutant strain	Provides ~40 % protection against s.c. challenge with virulent strain *Y. pestis* 231	[280]

χ10004 (pCD1Ap) Δ*relA* Δ*spoT*	5.8×10^5 CFU for s.c. infection in female Swiss Webster mice	s.c. immunization with 2.5×10^4 CFU of mutant strain	A single s.c. vaccination could provide complete protection against s.c. challenge with 1.5×10^5 CFU of virulent strain and 60 % protection against i.n. challenge with 2.0×10^4 CFU of *Y. pestis* KIM5+	[117]
Y. pestis CO92 Δ*pgm* Δ*smpB-ssrA*	>10^6 CFU by i.v. infection and >10^8 CFU by i.n. infection in C57BL/6 mice	i.v. and i.n. immunization with 10^4 CFU of mutant strain	Provides complete protection against i.n. challenge with 2×10^5 CFU of *Y. pestis* CO92Δ*pgm*	[143]
χ10010 (pCD1Ap) Δ*crp*	>3×10^7 CFU by s.c. infection and >10^5 CFU by i.n. infection female Swiss Webster mice	s.c. immunization with 3.8×10^7 CFU of mutant strain	Provides 80 % protection against s.c. challenge with 10^7 CFU of virulent strain and no protection against i.n. challenge with 1×10^4 CFU of *Y. pestis* KIM5+	[118]
χ10017 (pCD1Ap) ΔP_{crp}::*araC* P_{BAD} *crp*	>4.3×10^5 CFU by s.c. infection and ~10^5 CFU by i.n. infection in female Swiss Webster mice	s.c. immunization with 3.0×10^4 CFU of mutant strain	Provides complete protection against s.c. challenge with 10^5 CFU of virulent strain and 70 % protection against i.n. challenge with 1×10^4 CFU of *Y. pestis* KIM5+	[118]
Y. pestis GB Δ*guaBA*	>7×10^4 CFU by s.c. and i.v. infection in female BALB/c mice	s.c. and i.n. immunization with 7×10^4 CFU of mutant strain	Provides 100 % protection against s.c. challenge with ~66 LD_{50} of the wild-type *Y. pestis* strain GB	[281]
χ10015 (pCD1Ap) Δ*lpxP*::P_{lpxL} *lpxL*	~10^7 CFU by s.c. infection and 2.7×10^3 CFU by i.n. infection in female Swiss Webster mice	s.c. immunization with 2.2×10^6 CFU of χ10015(pCD1Ap)	Provides complete protection against s.c. challenge with 3.57×10^7 CFU of virulent strain and 90 % protection against i.n. challenge with 1.24×10^4 CFU of *Y. pestis* KIM5+	[128]
χ10027 (pCD1Ap) Δ*lpxP32*::P_{lpxL} *lpxL* Δ*lacI23*::P_{lpp} *lpxE*	LD_{50}s: 2.3×10^4 CFU for s.c. infection and 7.8×10^4 CFU for i.n. infection	ND	ND	[127]

(continued)

Table 12.1 (continued)

Y. pestis mutant	LD_{50}	Immunization	Protective efficacy	References
χ10028 (pCD1Ap) *pgm*⁻ Δ*lpxP*::P_{lpxL} *lpxL*	~10^8 CFU by s.c. infection and 5×10^3 CFU by i.n. infection in female Swiss Webster mice	s.c. immunization 1.4×10^7 CFU of χ10028(pCD1Ap)	Provides 80 % protection against s.c. challenge with 3.57×10^7 CFU of virulent strain and 60 % protection against i.n. challenge with 1.24×10^4 CFU of *Y. pestis* KIM5+	[128]
χ10030 (pCD1Ap) ΔP_{crp} TT *araC* P_{BAD} *crp* Δ*lpxP*::P_{lpxL} *lpxL*	>10^8 CFU by s.c. infection and >10^6 CFU by i.n. infection in female Swiss Webster mice	s.c. immunization 1.4×10^7 CFU of χ10030(pCD1Ap)	Provides complete protection against s.c. challenge with 3.57×10^7 CFU of virulent strain and 80 % protection against i.n. challenge with 1.24×10^4 CFU of *Y. pestis* KIM5+	[128]
Δ*yscN Y. pestis* CO92	>4.44 × 10^6 CFU by s.c. infection in female Swiss Webster mice	s.c. immunization with 4.98×10^7 CFU of Δ*yscN Y. pestis* CO92 and s.c. booster at 28 days with 3.22×10^7 CFU	Provides 90 % protection against s.c. challenge with 180 CFU of virulent *Y. pestis* CO92	[133]
Δ*lpp* Δ*msbB Y. pestis* CO92	LD50s: >30 CFU by s.c. infection and ~ 10^3 CFU by i.n. infection in female Swiss Webster mice	Survival animals rechallenged by *Y. pestis* CO92	Provide partial protection against s.c. and i.n. challenge	[134]
Δ*yscB Y. pestis* biovar *Microtus* strain 201	LD_{50} of the Δ*yscB* mutant in BALB/c mice is estimated to be more than 10^6 CFU, which is more than 400,000-fold higher than 3 CFU of the wild-type strain	s.c. immunization with 1.63×10^4 CFU of Δ*yscB* mutant	The Δ*yscB* mutant provided 87.5 % or complete protection by the subcutaneous or intranasal challenge with 1.24×10^6 CFU of virulent *Y. pestis* strain 141	[135]
Y. pestis biovar *Microtus* strain 201	Avirulent to humans or primates	s.c. immunization with 1.4×10^{10} CFU of *Y. pestis* strain 201	The immunization provided 83.3 % protection by the subcutaneous challenge with 1.74×10^9 CFU of virulent *Y. pestis* strain 141 for rhesus macaques	[136]

ΔlppΔmsbB ΔrbsA Y. pestis CO92	20-50 LD_{50} in a mouse model of pneumonic plague	Survival animals rechallenged by *Y. pestis* CO92	The triple mutant-infected mice at 50 LD_{50} were 90 % protected upon subsequent challenge with 12 LD_{50} of *Y. pestis* CO92	[137]
Δail Δlpp ΔmsbB Y. pestis CO92	LD_{50} of the triple mutant (*Δail Δlpp ΔmsbB*) was equal to 6800 LD_{50} of *Y. pestis* CO92	Survival animals rechallenged by *Y. pestis* CO92	Partially protection against intranasal challenge with the bioluminescent *Y. pestis* CO92 strain (20 to 28 LD_{50}s)	[138]
V674pF1	LD_{50} of the V674pF1 strain in OF1 female mice is more than 10^9 CFU	Oral immunization with 10^8 or 10^9 CFU of V674pF1 strain	A very high challenge dose of 10^7 CFU CO92 ($3300 \times LD_{50}$) was used to mimic a severe contamination, mice vaccinated with 10^8 CFU of V674pF1 showed 80 % protection, and this protection reached 94 % when a vaccine dose of 10^9 CFU was administered	[144]
χ10057 *Δasd-206 ΔmsbB868*::P_{msbB} *msbB(EC)* ΔP_{crp21}::TT *araC* P_{BAD} *crp*	LD_{50} of the χ10057 strain in Swiss Webster mice is more than 10^9 CFU	Oral immunization with 10^9 CFU of χ10057(pYA5199) strain	Provide 83 % protection against i.n. challenge with 1.3×10^4 CFU of virulent *Y. pestis* KIM6+ (pCD1Ap) strain	[145]

Note: *Y. pestis* GB, LD_{50} (s.c.) = 1 CFU [86]; *Y. pestis* KIM5+, LD_{50} (s.c.) < 10 CFU, LD_{50} (i.n.) ~100 CFU [128]; *Y. pestis* Kimberley53, LD_{50} (s.c.) 1 ~ 3 CFU, LD_{50} (i.n.) = 550 CFU [131]; *Y. pestis* 231, LD_{50} (s.c.) < 10 CFU [282]; *Y. pestis* CO92, LD_{50} (s.c.) = 1.9 CFU, LD_{50} (i.n.) ~250 CFU, LD_{50} (aerosol) ~2100 CFU [63, 283–284]; *Y. pestis* CO92Δ*pgm* strain, LD_{50} (i.n.) = 2×10^4 CFU [143]. CFUs, colony-forming units; LD_{50}, 50 % lethal dose; s.c., subcutaneous; i.n., intranasal; i.p., intraperitoneal and evaluated the role of *Yersinia* outer proteins (Yops) in conferring protective immunity against plague in mice injected with above these proteins. Most Yop-vaccinated animals succumbed to infection with either wild-type encapsulated *Y. pestis* or a virulent, nonencapsulated isogenic variant. Vaccination with YpkA significantly prolonged the mean survival time but did not increase the overall survival of mice challenged with the nonencapsulated strain. Only immunization with YopD provided significant protection for mice against challenge with the nonencapsulated *Y. pestis* strain [44].

Straley's group firstly found that *Y. pestis* YadB and YadC, two new members of the Oca (oligomeric coiled-coil adhesins) family of proteins [45, 46], have the ability to form trimers and correlate with the invasion of *Y. pestis* into epithelioid cells [47]. Loss of *yadBC* caused a modest loss of invasiveness for epithelioid cells and a subtle decrease in virulence for bubonic plague but not for pneumonic plague in mice [47, 48]. But immunization with the GST-YadC$_{137–409}$ protein, which fused YadC aa 137–409 to C terminal of glutathione S-transferase (GST), provided partial protection against F1^{-} *Y. pestis* challenge in mice and was found to stimulate mixed Th1/Th2 responses [49]. However, Sun et al. showed that YadC810 protein immunization could not provide any protection against subcutaneous and intranasal challenge of virulent *Y. pestis* CO92 [50]. The explanations for this contrary result are (1) the higher challenge dose we used and (2) the difference between YadC protein (aa 32–551) used for immunization in our studies and the YadC protein (aa 137–422) used by Murphy et al. [49]. The variation of amino acid sequences might change the configuration of the YadC protein in the two cases [50]. YscF is a surface-localized protein that is required both to secrete Yops and to translocate toxins into eukaryotic cells [51–54], which suggested that YscF was required for virulence and might be a potential protective antigen. Matson et al. showed that a robust antibody response to YscF primed by immunization was able to afford significant protection to immunized mice following challenge with *Y. pestis* [55].

Yang's group employed the high-throughput screening to identify new protective antigens of *Y. pestis*. Total 261 genes from *Y. pestis* were selected on the basis of bioinformatics analysis and were expressed in *Escherichia coli* BL21(DE3). After purification, 101 proteins were qualified for examination of their abilities to induce the production of gamma interferon in mice immunized with live vaccine EV76 by enzyme-linked immunospot assay. Thirty-four proteins were found to stimulate strong T-cell responses. The protective efficiencies for 24 of them were preliminarily evaluated in mice. In addition to LcrV, nine proteins (YPO0606, YPO1914, YPO0612, YPO3119, YPO3047, YPO1377, YPCD1.05c, YPO0420, and YPO3720) provided partial protection against challenge with a low dose (20 times the 50 % lethal dose [20× LD_{50}]) of *Y. pestis* 201, but only YPO0606 could partially protect mice from infection with *Y. pestis* 201 at a higher challenge dosage (200× LD_{50}) [41].

Recently, Chopra's group [56] indicated that immunizations with OmpA, Ail/OmpX, Pla, and F1-V by intramuscular (i.m.) route induced different amounts of antibody titers against the above antigens. The titers of anti-Pla and anti-OmpA antibodies were the lowest, while the anti-Ail/OmpX antibody titers were similar to that of F1-V antigen. Mice immunized with the F1-V antigen were completely protected by s.c. challenge with 500 LD_{50} Δ*caf1* mutant strain of *Y. pestis* CO92 which corresponded to approximately 75–100 LD_{50} of the WT *Y. pestis* CO92 [57]. Antibodies to both OmpA and Ail/OmpX provided protection to mice resulting in 40–50 % mice survival, respectively, while antibodies to Pla did not provide any protection to mice. All of the unimmunized, naive control mice died by day 20, although 90 % of mice immunized with Pla antigen died by day 10. Comparing with the protective effects of Ail/OmpX alone, they did not observe any additive or synergistic effect on protection against bubonic plague in mice immunized with Ail/OmpX, OmpA, and Pla cocktail. In a pneumonic plague mouse model, F1-V immunization clearly provided complete protection for mice against i.n. challenge. Interestingly, immunization with Pla provided 60 % protection to animals when challenged with 15 LD_{50} [7500 colony-forming units (CFUs)] of Δ*caf1* mutant of CO92 despite Pla immunization not protecting against a bubonic plague challenge. Neither Ail/OmpX nor OmpA immunizations protected mice from developing pneumonic plague despite immunizations with both of the aforementioned antigens providing protection against bubonic plague in mice. In rat model, immunization with OmpA, Ail/OmpX, Pla, and F1-V in rats also produced somewhat comparable antibody titers. The F1-V immunization provided complete protection for rats against i.n. challenge with 8.5 LD_{50} of wild-type CO92, and Ail/OmpX immunization provided 50 % protection, while OmpA and Pla immunization failed to provide any protection. In addition, immunization with Ail/OmpX, Pla, or OmpA did not provide any protection to rats against s.c. challenge with 7 LD_{50} of *Y. pestis* CO92. Although results indicated that none of immunization with above antigens provided better protection against *Y. pestis* challenge in mice than that with F1 and/or LcrV antigens, these antigens may be useful to be the combination with F1 and/or LcrV to augment protective immunity of such subunit vaccines. In the following, we will emphasize on summarizing the progresses in F1 and/or LcrV subunit vaccines.

Y. pestis produces a specific capsule composed of the fraction I, the biosynthesis of which is temperature dependent. The monomer of the protein capsule is 17.5 kDa, i.e., F1 antigen, confers resistance to phagocytosis [58], has good immunogenic properties in *Y. pestis*, and is secreted onto the bacterial surface by the Caf1 system, which consists of Caf1A as an anchor, Caf1M as a chaperone, and several Caf1 (F1) structural proteins [59, 60]. Meyer et al. indicated that the anti-F1 human serum afforded significant passive protection in mice [61]. Mice immunized with F1 antigen purified from the *E. coli* recombinant [62, 63] and *Y. pestis* [63] were protected by lethal *Y. pestis* challenge. However, protective efficacy against plague challenge correlated with titer of F1 antibody, which was a discrepancy in these studies [62, 63]. Immunization with one dose of F1 formulated in poly(lactide-co-glycolide) (PLG) microparticles and liposomes induced high F1 antibody titers in mice and provided great protection against 10^5 CFU of *Y. pestis* GB strain [64].

Since F1-negative strains have been isolated from natural sources and caused experimental fatal disease [65], Davis et al. indicated that African green monkeys that inhaled F1-negative and/or F1-positive strains of *Y. pestis* died at 4–10 days postexposure and had lesions consistent with primary pneumonic plague [66].

Moreover, Quenee et al. also confirmed that Δ*caf1* *Y. pestis* was not only fully virulent in animal models of bubonic and pneumonic plague but also broke through immune responses generated with live, attenuated strains or F1 subunit vaccines [67]. The same group indicated that immunization of mice and guinea pigs with the recombinant F1 generated robust humoral immune responses. Sixty percent of immunized mice survived pneumonic plague challenge with 100 MLD *Y. pestis* CO92 [68], but only 12.5 % guinea pigs survived pneumonic plague challenge with 250 MLD *Y. pestis* CO92 [69]. These studies suggested that plague subunit vaccines shouldn't be solely based on the F1 antigen, although F1 antigen is a good protective immunogen.

LcrV is a multifunctional virulence protein encoded on these 70-kb pCD1 plasmids, which codes for a virulence-associated type III secretion system (T3SS) necessary for the translocation of Yops into eukaryotic target cells [70] and is the core of the *Yersinia* pathogenicity machinery that targets cells of the immune system [71, 72]. LcrV is exported to the bacterial surface by the T3SS, localizes to the tip of the T3SS needle structure, and can be secreted into the extracellular milieu [73–76]. LcrV also interacts with the Ysc gate protein LcrG [70, 77] and cooperates with YopB and D to form a channel or translocon for delivering Yops into eukaryotic cells [75, 78].

LcrV as a primary antigen or passive anti-LcrV antibodies are demonstrated to protect bubonic or pneumonic *Y. pestis* infection in many research articles and reviews [11, 17, 18, 69, 79–84]. Anderson et al. demonstrated that immunization with the recombinant full-length LcrV antigen protected mice from lethal bubonic and pneumonic challenge by a wild-type *Y. pestis* CO92 ($F1^+$ strain) or by the isogenic $F1^-$ strain C12 [80]. As described above, vaccination of mice with F1 partially protects mice and rats from s.c. challenge with *Y. pestis* and macaques against pneumonic plague by passive transfer of sera collected from F1-vaccinated rabbits [67], whereas there is some evidence that F1 has adjuvantizing activity on the immune response to the coadministered V antigen and that this effect is specific and not due to LPS contamination [85]. Thus, including F1 into LcrV antigen combination might augment protective immunity against *Y. pestis* strains.

Williamson et al. showed that co-immunization with the purified culture-derived F1 and the recombinant LcrV subunits afforded a greater level of protection than with either subunit alone [86]. Also they showed that the antibody titers to F1 and V were correlated with protection [87] and immunization with a single dose of Alhydrogel-adsorbed F1 + V vaccine afforded great protection against aerosolized *Y. pestis* challenge [85]. Their studies demonstrated the potential of the combined F1 + V vaccine to be developed as a human prophylactic for pneumonic plague. The Defence Science and Technology Laboratory (Dstl) at Porton Down (United Kingdom) has developed the subunit vaccine rF1 + rV, which is comprised of purified recombinant F1 (rF1) as well as recombinant LcrV (rV, derived from GST-LcrV) conjugated with Alhydrogel [85, 88]. Avecia Biologics Ltd. initially licensed the technology to manufacture rF1 + rV vaccine in GMP level for phase I and II trials. Then, PharmAthene Inc. acquired an exclusive license for rF1 + rV, which is now manufactured as RypVaxTM, and continues vaccine development toward FDA licensure. In a phase I clinical trial, GMP manufactured rF1+rV (RypVax) was administered to 24 healthy adult males in a double-blind, ascending dose design study, where groups of six individuals received vaccine at dose levels of 5, 10, 20, or 40 μg protein in a volume of 0.5 ml, administered intramuscularly with two doses, prime (day 1) and booster (day 21). RypVax-immunized volunteers developed rF1- and rV-specific antibodies on day 14 and increased titers after the booster injection [89]. PharmAthene Inc. conducted three phase I trials and reported vaccine safety in humans. Levels of antibodies varied considerably among members of each vaccine group. A phase II efficacy trial for RypVax was launched in 2003; however experimental tools and generated data are currently not available [84].

Since the US Department of Homeland Security and Department of Defense require an effective vaccine to protect state and local emergency response and rescue teams, as well as scientists and members of the World Health Organization during laboratory and fieldwork with *Y. pestis* [90], the US Army Medical Research Institute of Infectious Diseases (USAMRIID) developed the rF1-V fusion protein as a vaccine. The rF1-V vaccine protected experimental mice against pneumonic as well as bubonic plague, caused by either an $F1^+$ or $F1^-$ strain of *Y. pestis*, and provided better protection than F1 or V alone against the $F1^+$ strain [91]. Moreover, vaccination with F1-V fusion antigen provided similar protective efficacy against *Y. pestis* challenge as vaccination with F1+V combination [90, 91]. Under contract with the US Department of Defense (DOD), DynPort Vaccine Company developed rF1-V toward FDA licensure. The progress of DynPort Vaccine Company toward licensure of the rF1-V vaccine was posted on the National Institutes of Health clinical trial website (www.clinicaltrials.gov). Two clinical trials were reported for the rF1-V plague vaccine. The first trial was designed as a phase I, open-label, dose-escalating study for safety, tolerability, and immunogenicity in healthy volunteers. The second trial was designed as a phase II, dose-blinded, block-randomized, multicenter study to select dosage and schedule of rF1V, examining immune responses up to 210 days, with additional immunogenicity and safety/reactogenicity data collection through 540 days. Both trials have been completed; however results are not yet available. According to a DynPort Vaccine Company press release, 44 healthy volunteers between the ages of 18 and 40 were enrolled in the phase I trial and no vaccine-related serious adverse events were identified. The phase II trial tested 400 healthy volunteers between the ages of 18 and 55 [84].

LcrV was reported to trigger the release of interleukin-10 by host immune cells through its interaction with TLR2/CD14 [92–94] and also to suppress tumor necrosis factor alpha (TNF-α) and interferon gamma (IFN-γ) production [82, 95]. However, several other groups demonstrated that *Y. pestis* LcrV could not efficiently activate TLR2 signaling and that TLR2-mediated immunomodulation did not play a major role in pathogenesis of plague [96–98]. Arguments about immunomodulatory features of LcrV need to be resolved in further study, but potential concerns have been dispelled in the vaccine design. Schneewind's group from the University of Chicago developed a nontoxigenic LcrV vaccine, rV10, a variant with a deletion of LcrV residues 271–300, removing potential epitope of interaction with TLR2/CD14. Immunization with recombinant purified rV10 conjugated with Alhydrogel elicited immune responses that protected mice against a lethal challenge with the fully virulent *Y. pestis* strain CO92 [99, 100]. Compared to rLcrV immunization, rV10 immunization provided equal levels of vaccine protection to mice [99]. The rV10 displayed a reduced ability to release interleukin-10 or prevent the release of tumor necrosis factor alpha from lipopolysaccharide-stimulated primary macrophages [100]. Immunization with rV10 also provided great protection against bubonic and pneumonic plague challenge in rats, guinea pigs, and nonhuman primates [101]. Eighty-seven percent of guinea pigs immunized with the rV10 plague vaccine survived pneumonic plague challenge with 250 MLD *Y. pestis* CO92, while only 50 % of guinea pigs immunized with rLcrV survived this challenge [69]. Additionally, guinea pigs vaccinated with rV10 or F1-V had similar survival by the intranasal challenge with 1000 MLD *Y. pestis* CAC1 (*caf1A*::IS*1541*) [101]. In an NIAID-supported program, the University of Chicago investigators use GMP manufactured rV10 to demonstrate preclinical efficacy in animals [68]. The rV10 vaccine is currently undergoing US Food and Drug Administration (FDA) pre-Investigational New Drug (pre-IND) authorization review for a future phase I trial [84].

Additionally, Yang and Wang's group developed a new subunit vaccine consisting of F1 and rV270 (a recombinant LcrV variant lacking amino acids 271–326 to reduce its potential immunosuppressive activity) being evaluated in different animal models including mouse, guinea pigs, rabbits, and Chinese-origin rhesus macaque [102–106]. Mice immunized with the subunit vaccines and EV76 vaccine achieved complete protection against challenge with 10^6 CFU of virulent *Y. pestis* strain 141. Immunization with the subunit vaccine and EV76 vaccine provided good protection against challenge with the similar CFU of *Y. pestis* in guinea pigs and rabbits, respectively. Immunization with subunit vaccines stimulated significantly higher anti-rV270 and anti-F1 IgG titers in mice than in guinea pigs and rabbits, and guinea pigs developed significantly higher IgG titers than the rabbits, but the anti-F1 response in guinea pigs was more variable than in mice and rabbits. All the above animals immunized with EV76 developed a negligible IgG titer to rV270 antigen. Analysis of IgG subclasses demonstrated that subunit vaccines induced strong predominant IgG1 responses, whereas those receiving EV76 immunization primed IgG1 and IgG2a balanced responses [102]. They further compared immunogenicity of the subunit vaccines SV1 (20 μg F1 + 10 μg rV270) and SV2 (200 μg F1 + 100 μg rV270) and EV76 in Chinese-origin rhesus macaques, *Macaca mulatta*. Similar to other animal models described above, the macaques immunized with SV1 or SV2 developed higher anti-rV270 IgG titer, while animals immunized with EV76 elicited a negligible IgG to rV270 antigen. No significant antibody titer differences were observed between SV1 and SV2 immunized groups. Also there were no statistical differences for CD4/CD8 ratios, IL-4, and CD69 levels between the three-vaccine immunized groups. However, the EV76 immunized animals produced a significant higher level of IL-12 than the subunit immunized groups, indicating that EV76 had an advantage over SV in respect of cellular immunity. Immunization with SV and EV76 provided a similar protective level against s.c. challenge with 6×10^6 CFU of *Y. pestis* in Chinese-origin rhesus macaques [103]. They also evaluated the long-term immune responses, transmission modes of maternal antibodies, and protective efficacy in mice vaccinated with the subunit vaccine SV1. Their studies demonstrated that antibodies to F1 and rV270 were detectable over a period of 518 days, and the complete protection against 10^6 CFU of *Y. pestis* 141 by s.c. challenge was achieved up to day 518 after primary immunization [106]. The rV270- and F1-specific antibodies could be transmitted to newborn mice from their mothers until 10 to 14 weeks of age. There was no difference in antibody titers between the parturient mice immunized with SV1 (PM-S) and the cesarean-section newborns from the PM-S or between the lactating mice immunized with SV1 (LM-S) and the cross-fostered mice (CFM) during 3 weeks of lactation. Seventy-two percent of newborn mice survived against s.c. challenge with 4800 CFU of *Y. pestis* strain 141 at 6-week age, but none of these mice survived against 5700 CFU of *Y. pestis* challenge at 14-week age. Eighty-four percent of CFM could survive against 5000 CFU of *Y. pestis* challenge at 7-week age. Their study showed that maternal antibodies induced by SV1 subunit vaccine in mother mice could be transferred to their offspring (newborn mice) by both placenta and lactation. Passive antibodies from the immunized mothers could persist for 3 months in newborn mice and provided early protection for newborn mice [105].

12.3 Live Attenuated *Yersinia* Vaccines

Attenuated *Y. pestis* strains that effectively protected albino mice against experimental plague were developed in 1895 by Yersin and in 1903–1904 by Kolle and Otto but were not tested in humans owing to fears of reversion to virulence. The first vaccination of humans with live plague vaccine was done in Manila, Philippines, in 1907, but reliable evidence of its efficacy was not obtained as there were no plague cases in the city at that time [107]. Subsequently, the EV76 strain, a spontaneous *pgm* mutant, was developed from the EV strain isolated by Girard and Robic from a human case of bubonic plague in Madagascar in 1926 [108]. In 1936, a subculture of the EV76 vaccine strain was established at the NIIEG (designation based on the Russian abbreviation of the Scientific Research Institute of Epidemiology and Hygiene, Kirov, Russian Federation) in the former USSR [108]. This strain was employed for the development of the live vaccine designated as *Vaccinum pestosum vivum siccum*, which is manufactured in the former USSR from 1940 [24, 109]. The EV NIIEG strain has been used as a live plague vaccine for the protection of plague researchers and people living in territories endemic for plague in the countries of the former USSR and some Asian countries and is still in use today [18, 109]. Nevertheless, a single dose of the EV NIIEG live vaccine conferred a prompt (day 7 postvaccination) and pronounced immunity in vaccinees lasting for 10–12 months against bubonic and, to some extent, pneumonic plague [11, 109].

However, EV76 vaccine strain can cause disease in some nonhuman primates, raising questions about its suitability as a human vaccine [110]. The live Pgm^- strain conferred greater protection against bubonic and pneumonic plague than killed vaccines in animals, but it sometimes caused local and systemic reactions [19, 21, 110, 111]. In addition, a live Pgm⁻ strain retains virulence when administered by the intranasal (i.n.) and intravenous (i.v.) routes [17, 110, 112]. Variable virulence of the live vaccine strains in animal models and reactogenicity in humans has prevented this vaccine from gaining worldwide acceptance, especially in the United States and Europe [14, 113]. Although licensing live attenuated *Y. pestis* as a vaccine will undoubtedly be a long and arduous process, it does not extinguish researchers' passion to explore new attenuated *Y. pestis* mutants as vaccines. Table 12.1 shows lists of recent developments of live, rationally attenuated *Y. pestis* mutants as vaccines against plague.

In *Salmonella*, Δ*relA* Δ*spoT* mutants are attenuated [114] and *crp* mutants are attenuated and immunogenic [115]. It has also been established that *Y. pestis crp* mutants are attenuated for virulence [116]. In our laboratory, we examined the vaccine potential of *Y. pestis* Δ*relA* Δ*spoT* [117] and Δ*crp* [118] mutants. Mice vaccinated subcutaneously (s.c.) with 2.5×10^4 CFU of the Δ*relA* Δ*spoT* mutant, which developed high anti-*Y. pestis* serum IgG titers, were completely protected against s.c. challenge with 1.5×10^5 CFU of virulent *Y. pestis* and partially protected (60 % survival) against pulmonary challenge with 2.0×10^4 CFU of virulent *Y. pestis* [117]. Results indicate that ppGpp represents an important virulence determinant in *Y. pestis*, and the Δ*relA* Δ*spoT* mutant strain is a promising vaccine candidate to provide protection against plague.

The Δ*crp* mutant was completely attenuated (s.c. $LD_{50}>10^7$ CFU) and partially protective against bubonic plague but not protective against pneumonic plague [118]. The strategy of regulated delayed attenuation was developed in *Salmonella*, in which the virulence gene expression of bacteria is dependent on the presence of sugars (arabinose, mannose, or rhamnose). When cells are grown in the presence of sugar, the virulence gene is expressed. Once the cells invade host tissues where free arabinose is not available, virulence gene expression ceases and the cells become attenuated [119]. This strategy was applied to *Y. pestis*, constructing a strain with *crp* under transcriptional control of the *araC* P_{BAD} promoter [118]. The resulting strain was partially attenuated ($LD_{50}=4.3\times10^5$ CFU) and protective against both bubonic and pneumonic plague [118].

One strategy used by *Y. pestis* to evade the host immune system is to produce lipid A that is not recognized by toll-like receptor 4 (TLR4). At environment temperature (26 °C), *Y. pestis* adds palmitoleate to Kdo_2-lipid IV_A precursor by palmitoleoyltransferase encoded by a temperature-regulated gene, *lpxP* at 26 °C, then stepwise adds myristate by myristoyltransferase encoded by *lpxM* gene to form hexa-acylated lipid A, while at 37 °C, the body temperature of mammalian hosts, the palmitoleoyltransferase encoded by *lpxP* is activated, resulting in tetra-acylated lipid A [120, 121], which is not recognized by TLR4 [122] that preferentially recognizes hexa-acylated lipid A [123–125]. In 2006, Montminy et al. reported that a *Y. pestis* strain engineered to produce hexa-acylated lipid A at 37 °C by constitutive expression of the *E. coli lpxL* gene from a multicopy plasmid is attenuated [126]. Based on those knowledge, we constructed a strain χ10015(pCD1Ap) (Δ*lpxP32*::P_{lpxL} *lpxL*), which expresses *E. coli lpxL* from the chromosome of *Y. pestis*

KIM6+(pCD1Ap), providing greater genetic stability than plasmid expression. The χ10015(pCD1Ap) was highly attenuated by s.c. administration, but χ10015(pCD1Ap) stimulated a strong inflammatory reaction, which results in mice being sick and looking ruffled in early infection stage, and also retained virulence via intranasal infection. Heterologous expression of the lipid A 1-phosphatase, LpxE, from *Francisella tularensis* in *Y. pestis* yields predominantly 1-dephosphorylated lipid A that might reduce hyper-inflammation of χ10015(pCD1Ap) and the virulence of χ10015(pCD1Ap) by i.n. infection. Results indicated that expression of LpxE on top of LpxL provided no significant reduction in virulence of *Y. pestis* in mice when it was administered intranasally, but actually reduced LD_{50} by three orders of magnitude when the strain was administered subcutaneously [127].

The strain, χ10030(pCD1Ap), produces hexa-acylated lipid A at 37 °C and carries the arabinose-regulated *crp* gene [128]. Our results demonstrated an increase in the LD_{50} of χ10030(pCD1Ap) by s.c. and i.n. inoculation of more than 1.5×10^7 and 3.4×10^4-fold, respectively, in Swiss Webster mice, compared to the wild-type virulent *Y. pestis* KIM6+(pCD1Ap) strain. Both s.c. and i.n. immunization with strain χ10030(pCD1Ap) induced significant protection against both bubonic and pneumonic plague with minimal reactogenicity in mice, attributes consistent with our goal of designing a live safe *Y. pestis* vaccine. However, this strain was still able to induce IL-10 early in infection, a known strategy used by *Y. pestis* to evade detection by the host [73]. Also, due to safety concerns surrounding a live plague vaccine, we consider it prudent to identify and include an attenuating deletion mutation in our final vaccine. Therefore, we plan to enhance the safety and efficacy of χ10030(pCD1Ap) by including a yet to be identified deletion mutation and eliminating its ability to elicit IL-10 early in infection.

Other mutations that affect genes specific for *Yersinia* have also been examined as a basis for attenuating *Y. pestis*. Of note, a *Y. pestis* Δ*yopH* mutant is attenuated and provides a high level of protection against bubonic and pneumonic plague in mice [129]. Studies showed that Δ*pcm* and Δ*nlpD* mutants were attenuated and elicited protective immunity in mice [130, 131], but the immunization with Δ*nlpD* mutants failed to protect guinea pigs [132].

YscN, an ATPase of *Y. pestis*, has a critical role for virulence factor delivery. Bozue et al. indicated that introduction of the *yscN* gene into the *Y. pestis* CO92 led to attenuation following s.c. mice challenges. No mice succumbed to challenge with 4.44×10^4 or 4.44×10^6 CFU of the Δ*yscN* mutant by s.c. route. The attenuation of the *Y. pestis* Δ*yscN* strain suggested the possible use of the strain as a live vaccine. Immunization administered s.c. twice with 10^7 CFU of the *Y. pestis* Δ*yscN* strain in mice provided 90 % protection against s.c. challenge with 180 CFU of the wild-type CO92 strain [133].

Sha et al. showed that the Δ*lpp* Δ*msbB* double mutant *Y. pestis* CO92 strain was grossly compromised in its ability to disseminate to distal organs in mice and in evoking cytokines/chemokines in infected animal tissues. Additionally, mice that survived challenge with the Δ*lpp* Δ*msbB* double mutant, but not the Δ*lpp* or Δ*msbB* single mutant, in a pneumonic plague model were significantly protected against a subsequent lethal wild-type CO92 rechallenge. Thus, the Δ*lpp* Δ*msbB* double mutant might provide a new live attenuated background vaccine candidate strain [134]. Identification of other attenuating mutations that target unique *Y. pestis* virulence genes will be of significant interest for developing safe attenuated *Y. pestis* vaccines.

Zhang et al. constructed the Δ*yscB* mutant *Y. pestis* biovar *Microtus* strain 201 that is avirulent to humans, but virulent to mice. The evaluation of virulence, immunogenicity, and protective efficacy of the Δ*yscB* mutant showed that the Δ*yscB* mutant was severely attenuated, elicited a higher F1-specific antibody titer, and provided protective efficacy against bubonic and pneumonic challenge with *Y. pestis* 141 strain (Antigua biovar) in mouse model. The Δ*yscB* mutant could induce the secretion of both Th1-associated cytokines (IFN-γ, IL-2, and TNF-α) and Th2-associated cytokines (IL-4 and IL-10) [135]. The same group evaluated the protective efficacy of the *Y. pestis Microtus* strain 201 as a live attenuated plague vaccine candidate. Their results showed that this strain was highly attenuated by subcutaneous route, elicited an F1-specific antibody titer similar to the EV76, and provided a similar protective efficacy with the EV76 against bubonic plague in Chinese-origin rhesus macaques. The immunization with *Y. pestis Microtus* strain 201 induced elevated secretion of both Th1-associated cytokines (IFN-γ, IL-2, and TNF-α) and Th2-associated cytokines (IL-4, IL-5, and IL-6), as well as chemokines MCP-1 and IL-8. However, the protected animals developed skin ulcer at challenge site with different severities in most of the 201-immunized and some of the EV-immunized monkeys [136].

Recently, Chopra's group employed high-throughput signature-tagged mutagenic means to identify novel virulence factors from *Y. pestis* CO92. In this study, they found *rbsA* that codes for a putative sugar transport system ATP-binding protein, and *vasK*, a component of the type VI secretion system, exhibited attenuation at 11–12 LD_{50} in a mouse model of pneumonic plague. Combining Δ*rbsA* and Δ*vasK* genes into either the Δ*lpp* single or the Δ*lpp*Δ*msbB* double mutant augmented the attenuation to provide 90–100 % survivability to mice in a pneumonic plague model at 20–50 LD_{50}s. The Δ*lpp*Δ*msbB* Δ*rbsA* triple mutant-infected mice at 50 LD_{50} were 90 % protected upon subsequent challenge with 12 LD_{50} of *Y. pestis* CO92 [137]. They also evaluated whether the deletion of *ail* gene affected virulence of *Y. pestis* CO92. Results indicated that the *ail* single mutant was slightly attenuated compared to the WT bacterium in a mouse model of pneumonic plague; however combining Δ*ail* into Δ*lpp* single mutant strain and Δ*lpp* Δ*msbB* double mutant strain increased their attenuation. LD_{50} of the triple mutant (Δ*ail* Δ*lpp* Δ*msbB*) was equal to 6800 LD_{50} of *Y. pestis* CO92. The mutant-infected animals developed balanced Th1- and Th2-based immune responses based on antibody isotypes. The triple mutant was cleared from mouse organs rapidly, with concurrent decreases in the production of various cytokines and histopathological lesions. Animals surviving from infection with the triple mutant were partially protected against subsequently challenged on day 24 with the bioluminescent *Y. pestis* CO92 strain (20–28 LD_{50}s) by intranasal route; however efficient clearing of the invading pathogen was visualized in real time by in vivo imaging [138].

Y. pseudotuberculosis, a recent ancestor of *Y. pestis* [139], is much less virulent and typically causes an enteric disease that is rarely fatal. Its lifestyle as an enteric pathogen should facilitate its use as an oral vaccine. With the exception of two additional plasmids carried by *Y. pestis* (pPCP1 and pMT1), the two species share >95 % genetic identity and a common virulence plasmid with a conserved colinear backbone [140]. Based on these similarities, the use of avirulent *Y. pseudotuberculosis* strains as a plague vaccine has been explored. Oral immunization with attenuated *Y. pseudotuberculosis* strains stimulates cross-immunity to *Y. pestis* and provides partial protection against pulmonary challenge with *Y. pestis* [141–143]. While protection was not stellar, it was significant, demonstrating the feasibility of using this approach. Derbise et al. showed that an encapsulated *Y. pseudotuberculosis* IP32953 was generated by cloning the *Y. pestis* F1-encoding *caf* operon and expressing it in the attenuated strain. The new V674pF1 strain produced the F1 capsule in vitro and in vivo. Oral inoculation of V674pF1 allowed the colonization of the gut without lesions to Peyer's patches and the spleen. Vaccination induced both humoral and cellular components of immunity, at the systemic (IgG and Th1 cells) and the mucosal levels (IgA and Th17 cells). A single oral dose conferred 100 % protection against a lethal pneumonic plague challenge (33 × LD_{50} of the fully virulent *Y. pestis* CO92 strain) and 94 % against a high challenge dose (3300 × LD_{50}). Both F1 and other *Yersinia* antigens were recognized and V674pF1 efficiently protected against an F1-negative *Y. pestis* [144].

Recently, Sun et al. constructed a *Y. pseudotuberculosis* mutant strain with arabinose-dependent regulated delayed shutoff of *crp* expression (*araC* P_{BAD} *crp*) and replacement of the *msbB* gene with the *E. coli msbB* gene to attenuate it. Then, we inserted the *asd* mutation into this construction to form χ10057 (Δ*asd-206* Δ*msbB868*:: P_{msbB} *msbB* $_{(EC)}$ ΔP_{crp21}::TT *araC* P_{BAD} *crp*) for adapting with a balanced-lethal Asd^+ plasmid to facilitate antigen synthesis. A hybrid protein composed of YopE (1-138aa) and fused with full-length LcrV of *Y. pestis* ($YopE_{Nt138}$-LcrV) was synthesized in χ10057 harboring an Asd^+ plasmid (pYA5199, *$yopE_{Nt138}$-lcrV*) and could be secreted through type III secretion system (T3SS) in vitro and in vivo. Animal studies indicated that mice orally immunized with χ10057(pYA5199) developed similar titers of IgG response to whole-cell lysates of *Y. pestis* (YpL) and LcrV as χ10057(pYA3332, empty plasmid). The χ10057(pYA5199) induced higher level of protection (83 % survival) against intranasal (i.n.) challenge with ~130 LD_{50} (1.3 × 10^4 CFU) of *Y. pestis* KIM6+ (pCD1Ap) than induced by χ10057(pYA3332) (40 % survival). Splenocytes from mice vaccinated with χ10057(pYA5199) produced significant levels of IFN-γ, TNF-α, and IL-17 after restimulation with LcrV and YpL antigens [145].

Additionally, a *Y. pseudotuberculosis* mutant strain combined with chromosome insertion of *caf1R-caf1A-caf1M-caf1* operon and deletions of *yopJ* and *yopK*, χ10068 [pYV-ω2 (Δ*yopJ315*Δ*yopK108*) Δ*lacZ044*:: *caf1R-caf1M-caf1A-caf1*], was constructed. Results indicated that gene insertion and deletion did not affect the growth rate of χ10068 compared to wild-type *Y. pseudotuberculosis* cultured at 26 °C, and also F1 antigen in χ10068 was synthesized at 37 °C (mammal temperature), not at regular culture temperature (26 °C). Immunization with χ10068 primed both antibody responses and specific T-cell responses to F1 and YpL. A single dose of oral immunization with χ10068 provided 70 % protection against a subcutaneous (s.c.) challenge with ~2.6 × 10^5 LD_{50} of *Y. pestis* KIM6+ (pCD1Ap) and 90 % protection against an intranasal (i.n.) challenge with ~500 LD_{50} of *Y. pestis* KIM6+ (pCD1Ap) in mice (manuscript in preparation). As a naturally occurring enteric pathogen, live attenuated *Y. pseudotuberculosis*-based vaccines may be used as an oral vaccine delivered by baits to wild animals, which might reduce the transmission of sylvatic plague to humans by controlling it or eradicating it in its natural rodent hosts.

12.4 Live Vectored Plague Vaccines

In the process of attenuation, an infectious agent is altered so that it becomes harmless or less virulent, while retaining its ability to interact with the host and stimulate a protective immune response [146]. There are many examples of successful live attenuated vaccines delivered by injection, including the current bacterial vaccine for tuberculosis (BCG) [147] and viral vaccines for measles, mumps, rubella, chicken pox, and yellow fever [148]. Rabies vaccines are now available in two different attenuated forms, one for use in humans and one for animals [149]. There are also a number of mucosally delivered live vaccines. These include oral vaccines against poliovirus [150], cholera [151], rotavirus [152], and typhoid fever [153] and the nasally delivered vaccines against influenza [154–156].

Most pathogens gain entry to the host via mucosal surfaces [157, 158]. Thus, parenterally administered vaccines, which may be limited in their capacity to induce mucosal immune responses, may not be the most appropriate form of vaccination for many infections. In contrast, mucosally delivered vaccines have the potential for inducing both systemic and mucosal immunity. Ideally delivered by the oral or intranasal (i.n.) route, such vaccines also offer the advantage of being easier and safer to administer than needle-based delivery [159]. Therefore, live attenuated vaccines have advantages over subunit vaccines as they are typically taken orally, still inducing strong mucosal and durable immunity [24, 158, 160]. In addition, they are often less expensive to manufacture than subunit vaccines. The major disadvantages of live vaccines include inadequate attenuation, particularly in the case of immunocompromised individuals and the potential to revert to virulence. However, application of modern molecular techniques in conjunction with a detailed understanding of the virulence attributes of the delivery vector or, in some cases, of the pathogen itself prior to attenuation makes the latter unlikely in a well-characterized rationally attenuated vaccine. Thus, the development of live vaccines against plague at this time represents an underutilized strategy for preventing this disease.

12.4.1 Live Bacterially Vectored Plague Vaccines

The commensal, nonpathogenic bacterium *Lactococcus lactis* has been used to deliver LcrV [161, 162] with some success. However, most of the studies examining the use of live bacterially vectored vaccines for plague, including work in our laboratory, have focused on exploiting live attenuated *Salmonella* to deliver *Y. pestis* antigens. Live attenuated *Salmonella* have attracted considerable attention as vectors for the delivery of a variety of heterologous vaccine antigens. After delivery by the oral route, the bacteria enter the intestinal subepithelium via M cells and are trafficked via mesenteric lymph nodes to fixed macrophages in the spleen and liver [163–165]. This colonization pathway results in the induction of mucosal and systemic immune responses. Table 12.2 summarizes a number of recent studies utilizing live attenuated *Salmonella* vaccines to deliver *Y. pestis* antigens.

With a few exceptions, all the studies listed in Table 12.3 used *Salmonella* to deliver F1, LcrV, or both. Titball's group has done numerous studies in this area, constructing strains that produce F1-V fusion protein [166], LcrV [167], and F1 capsule on the surface of the cell [291]. Pascual's group took the effort one step further and constructed a *Salmonella* strain that produced F1 as an extracellular capsule and LcrV as a soluble cytoplasmic protein [291]. Sizemore et al. demonstrated that attenuated *S. typhimurium* strains expressing cytoplasmically localized F1-V and V antigen antigens were more immunogenic than strains that secreted or localized plague antigens to the outer membrane [168]. In all of these studies, *S. typhimurium* vaccine strains synthesizing F1 and/or LcrV or fragments of LcrV were demonstrated to elicit humoral and/or cellular immunity against the vectored antigen and to provide some level of protective immunity against either subcutaneous or intranasal challenge with *Y. pestis*. Interestingly, some authors noted that immunization with attenuated *Salmonella* alone (no *Y. pestis* antigens) could provide a low level of protection [169–171], indicating that the use of *Salmonella* as a plague vaccine may provide an additional benefit.

Studies have also described *S. typhi* constructs as candidates for human vaccines. In one study, an *S. typhi* strain synthesizing capsular F1 was demonstrated to elicit protective immunity when used to intranasally immunize mice [172]. A similar vaccine strain was administered intranasally to 7-day old mice [173]. Immunized mice developed mucosal antibody and IFN-γ-secreting cells and were efficiently primed for a later injection of F1 plus alum adjuvant. The *Salmonella* vaccine provided more potent priming than an F1 plus alum prime, demonstrating the potential for using a *Salmonella*-vectored plague vaccine in a prime boost scenario. Recently, Galen et al. used a live attenuated *S. typhi* strain to create a bivalent mucosal plague vaccine that produces both the protective F1 capsular antigen of *Y. pestis* and the LcrV protein required for secretion of virulence effector proteins. To reduce metabolic burden associated with the co-expression of F1 and LcrV within the live vector, we balanced the expression of both antigens by combining plasmid-based expression of F1 with chromosomal expression of LcrV from three independent loci. The serum antibody responses to LPS induced by the optimized bivalent vac-

Table 12.2 Evaluation of *Salmonella* vaccine strains expressing antigen(s) of *Y. pestis*

Strain	Genotype	Antigen(s)	Immunization route	Protective efficacy	References
S. typhimurium					
SL3261	*aroA*	F1	i.v. or oral	i.v. immunization provides complete protection/ oral immunization provides 90 % protection against 50 LD_{50}s of *Y. pestis* GB strain s.c. challenge	[285]
SL3261	*aroA*	F1/V fusion	i.v.	Provides 85 % protection against 50 LD_{50}s of *Y. pestis* GB strain s.c. challenge	[166]
SL3261	*aroA*	F1 expressed under different promoters	Oral	ND (serum antibody responses to F1 reported)	[286]
SL3261	*aroA*	LcrV	Oral	Provides 30 % protection against 97 LD_{50}s of *Y. pestis* GB strain s.c. challenge	[167]
SL3261 SLDAPD	*aroA*	F1	Oral	Oral immunization provides complete protection against 8.8×10^4 LD_{50}s of *Y. pestis* GB strain s.c. challenge	[287]
H683	*asd aroA*	F1 and/or LcrV	Oral	90 % protection against ~1000LD_{50} *Y. pestis* 195/P s.c. challenge and 90 % protection against ~100LD_{50} *Y. pestis* Madagascar	[291]
χ8501	*hisG Δcrp-28 ΔasdA16*	F1 or LcrV	i.n. or oral	3 doses of strain synthesizing F1 or LcrV provided 80 % or 60 % protection against 2×10^3 CFU of *Y. pestis* (Yokohama-R strain) i.p. challenge, respectively	[288]
χ8501	*hisG Δcrp-28 ΔasdA16*	Truncated LcrV (aa131-aa327) and Psn	Oral	Truncated LcrV and Psn immunization provide 80 % protection against 3000 CFU of *Y. pestis* CO92 and 75 % protection against 1300 CFU of *Y. pestis* CO92 by s.c. challenge	[170]
χ9447	Complex genotype[a] Strain displays "regulated in vivo lysis"	LcrV196 on "runaway-like replication plasmid"	Oral	Oral immunization provides 87.5 % survival post-s.c. challenge with 4.49×10^2 CFU or 5.63×10^3 CFU of *Y. pestis* CO92. Oral immunization shows great efficacy with 75 % and 50 % survival post-i.n. challenge with 4.1×10^3 CFU or 4.4×10^4 CFU of *Y. pestis* CO92, respectively	[289]

χ9558	Complex genotype[b]	PsaA	Oral	No protection against s.c. challenge with 4.49×10^2 CFU of *Y. pestis* CO92 and 25 % protection against i.n. challenge with 4.1×10^3 CFU of *Y. pestis* CO92	[181]
χ8501	*hisG* Δ*crp-28* Δ*asdA16*	Psn, HmuR, LcrV196	Oral	Immunization with the strain expressing Psn or LcrV196 afforded nearly full protection against s.c challenge and partial protection against i.n. challenge. While immunization with the strain expressing LcrV5214 or HmuR no protection against s.c. or i.n challenge	[171]
χ9558	Complex genotype[b]	LcrV5214			
S. Typhi					
BRD1116	*aroA aroC htrA*	F1	i.n.	Intranasal immunization provides 65 % protection against 113 CFU of *Y. pestis* GB strain s.c. challenge	[172]
ACAM948CVD (CVD 9080)	*htrA*, *aroC*, *aroD*	F1	i.n.	~80 % protection against i.v. challenge with 10,000 CFU of *Y. pestis* EV76 in 0.2 ml of sterile PBS. $FeCl_2$ (40 μg/mouse) was administered i.p. immediately	[173]
CVD 910FV	*guaBA* and *htrA*	F1, LcrV	i.n.	Mice receiving the optimized bivalent vaccine were fully protected against lethal pulmonary challenge	[290]

Note: *Y. pestis* GB, LD_{50} (s.c.) = 1 CFU [86]; *Y. pestis* 195/P, LD_{50} (s.c.) = 1 CFU [291]; *Y. pestis* Madagascar, LD_{50} (i.n.) ~100 CFU [291]; *Y. pestis* (Yokohama-R strain), LD_{50} (i.p.) = 15 CFU [292]; *Y. pestis* EV76, LD_{50} (i.v.) =56 CFU [173]; *Y. pestis* CO92, LD_{50} (s.c.) =1.9 CFU, LD_{50} (i.n.) ~250 CFU, LD_{50} (aerosol) ~2100 CFU [63, 283, 284]

CFUs colony-forming units, LD_{50} 50 % lethal dose, *s.c.* subcutaneous, *i.n.* intranasal, *i.v.* intravenous, *i.p.* intraperitoneal, *ND* no detected

[a]Δ*pmi-2426* Δ(*gmd-fcl*)*-26* ΔP_{fur81}::TT *araC* P_{BAD} *fur* ΔP_{crp527}::TT *araC* P_{BAD} *crp* Δ*asdA21*::TT *araC* P_{BAD} *c2* Δ*araE25* Δ*araBAD23* Δ*relA198*::*araC* P_{BAD} *lacI* TT ΔP_{murA7}::TT *araC* P_{BAD} *murA* Δ*endA2311*

[b]Δ*pmi-2426* Δ(*gmd-fcl*)*-26* ΔP_{fur81}::TT *araC* P_{BAD} *fur* ΔP_{crp527}::TT *araC* P_{BAD} *crp* Δ*asdA27*::TT *araC* P_{BAD} *c2* Δ*araE25* Δ*araBAD23* Δ*relA198*::*araC* P_{BAD} *lacI* TT Δ*sopB1925* Δ*agfBAC811*

Table 12.3 Immunogenicity and protective efficacy of subunit antigens of *Y. pestis*

Antigen	Function	Immunization route	Protective efficacy	References
LPS	Lipopolysaccharide	s.c.	Bubonic – not protective	[33]
pH 6 antigen	Fimbrial adhesion	s.c.	Provided a significant protection (70 %) against an intranasal infection with *Y. pestis* KIM5 (Pgm-) in the iron dextran-treated mouse model and no any protection against s.c. challenge with fully virulent strains 231 and I-1996	[42, 177]
YopD	Type III system – translocation Yop	s.c. prime and s.c. and i.p. booster	No protection against encapsulated CO92 by s.c. challenge, partial protection against nonencapsulated CO92 by s.c. challenge	[44]
YopE	Type III system – cytotoxin effector Yop	s.c. prime and s.c. and i.p. booster	No protection against encapsulated and nonencapsulated CO92 by s.c. challenge	[44, 277]
YopH	Type III system – PTPase effector Yop	s.c. prime and s.c. and i.p. booster	No protection against encapsulated and nonencapsulated CO92 by s.c. challenge	[44]
YopK	Type III system – regulates Yop release	s.c. prime and s.c. and i.p. booster	No protection against encapsulated and nonencapsulated CO92 by s.c. challenge	[44, 277]
YopM	Type III system – effector Yop	s.c. prime and s.c. and i.p. booster	No protection against encapsulated and nonencapsulated CO92 by s.c. challenge	[44, 278]
YopN	Type III system – regulates Yop release	s.c. prime and s.c. and i.p. booster	No protection against encapsulated and nonencapsulated CO92 by s.c. challenge	[44, 277]
YpkA	Type III system – Ser/Thr kinase effector Yop	s.c. prime and s.c. and i.p. booster	No protection against encapsulated and nonencapsulated CO92 by s.c. challenge, but significantly prolonged mean survival time	[44]
YadC	A member of the oligomeric coiled-coil adhesins	s.c.	Partial protection against $F1^-$ *Y. pestis* challenge in mice	[49]
YscF	*Yersinia* secretory factor F	i.p.	Partial protection against intravenous challenge with *Y. pestis* KIM5	[55]
YPO0606	Autotransporter protein YapF	i.m.	Partial protection against s.c. challenge with *Y. pestis* 201 strain	[41]
YPO1914	Inner membrane ABC transporter YbtQ	i.m.	Partial protection against s.c. challenge with *Y. pestis* 201 strain	[41]
YPO0612	Putative sugar-binding protein	i.m.	Partial protection against s.c. challenge with *Y. pestis* 201 strain	[41]

YPO3119	Heat shock protein HtpG	i.m.	Partial protection against s.c. challenge with *Y. pestis* 201 strain	[41]
YPO3047	Putative sulfatase YdeN	i.m.	Partial protection against s.c. challenge with *Y. pestis* 201 strain	[41]
YPO1377	Putative outer membrane lipoprotein carrier protein	i.m.	Partial protection against s.c. challenge with *Y. pestis* 201 strain	[41]
YPCD1.05c	YopE chaperone SycE/YerA, YopE-targeting protein	i.m.	Partial protection against s.c. challenge with *Y. pestis* 201 strain	[41]
YPO0420	Putative lipoprotein	i.m.	Partial protection against s.c. challenge with *Y. pestis* 201 strain	[41]
YPO3720	Hemolysin activator protein	i.m.	Partial protection against s.c. challenge with *Y. pestis* 201 strain	[41]
Ail/OmpX	Outer membrane protein X	i.m.	Partial protection against bubonic plague, no protection against pneumonic plague in mice model	[56]
OmpA	Outer membrane protein A	i.m.	Partial protection against bubonic plague, no protection against pneumonic plague in mice model	[56]
Pla	The plasminogen activator	i.m.	Pla-immunized mice were 60 % protected against i.n. challenge of Δ*caf1 Y. pestis* CO92 at a 15 LD_{50}, but no protection against s.c. challenge of Δ*caf1 Y. pestis* CO92 at a 500 LD_{50}	[56]
F1	Surface capsule	s.c.	Bubonic and pneumonic, protective	[56, 61–64, 86]
LcrV	Type III system – part of the injectosome	i.m.	Bubonic and pneumonic, protective	[11, 17, 18, 69, 79–84]
F1-V		i.m.	Great protection against bubonic and pneumonic plague	[56, 101]
V10		i.m.	Great protection against bubonic and pneumonic plague	[101]
V10-2		i.m.	Great protection against bubonic and pneumonic plague	[101]
F1 + rV270		i.m.	Great protection against bubonic and pneumonic plague in mice, guinea pigs, rabbits, and rhesus macaques	

cine were indistinguishable from those elicited by the parent strain, suggesting adequate immunogenic capacity maintained through preservation of bacterial fitness. Importantly, mice receiving the optimized bivalent vaccine were fully protected against lethal pulmonary challenge [174].

Our philosophy with regard to *Salmonella*-vectored vaccines for plague is that F1 and LcrV, while highly effective in laboratory models, may not be sufficient to protect against all strains of *Y. pestis*. For example, nonencapsulated (F1-negative) *Y. pestis* mutants can cause chronic, lethal infections in laboratory rats and mice [175, 176]. However, the relevance of these observations has been brought into question by a recent study showing that the impact of the F1 capsule on *Y. pestis* virulence depends on the strain and genotype of mouse used for testing [177]. On the other hand, this concern appears to be relevant to humans as an F1-negative strain of *Y. pestis* has been implicated in an acute fatal human infection [178]. Additionally, there are known polymorphisms of LcrV [179] and F1 [180] that may influence protective efficacy. Therefore, using only two antigens for presentation by *Salmonella* might be insufficient to combat weaponized or naturally occurring *Y. pestis*, leading us to evaluate additional antigens. In addition to LcrV, our group has used *Salmonella* to vector three other *Y. pestis* antigens, Psn [171], HmuR [171], PsaA, also called pH 6 antigen [34], which forms a fibrillar structure on the *Y. pestis* cell surface [181] and YadC, a member of the oligomeric coiled-coil adhesins [50]. Psn and HmuR are outer membrane proteins involved in iron acquisition [182, 183]. The role of PsaA in virulence is not clear [35, 38, 39], but available data indicates it may serve as an adhesin [184] and an antiphagocytic factor [185]. We demonstrated that *Salmonella* delivering Psn elicited significant protective immunity against subcutaneous challenge [171]. We observed partial protection against intranasal challenge, although this did not achieve statistical significance. PsaA was highly immunogenic, eliciting strong serum IgG and mucosal IgA antibodies. However, immunized mice were not protected from subcutaneous challenge, and similar to what we observed with Psn, some immunized mice were protected from intranasal challenge, but the result was not statistically significant [181]. When delivered by our *Salmonella* strains, HmuR was poorly immunogenic and did not confer protection against either challenge route [171]. Mice immunized with *Salmonella* synthesizing YadC or YadC810 are afforded 50 % protection but no protection by immunization with the *Salmonella* strain synthesizing YadBC by s.c. challenge with ~230 LD_{50} of *Y. pestis* CO92. None of these provided protection against i.n. challenge with ~31 LD_{50} of *Y. pestis* CO92 [50]. Recently, we optimized expression of three antigens (LcrV196, F1, and Psn) in our newly improved *Salmonella* strain. Oral immunization with the *Salmonella* strain delivering three antigens provided complete protection against s.c. challenge with 5700 CFU of *Y. pestis* CO92 and 60 % protection against i.n. challenge with 5000 CFU of *Y. pestis* CO92 (manuscript in preparation).

12.4.2 Virally Vectored Live Plague Vaccines

Replication-deficient adenovirus (Ad) vectors are excellent candidates for vaccine platforms as they transfer genes effectively to antigen-presenting cells (APCs) in vivo, with consequent activation of APCs, thus conveying immune adjuvant properties and inducing strong, rapid humoral and cellular immune responses against the transgene product [186]. Crystal's group developed a replication-deficient adenovirus (Ad) gene transfer vector encoding V antigen and demonstrated that a single injection of the recombinant virus elicited strong anti-LcrV serum antibody responses, LcrV-specific $CD4^+$ and $CD8^+$ responses, and protective immunity against an intranasal *Y. pestis* challenge [187, 188]. In a subsequent study, they fused either F1 or LcrV to the C terminus of adenovirus capsid protein, IX. Both constructs elicited strong humoral immunity in mice immunized intramuscularly with greater efficacy than an injection of adjuvanted purified V or F1 [188]. In addition, they also employed adenovirus (Ad) to deliver monoclonal antibodies (MAbs) specific for the *Y. pestis* LcrV antigen, which provided good protection for immunized mice against intranasal challenge with 363 LD_{50} of *Y. pestis* CO92 [189, 190].

Rose's group devised a vaccine utilizing recombinant vesicular stomatitis virus (VSV) vectors expressing the *Y. pestis lcrV* gene [191]. Two intranasal doses elicited high titers of anti-LcrV IgG and protected immunized mice against intranasal challenge. In a follow-up study, the virus was modified to encode a secreted form of LcrV [192]. A single intramuscular dose of 10^9 PFU was sufficient to protect 90 % of the immunized mice from a lethal *Y. pestis* challenge. The secreted LcrV was a more potent vaccine than the previous vaccine that encoded the nonsecreted form, and the authors showed that a high level of protection was dependent on $CD4^+$ but not $CD8^+$ cells and correlated with increased anti-LcrV antibody and a bias toward IgG2a and away from IgG1 isotypes [191, 192]. In addition, vaccinia virus (VACV) is the vaccine for smallpox and a widely used vaccine vector for infectious diseases and cancers [193]. Several groups demonstrated a vaccinia viral vector expressing either *lcrV* or *caf1* (gene encoding F1) as vaccines which are highly immunogenic in BALB/c mice and safe in immunocompromised SCID mice [193–195].

Barton et al. reported that latent infection of mice with either murine gammaherpesvirus 68 or murine cytomegalovirus results in an increased resistance to both intranasal and subcutaneous infections with either *Listeria monocytogenes* or *Y. pestis* [196]. Latency-induced protection is not antigen specific but involves prolonged production of the antiviral cytokine interferon-γ and systemic activation of macrophages, which upregulates the basal activation state of innate immunity against lethal challenge of plague [196]. This observation might be translated into a proactive approach to provide immunity against plague or other pathogens.

A number of reports described studies to develop viral-vectored bait vaccines to be used to control environmental sources of plague. One group constructed a recombinant vaccinia virus to direct synthesis of an F1-V fusion protein with promising results [194, 195]. Orally immunized mice developed high serum antibody titers against the F1-V antigen and achieved 90 % protection against a challenge of 10

LD_{50} of *Y. pestis* [194, 195]. Researchers at the US Geological Survey's National Wildlife Health Center have been developing a recombinant raccoon poxvirus (RCN) that directs synthesis of the F1 antigen (herein designated RCN-F1) as a bait vaccine to protect Prairie dogs (*Cynomys* spp.). Prairie dogs are highly susceptible to *Y. pestis*. In initial studies, the vaccine protected mice from virulent plague challenge [197], and black-tailed prairie dogs (*Cynomys ludovicianus*) vaccinated intramuscularly with RCN-F1 survived subcutaneous challenge with virulent *Y. pestis* [198]. To provide a more practical approach for field vaccination, the RCN-F1 vaccine was incorporated into palatable, edible bait and offered to black-tailed prairie dogs. Antibody titers against *Y. pestis* F1 antigen increased significantly in vaccinated animals, and their survival was significantly higher upon challenge with *Y. pestis* than that of negative controls [198, 199], demonstrating that oral bait immunization of prairie dogs can provide protection against plague.

12.5 Other Vaccines for Plague

DNA Vaccines Recently, a novel methodology of DNA vaccines has been developed in which genes encoding protein antigens are delivered into host cells for enabling antigen production to occur in vivo. There are several advantages about DNA vaccines, such as ease of construction, low cost of mass production, high levels of temperature stability, and the ability to elicit both humoral and cell-mediated immune responses [200, 201]. Bennett et al. first reported that a DNA vaccine vector encoding a fusion of the *Y. pestis* V antigen and glutathione *S*-transferase (GST) under the CMV promoter could induce V antigen-specific antibody in mice [202], which suggested that there was potential for the development of a DNA vaccine against plague. Grosfeld et al. constructed three plasmids expressing the full-length F1, F1 devoid of its putative signal peptide (deF1), and F1 fused to the signal-bearing E3 polypeptide of Semliki Forest virus (E3/F1). Among them, intramuscular vaccination of mice with the plasmid expressing deF1 induced the most effective in eliciting anti-F1 antibodies. Immunization with deF1 DNA conferred complete protection against s.c. challenge with 4000 LD_{50} of the virulent *Y. pestis* Kimberley53 strain [203]. Garmory et al. reported that immunization with the plasmid containing CMV-TE eukaryotic promoters for driving expression of V antigen induced higher IgG2a titers than other five different eukaryotic promoters, but alteration of the codon usage of the *lcrV* gene was not found to improve the anti-LcrV antibody responses [204]. The comparison of DNA vaccines delivered via intramuscular injection with gene gun administration indicated that gene gun delivery induced significantly higher antibody responses to F1 or LcrV and also afforded the highest level of protection against *Y. pestis* challenge [203, 204].

Williamson et al. reported that mice primed with a combination of plasmid DNA encoding either protective antigen of *Bacillus anthracis* or LcrV antigen of *Y. pestis* and boosted with a combination of the recombinant proteins were fully protected

(6/6) against challenge with *Y. pestis*. However, mice primed only with plasmid DNA encoding the V antigen and boosted with rV were partially protected (3/6) against *Y. pestis* challenge, or mice primed and boosted with plasmid DNA encoding the V antigen were poorly protected (1/6) against *Y. pestis* challenge. This protective enhancement may be due to the effect of CpG motifs known to be present in the plasmid DNA construct encoding protective antigen of *B. anthracis* [205]. Recently, Albrecht et al. evaluated the efficacy of multi-agent DNA vaccines consisting of a truncated gene encoding *B. anthracis* lethal factor (LFn) fused to either *Y. pestis* V antigen (V) or *Y. pestis* F1. Mice immunized with above DNA vaccine by gene gun and developed predominantly IgG1 responses to LFn, V, and F1, respectively, were fully protected against a lethal aerosolized *B. anthracis* spore challenge but were partially protected against a lethal aerosolized *Y. pestis* [206]. In addition, Wang et al. demonstrated that a novel DNA vaccine expressing a modified V antigen (LcrV) of *Y. pestis*, with a human tissue plasminogen activator (tPA) signal sequence, elicited strong V-specific antibody responses in BALB/c mice. The tPA-V DNA vaccine provided better protection against intranasal challenge with lethal doses of *Y. pestis* than a DNA vaccine only expressing the wild-type V antigen in mice. Additionally, oligomers formed spontaneously by tPA-V primed a higher IgG2a anti-V antibody response in immunized mice, which tends to induce Th1-type cellular immune response [207]. The same group found that an LcrV DNA vaccine was able to elicit CD8+ T-cell immune responses against specific epitopes of LcrV antigen and induced protective immunity against i.n. challenge with *Y. pestis* [208].

Mixture of IL-12 with protective antigens might enhance vaccine efficacy, because IL-12 has a central function in initiating and regulating cellular immune responses by stimulating gamma interferon (IFN-γ) production in both natural killer (NK) cells and helper T cells [209, 210]. Therefore, Yamanaka et al. construct two bicistronic plasmids encoding an F1-V fusion protein and IL-12 with different copy numbers to produce high or low level of IL-12. Animal experiments indicated that mice immunized with IL-12(Low)/F1-V vaccine were provided the best protective efficacy (80 % survival) against pneumonic challenge of *Y. pestis* compared to mice immunized with IL-12(Low)/F1, IL-12(Low)/V, or IL-12(Low) vector DNA vaccines. However improved expression of IL-12 resulted in lost efficacy when using the IL-12(High)/F1-V DNA vaccine. Although there were differences in the amount of IL-12 produced by the two F1-V DNA vaccines, antibody responses and Th-cell responses to F1- and V-Ags were similar [209, 211].

Nanovaccines In order to improve efficacy of subunit vaccines, nanotechnology platforms have recently been incorporated into vaccine development to overcome certain concerns about vaccines, such as the weak immunogenicity, intrinsic instability in vivo, toxicity, and the need for multiple administrations. Nanocarrier-based delivery systems facilitate uptake of nanovaccines by phagocytic cells, the gut-associated lymphoid tissue, and the mucosa-associated lymphoid tissue, leading to efficient antigen recognition and presentation and offering an opportunity to enhance the humoral and cellular immune responses [212]. In addition, modifying the surfaces of nanocarriers with a variety of targeting moieties allows the delivery of antigens to specific cell surface receptors, thereby stimulating specific and selective immune responses [213].

Zeng et al. firstly demonstrated that intranasal mucosal vaccination of mice with nanostructural and single-molecule force bases of *Y. pestis* V antigen fused with protein anchor (V-PA) loaded on gram-positive enhancer matrix (GEM) vaccine particles elicited robust antigen-specific immune response. This study indicated that high-density, high-stability, specific, and immunological pH-responsive loading of immunogen nanoclusters on vaccine particles could readily be presented to the immune system for induction of strong antigen-specific immune responses [214]. Hoeprich's group tried to immobilize hexa-His-tagged LsrB, a *Y. pestis* transport protein, onto nickel-chelating nanolipoprotein particles (NiNLPs) [215]. Then, they employed a nanolipoprotein particle (NLP)-based vaccine delivery platform to co-deliver both subunit antigens and amphipathic adjuvants such as monophosphoryl lipid A and cholesterol-modified CpG oligodeoxynucleotides, which can bind His-tagged protein antigens. Immunization with this co-delivery platform primed 5–10 times higher antibody titers against His-tagged influenza hemagglutinin 5 and *Y. pestis* LcrV antigens in mice than with coadministration formulations and nonadjuvanted NiNLPs. This study indicated that colocalized delivery of adjuvant and antigen could induce significantly greater immune response in mice than coadministered formulations [216].

Narasimhan's group developed a novel biodegradable polyanhydride nanoparticle encapsulated with F1-V vaccine. Immunization with the nanoparticle-based vaccine induced higher titer and higher avidity anti-F1-V IgG1 antibody that persisted for at least 23 weeks postvaccination in mice than immunization with the recombinant protein F1-V alone and MPLA-adjuvanted F1-V. The single-dose intranasal immunization with nanoparticle-based F1-V vaccine induced long-lived protective immunity against pneumonic plague. After intranasal challenge, no *Y. pestis* were recovered from the lungs, livers, or spleens of mice vaccinated with the nanoparticle-based F1-V vaccine [217, 218]. They also compared the deposition within the lung and internalization by phagocytic cells of F1-V-encapsulated polyanhydride nanoparticles with that of soluble F1-V alone or F1-V adjuvanted with monophosphoryl lipid A (MPLA). Results demonstrated that encapsulation of F1-V into polyanhydride nanoparticles prolonged its presence, while MPLA-adjuvanted F1-V is undetectable within 48 h. Moreover, the inflammation induced by the nanovaccine is mild compared with the marked inflammation induced by the MPLA-adjuvanted F1-V [219]. Further, they investigated the effect of nanoparticle chemistry and its attributes on the kinetics and maturation of the antigen-specific serum antibody response. Results demonstrated that decoration of polyanhydride nanoparticle chemistry facilitated improving antibody titers, avidity, and epitope specificity. Their studies indicated that immunization with nanoparticle encapsulated with subunit vaccine formulations could induce long-lasting and mature antibody responses, which can be used for the rational design of effective vaccine [220].

Gregory et al. decorated the 15-nm gold nanoparticles (AuNPs) with *Y. pestis* F1 antigen using N-hydroxysuccinimide and N-(3-dimethylaminopropyl)-N′-ethylcarbodiimide hydrochloride coupling chemistry. Compared with mice vaccinated with AuNP-F1 in PBS or unconjugated F1 antigen in PBS or Alhydrogel, mice vaccinated with AuNP-F1 in Alhydrogel generated the highest IgG antibody response to F1 antigen [221].

Rao's group employed the bacteriophage T4 DNA nanoparticles carrying reporter genes, vaccine candidates, functional enzymes, and targeting ligands that were efficiently delivered into cells or targeted to antigen-presenting dendritic cells. Mice vaccinated with a single dose of F1-V plague vaccine containing both gene and protein in the T4 head elicited robust antibody and cellular immune responses [222]. Based on this work, they delivered F1mut-V fusion protein by phage T4 nanoparticle, in which F1 eliminated polymerization by transplanting the NH_2-terminal β-strand of F1 to the COOH terminus, but the T-cell epitopes of F1 were retained. The F1mut-V was displayed on phage T4 nanoparticle via the small outer capsid protein, Soc. The immunization with purified F1mut-V monomer adjuvanted Alhydrogel or the T4-decorated F1mut-V without any adjuvant induced robustly immunogenic responses in mice. Vaccination with either the purified F1mut-V mixed with Alhydrogel or T4-decorated F1mut-V without adjuvant provided complete protection to mice and rats against intranasal challenge with high doses of *Y. pestis* CO92. This novel delivery platform might generate new-type vaccines and genetic therapies [223].

New Adjuvanted Plague Vaccines Adjuvants are compounds that enhance the specific immune response against co-inoculated antigens. Thus, antigens mixed with adjuvants are required to achieve the generation of a strong immune response [224]. Jones et al. indicated that intranasal immunization with F1-V formulated with a proteasome-based adjuvant (Protollin) elicited high titers of anti-F1-V IgA in the lungs of mice, whereas intranasal immunization with F1-V alone or intramuscular immunization with Alhydrogel adjuvanted F1-V did not, and also induced higher serum titers of anti-F1-V IgG than those induced by intramuscular Alhydrogel adjuvanted F1-V, which provided 100 % and 80 % protection against aerosol challenge with 170 LD_{50} and against 255 LD_{50} of *Y. pestis*, respectively [225]. This study suggested that Protollin might be a more effective adjuvant than Alhydrogel to induce potent immune responses.

Several studies demonstrated that interleukin-12 (IL-12) could be used as a highly effective vaccine adjuvant against bacterial and viral infections [226–233]. Kumar et al. showed that intranasal vaccination with inactivated *Y. pestis* CO92 (iYp) adjuvanted with IL-12 provided complete protection for mice against an i.n. challenge with a lethal dose of *Y. pestis* CO92. Survival of the vaccinated mice correlated with levels of systemic and lung antibodies, and immunization with iYp adjuvanted with IL-12 reduced pulmonary pathology, proinflammatory cytokines, and the presence of lung lymphoid cell aggregates after *Y. pestis* challenge. Protection against pneumonic plague was partially dependent upon Fc receptors and could be transferred to naïve mice with immune mouse serum and was not dependent upon complement. Interestingly, depletion of CD4 and/or CD8 T cells from vaccinated mice before challenge did not affect their survival. This study suggested the safety, immunogenicity, and protective efficacy of i.n. administered iYp plus IL-12 in a mouse model of pneumonic plague [234].

Do et al. investigated a novel approach based on targeting of dendritic cells using the DEC-205/CD205 receptor (DEC) via the intranasal route as a way to improve mucosal cellular immunity to the vaccine. Intranasal administration of *Y. pestis* LcrV (V) protein fused to anti-DEC antibody together with poly IC as an adjuvant induced high frequencies of IFN-γ-secreting $CD4^+$ T cells in the airway and lung as well as pulmonary IgG and IgA antibodies. Anti-DEC/LcrV was more efficient to induce IFN-γ/TNF-α/IL-2-secreting polyfunctional $CD4^+$ T cells when compared to nontargeted soluble protein vaccine. In addition, the intranasal route of immunization with anti-DEC/LcrV was associated with improved survival upon pulmonary challenge with the virulent *Y. pestis* CO92. Taken together, these data indicate that targeting dendritic cells via the mucosal route is a potential new avenue for the development of a mucosal vaccine against pneumonic plague [235].

Dinc et al. evaluated the efficacy of a novel SA-4-1BBL costimulatory molecule as a Th1 adjuvant to improve cellular responses generated by the rF1-V vaccine. They found that rF1-V recombinant antigen adjuvanted with SA-4-1BBL had better efficacy than with alum in generating CD4+ and CD8+ T cells producing TNF-α and IFN-γ for Th1 responses. However, SA-4-1BBL as a single adjuvant did not generate a significant antibody response against rF1-V. SA-4-1BBL in combination with alum did not increase antibody titers to F1 and LcrV, but significantly increased the ratio of Th1-regulated IgG2c to the Th2-regulated IgG1 in C57BL/6 mice. Protective experiment indicated that a single vaccination with rF1-V adjuvanted with SA-4-1BBL + alum provided better protection against bubonic challenge with *Y. pestis* CO92 than vaccines containing individual adjuvants [236]. The results suggested that SA-4-1BBL as an adjuvant generated a more balanced Th1 cellular and humoral immune response and might be employed to deal with other pathogens.

Plant-Based Plague Vaccine Plants are emerging as an economical alternative to fermentation-based expression systems for producing vaccine antigens from bacteria, viruses, and parasites. Several vaccines in phase I human clinical trials accomplished with plant-made technology have been reviewed in recent paper [237]. The tobacco cell-derived vaccine against the Newcastle disease virus was the first licensed plant-derived vaccine [238]. ZMapp used to fight for Ebola virus infection was a plant-made antibody as a hot spot recently reported by scientific news [239].

Santi et al. used *Nicotiana benthamiana* by using a deconstructed tobacco mosaic virus-based system to rapidly synthesize high levels of the plague antigens F1, V, and fusion protein F1-V. Subcutaneous immunization with these plant-derived purified antigens F1, V, and fusion protein F1-V to guinea pigs generated systemic immune responses and provided protection against an aerosol challenge of virulent *Y. pestis* CO92 [240]. Alvarez et al. employed tomato to deliver *Y. pestis* F1-V antigen as oral vaccine to facilitate antigen delivery and induce mucosal immune response [241]. Mice were primed subcutaneously with bacterially produced F1-V and boosted orally with transgenic tomato fruit. Analysis of antibody responses indicated that the F1- and V-specific IgG1 concentrations were significantly higher in mice boosted with the transgenic tomato fruit than in mice boosted with W.T. (non-transgenic tomato fruit), and F1- and V-specific mucosal IgA was elicited only in mice boosted with oral transgenic F1-V tomato [241]. Also they found that genetically modified tomato with the highest P19 protein levels was correlated with the highest F1-V antigen accumulation [242]. On the other hand, they tried to use Zera technology to induce protein body formation in non-seed tissues. Zera (gamma-Zein ER-accumulating domain) is the N-terminal proline-rich domain of gamma-zein that is sufficient to induce the assembly of protein bodies (PB) formation in the rough endoplasmic reticulum (ER)-derived organelles. Their studies demonstrated that Zera-F1-V protein accumulation was at least 3x higher than F1-V without Zera fusion in three different host plant systems: *N. benthamiana*, *Medicago sativa* (alfalfa), and *Nicotiana tabacum* NT1 cells [243].

Yusibov's group reported that the F1, LcrV, and F1-LcrV antigens of *Y. pestis* fused with a thermostable carrier molecule, lichenase (LicKM), from *Clostridium thermocellum* were synthesized in *N. benthamiana*. Subcutaneous immunization with the purified antigens as vaccines from *N. benthamiana* in cynomolgus macaques induced high titers of serum IgG, mainly of the IgG1 isotype, against both F1 and LcrV. Challenge study indicated that the LcrV-F1 plant-produced vaccine conferred complete protection against aerosolized *Y. pestis* [244, 245]. This study clearly demonstrates the efficacy of a plant-produced plague vaccine candidate in a primate model. Additionally, Del Prete et al. showed that F1, V, and F1-V fusion protein produced in *N. benthamiana* administered to guinea pigs resulted in immunity and protection against an aerosol challenge of virulent *Y. pestis*. The plant-derived F1, V, and F1-V antigens could engage in TLR2 signaling and activated IL-6 and CXCL8 production in monocytes, but did not affect the production of IL-12, IL-10, IL-1β, and CXCL10. Native F1 antigen and recombinant plant-derived F1 (rF1) and rF1-V all induced similar specific T-cell responses [246].

The engineering of chloroplasts for the production of vaccines and biopharmaceuticals has ushered in a new era in biotechnology [247]. Arlen et al. firstly reported high expression of the plague F1-V fusion antigen in chloroplasts as an oral vaccine that provided protection against aerosol challenge of *Y. pestis* CO92 [248]. Rosales-Mendoza et al. synthesized F1-V fusion antigen as an oral vaccine in lettuce [249] or carrot [250] via *Agrobacterium*-mediated transformation. An ELISA analysis confirmed that the expected antigenic F1-V protein was successfully expressed in transgenic lines. Mice immunized subcutaneously with lettuce or carrot expressing the F1-V antigen developed systemic humoral responses [249, 250].

12.6 Perspectives

Y. pestis began to be used as a biological weapon at least 700 years ago and is today considered one of the more likely bioweapons, owing to its extreme virulence, its low infectious dose, and the ease of its transmission. Pneumonic plague caused by inhaling *Y. pestis* has a short incubation period and progresses rapidly to a fatal infection, and victims often become sources of secondary infections as indicated by historical plague epidemics [2, 251]. All these factors stimulate studies on the development of different plague vaccines including live recombinant, purified subunit, recombinant subunit, DNA, chemical fractions, plant-based vaccine.

Currently, two subunit vaccines based on rF1 and rV antigens have passed through phase I and II clinical trials and into the licensing process. Although direct determination of efficacy is not possible due to ethical considerations, human immune responses to subunit plague vaccine have shown good correlation with macaque and mouse immune responses [89]. The third one, rV10 vaccine is currently undergoing US Food and Drug Administration (FDA) pre-Investigational New Drug (pre-IND) authorization review for a future phase I trial [84]. The rV10 vaccine provided complete protection for guinea pigs against intranasal challenge with 1000 MLD *Y. pestis* CAC1 (*caf1A*::IS*1541*) [101]. Passive transfer of rLcrV- or rV10-specific antibodies to BALB/c mice provided protection for them intravenously challenged with *Y. pestis* or *Y. pestis* strains expressing polymorphic *lcrV* KIM D27 (KIMD27) or *lcrV* WA-314 (NCM4) [252]. Also, plague molecular microencapsulated vaccine based on rF1 and rV antigens in Russia has passed through phase I clinical trials (http://www.niigpk.ru/science/klinicheskie-issledovaniya/zavershennye-klinicheskie-issledovaniya). Thus, subunit vaccines are the most promising prospects for human use in current situations.

The live attenuated *Y. pestis* vaccines, EV76 derivatives, can induce great protection against bubonic and pneumonic plague, but these vaccines are not favored in the United States and Europe due to safety concerns [14, 113]. However, live, attenuated *Y. pestis* strains can be altered rationally to become safe vaccines eliciting both humoral and cellular immune responses providing stronger protection against *Y. pestis* than vaccines based on only one or two antigens. Thus, the development of new, improved, live attenuated *Y. pestis* vaccines should be encouraged. In addition, since nearly 30 years ago, live attenuated mutants of *Salmonella* have attracted considerable attention as vectors for the delivery of a variety of heterologous vaccine antigens [253–258]. Much progress was achieved in developing safe efficacious live attenuated *Salmonella* vaccines for poultry [259], swine [260], cattle [261], and humans [262–267]. The vaccines for poultry and swine are licensed and used internationally to prevent infection with a broad host range and host-species adapted *Salmonella* serotypes. *Salmonella*-based vaccine administration is needle-free and easier and less expensive to manufacture (~10 cents/dose for human vaccines) than subunit vaccines. Several groups are endeavored to develop a vastly improved array of means to enhance the safety, efficacy, and utility of *Salmonella* antigen delivery technologies [174, 268, 269]. Use of the *Salmonella* vector system for delivery of multiple *Yersinia* antigens also has good prospects against plague.

Y. pestis is now endemic in rodent populations in many regions around the world making it difficult to control [270]. Bubonic plague is primarily a disease of rodents that is spread by fleas in nature; humans are occasionally infected either by fleabite or by inhalational exposure, usually through a secondary host, for example, a wild rabbit or prairie dog or domestic cat, or, rarely, through another infected person [2]. Plague seroprevalence also indicated that wild animals are involved in the persistence and transmission of *Y. pestis* [271–276]. A viable alternative strategy is to immunize targeted wild rodent populations against plague, especially those living in close contact with humans. This approach would directly address the source of *Y. pestis* and prevent its spread into humans. Importantly, it would also provide a means to potentially reduce plague epidemic in rodent populations at treated areas, which is not possible today with existing tools. As enteric pathogens, *S. typhimurium*- or *Y. pseudotuberculosis*-based vaccines can be oral administration and less expensive to manufacture than subunit vaccines. Therefore, palatable baits containing live *S. typhimurium*- or *Y. pseudotuberculosis*-based vaccines for herd immunization might reduce plague epidemics.

References

1. Wagner DM, Klunk J, Harbeck M, Devault A, Waglechner N, Sahl JW, Enk J, Birdsell DN, Kuch M, Lumibao C, et al. *Yersinia pestis* and the Plague of Justinian 541–543 AD: a genomic analysis. Lancet Infect Dis. 2014;14(4):319–26.
2. Perry RD, Fetherston JD. *Yersinia pestis*–etiologic agent of plague. Clin Microbiol Rev. 1997;10(1):35–66.
3. Schrag SJ, Wiener P. Emerging infectious disease: what are the relative roles of ecology and evolution? Trends Ecol Evol. 1995;10(8):319–24.
4. Duplantier J-M, Duchemin J-B, Chanteau S, Carniel E. From the recent lessons of the Malagasy foci towards a global understanding of the factors involved in plague reemergence. Vet Res. 2005;36(3):437–53.
5. Stenseth NC, Atshabar BB, Begon M, Belmain SR, Bertherat E, Carniel E, Gage KL, Leirs H, Rahalison L. Plague: past, present, and future. PLoS Med. 2008;5(1):e3.
6. Inglesby TV, Dennis DT, Henderson DA, Bartlett JG, Ascher MS, Eitzen E, Fine AD, Friedlander AM, Hauer J, Koerner JF, et al. Plague as a biological weapon: medical and public health management. Working Group on Civilian Biodefense. JAMA. 2000;283(17):2281–90.
7. Guiyoule A, Rasoamanana B, Buchrieser C, Michel P, Chanteau S, Carniel E. Recent emergence of new variants of *Yersinia pestis* in Madagascar. J Clin Microbiol. 1997;35(11):2826–33.
8. Guiyoule A, Gerbaud G, Buchrieser C, Galimand M, Rahalison L, Chanteau S, Courvalin P, Carniel E. Transferable plasmid-mediated resistance to streptomycin in a clinical isolate of *Yersinia pestis*. Emerg Infect Dis. 2001;7(1):43–8.
9. Galimand M, Guiyoule A, Gerbaud G, Rasoamanana B, Chanteau S, Carniel E, Courvalin P. Multidrug resistance in *Yersinia pestis* mediated by a transferable plasmid. N Engl J Med. 1997;337(10):677–80.
10. Smiley ST. Current challenges in the development of vaccines for pneumonic plague. Expert Rev Vaccines. 2008;7(2):209–21.
11. Feodorova VA, Corbel MJ. Prospects for new plague vaccines. Expert Rev Vaccines. 2009;8(12):1721–38.
12. Hawgood BJ. Waldemar Mordecai Haffkine, CIE (1860–1930): prophylactic vaccination against cholera and bubonic plague in British India. J Med Biogr. 2007;15(1):9–19.
13. Cavanaugh DC. K F Meyer's work on plague. J Infect Dis. 1974;129(Suppl):S10–2.
14. Meyer KF. Effectiveness of live or killed plague vaccines in man. Bull World Health Organ. 1970;42(5):653–66.
15. Cavanaugh DC, Elisberg BL, Llewellyn CH, Marshall Jr JD, Rust Jr JH, Williams JE, Meyer KF. Plague immunization. V. Indirect evidence for the efficacy of plague vaccine. J Infect Dis. 1974;129(Suppl):S37–40.

16. Butler T. Plague and other *Yersinia* infections. New York: Plenum Press; 1983.
17. Smiley ST. Immune defense against pneumonic plague. Immunol Rev. 2008;225:256–71.
18. Titball RW, Williamson ED. *Yersinia pestis* (plague) vaccines. Expert Opin Biol Ther. 2004;4(6):965–73.
19. Meyer KF, Cavanaugh DC, Bartelloni PJ, Marshall Jr JD. Plague immunization. I. Past and present trends. J Infect Dis. 1974;129(Suppl):S13–8.
20. Jones SM, Griffin KF, Hodgson I, Williamson ED. Protective efficacy of a fully recombinant plague vaccine in the guinea pig. Vaccine. 2003;21(25–26):3912–8.
21. Russell P, Eley SM, Hibbs SE, Manchee RJ, Stagg AJ, Titball RW. A comparison of Plague vaccine, USP and EV76 vaccine induced protection against *Yersinia pestis* in a murine model. Vaccine. 1995;13(16):1551–6.
22. Demeure C, editor. Live vaccines against plague and *pseudotuberculosis*. Wymondham: Caister Academic Press; 2012.
23. Williamson ED, Oyston PCF, editors. Acellular vaccines against plague. Wymondham: Caister Academic Press; 2012.
24. Dentovskaya SV, Kopylov PK, Ivanov SA, Ageev SA, Anisimov AP. Molecular bases of vaccine-prevention of plague. Mol Genet Microbiol. 2013;28(3):87–98.
25. Pier GB, Meluleni G, Goldberg JB. Clearance of *Pseudomonas aeruginosa* from the murine gastrointestinal-tract is effectively mediated by O-antigen-specific circulating antibodies. Infect Immun. 1995;63(8):2818–25.
26. Roland K, Karaca K, Sizemore D. Expression of *Escherichia coli* antigens in *Salmonella typhimurium* as a vaccine to prevent airsacculitis in chickens. Avian Dis. 2004;48(3):595–605.
27. de Xu Q, Cisar JO, Osorio M, Wai TT, Kopecko DJ. Core-linked LPS expression of *Shigella dysenteriae* serotype 1 O-antigen in live *Salmonella* Typhi vaccine vector Ty21a: preclinical evidence of immunogenicity and protection. Vaccine. 2007;25(33):6167–75.
28. Conlan JW, Shen H, Webb A, Perry MB. Mice vaccinated with the O-antigen of *Francisella tularensis* LVS lipopolysaccharide conjugated to bovine serum albumin develop varying degrees of protective immunity against systemic or aerosol challenge with virulent type A and type B strains of the pathogen. Vaccine. 2002;20(29–30):3465–71.
29. Fulop M, Manchee R, Titball R. Role of lipopolysaccharide and a major outer membrane protein from *Francisella tularensis* in the induction of immunity against tularemia. Vaccine. 1995;13(13):1220–5.
30. Fulop M, Mastroeni P, Green M, Titball RW. Role of antibody to lipopolysaccharide in protection against low- and high-virulence strains of *Francisella tularensis*. Vaccine. 2001;19(31):4465–72.
31. Sandstrom G, Tarnvik A, Wolf-Watz H, Lofgren S. Antigen from *Francisella tularensis*: nonidentity between determinants participating in cell-mediated and humoral reactions. Infect Immun. 1984;45(1):101–6.
32. Thomas RM, Titball RW, Oyston PC, Griffin K, Waters E, Hitchen PG, Michell SL, Grice ID, Wilson JC, Prior JL. The immunologically distinct O antigens from *Francisella tularensis* subspecies tularensis and *Francisella novicida* are both virulence determinants and protective antigens. Infect Immun. 2007;75(1):371–8.
33. Prior JL, Hitchen PG, Williamson ED, Reason AJ, Morris HR, Dell A, Wren BW, Titball RW. Characterization of the lipopolysaccharide of *Yersinia pestis*. Microb Pathog. 2001;30(2):49–57.
34. Ben-Efraim S, Aronson M, Bichowsky-Slomnicki L. New antigen component of *Pasteurella pestis* formed under specified conditions of pH and temperature. J Bacteriol. 1961;81(5):704–14.
35. Cathelyn JS, Crosby SD, Lathem WW, Goldman WE, Miller VL. RovA, a global regulator of *Yersinia pestis*, specifically required for bubonic plague. Proc Natl Acad Sci U S A. 2006;103(36):13514–9.

36. Chauvaux S, Rosso ML, Frangeul L, Lacroix C, Labarre L, Schiavo A, Marceau M, Dillies MA, Foulon J, Coppee JY, et al. Transcriptome analysis of *Yersinia pestis* in human plasma: an approach for discovering bacterial genes involved in septicaemic plague. Microbiology. 2007;153(Pt 9):3112–24.
37. Liu H, Wang H, Qiu J, Wang X, Guo Z, Qiu Y, Zhou D, Han Y, Du Z, Li C, et al. Transcriptional profiling of a mice plague model: insights into interaction between *Yersinia pestis* and its host. J Basic Microbiol. 2009;49(1):92–9.
38. Lindler LE, Klempner MS, Straley SC. *Yersinia pestis* pH 6 antigen: genetic, biochemical, and virulence characterization of a protein involved in the pathogenesis of bubonic plague. Infect Immun. 1990;58(8):2569–77.
39. Anisimov AP, Bakhteeva IV, Panfertsev EA, Svetoch TE, Kravchenko TB, Platonov ME, Titareva GM, Kombarova TI, Ivanov SA, Rakin AV, et al. The subcutaneous inoculation of pH 6 antigen mutants of *Yersinia pestis* does not affect virulence and immune response in mice. J Med Microbiol. 2009;58(Pt 1):26–36.
40. Li B, Jiang L, Song Q, Yang J, Chen Z, Guo Z, Zhou D, Du Z, Song Y, Wang J, et al. Protein microarray for profiling antibody responses to *Yersinia pestis* live vaccine. Infect Immun. 2005;73(6):3734–9.
41. Li B, Zhou L, Guo J, Wang X, Ni B, Ke Y, Zhu Z, Guo Z, Yang R. High-throughput identification of new protective antigens from a *Yersinia pestis* live vaccine by enzyme-linked immunospot assay. Infect Immun. 2009;77(10):4356–61.
42. Galvan EM, Nair MK, Chen H, Del Piero F, Schifferli DM. Biosafety level 2 model of pneumonic plague and protection studies with F1 and Psa. Infect Immun. 2010;78(8):3443–53.
43. Benner GE, Andrews GP, Byrne WR, Strachan SD, Sample AK, Heath DG, Friedlander AM. Immune response to Yersinia outer proteins and other *Yersinia pestis* antigens after experimental plague infection in mice. Infect Immun. 1999;67(4):1922–8.
44. Andrews GP, Strachan ST, Benner GE, Sample AK, Anderson Jr GW, Adamovicz JJ, Welkos SL, Pullen JK, Friedlander AM. Protective efficacy of recombinant Yersinia outer proteins against bubonic plague caused by encapsulated and nonencapsulated *Yersinia pestis*. Infect Immun. 1999;67(3):1533–7.
45. Roggenkamp A, Ackermann N, Jacobi CA, Truelzsch K, Hoffmann H, Heesemann J. Molecular analysis of transport and oligomerization of the *Yersinia enterocolitica* adhesin YadA. J Bacteriol. 2003;185(13):3735–44.
46. Ackermann N, Tiller M, Anding G, Roggenkamp A, Heesemann J. Contribution of trimeric autotransporter C-terminal domains of oligomeric coiled-coil adhesin (Oca) family members YadA, UspA1, EibA, and Hia to translocation of the YadA passenger domain and virulence of *Yersinia enterocolitica*. J Bacteriol. 2008;190(14):5031–43.
47. Forman S, Wulff CR, Myers-Morales T, Cowan C, Perry RD, Straley SC. *yadBC* of *Yersinia pestis*, a new virulence determinant for bubonic plague. Infect Immun. 2008;76(2):578–87.
48. Partial Retraction. *yadBC* of *Yersinia pestis*, a new virulence determinant for bubonic plague. Infect Immun. 2013;81(2):619.
49. Murphy BS, Wulff CR, Garvy BA, Straley SC. *Yersinia pestis* YadC: a novel vaccine candidate against plague. Adv Exp Med Biol. 2007;603:400–14.
50. Sun W, Olinzock J, Wang S, Sanapala S, Curtiss 3rd R. Evaluation of YadC protein delivered by live attenuated *Salmonella* as a vaccine against plague. Pathog Dis. 2014;70(2):119–31.
51. Allaoui A, Schulte R, Cornelis GR. Mutational analysis of the *Yersinia-Enterocolitica* Virc Operon – characterization of Ysce, Yscf, Yscg, Ysci, Yscj, Ysck required for Yop secretion and Ysch Encoding Yopr. Mol Microbiol. 1995;18(2):343–55.
52. Haddix PL, Straley SC. Structure and regulation of the *Yersinia pestis* yscBCDEF operon. J Bacteriol. 1992;174(14):4820–8.
53. Hoiczyk E, Blobel G. Polymerization of a single protein of the pathogen *Yersinia enterocolitica* into needles punctures eukaryotic cells. Proc Natl Acad Sci U S A. 2001;98(8):4669–74.

54. Marenne MN, Journet L, Mota LJ, Cornelis GR. Genetic analysis of the formation of the Ysc-Yop translocation pore in macrophages by *Yersinia enterocolitica*: role of LcrV. YscF and YopN. Microb Pathog. 2003;35(6):243–58.
55. Matson JS, Durick KA, Bradley DS, Nilles ML. Immunization of mice with YscF provides protection from *Yersinia pestis* infections. BMC Microbiol. 2005;5:38.
56. Erova TE, Rosenzweig JA, Sha J, Suarez G, Sierra JC, Kirtley ML, van Lier CJ, Telepnev MV, Motin VL, Chopra AK. Evaluation of protective potential of *yersinia pestis* outer membrane protein antigens as possible candidates for a new-generation recombinant plague vaccine. Clin Vaccine Immunol. 2013;20(2):227–38.
57. Sha J, Endsley JJ, Kirtley ML, Foltz SM, Huante MB, Erova TE, Kozlova EV, Popov VL, Yeager LA, Zudina IV, et al. Characterization of an F1 deletion mutant of *yersinia pestis* CO92, pathogenic role of F1 antigen in bubonic and pneumonic plague, and evaluation of sensitivity and specificity of F1 antigen capture-based dipsticks. J Clin Microbiol. 2011;49(5):1708–15.
58. Rodrigues CG, Carneiro CM, Barbosa CT, Nogueira RA. Antigen F1 from *Yersinia pestis* forms aqueous channels in lipid bilayer membranes. Braz J Med Biol Res. 1992;25(1): 75–9.
59. Galyov EE, Karlishev AV, Chernovskaya TV, Dolgikh DA, Smirnov O, Volkovoy KI, Abramov VM, Zav'yalov VP. Expression of the envelope antigen F1 of *Yersinia pestis* is mediated by the product of *caf1M* gene having homology with the chaperone protein PapD of *Escherichia coli*. FEBS Lett. 1991;286(1–2):79–82.
60. Karlyshev AV, Galyov EE, Smirnov O, Guzayev AP, Abramov VM, Zav'yalov VP. A new gene of the *f1* operon of *Y. pestis* involved in the capsule biogenesis. FEBS Lett. 1992;297(1–2):77–80.
61. Meyer KF, Hightower JA, McCrumb FR. Plague immunization. VI. Vaccination with the fraction I antigen of *Yersinia pestis*. J Infect Dis. 1974;129(Suppl):S41–5.
62. Simpson WJ, Thomas RE, Schwan TG. Recombinant capsular antigen (fraction 1) from *Yersinia pestis* induces a protective antibody response in BALB/c mice. AmJTrop Med Hyg. 1990;43(4):389–96.
63. Andrews GP, Heath DG, Anderson Jr GW, Welkos SL, Friedlander AM. Fraction 1 capsular antigen (F1) purification from *Yersinia pestis* CO92 and from an *Escherichia coli* recombinant strain and efficacy against lethal plague challenge. Infect Immun. 1996;64(6):2180–7.
64. Reddin KM, Easterbrook TJ, Eley SM, Russell P, Mobsby VA, Jones DH, Farrar GH, Williamson ED, Robinson A. Comparison of the immunological and protective responses elicited by microencapsulated formulations of the F1 antigen from *Yersinia pestis*. Vaccine. 1998;16(8):761–7.
65. Meka-Mechenko TV. F1-negative natural *Y. pestis* strains. Adv Exp Med Biol. 2003;529:379–81.
66. Davis KJ, Fritz DL, Pitt ML, Welkos SL, Worsham PL, Friedlander AM. Pathology of experimental pneumonic plague produced by fraction 1-positive and fraction 1-negative *Yersinia pestis* in African green monkeys (*Cercopithecus aethiops*). Arch Pathol Lab Med. 1996;120(2):156–63.
67. Quenee LE, Cornelius CA, Ciletti NA, Elli D, Schneewind O. *Yersinia pestis* caf1 variants and the limits of plague vaccine protection. Infect Immun. 2008;76(5):2025–36.
68. Cornelius CA, Quenee LE, Overheim KA, Koster F, Brasel TL, Elli D, Ciletti NA, Schneewind O. Immunization with recombinant V10 protects *cynomolgus macaques* from lethal pneumonic plague. Infect Immun. 2008;76(12):5588–97.
69. Quenee LE, Ciletti N, Berube B, Krausz T, Elli D, Hermanas T, Schneewind O. Plague in Guinea pigs and its prevention by subunit vaccines. Am J Pathol. 2011;178(4):1689–700.
70. Cornelis GR, Boland A, Boyd AP, Geuijen C, Iriarte M, Neyt C, Sory MP, Stainier I. The virulence plasmid of *Yersinia*, an antihost genome. Microbiol Mol Biol Rev. 1998;62(4):1315–52.
71. Cornelis GR. *Yersinia* type III secretion: send in the effectors. J Cell Biol. 2002;158(3): 401–8.

72. Perry RD, Straley SC, Fetherston JD, Rose DJ, Gregor J, Blattner FR. DNA sequencing and analysis of the low-Ca^{2+} -response plasmid pCD1 of *Yersinia pestis* KIM5. Infect Immun. 1998;66(10):4611–23.
73. Brubaker RR. Interleukin-10 and inhibition of innate immunity to Yersiniae: roles of Yops and LcrV (V antigen). Infect Immun. 2003;71(7):3673–81.
74. Cornelis GR. The type III secretion injectisome. Nat Rev Microbiol. 2006;4(11):811–25.
75. Mueller CA, Broz P, Cornelis GR. The type III secretion system tip complex and translocon. Mol Microbiol. 2008;68(5):1085–95.
76. Ivanov MI, Hill J, Bliska JB. Direct neutralization of type III effector translocation by the variable region of a monoclonal antibody to *Yersinia pestis* LcrV. Clin Vaccine Immunol. 2014;21(5):667–73.
77. Skrzypek EaS SC. LcrG, a secreted protein involved in negative regulation of the low-calcium response in *Yersinia pestis*. J Bacteriol. 1993;175:3520–8.
78. Sarker MR, Neyt C, Stainier I, Cornelis GR. The *Yersinia* Yop virulon: LcrV is required for extrusion of the translocators YopB and YopD. J Bacteriol. 1998;180(5):1207–14.
79. Oyston PC, Williamson ED. Prophylaxis and therapy of plague. Expert Rev Anti-Infe. 2013;11(8):817–29.
80. Anderson Jr GW, Leary SE, Williamson ED, Titball RW, Welkos SL, Worsham PL, Friedlander AM. Recombinant V antigen protects mice against pneumonic and bubonic plague caused by F1-capsule-positive and -negative strains of *Yersinia pestis*. Infect Immun. 1996;64(11):4580–5.
81. Leary SE, Williamson ED, Griffin KF, Russell P, Eley SM, Titball RW. Active immunization with recombinant V antigen from *Yersinia pestis* protects mice against plague. Infect Immun. 1995;63(8):2854–8.
82. Motin VL, Nakajima R, Smirnov GB, Brubaker RR. Passive immunity to yersiniae mediated by anti-recombinant V antigen and protein A-V antigen fusion peptide. Infect Immun. 1994;62(10):4192–201.
83. Une T, Brubaker RR. Roles of V antigen in promoting virulence and immunity in yersiniae. J Immunol. 1984;133(4):2226–30.
84. Quenee LE, Schneewind O. Plague vaccines and the molecular basis of immunity against *Yersinia pestis*. Hum Vaccin. 2009;5(12):817–23.
85. Williamson ED, Eley SM, Stagg AJ, Green M, Russell P, Titball RW. A single dose sub-unit vaccine protects against pneumonic plague. Vaccine. 2000;19(4–5):566–71.
86. Williamson ED, Eley SM, Griffin KF, Green M, Russell P, Leary SE, Oyston PC, Easterbrook T, Reddin KM, Robinson A, et al. A new improved sub-unit vaccine for plague: the basis of protection. FEMS Immunol Med Microbiol. 1995;12(3–4):223–30.
87. Williamson ED, Vesey PM, Gillhespy KJ, Eley SM, Green M, Titball RW. An IgG1 titre to the F1 and V antigens correlates with protection against plague in the mouse model. Clin Exp Immunol. 1999;116(1):107–14.
88. Williamson ED, Flick-Smith HC, Waters E, Miller J, Hodgson I, Le Butt CS, Hill J. Immunogenicity of the rF1+rV vaccine for plague with identification of potential immune correlates. Microb Pathog. 2007;42(1):11–21.
89. Williamson ED, Flick-Smith HC, Lebutt C, Rowland CA, Jones SM, Waters EL, Gwyther RJ, Miller J, Packer PJ, Irving M. Human immune response to a plague vaccine comprising recombinant F1 and V antigens. Infect Immun. 2005;73(6):3598–608.
90. Powell BS, Andrews GP, Enama JT, Jendrek S, Bolt C, Worsham P, Pullen JK, Ribot W, Hines H, Smith L, et al. Design and testing for a nontagged F1-V fusion protein as vaccine antigen against bubonic and pneumonic plague. Biotechnol Prog. 2005;21(5):1490–510.
91. Heath DG, Anderson Jr GW, Mauro JM, Welkos SL, Andrews GP, Adamovicz J, Friedlander AM. Protection against experimental bubonic and pneumonic plague by a recombinant capsular F1-V antigen fusion protein vaccine. Vaccine. 1998;16(11–12):1131–7.
92. Sing A, Roggenkamp A, Geiger AM, Heesemann J. *Yersinia enterocolitica* evasion of the host innate immune response by V antigen-induced IL-10 production of macrophages is abrogated in IL-10-deficient mice. J Immunol. 2002;168(3):1315–21.

93. Sing A, Rost D, Tvardovskaia N, Roggenkamp A, Wiedemann A, Kirschning CJ, Aepfelbacher M, Heesemann J. Yersinia V-antigen exploits toll-like receptor 2 and CD14 for interleukin 10-mediated immunosuppression. J Exp Med. 2002;196(8):1017–24.
94. Reithmeier-Rost D, Bierschenk S, Filippova N, Schroder-Braunstein J, Sing A. Yersinia V antigen induces both TLR homo- and heterotolerance in an IL-10-involving manner. Cell Immunol. 2004;231(1–2):63–74.
95. Nedialkov YA, Motin VL, Brubaker RR. Resistance to lipopolysaccharide mediated by the *Yersinia pestis* V antigen-polyhistidine fusion peptide: amplification of interleukin-10. Infect Immun. 1997;65(4):1196–203.
96. Pouliot K, Pan N, Wang S, Lu S, Lien E, Goguen JD. Evaluation of the role of LcrV-Toll-like receptor 2-mediated immunomodulation in the virulence of *Yersinia pestis*. Infect Immun. 2007;75(7):3571–80.
97. Reithmeier-Rost D, Hill J, Elvin SJ, Williamson D, Dittmann S, Schmid A, Wilharm G, Sing A. The weak interaction of LcrV and TLR2 does not contribute to the virulence of Yersinia pestis. Microbes Infect. 2007;9(8):997–1002.
98. Sun W, Curtiss 3rd R. Amino acid substitutions in LcrV at putative sites of interaction with toll-like receptor 2 do not affect the virulence of *Yersinia pestis*. Microb Pathog. 2012;53(5–6):198–206.
99. DeBord KL, Anderson DM, Marketon MM, Overheim KA, DePaolo RW, Ciletti NA, Jabri B, Schneewind O. Immunogenicity and protective immunity against bubonic plague and pneumonic plague by immunization of mice with the recombinant V10 antigen, a variant of LcrV. Infect Immun. 2006;74(8):4910–4.
100. Overheim KA, Depaolo RW, Debord KL, Morrin EM, Anderson DM, Green NM, Brubaker RR, Jabri B, Schneewind O. LcrV plague vaccine with altered immunomodulatory properties. Infect Immun. 2005;73(8):5152–9.
101. Quenee LE, Ciletti NA, Elli D, Hermanas TM, Schneewind O. Prevention of pneumonic plague in mice, rats, guinea pigs and non-human primates with clinical grade rV10, rV10-2 or F1-V vaccines. Vaccine. 2011;29(38):6572–83.
102. Qi Z, Zhou L, Zhang Q, Ren L, Dai R, Wu B, Wang T, Zhu Z, Yang Y, Cui B, et al. Comparison of mouse, guinea pig and rabbit models for evaluation of plague subunit vaccine F1+rV270. Vaccine. 2010;28(6):1655–60.
103. Qiu Y, Liu Y, Qi Z, Wang W, Kou Z, Zhang Q, Liu G, Liu T, Yang Y, Yang X, et al. Comparison of immunological responses of plague vaccines F1+rV270 and EV76 in Chinese-origin rhesus macaque, *Macaca mulatta*. Scand J Immunol. 2010;72(5):425–33.
104. Tian G, Qiu Y, Qi Z, Wu X, Zhang Q, Bi Y, Yang Y, Li Y, Yang X, Xin Y, et al. Histopathological observation of immunized rhesus macaques with plague vaccines after subcutaneous infection of Yersinia pestis. Plos One. 2011;6(4):e19260.
105. Qi ZZ, Zhao HH, Zhang QW, Bi YJ, Ren LL, Zhang XC, Yang HQ, Yang XY, Wang Q, Li CX, et al. Acquisition of maternal antibodies both from the placenta and by lactation protects mouse offspring from *Yersinia pestis* challenge. Clin Vaccine Immunol. 2012;19(11): 1746–50.
106. Wang Z, Zhou L, Qi Z, Zhang Q, Dai R, Yang Y, Cui B, Wang H, Yang R, Wang X. Long-term observation of subunit vaccine F1-rV270 against *Yersinia pestis* in mice. Clin Vaccine Immunol. 2010;17(1):199–201.
107. Jackson S, Burrows TW. The virulence-enhancing effect of iron on non-pigmented mutants of virulent strains of Pasteurella pestis. Br J Exp Pathol. 1956;37:577–83.
108. Girard G. Immunity in plague. Acquisitions supplied by 30 years of work ont the "*Pasteurella pestis* EV" (Girard and Robic) strain. Biol Med (Paris). 1963;52:631–731.
109. Saltykova RA, Faibich MM. Experience from a 30-year study of the stability of the properties of the plague vaccine strain EV in the USSR. Zh Mikrobiol Epidemiol Immunobiol. 1975;6:3–8.
110. Meyer KF, Smith G, Foster L, Brookman M, Sung M. Live, attenuated *Yersinia pestis* vaccine: virulent in nonhuman primates, harmless to guinea pigs. J Infect Dis. 1974;129(Suppl): S85–120.

111. Hallett AF, Isaacson M, Meyer KF. Pathogenicity and immunogenic efficacy of a live attenuated plaque vaccine in vervet monkeys. Infect Immun. 1973;8(6):876–81.
112. Une T, Brubaker RR. In vivo comparison of avirulent Vwa- and Pgm- or Pstr phenotypes of yersiniae. Infect Immun. 1984;43(3):895–900.
113. Liu S, Bayles DO, Mason TM, Wilkinson BJ. A cold-sensitive *Listeria monocytogenes* mutant has a transposon insertion in a gene encoding a putative membrane protein and shows altered (p)ppGpp levels. Appl Environ Microbiol. 2006;72(6):3955–9.
114. Na HS, Kim HJ, Lee HC, Hong Y, Rhee JH, Choy HE. Immune response induced by *Salmonella typhimurium* defective in ppGpp synthesis. Vaccine. 2006;24(12):2027–34.
115. Curtiss III R, Kelly SM. *Salmonella typhimurium* deletion mutants lacking adenylate cyclase and cyclic AMP receptor protein are avirulent and immunogenic. Infect Immun. 1987;55(12):3035–43.
116. Zhan L, Han Y, Yang L, Geng J, Li Y, Gao H, Guo Z, Fan W, Li G, Zhang L, et al. The cyclic AMP receptor protein, CRP, is required for both virulence and expression of the minimal CRP regulon in *Yersinia pestis* biovar microtus. Infect Immun. 2008;76(11):5028–37.
117. Sun W, Roland KL, Branger CG, Kuang X, Curtiss 3rd R. The role of *relA* and *spoT* in *Yersinia pestis* KIM5+ pathogenicity. Plos One. 2009;4(8):e6720.
118. Sun W, Roland KL, Kuang X, Branger CG, Curtiss 3rd R. *Yersinia pestis* with regulated delayed attenuation as a vaccine candidate to induce protective immunity against plague. Infect Immun. 2010;78(3):1304–13.
119. Curtiss III R, Wanda SY, Gunn BM, Zhang X, Tinge SA, Ananthnarayan V, Mo H, Wang S, Kong W. *Salmonella enterica* serovar Typhimurium strains with regulated delayed attenuation in vivo. Infect Immun. 2009;77(3):1071–82.
120. Knirel YA, Anisimov AP. Lipopolysaccharide of *Yersinia pestis*, the cause of plague: structure, genetics. Biological properties. Acta Nat. 2012;4(3):46–58.
121. Rebeil R, Ernst RK, Jarrett CO, Adams KN, Miller SI, Hinnebusch BJ. Characterization of late acyltransferase genes of *Yersinia pestis* and their role in temperature-dependent lipid A variation. J Bacteriol. 2006;188(4):1381–8.
122. Poltorak A, Ricciardi-Castagnoli P, Citterio S, Beutler B. Physical contact between lipopolysaccharide and toll-like receptor 4 revealed by genetic complementation. Proc Natl Acad Sci U S A. 2000;97(5):2163–7.
123. Hoshino K, Takeuchi O, Kawai T, Sanjo H, Ogawa T, Takeda Y, Takeda K, Akira S. Cutting edge: Toll-like receptor 4 (TLR4)-deficient mice are hyporesponsive to lipopolysaccharide: evidence for TLR4 as the Lps gene product. J Immunol. 1999;162(7):3749–52.
124. Lien E, Means TK, Heine H, Yoshimura A, Kusumoto S, Fukase K, Fenton MJ, Oikawa M, Qureshi N, Monks B, et al. Toll-like receptor 4 imparts ligand-specific recognition of bacterial lipopolysaccharide. J Clin Invest. 2000;105(4):497–504.
125. Poltorak A, He X, Smirnova I, Liu MY, Van Huffel C, Du X, Birdwell D, Alejos E, Silva M, Galanos C, et al. Defective LPS signaling in C3H/HeJ and C57BL/10ScCr mice: mutations in Tlr4 gene. Science. 1998;282(5396):2085–8.
126. Montminy SW, Khan N, McGrath S, Walkowicz MJ, Sharp F, Conlon JE, Fukase K, Kusumoto S, Sweet C, Miyake K, et al. Virulence factors of *Yersinia pestis* are overcome by a strong lipopolysaccharide response. Nat Immunol. 2006;7(10):1066–73.
127. Sun W, Six DA, Reynolds CM, Chung HS, Raetz CR, Curtiss 3rd R. Pathogenicity of *Yersinia pestis* synthesis of 1-dephosphorylated lipid A. Infect Immun. 2013;81(4):1172–85.
128. Sun W, Six D, Kuang XY, Roland KL, Raetz CRH, Curtiss 3rd R. A live attenuated strain of *Yersinia pestis* KIM as a vaccine against plague. Vaccine. 2011;29:2986–98.
129. Bubeck SS, Dube PH. *Yersinia pestis* CO92 delta *yopH* is a potent live, attenuated plague vaccine. Clin Vaccine Immunol. 2007;14(9):1235–8.
130. Flashner Y, Mamroud E, Tidhar A, Ber R, Aftalion M, Gur D, Lazar S, Zvi A, Bino T, Ariel N, et al. Generation of *Yersinia pestis* attenuated strains by signature-tagged mutagenesis in search of novel vaccine candidates. Infect Immun. 2004;72(2):908–15.

131. Tidhar A, Flashner Y, Cohen S, Levi Y, Zauberman A, Gur D, Aftalion M, Elhanany E, Zvi A, Shafferman A, et al. The *NlpD* lipoprotein is a novel *Yersinia pestis* virulence factor essential for the development of plague. Plos One. 2009;4(9):e7023.
132. Dentovskaya SV, Ivanov SA, Kopylov P, Shaikhutdinova RZ, Platonov ME, Kombarova TI, Gapel'chenkova TV, Balakhonov SV, Anisimov AP. Selective protective potency of *Yersinia pestis* Delta*nlpD* mutants. Acta Nat. 2015;7(1):102–8.
133. Bozue J, Cote CK, Webster W, Bassett A, Tobery S, Little S, Swietnicki W. A *Yersinia pestis* YscN ATPase mutant functions as a live attenuated vaccine against bubonic plague in mice. Fems Microbiol Lett. 2012;332(2):113–21.
134. Sha J, Kirtley ML, van Lier CJ, Wang S, Erova TE, Kozlova EV, Cao A, Cong Y, Fitts EC, Rosenzweig JA, et al. Deletion of the Braun lipoprotein-encoding gene and altering the function of lipopolysaccharide attenuate the plague bacterium. Infect Immun. 2013;81(3): 815–28.
135. Zhang X, Qi Z, Du Z, Bi Y, Zhang Q, Tan Y, Yang H, Xin Y, Yang R, Wang X. A live attenuated strain of *Yersinia pestis* Delta*yscB* provides protection against bubonic and pneumonic plagues in mouse model. Vaccine. 2013;31(22):2539–42.
136. Zhang Q, Wang Q, Tian G, Qi Z, Zhang X, Wu X, Qiu Y, Bi Y, Yang X, Xin Y, et al. *Yersinia pestis* biovar Microtus strain 201, an avirulent strain to humans, provides protection against bubonic plague in rhesus macaques. Hum Vaccin Immunother. 2014;10(2):368–77.
137. Ponnusamy D, Fitts EC, Sha J, Erova TE, Kozlova EV, Kirtley ML, Tiner BL, Andersson JA, Chopra AK. High-throughput signature-tagged mutagenic approach to identify novel virulence factors of *Yersinia pestis* CO92 in a mouse model of infection. Infect Immun. 2015;83(5):2065–81.
138. Tiner BL, Sha J, Kirtley ML, Erova TE, Popov VL, Baze WB, van Lier CJ, Ponnusamy D, Andersson JA, Motin VL, et al. Combinational deletion of three membrane protein-encoding genes highly attenuates *Yersinia pestis* while retaining immunogenicity in a mouse model of pneumonic plague. Infect Immun. 2015;83(4):1318–38.
139. Achtman M, Zurth K, Morelli G, Torrea G, Guiyoule A, Carniel E. *Yersinia pestis*, the cause of plague, is a recently emerged clone of *Yersinia pseudotuberculosis*. Proc Natl Acad Sci U S A. 1999;96(24):14043–8.
140. Chain PS, Carniel E, Larimer FW, Lamerdin J, Stoutland PO, Regala WM, Georgescu AM, Vergez LM, Land ML, Motin VL, et al. Insights into the evolution of *Yersinia pestis* through whole-genome comparison with *Yersinia pseudotuberculosis*. Proc Natl Acad Sci U S A. 2004;101(38):13826–31.
141. Taylor VL, Titball RW, Oyston PC. Oral immunization with a *dam* mutant of *Yersinia pseudotuberculosis* protects against plague. Microbiology. 2005;151(Pt 6):1919–26.
142. Blisnick T, Ave P, Huerre M, Carniel E, Demeure CE. Oral vaccination against bubonic plague using a live avirulent *Yersinia pseudotuberculosis* strain. Infect Immun. 2008;76(8):3808–16.
143. Okan NA, Mena P, Benach JL, Bliska JB, Karzai AW. The *smpB-ssrA* mutant of *Yersinia pestis* functions as a live attenuated vaccine to protect mice against pulmonary plague infection. Infect Immun. 2010;78(3):1284–93.
144. Derbise A, Cerda Marin A, Ave P, Blisnick T, Huerre M, Carniel E, Demeure CE. An encapsulated *Yersinia pseudotuberculosis* is a highly efficient vaccine against pneumonic plague. Plos Neglect Trop D. 2012;6(2):e1528.
145. Sun W, Sanapala S, Henderson JC, Sam S, Olinzock J, Trent MS, Curtiss 3rd R. LcrV delivered via type III secretion system of live attenuated *Yersinia pseudotuberculosis* enhances immunogenicity against pneumonic plague. Infect Immun. 2014;82(10):4390–404.
146. Badgett MR, Auer A, Carmichael LE, Parrish CR, Bull JJ. Evolutionary dynamics of viral attenuation. J Virol. 2002;76(20):10524–9.
147. Behr MA. BCG–different strains, different vaccines? Lancet Infect Dis. 2002;2(2):86–92.
148. Male D, Brostoff J, Roth D and Roitt V. Immunology, 7th ed. Philadelphia: Elsevier; 2006.
149. Rabies. www.immunizationinfo.org/vaccines/rabies. 2010.

150. Modlin JF. Poliomyelitis in the United States: the final chapter? JAMA. 2004;292(14): 1749–51.
151. Levine MM. Immunization against bacterial diseases of the intestine. J Pediatr Gastroenterol Nutr. 2000;31(4):336–55.
152. Kapikian AZ, Hoshino Y, Chanock RM, Perez-Schael I. Efficacy of a quadrivalent rhesus rotavirus-based human rotavirus vaccine aimed at preventing severe rotavirus diarrhea in infants and young children. J Infect Dis. 1996;174 Suppl 1:S65–72.
153. Germanier R, Fuer E. Isolation and characterization of Gal E mutant Ty 21a of *Salmonella typhi*: a candidate strain for a live, oral typhoid vaccine. J Infect Diseas. 1975;131(5):553–8.
154. Minne A, Jaworska J, Gerhold K, Ahrens B, Avagyan A, Vanbever R, Matricardi PM, Schmidt AC, Hamelmann E. Intranasal delivery of whole influenza vaccine prevents subsequent allergen-induced sensitization and airway hyper-reactivity in mice. Clin Exp Allergy. 2007;37(8):1250–8.
155. Lin K. Intranasal influenza vaccine may be a safe, effective option for many children. J Pediatr. 2007;151(1):102–3.
156. Vinella D, Albrecht C, Cashel M, D'Ari R. Iron limitation induces SpoT-dependent accumulation of ppGpp in Escherichia coli. Mol Microbiol. 2005;56(4):958–70.
157. Medina E, Guzman CA. Modulation of immune responses following antigen administration by mucosal route. FEMS Immunol Med Microbiol. 2000;27(4):305–11.
158. Neutra MR, Kozlowski PA. Mucosal vaccines: the promise and the challenge. Nat Rev Immunol. 2006;6(2):148–58.
159. Garmory HS, Leary SE, Griffin KF, Williamson ED, Brown KA, Titball RW. The use of live attenuated bacteria as a delivery system for heterologous antigens. J Drug Target. 2003; 11(8–10):471–9.
160. Atkins HS, Morton M, Griffin KF, Stokes MG, Nataro JP, Titball RW. Recombinant *Salmonella* vaccines for biodefence. Vaccine. 2006;24(15):2710–7.
161. Ramirez K, Ditamo Y, Rodriguez L, Picking WL, van Roosmalen ML, Leenhouts K, Pasetti MF. Neonatal mucosal immunization with a non-living, non-genetically modified Lactococcus lactis vaccine carrier induces systemic and local Th1-type immunity and protects against lethal bacterial infection. Mucosal Immunol. 2010;3(2):159–71.
162. Foligne B, Dessein R, Marceau M, Poiret S, Chamaillard M, Pot B, Simonet M, Daniel C. Prevention and treatment of colitis with *Lactococcus lactis* secreting the immunomodulatory *Yersinia* LcrV protein. Gastroenterology. 2007;133(3):862–74.
163. Carter PB, Collins FM. The route of enteric infection in normal mice. J Exp Med. 1974;139(5):1189–203.
164. Giannasca PJ, Neutra MR. Interactions of microorganisms with intestinal M cells: mucosal invasion and induction of secretory immunity. Infect Agents Dis. 1993;2(4):242–8.
165. Nix RN, Altschuler SE, Henson PM, Detweiler CS. Hemophagocytic macrophages harbor *Salmonella enterica* during persistent infection. PLoS Pathog. 2007;3(12):e193.
166. Leary SE, Griffin KF, Garmory HS, Williamson ED, Titball RW. Expression of an F1/V fusion protein in attenuated *Salmonella typhimurium* and protection of mice against plague. Microb Pathog. 1997;23(3):167–79.
167. Garmory HS, Griffin KF, Brown KA, Titball RW. Oral immunisation with live *aroA* attenuated *Salmonella enterica* serovar Typhimurium expressing the *Yersinia pestis* V antigen protects mice against plague. Vaccine. 2003;21(21–22):3051–7.
168. Sizemore DR, Warner EA, Lawrence JA, Thomas LJ, Roland KL, Killeen KP. Construction and screening of attenuated Delta*phoP/Q Salmonella typhimurium* vectored plague vaccine candidates. Human Vaccin Immunother. 2012;8(3):371–83.
169. Titball RW, Howells AM, Oyston PC, Williamson ED. Expression of the *Yersinia pestis* capsular antigen (F1 antigen) on the surface of an *aroA* mutant of *Salmonella* typhimurium induces high levels of protection against plague. Infect Immun. 1997;65(5):1926–30.

170. Branger CG, Fetherston JD, Perry RD, Curtiss III R. Oral vaccination with different antigens from *Yersinia pestis* KIM delivered by live attenuated *Salmonella* typhimurium elicits a protective immune response against plague. Adv Exp Med Biol. 2007;603:387–99.
171. Branger CG, Sun W, Torres-Escobar A, Perry R, Roland KL, Fetherston J, Curtiss 3rd R. Evaluation of Psn, HmuR and a modified LcrV protein delivered to mice by live attenuated *Salmonella* as a vaccine against bubonic and pneumonic *Yersinia pestis* challenge. Vaccine. 2010;29(2):274–82.
172. Morton M, Garmory HS, Perkins SD, O'Dowd AM, Griffin KF, Turner AK, Bennett AM, Titball RW. A *Salmonella enterica* serovar Typhi vaccine expressing *Yersinia pestis* F1 antigen on its surface provides protection against plague in mice. Vaccine. 2004;22(20):2524–32.
173. Ramirez K, Capozzo AV, Lloyd SA, Sztein MB, Nataro JP, Pasetti MF. Mucosally delivered *Salmonella* typhi expressing the *Yersinia pestis* F1 antigen elicits mucosal and systemic immunity early in life and primes the neonatal immune system for a vigorous anamnestic response to parenteral F1 boost. J Immunol. 2009;182(2):1211–22.
174. Galen JE, Wang JY, Carrasco JA, Lloyd SA, Mellado-Sanchez G, Diaz-McNair J, Franco O, Buskirk AD, Nataro JP, Pasetti MF. A bivalent typhoid live vector vaccine expressing both chromosome- and plasmid-encoded *Yersinia pestis* antigens fully protects against murine lethal pulmonary plague infection. Infect Immun. 2015;83(1):161–72.
175. Williams JE, Cavanaugh DC. Chronic infections in laboratory rodents from inoculation of nonencapsulated plague bacilli (*Yersinia pestis*). Experientia. 1983;39(4):408–9.
176. Williams JE, Cavanaugh DC. Potential for rat plague from nonencapsulated variants of the plague bacillus (*Yersinia pestis*). Experientia. 1984;40(7):739–40.
177. Weening EH, Cathelyn JS, Kaufman G, Lawrenz MB, Price P, Goldman WE, Miller VL. The dependence of the *Yersinia pestis* capsule on pathogenesis is influenced by the mouse background. Infect Immun. 2011;79(2):644–52.
178. Winter CC, Cherry WB, Moody MD. An unusual strain of *Pasteurella pestis* isolated from a fatal human case of plague. Bull World Health Organ. 1960;23:408–9.
179. Anisimov AP, Dentovskaya SV, Panfertsev EA, Svetoch TE, Kopylov PK, Segelke BW, Zemla A, Telepnev MV, Motin VL. Amino acid and structural variability of *Yersinia pestis* LcrV protein. Infect Genet Evol. 2009;10(1):137–45.
180. Revazishvili T, Rajanna C, Bakanidze L, Tsertsvadze N, Imnadze P, O'Connell K, Kreger A, Stine OC, Morris Jr JG, Sulakvelidze A. Characterisation of *Yersinia pestis* isolates from natural foci of plague in the Republic of Georgia, and their relationship to *Y. pestis* isolates from other countries. Clin Microbiol Infect. 2008;14(5):429–36.
181. Torres-Escobar A, Juarez-Rodriguez MD, Branger CG, Curtiss 3rd R. Evaluation of the humoral immune response in mice orally vaccinated with live recombinant attenuated *Salmonella enterica* delivering a secreted form of *Yersinia pestis* PsaA. Vaccine. 2010;28(36):5810–6.
182. Fetherston JD, Lillard Jr JW, Perry RD. Analysis of the pesticin receptor from *Yersinia pestis*: role in iron-deficient growth and possible regulation by its siderophore. J Bacteriol. 1995;177(7):1824–33.
183. Hornung JM, Jones HA, Perry RD. The hmu locus of *Yersinia pestis* is essential for utilization of free haemin and haem–protein complexes as iron sources. Mol Microbiol. 1996;20(4):725–39.
184. Yang Y, Merriam JJ, Mueller JP, Isberg RR. The *psa* locus is responsible for thermoinducible binding of *Yersinia pseudotuberculosis* to cultured cells. Infect Immun. 1996;64(7):2483–9.
185. Huang XZ, Lindler LE. The pH 6 antigen is an antiphagocytic factor produced by *Yersinia pestis* independent of Yersinia outer proteins and capsule antigen. Infect Immun. 2004;72(12):7212–9.
186. Wilson JM. Adenoviruses as gene-delivery vehicles. N Engl J Med. 1996;334(18):1185–7.
187. Chiuchiolo MJ, Boyer JL, Krause A, Senina S, Hackett NR, Crystal RG. Protective immunity against respiratory tract challenge with *Yersinia pestis* in mice immunized with an adenovirus-based vaccine vector expressing V antigen. J Infect Dis. 2006;194(9):1249–57.

188. Boyer JL, Sofer-Podesta C, Ang J, Hackett NR, Chiuchiolo MJ, Senina S, Perlin D, Crystal RG. Protective immunity against a lethal respiratory *Yersinia pestis* challenge induced by V antigen or the F1 capsular antigen incorporated into adenovirus capsid. Hum Gene Ther. 2010;21(7):891–901.
189. Sofer-Podesta C, Ang J, Hackett NR, Senina S, Perlin D, Crystal RG, Boyer JL. Adenovirus-mediated delivery of an anti-V antigen monoclonal antibody protects mice against a lethal *Yersinia pestis* challenge. Infect Immun. 2009;77(4):1561–8.
190. Van Blarcom TJ, Sofer-Podesta C, Ang J, Boyer JL, Crystal RG, Georgiou G. Affinity maturation of an anti-V antigen IgG expressed in situ through adenovirus gene delivery confers enhanced protection against *Yersinia pestis* challenge. Gene Ther. 2010;17(7):913–21.
191. Palin A, Chattopadhyay A, Park S, Delmas G, Suresh R, Senina S, Perlin DS, Rose JK. An optimized vaccine vector based on recombinant vesicular stomatitis virus gives high-level, long-term protection against *Yersinia pestis* challenge. Vaccine. 2007;25(4):741–50.
192. Chattopadhyay A, Park S, Delmas G, Suresh R, Senina S, Perlin DS, Rose JK. Single-dose, virus-vectored vaccine protection against *Yersinia pestis* challenge: CD4+ cells are required at the time of challenge for optimal protection. Vaccine. 2008;26(50):6329–37.
193. Embry A, Meng X, Cantwell A, Dube PH, Xiang Y. Enhancement of immune response to an antigen delivered by vaccinia virus by displaying the antigen on the surface of intracellular mature virion. Vaccine. 2011;29(33):5331–9.
194. Brewoo JN, Powell TD, Stinchcomb DT, Osorio JE. Efficacy and safety of a modified vaccinia Ankara (MVA) vectored plague vaccine in mice. Vaccine. 2010;28:5891–9.
195. Bhattacharya D, Mecsas J, Hu LT. Development of a vaccinia virus based reservoir-targeted vaccine against *Yersinia pestis*. Vaccine. 2010;28(48):7683–9.
196. Barton ES, White DW, Cathelyn JS, Brett-McClellan KA, Engle M, Diamond MS, Miller VL, Virgin HW. Herpesvirus latency confers symbiotic protection from bacterial infection. Nature. 2007;447(7142):326–9.
197. Osorio JE, Powell TD, Frank RS, Moss K, Haanes EJ, Smith SR, Rocke TE, Stinchcomb DT. Recombinant raccoon pox vaccine protects mice against lethal plague. Vaccine. 2003;21(11–12):1232–8.
198. Mencher JS, Smith SR, Powell TD, Stinchcomb DT, Osorio JE, Rocke TE. Protection of black-tailed prairie dogs (*Cynomys ludovicianus*) against plague after voluntary consumption of baits containing recombinant raccoon poxvirus vaccine. Infect Immun. 2004;72(9):5502–5.
199. Rocke TE, Smith SR, Stinchcomb DT, Osorio JE. Immunization of black-tailed prairie dog against plague through consumption of vaccine-laden baits. J Wildl Dis. 2008;44(4):930–7.
200. Reyes-Sandoval A, Ertl HC. DNA vaccines. Curr Mol Med. 2001;1(2):217–43.
201. Gurunathan S, Klinman DM, Seder RA. DNA vaccines: immunology, application, and optimization*. Annu Rev Immunol. 2000;18:927–74.
202. Bennett AM, Phillpotts RJ, Perkins SD, Jacobs SC, Williamson ED. Gene gun mediated vaccination is superior to manual delivery for immunisation with DNA vaccines expressing protective antigens from *Yersinia pestis* or Venezuelan Equine Encephalitis virus. Vaccine. 1999;18(7–8):588–96.
203. Grosfeld H, Cohen S, Bino T, Flashner Y, Ber R, Mamroud E, Kronman C, Shafferman A, Velan B. Effective protective immunity to *Yersinia pestis* infection conferred by DNA vaccine coding for derivatives of the F1 capsular antigen. Infect Immun. 2003;71(1):374–83.
204. Garmory HS, Freeman D, Brown KA, Titball RW. Protection against plague afforded by immunisation with DNA vaccines optimised for expression of the *Yersinia pestis* V antigen. Vaccine. 2004;22(8):947–57.
205. Williamson ED, Bennett AM, Perkins SD, Beedham RJ, Miller J, Baillie LW. Co-immunisation with a plasmid DNA cocktail primes mice against anthrax and plague. Vaccine. 2002;20(23–24):2933–41.
206. Albrecht MT, Eyles JE, Baillie LW, Keane-Myers AM. Immunogenicity and efficacy of an anthrax/plague DNA fusion vaccine in a mouse model. Fems Immunol Med Microbiol. 2012;65(3):505–9.

207. Wang S, Heilman D, Liu F, Giehl T, Joshi S, Huang X, Chou TH, Goguen J, Lu S. A DNA vaccine producing LcrV antigen in oligomers is effective in protecting mice from lethal mucosal challenge of plague. Vaccine. 2004;22(25–26):3348–57.
208. Wang S, Goguen JD, Li F, Lu S. Involvement of CD8+ T cell-mediated immune responses in LcrV DNA vaccine induced protection against lethal *Yersinia pestis* challenge. Vaccine. 2011;29(39):6802–9.
209. Yamanaka H, Hoyt T, Yang X, Golden S, Bosio CM, Crist K, Becker T, Maddaloni M, Pascual DW. A nasal interleukin-12 DNA vaccine coexpressing *Yersinia pestis* F1-V fusion protein confers protection against pneumonic plague. Infect Immun. 2008;76(10):4564–73.
210. Gately MK, Renzetti LM, Magram J, Stern AS, Adorini L, Gubler U, Presky DH. The interleukin-12/interleukin-12-receptor system: role in normal and pathologic immune responses. Annu Rev Immunol. 1998;16:495–521.
211. Yamanaka H, Hoyt T, Bowen R, Yang X, Crist K, Golden S, Maddaloni M, Pascual DW. An IL-12 DNA vaccine co-expressing *Yersinia pestis* antigens protects against pneumonic plague. Vaccine. 2009;27(1):80–7.
212. Nasir A. Nanotechnology in vaccine development: a step forward. J Invest Dermatol. 2009;129(5):1055–9.
213. Plebanski M, Xiang SD. Nanotechnology and vaccine development: methods to study and manipulate the interaction of nanoparticles with the immune system. Methods. 2013;60(3):225.
214. Zeng G, Chen J, Zhong L, Wang R, Jiang L, Cai J, Yan L, Huang D, Chen CY, Chen ZW. NSOM- and AFM-based nanotechnology elucidates nano-structural and atomic-force features of a *Y. pestis* V immunogen-containing particle vaccine capable of eliciting robust response. Proteomics. 2009;9(6):1538–47.
215. Blanchette CD, Fischer NO, Corzett M, Bench G, Hoeprich PD. Kinetic analysis of his-tagged protein binding to nickel-chelating nanolipoprotein particles. Bioconjug Chem. 2010;21(7):1321–30.
216. Fischer NO, Rasley A, Corzett M, Hwang MH, Hoeprich PD, Blanchette CD. Colocalized delivery of adjuvant and antigen using nanolipoprotein particles enhances the immune response to recombinant antigens. J Am Chem Soc. 2013;135(6):2044–7.
217. Ulery BD, Kumar D, Ramer-Tait AE, Metzger DW, Wannemuehler MJ, Narasimhan B. Design of a protective single-dose intranasal nanoparticle-based vaccine platform for respiratory infectious diseases. Plos One. 2011;6(3):e17642.
218. Ulery BD, Petersen LK, Phanse Y, Kong CS, Broderick SR, Kumar D, Ramer-Tait AE, Carrillo-Conde B, Rajan K, Wannemuehler MJ, et al. Rational design of pathogen-mimicking amphiphilic materials as nanoadjuvants. Sci Rep. 2011;1:198.
219. Ross KA, Haughney SL, Petersen LK, Boggiatto P, Wannemuehler MJ, Narasimhan B. Lung deposition and cellular uptake behavior of pathogen-mimicking nanovaccines in the first 48 hours. Adv Healthcare Mater. 2014;3(7):1071–7.
220. Haughney SL, Ross KA, Boggiatto PM, Wannemuehler MJ, Narasimhan B. Effect of nanovaccine chemistry on humoral immune response kinetics and maturation. Nanoscale. 2014;6(22):13770–8.
221. Gregory AE, Williamson ED, Prior JL, Butcher WA, Thompson IJ, Shaw AM, Titball RW. Conjugation of *Y. pestis* F1-antigen to gold nanoparticles improves immunogenicity. Vaccine. 2012;30(48):6777–82.
222. Tao P, Mahalingam M, Marasa BS, Zhang ZH, Chopra AK, Rao VB. In vitro and in vivo delivery of genes and proteins using the bacteriophage T4 DNA packaging machine. Proc Natl Acad Sci U S A. 2013;110(15):5846–51.
223. Tao P, Mahalingam M, Kirtley ML, van Lier CJ, Sha J, Yeager LA, Chopra AK, Rao VB. Mutated and bacteriophage T4 nanoparticle arrayed F1-V immunogens from *Yersinia pestis* as next generation plague vaccines. PLoS Pathog. 2013;9(7):e1003495.
224. Petrovsky N, Aguilar JC. Vaccine adjuvants: current state and future trends. Immunol Cell Biol. 2004;82(5):488–96.
225. Jones T, Adamovicz JJ, Cyr SL, Bolt CR, Bellerose N, Pitt LM, Lowell GH, Burt DS. Intranasal Protollin/F1-V vaccine elicits respiratory and serum antibody responses and protects mice against lethal aerosolized plague infection. Vaccine. 2006;24(10):1625–32.

226. Arulanandam BP, Lynch JM, Briles DE, Hollingshead S, Metzger DW. Intranasal vaccination with pneumococcal surface protein A and interleukin-12 augments antibody-mediated opsonization and protective immunity against *Streptococcus pneumoniae* infection. Infect Immun. 2001;69(11):6718–24.
227. Baron SD, Singh R, Metzger DW. Inactivated *Francisella tularensis* live vaccine strain protects against respiratory tularemia by intranasal vaccination in an immunoglobulin A-dependent fashion. Infect Immun. 2007;75(5):2152–62.
228. Metzger DW, Buchanan JM, Collins JT, Lester TL, Murray KS, Van Cleave VH, Vogel LA, Dunnick WA. Enhancement of humoral immunity by interleukin-12. Ann N Y Acad Sci. 1996;795:100–15.
229. McKnight AJ, Zimmer GJ, Fogelman I, Wolf SF, Abbas AK. Effects of IL-12 on helper T cell-dependent immune responses in vivo. J Immunol. 1994;152(5):2172–9.
230. Germann T, Bongartz M, Dlugonska H, Hess H, Schmitt E, Kolbe L, Kolsch E, Podlaski FJ, Gately MK, Rude E. Interleukin-12 profoundly up-regulates the synthesis of antigen-specific complement-fixing IgG2a, IgG2b and IgG3 antibody subclasses in vivo. Eur J Immunol. 1995;25(3):823–9.
231. Bliss J, VanCleave V, Murray K, Wiencis A, Ketchum M, Maylor R, Haire T, Resmini C, Abbas AK, Wolf SF. IL-12, as an adjuvant, promotes a T helper 1 cell, but does not suppress a T helper 2 cell recall response. J Immunol. 1996;156(3):887–94.
232. Arulanandam BP, Mittler JN, Lee WT, O'Toole M, Metzger DW. Neonatal administration of IL-12 enhances the protective efficacy of antiviral vaccines. J Immunol. 2000;164(7):3698–704.
233. Arulanandam BP, O'Toole M, Metzger DW. Intranasal interleukin-12 is a powerful adjuvant for protective mucosal immunity. J Infect Dis. 1999;180(4):940–9.
234. Kumar D, Kirimanjeswara G, Metzger DW. Intranasal administration of an inactivated *Yersinia pestis* vaccine with interleukin-12 generates protective immunity against pneumonic plague. Clin Vaccine Immunol. 2011;18(11):1925–35.
235. Do Y, Didierlaurent AM, Ryu S, Koh H, Park CG, Park S, Perlin DS, Powell BS, Steinman RM. Induction of pulmonary mucosal immune responses with a protein vaccine targeted to the DEC-205/CD205 receptor. Vaccine. 2012;30(45):6359–67.
236. Dinc G, Pennington JM, Yolcu ES, Lawrenz MB, Shirwan H. Improving the Th1 cellular efficacy of the lead *Yersinia pestis* rF1-V subunit vaccine using SA-4-1BBL as a novel adjuvant. Vaccine. 2014;32(39):5035–40.
237. Alvarez ML, Cardineau GA. Prevention of bubonic and pneumonic plague using plant-derived vaccines. Biotechnol Adv. 2010;28(1):184–96.
238. Release UN: USDA issues license for plant-cell-produced Newcastle disease vaccine for chickens. January, 2006. Available from http://www.aphis.usda.gov/newsroom/content/2006/01/ndvaccine.shtml. In; 2006.
239. Rybicki EP. Plant-based vaccines against viruses. Virol J. 2014;11(1):205.
240. Santi L, Giritch A, Roy CJ, Marillonnet S, Klimyuk V, Gleba Y, Webb R, Arntzen CJ, Mason HS. Protection conferred by recombinant *Yersinia pestis* antigens produced by a rapid and highly scalable plant expression system. Proc Natl Acad Sci U S A. 2006;103(4):861–6.
241. Alvarez ML, Pinyerd HL, Crisantes JD, Rigano MM, Pinkhasov J, Walmsley AM, Mason HS, Cardineau GA. Plant-made subunit vaccine against pneumonic and bubonic plague is orally immunogenic in mice. Vaccine. 2006;24(14):2477–90.
242. Alvarez ML, Pinyerd HL, Topal E, Cardineau GA. P19-dependent and P19-independent reversion of F1-V gene silencing in tomato. Plant Mol Biol. 2008;68(1–2):61–79.
243. Alvarez ML, Topal E, Martin F, Cardineau GA. Higher accumulation of F1-V fusion recombinant protein in plants after induction of protein body formation. Plant Mol Biol. 2010;72(1–2):75–89.
244. Mett V, Lyons J, Musiychuk K, Chichester JA, Brasil T, Couch R, Sherwood R, Palmer GA, Streatfield SJ, Yusibov V. A plant-produced plague vaccine candidate confers protection to monkeys. Vaccine. 2007;25(16):3014–7.

245. Chichester JA, Musiychuk K, Farrance CE, Mett V, Lyons J, Yusibov V. A single component two-valent LcrV-F1 vaccine protects non-human primates against pneumonic plague. Vaccine. 2009;27(25–26):3471–4.
246. Del Prete G, Santi L, Andrianaivoarimanana V, Amedei A, Domarle O, D'Elios MM, Arntzen CJ, Rahalison L, Mason HS. Plant-derived recombinant F1, V, and F1-V fusion antigens of Yersinia pestis activate human cells of the innate and adaptive immune system. Int J Immunopathol Pharmacol. 2009;22(1):133–43.
247. Daniell H. Production of biopharmaceuticals and vaccines in plants via the chloroplast genome. Biotechnol J. 2006;1(10):1071–9.
248. Arlen PA, Singleton M, Adamovicz JJ, Ding Y, Davoodi-Semiromi A, Daniell H. Effective plague vaccination via oral delivery of plant cells expressing F1-V antigens in chloroplasts. Infect Immun. 2008;76(8):3640–50.
249. Rosales-Mendoza S, Soria-Guerra RE, Moreno-Fierros L, Alpuche-Solis AG, Martinez-Gonzalez L, Korban SS. Expression of an immunogenic F1-V fusion protein in lettuce as a plant-based vaccine against plague. Planta. 2010;232(2):409–16.
250. Rosales-Mendoza S, Soria-Guerra RE, Moreno-Fierros L, Han Y, Alpuche-Solis AG, Korban SS. Transgenic carrot tap roots expressing an immunogenic F1-V fusion protein from Yersinia pestis are immunogenic in mice. J Plant Physiol. 2011;168(2):174–80.
251. Straley SC, Starnbach MN. Yersinia: strategies that thwart immune defenses. In: Cunningham MW, Fujinami RS, editors. Effects of microbes on the immune system. Philadelphia: Lippincott Williams and Wilkins; 2000.
252. Miller NC, Quenee LE, Elli D, Ciletti NA, Schneewind O. Polymorphisms in the lcrV gene of *Yersinia enterocolitica* and their effect on plague protective immunity. Infect Immun. 2012;80(4):1572–82.
253. Formal SB, Baron LS, Kopecko DJ, Washington O, Powell C, Life CA. Construction of a potential bivalent vaccine strain: introduction of *Shigella sonnei* form I antigen genes into the *galE Salmonella typhi* Ty21a typhoid vaccine strain. Infect Immun. 1981;34(3):746–50.
254. Gonzalez C, Hone D, Noriega FR, Tacket CO, Davis JR, Losonsky G, Nataro JP, Hoffman S, Malik A, Nardin E, et al. *Salmonella typhi* vaccine strain CVD 908 expressing the circumsporozoite protein of *Plasmodium falciparum*: strain construction and safety and immunogenicity in humans. J Infect Diseas. 1994;169(4):927–31.
255. Hone DM, Lewis GK, Beier M, Harris A, McDaniels T, Fouts TR. Expression of human immunodeficiency virus antigens in an attenuated *Salmonella typhi* vector vaccine. Dev Biol Stand. 1994;82:159–62.
256. Nardelli-Haefliger D, Kraehenbuhl JP, Curtiss 3rd R, Schodel F, Potts A, Kelly S, De Grandi P. Oral and rectal immunization of adult female volunteers with a recombinant attenuated *Salmonella typhi* vaccine strain. Infect Immun. 1996;64(12):5219–24.
257. Tacket CO, Kelly SM, Schodel F, Losonsky G, Nataro JP, Edelman R, Levine MM, Curtiss 3rd R. Safety and immunogenicity in humans of an attenuated *Salmonella* typhi vaccine vector strain expressing plasmid-encoded *hepatitis* B antigens stabilized by the Asd-balanced lethal vector system. Infect Immun. 1997;65(8):3381–5.
258. Viret JF, Cryz SJ, Lang AB, Favre D. Molecular-cloning and characterization of the genetic-determinants that express the complete Shigella serotype-D (*Shigella-sonnei*) lipopolysaccharide in heterologous live attenuated vaccine strains. Mol Microbiol. 1993;7(2):239–52.
259. Cerquetti MC, Gherardi MM. Orally administered attenuated *Salmonella enteritidis* reduces chicken cecal carriage of virulent Salmonella challenge organisms. Vet Microbiol. 2000;76(2):185–92.
260. Barrow PA, Page K, Lovell MA. The virulence for gnotobiotic pigs of live attenuated vaccine strains of *Salmonella enterica* serovars Typhimurium and Enteritidis. Vaccine. 2001;19(25–26):3432–6.
261. Van der Walt ML, Vorster JH, Steyn HC, Greeff AS. Auxotrophic, plasmid-cured *Salmonella enterica* serovar Typhimurium for use as a live vaccine in calves. Vet Microbiol. 2001; 80(4):373–81.

262. Tacket CO, Galen J, Sztein MB, Losonsky G, Wyant TL, Nataro J, Wasserman SS, Edelman R, Chatfield S, Dougan G, et al. Safety and immune responses to attenuated *Salmonella enterica* serovar Typhi oral live vector vaccines expressing tetanus toxin fragment C. Clin Immunol. 2000;97(2):146–53.
263. Lowe DC, Savidge TC, Pickard D, Eckmann L, Kagnoff MF, Dougan G, Chatfield SN. Characterization of candidate live oral *Salmonella typhi* vaccine strains harboring defined mutations in *aroA*, *aroC*, and *htrA*. Infect Immun. 1999;67(2):700–7.
264. Khan S, Chatfield S, Stratford R, Bedwell J, Bentley M, Sulsh S, Giemza R, Smith S, Bongard E, Cosgrove CA, et al. Ability of SPI2 mutant of *S. typhi* to effectively induce antibody responses to the mucosal antigen enterotoxigenic *E. coli* heat labile toxin B subunit after oral delivery to humans. Vaccine. 2007;25(21):4175–82.
265. Hohmann EL, Oletta CA, Killeen KP. Miller SI: *phoP/phoQ*-deleted *Salmonella typhi* (Ty800) is a safe and immunogenic single-dose typhoid fever vaccine in volunteers. J Infect Dis. 1996;173(6):1408–14.
266. Hindle Z, Chatfield SN, Phillimore J, Bentley M, Johnson J, Cosgrove CA, Ghaem-Maghami M, Sexton A, Khan M, Brennan FR, et al. Characterization of *Salmonella enterica* derivatives harboring defined *aroC* and Salmonella pathogenicity island 2 type III secretion system (*ssaV*) mutations by immunization of healthy volunteers. Infect Immun. 2002;70(7): 3457–67.
267. DiPetrillo MD, Tibbetts T, Kleanthous H, Killeen KP, Hohmann EL. Safety and immunogenicity of *phoP/phoQ*-deleted *Salmonella typhi* expressing *Helicobacter pylori* urease in adult volunteers. Vaccine. 1999;18(5–6):449–59.
268. Curtiss 3rd R, Xin W, Li Y, Kong W, Wanda SY, Gunn B, Wang S. New technologies in using recombinant attenuated *Salmonella* vaccine vectors. Crit Rev Immunol. 2010;30(3):255–70.
269. Wang S, Kong Q, Curtiss 3rd R. New technologies in developing recombinant attenuated *Salmonella* vaccine vectors. Microb Pathog. 2013;58:17–28.
270. Keeling MJ, Gilligan CA. Metapopulation dynamics of bubonic plague. Nature. 2000;407(6806):903–6.
271. Bevins SN, Tracey JA, Franklin SP, Schmit VL, Macmillan ML, Gage KL, Schriefer ME, Logan KA, Sweanor LL, Alldredge MW, et al. Wild felids as hosts for human plague. Western United States. Emerg Infect Dis. 2009;15(12):2021–4.
272. Poland JD, Barnes AM, Herman JJ. Human bubonic plague from exposure to a naturally infected wild carnivore. Am J Epidemiol. 1973;97(5):332–7.
273. Eisen RJ, Petersen JM, Higgins CL, Wong D, Levy CE, Mead PS, Schriefer ME, Griffith KS, Gage KL, Beard CB. Persistence of *Yersinia pestis* in soil under natural conditions. Emerg Infect Dis. 2008;14(6):941–3.
274. Griffin KA, Martin DJ, Rosen LE, Sirochman MA, Walsh DP, Wolfe LL, Miller MW. Detection of *Yersinia pestis* DNA in prairie dog-associated fleas by polymerase chain reaction assay of purified DNA. J Wildl Dis. 2010;46(2):636–43.
275. Wobeser G, Campbell GD, Dallaire A, McBurney S. Tularemia, plague, yersiniosis, and Tyzzer's disease in wild rodents and lagomorphs in Canada: a review. Can Vet J. 2009;50(12):1251–6.
276. Eisen RJ, Holmes JL, Schotthoefer AM, Vetter SM, Montenieri JA, Gage KL. Demonstration of early-phase transmission of *Yersinia pestis* by the mouse flea, *Aetheca wagneri* (Siphonaptera: Ceratophylidae), and implications for the role of deer mice as enzootic reservoirs. J Med Entomol. 2008;45(6):1160–4.
277. Leary SEC, Griffin KF, Galyov EE, Hewer J, Williamson ED, Holmstrom A, Forsberg A, Titball RW. Yersinia outer proteins (YOPS) E, K and N are antigenic but non-protective compared to V antigen, in a murine model of bubonic plague. Microb Pathog. 1999;26(3): 159–69.
278. Nemeth J, Straley SC. Effect of Yersinia pestis YopM on experimental plague. Infect Immun. 1997;65(3):924–30.
279. Robinson VL, Oyston PC, Titball RW. A *dam* mutant of *Yersinia pestis* is attenuated and induces protection against plague. Fems Microbiol Lett. 2005;252(2):251–6.

280. Feodorova VA, Pan'kina LN, Savostina EP, Sayapina LV, Motin VL, Dentovskaya SV, Shaikhutdinova RZ, Ivanov SA, Lindner B, Kondakova AN, et al. A *Yersinia pestis lpxM*-mutant live vaccine induces enhanced immunity against bubonic plague in mice and guinea pigs. Vaccine. 2007;25(44):7620–8.
281. Oyston PC, Mellado-Sanchez G, Pasetti MF, Nataro JP, Titball RW, Atkins HS. A *Yersinia pestis guaBA* mutant is attenuated in virulence and provides protection against plague in a mouse model of infection. Microb Pathog. 2010;48(5):191–5.
282. Feodorova VA, Devdariani ZL. Immunogeneity and structural organisation of some pLCR-encoded proteins of *Yersinia pestis*. J Med Microbiol. 2001;50(1):13–22.
283. Lathem WW, Crosby SD, Miller VL, Goldman WE. Progression of primary pneumonic plague: a mouse model of infection, pathology, and bacterial transcriptional activity. Proc Natl Acad Sci U S A. 2005;102(49):17786–91.
284. Agar SL, Sha J, Foltz SM, Erova TE, Walberg KG, Parham TE, Baze WB, Suarez G, Peterson JW, Chopra AK. Characterization of a mouse model of plague after aerosolization of *Yersinia pestis* CO92. Microbiology. 2008;154(Pt 7):1939–48.
285. Oyston PC, Williamson ED, Leary SE, Eley SM, Griffin KF, Titball RW. Immunization with live recombinant *Salmonella* typhimurium *aroA* producing F1 antigen protects against plague. Infect Immun. 1995;63(2):563–8.
286. Bullifent HL, Griffin KF, Jones SM, Yates A, Harrington L, Titball RW. Antibody responses to *Yersinia pestis* F1-antigen expressed in *Salmonella* typhimurium *aroA* from in vivo-inducible promoters. Vaccine. 2000;18(24):2668–76.
287. Garmory HS, Leckenby MW, Griffin KF, Elvin SJ, Taylor RR, Hartley MG, Hanak JA, Williamson ED, Cranenburgh RM. Antibiotic-free plasmid stabilization by operator-repressor titration for vaccine delivery by using live *Salmonella enterica* serovar Typhimurium. Infect Immun. 2005;73(4):2005–11.
288. Liu WT, Hsu HL, Liang CC, Chuang CC, Lin HC, Liu YT. A comparison of immunogenicity and protective immunity against experimental plague by intranasal and/or combined with oral immunization of mice with attenuated *Salmonella serova*r Typhimurium expressing secreted *Yersinia pestis* F1 and V antigen. FEMS Immunol Med Microbiol. 2007;51(1): 58–69.
289. Torres-Escobar A, Juarez-Rodriguez MD, Gunn BM, Branger CG, Tinge SA, Curtiss III R. Fine-tuning synthesis of *Yersinia pestis* LcrV from runaway-like replication balanced-lethal plasmid in a *Salmonella enterica* serovar typhimurium vaccine induces protection against a lethal *Y. pestis* challenge in mice. Infect Immun. 2010;78(6):2529–43.
290. Galen JE, Wang JY, Carrasco JA, Lloyd SA, Mellado-Sanchez G, Diaz-McNair J, Franco O, Buskirk AD, Nataro JP, Pasetti MF. A bivalent typhoid live vector vaccine expressing both chromosomal and plasmid-encoded *Y. pestis* antigens fully protects against murine lethal pulmonary plague infection. Infect Immun. 2014;83:161–72.
291. Yang X, Hinnebusch BJ, Trunkle T, Bosio CM, Suo Z, Tighe M, Harmsen A, Becker T, Crist K, Walters N, et al. Oral vaccination with *salmonella* simultaneously expressing *Yersinia pestis* F1 and V antigens protects against bubonic and pneumonic plague. J Immunol. 2007;178(2):1059–67.
292. Welkos S, O'Brien A. Determination of median lethal and infectious doses in animal model systems. Methods Enzymol. 1994;235:29–39.

Chapter 13
Bacteriophages of *Yersinia pestis*

Xiangna Zhao and Mikael Skurnik

Abstract Bacteriophage play many varied roles in microbial ecology and evolution. This chapter collates a vast body of knowledge and expertise on Yersinia pestis phages, including the history of their isolation and classical methods for their isolation and identification. The genomic diversity of Y. pestis phage and bacteriophage islands in the Y. pestis genome are also discussed because all phage research represents a branch of genetics. In addition, our knowledge of the receptors that are recognized by Y. pestis phage, advances in phage therapy for Y. pestis infections, the application of phage in the detection of Y. pestis, and clustered regularly interspaced short palindromic repeats (CRISPRs) sequences of Y. pestis from prophage DNA are all reviewed here.

Keywords *Yersinia pestis* • Bacteriophages • Genomic diversity • Receptors • Detection and treatment

13.1 Isolation of *Yersinia pestis*-Specific Bacteriophages

Bacteriophages (phages), which are viruses that infect and kill bacteria, were first discovered by Twort in 1915 and independently by d'Herelle in 1917 [1]. In the past 100 years, various *Y. pestis* phages have been isolated and studied. In 1919, d'Herelle suggested that a lytic phage might be used to treat plague [2, 3]. Then, in 1925, as the first bacteriophage therapy approach, d'Herelle treated four cases of bubonic plague using a highly virulent *Y. pestis* phage that had been isolated in 1920 from rat

X. Zhao (✉)
Institute of Disease Control and Prevention, Academy of Military Medical Sciences, No. Dongdajie, Fengtai, Beijing 100071, China
e-mail: xnazhao@163.com

M. Skurnik
Department of Bacteriology and Immunology, Medicum, Research Programs Unit, University of Helsinki, Helsinki 00014, Finland
e-mail: mikael.skurnik@helsinki.fi

R. Yang, A. Anisimov (eds.), *Yersinia pestis: Retrospective and Perspective*, Advances in Experimental Medicine and Biology 918,
DOI 10.1007/978-94-024-0890-4_13

feces in Indochina [4]. In 1927, Flu recovered phages that lyse *Y. pestis*, *Escherichia coli*, and *Shigella dysenteriae* from canal waters in Leiden, the Netherlands [2, 5].

Many phages are capable of lysing *Y. pestis*, and they are classified into four serovars based on their antigenic properties, morphology, virulence, genome structure, and specificity for *Y. pestis* [6]. Serovar 1 is the most common, and it comprises all known lytic plague diagnostic phages, including the d'Herelle and Pokrovskaya lytic phages, phiA1122, H (described by Molnar and Lawton), Y, Yepphi, Berlin, Yepe2, YpP-R, YpP-G, YpsP-G, and 21 other phages that were described by Smith and Burrows and that have been studied in the former Soviet Union [1, 2, 7–11]. These phages have isometric, hexagonal heads and short, noncontractile, conical tails, and they exhibit T7-like growth curves; thus, they belong to the family *Podoviridae*, morphotype C1, which include phages T3, T7, and phi II of *E. coli* and P22 of *Salmonella enterica* [2, 6, 12]. Serovar 2 phages are P2-like, and they account for the majority of *Y. pestis* temperate phages, including phage N, phage H reported by Novoseltsev (not the phage H in serovar 1), L-94, and L-413C. Their particle structures possess an isometric, polygonal head of similar size to those of serovar 1 (53–63 nm), but their tails are rather long (110–160 nm) and feature a sheath that is capable of contracting. Together with the common phages P1, P2, and P4, these phages belong to the family *Myoviridae*, morphotype A1 [2, 6, 13]. Serovar 3 consists of only one phage, lysogenic phage II [14], and serovar 4 is represented by the temperate phages Tal and 513 [15].

In 1929, Pokrovskaya recovered a lytic phage that is specific for *Y. pestis* from tissue from an infected ground squirrel (*Spermophilus* sp.), and later the Pokrovskaya phage became the best serovar 1 plague diagnostic phage in use in Russia and other countries of the Commonwealth of Independent States [2, 16]. In 1932, Sugino reported the isolation of *Y. pestis*-specific phages in Japan [2]. In 1933 in Senegal, Advier isolated a *Y. pestis* phage from blood from a patient with a clinical case of bubonic plague [2]. A stock of the Advier phage was obtained in 1945 by Gunnison and Lazarus from the Pasteur Institute of Dakar, Senegal [2]. In 1951, Gunnison and Lazarus demonstrated that the Advier phage could be used to rapidly differentiate *Y. pestis* and *Yersinia pseudotuberculosis*, and they renamed the phage as "P phage" [2, 17]. It is most likely that the Gunnison P phage was subsequently renamed phiA1122 [2]. PhiA1122 is used by the Centers for Disease Control and Prevention of the USA as a diagnostic agent to confirm the identification of *Y. pestis* isolates [2, 18, 19]. Indeed, in 1943, Girard suggested that phages could be used for the diagnosis and treatment of plague [2], and in 1957, he utilized phages to select pseudolysogenic mutants of *Y. pestis* [20]. In 1948, Gunnison and Lazarus reported the alteration of *Y. pestis* phage following successive passage in *Y. pseudotuberculosis* and in shigellae [21]. In 1953, Cavanaugh and Quan developed filter paper strips for the rapid identification of *Y. pestis*. These strips contained a specific, lyophilized phage that was isolated from sewage at the Hooper Foundation, University of California, Berkeley [2, 22]. Molnar and Lawton named the Cavanaugh and Quan phage as "phage H," and the properties of phage H were similar to those of the Gunnison P phage (phiA1122) [2, 9, 17, 23]. In 1973, Duberstein found that phage H appears to be related to the T3-T7 group of coliphages [24], and Summers reported that phage H and *E. coli* phage phi II are nearly identical [25]. Russian researchers have conducted various studies on plague phages, e.g., in 1963, Volosivets conducted a study of plague phages and plague-resistant *Y. pestis* mutants [26]. In 1964 and 1965, Shashaev described the biological properties of plague phages and studied the fate of plague- and *Y. pseudotuberculosis*-specific phages in the great gerbil (*Rhombomys opimus*) [26, 27]. In 1969, Arutiunov reported a new method for increasing the titer of plague-specific bacteriophages [28]. Novoseltsev conducted many phage-related studies [29, 30]. In 1967, he reported the temperate *Y. pestis* phages "H" and N [2, 6], and he conducted an electron microscopic study of *Yersinia* phages in 1971 [31]. In 1977, Novoseltsev reported the origin of phages that are moderately virulent for *Y. pestis*, and classified them as serotype 2, and in 1990, he reported lysogenic phage II and classified it as serovar 3 [14, 32]. In 1994, he reported the lysogenic phages Tal and 513 and classified them as serovar 4 [2]. In 1980, Dorozhko reported his studies of the lysogeny of *Y. pestis*-specific phages [33]. In 1985, Kunikowska explored the adsorption of *Y. pestis* phage and showed that group B human red blood cells exhibited the highest affinity receptors for *Y. pestis* phage [34]. In the former Soviet Union, the T7-related Pokrovskaya phage and the P2-related L-413 C phage were used for diagnostic purposes [2, 6]. In 1964, the phage L-413C was induced from the virulent *Y. pestis* strain 413 that had been isolated from a red-tailed gerbil (*Meriones erythrourus*) in a Central Asian desert plague focus [6, 13]. In 1985, Pak found that phage L-413C was not able to infect 27 *Y. pestis* strains, which indicated that these strains might be immune to the phage, perhaps being lysogens themselves [35]. Subsequently, the Pokrovskaya phage was recommended as an additional test for *Y. pestis* strains when lysis with the main diagnostic phage L-413C failed [6]. Recently, Skurnik has published the sequences of two *Y. pestis*-specific phages, phiD1 and phiEV-1, which were isolated from the sewage of Turku, Finland.

13.2 Isolation and Identification Protocols of *Y. pestis* Phages

Protocols for the isolation and identification of Y. pestis phage are similar to those of other phage. The application of microscopy and molecular genetic tools often requires that phage particles be extracted and concentrated within small volumes [36]. Protocols employed for the extraction of phage particles from a range of environments broadly follow a similar sequence of steps: (i) dissolution, (ii) centrifugation, (iii) sequential filtrations that first remove cells and then concentrate phages, and (iv) final purification by CsCl gradient ultracentrifugation [37, 38]. Here, methodological approaches for the isolation and concentration of Y. pestis phage from various samples are detailed. Owing to the physical differences between the samples, methodological approaches for the concentration of phage particles from water, soil, sediment and fecal samples differ [36]. Efficient recovery and concentration of phage from water samples requires the use of ultrafiltration membranes with a retention cut-off of at least 100 kD or smaller [36]. Obtaining a sample of phage particles from soil or aquatic sediment requires the extraction of phage particles from the porous media into a buffer solution [36]. Ideally, the buffer solution should compete and displace phage from sediment particles by disrupting the electrostatic and hydrophobic interactions between the phage and the porous media [36, 39]. For the extraction of phage from fecal samples, two methods based on tangential-flow filtration (TFF) and polyethylene glycol precipitation (PEG) have been reported [37]. Most bacterial cells carry prophage genomes either integrated into the host DNA or present as repressed plasmids [40]. For the isolation of prophage, two common methods have been reported that involve induction via a sublethal dose of mitomycin C or irradiation with ultraviolet light [40]. For example, the prophage YpfΦ was induced by adding 150 ng/ml of mitomycin to Y. pestis cultures [41]. After extraction, phages can be enumerated by the double agar overlay plaque assay [42], the direct plating plaque assay [43], the small drop plaque assay system [44] or using a standard benchtop flow cytometer [45]. For phage classification and characterization, negative staining of purified phage is the most important electron microscopy technique [46]. The principal stains used for this application are phosphotungstate and uranyl acetate, although both of these possess advantages and disadvantages [46]. No universal method for virus classification exists [47]. Phage classification is open-ended since new phage are discovered daily and the International Committee on Taxonomy of Viruses (ICTV) is far behind in its classification schedule [47]. For Y. pestis phages, most belong to the order Caudovirales. The host range of a bacteriophage is defined by the bacterial genera, species and strains that it can lyse, and is one of the defining biological characteristics of a particular bacterial phage [48]. For the identification of a new Y. pestis phage, a specific Y. pestis strain is used, onto which an environmental sample-such as sewage, water, soil or feces-is spread, forming plaques each of which should come from a single phage particle [48]. Some phage have a narrow host range, whereas for others the host range may be broad [48]. The first step in determining the host range for a given phage is to have a large array of bacteria to test. The second step is to spot test each of the phage on a selected set of potential hosts. The third step is to determine the efficiency of plating (EOP) of the phage on susceptible strains [48].

13.3 Genomic Diversity of *Y. pestis* Bacteriophages

Currently, the genomes of several phages that are capable of lysing *Y. pestis* have been fully sequenced, including the *Podoviridae* phages phiA1122, Yep-phi, Berlin, Yepe2, YpP-R, YpP-G, YpsP-G, and Yps-Y and the *Myoviridae* phages L-413C, PY100, YpsP-PST, and phiD1 [1, 2, 6, 7, 49, 50] (Table 13.1). Several phages are routinely used for plague diagnostics, including L-413C, phiA1122, Pokrovskaya, and Yep-phi [51]. All of the above-mentioned *Podoviridae* phages are T7 related, and their genomes are colinear with those of T7 and T3. They have capsids that possess icosahedral symmetry and a short tail. Comparative analyses support a very high degree of relatedness among the eight phages. Phage phiA1122 was used to

Table 13.1 Bacteriophages of *Y. pestis*

Bacteriophage	Group	Receptor	Sequence accession no.	References	Applications
PhiA1122	T7	Kdo/Ko of LPS	NC004777	[2, 60, 66, 68]	Plague diagnostics; reporter phage; qPCR detection
YpP-Y	T7	Hep(I)/ Glc of LPS	JQ965700	[51, 53, 96]	"Phage cocktail" for decontamination
YpP-R	T7	Beyond the LPS core	JQ965701, JX000007	[51, 53, 96]	*Y. pseudotuberculosis* diagnostics; "phage cocktail" for decontamination
YpsP-G (d'Herelle-m)	T7		JQ965703	[1, 51, 96]	*Y. pseudotuberculosis* diagnostics; "phage cocktail" for decontamination
Yep-phi	T7	LPS; OmpF; Ail	HQ333270	[7, 8]	Plague diagnostics
Pokrovskaya (Yepe2,YpP-G)	T7	Hep(II)/ Hep(III) of LPS	NC011038; JQ965702	[1, 7, 51]	Plague diagnostics; "phage cocktail" for decontamination
Berlin	T7		NC008694	[7]	
L-413C	P2	GlcNAc of LPS	NC004745	[51, 53, 96]	Plague diagnostics; qPCR detection
PhiJA1	T4	Kdo/Ko of LPS		[51, 53, 96]	
YpsP-PST	T4	Hep(II)/ Hep(III) of LPS	KF208315	[50, 51, 53, 96]	*Y. pseudotuberculosis* diagnostics; "phage cocktail" for decontamination
P2 *vir1*	P2	GlcNAc of LPS		[53]	
PY100	T1		AM076770	[49]	
$T7_{Yp}$	T7	Hep(I)/ Glc of LPS		[53]	
YpfΦ	*Inovirus*			[41, 61, 62]	
phiD1	*Myovirus*		HE956711	Mikael Skurnik	
phiEV-1	*Dwarf myovirus*			Mikael Skurnik	

Glc D-glucopyranose, *GlcNAc* N-acetyl-D-glucosamine, *Hep* L-*glycero*-D-*manno*-heptopyranose, *Kdo* 3-deoxy-D-*manno*-oct-2-ulopyranosonic acid, *Ko* D-*glycero*-D-*talo*-oct-2-ulopyranosonic acid differentiate *Y. pestis* and *Y. pseudotuberculosis*. At 37 °C, it grows on both species, but at 20 °C, it only grows on *Y. pestis* [2]. Phage phiA1122 is very closely related to phage T7, and the genome-wide level of nucleotide identity between phages phiA1122 and T7 is approximately 89 %; however, one-quarter of the phiA1122 genome is almost identical to the equivalent part of the T3 genome [2]. An analysis of sequence data suggested that phage T3 is a host range variant that arose from recombination between two yersiniophages: one of which is closely related to the *Yersinia enterocolitica*-specific phage phiYeO3-12, while the other might be phage phiA1122 or an extremely close relative [2, 52]. Comparative analyses revealed that phages phiA1122, YpP-R, YpP-Y, and YpsP-G are almost identical to each other; phage phiA1122 shares 98.5 % nucleotide sequence identity with phage YpP-R, 90.9 % with phage YpP-Y, and 97.9 % with phage YpsP-G. Phages YpP-R, YpsP-G, and YpsP-PST are diagnostic phages of *Y. pseudotuberculosis*, but they are also active against *Y. pestis* [51, 53]. Phage YpP-R is also designated as the Kotlyarova phage [53]. Phage YpsP-PST was isolated from pig feces in Guinea, and it belongs to the T4-like phages [1, 50]. Phages YpP-G and YpsP-G were components of diagnostic phage preparations that were produced in the former Soviet Union and used to identify *Y. pestis* [1]. Phage Yep-phi exclusively infects *Y. pestis* and is inactive toward other *Yersinia* species, irrespective of the growth temperature, and it is used as a plague diagnostic phage in China [8]. Phages Yep-phi, Berlin, Yepe2, and YpP-G are very similar to each other at the genome level. Phage Yep-phi shares 98.1 % nucleotide sequence identity with phage Berlin, 95.8 % with phage Yepe2, and 95.8 % with phage YpP-G. Comparatively, phages Yep-phi and phiA1122 have 60.2 % nucleotide sequence identity. Based on their genomic sequences, the eight related *Podoviridae* phages could be divided into two subgroups: the phiA1122 subgroup (phiA1122, YpP-R, YpP-Y, and YpsP-G) and the Yep-phi subgroup (Yep-phi, Berlin, Yepe2, and YpP-G) [1, 7]. Phage PY100 was isolated from manure at a pig farm in Germany, and it has a broad host range within the genus *Yersinia*, including the three human pathogenic species: *Y. enterocolitica*, *Y. pseudotuberculosis*, and *Y. pestis* [49]. Because its packaging mechanism is similar to that of *Salmonella* phage P22, it seems that phage PY100 can be used for transduction [49]. Phage L-413C is highly similar to phage P2 (GenBank accession no. AF063097) and the P2-like phage WΦ (GenBank accession no. AY135739). Its most striking feature is the mosaic structure of its tail fiber protein H, the amino- and carboxyl-termini of which display high similarity to the corresponding parts of the tail fiber proteins of phages P2 and WΦ; however, a large section in the middle of the H protein closely resembles the large subunit distal tail fiber of the T4-like phages T6 and RB32. The remaining region of the middle of the H protein is similar to those of a number of bacteriophage- or bacterially encoded phage tails. An analysis of sequence data indicated that phage L-413C is a member of the P2 group, and that it has experienced gene loss and replacement, as well as undergone recombination events in the gene encoding the tail fiber protein H, which led to a new host specificity [6]. None of the sequenced phages contain toxin or antibiotic resistance genes that might prevent their application for the treatment of plague.

13.3.1 Receptors for Y. pestis *Bacteriophages*

The interaction of phages with the bacterial surface (adsorption) is the first step during the infection process [54]. This step involves the recognition and binding of the phage receptor-binding protein to one or more cell envelope constituents, which eventually leads to the ejection of phage DNA from the capsid [55]. Detailed knowledge of the bacterial receptors that are recognized by the phages will increase our understanding of the infection process, and it will also help the implementation of efficient phage therapy to target pathogenic bacteria [56]. Phages tend to use structures that are exposed on the cell wall of the host bacteria as receptors, as they are easily accessible [57]. Lipopolysaccharide (LPS) is the major component of the outer membrane of Gram-negative bacteria, and it forms a surface structure that is exposed to the surrounding environment [58]. The LPS of *Y. pestis* lacks the O antigen, and it is composed of the core and lipid A moieties [59]. In 2011, Skurnik, using a series of *Y. pestis* mutants that had defects in genes involved in the synthesis of the LPS core, showed that the Hep/Glc (L-*glycero*-D-*manno*-heptose/D-glucopyranose)-Kdo/Ko (3-deoxy-D-*manno*-oct-2-ulopyranosonic acid/D-*glycero*-D-*talo*-oct-2-ulopyranosonic acid) region of the LPS cores of *Y. pestis* and *Y. pseudotuberculosis* is the receptor for phage phiA1122 [60]. Using an identical approach, Filippov identified the receptors for eight *Y. pestis* bacteriophages [53], and these phages (L-413C, P2 *vir1*, phiJA1, phiA1122, Pokrovskaya, $T7_{Yp}$, YpP-Y, and YpsP-PST) were demonstrated to employ six different components of the LPS core as a specific receptor [53] (Table 13.1). LPS mutants of *Y. pestis* are severely attenuated, as LPS is a critical virulence factor; therefore, the emergence of phage-resistant mutants, which is one of the major concerns of phage therapy, should not be a problem in such cases [53]. Furthermore, it is notable that not all *Y. pestis* phages employ LPS as their sole receptor, e.g., the receptor for phage YpP-R extends beyond the LPS core [53], and the outer membrane proteins Ail and OmpF of *Y. pestis* are involved, in addition to rough LPS, in the adsorption of Yep-phi [8]. This complex recognition specificity of phage Yep-phi could partly explain why no Yep-phi-resistant *Y. pestis* mutants have been isolated [8]. The receptor for the filamentous phage YpfΦ has not been identified. Deleting genes (*pilA*, *psaA*) that encode pili-like structures that are potential candidate receptors did not affect the susceptibility of *Y. pestis* to YpfΦ [61]. Based on the wide host range of YpfΦ, this phage may use several cell surface molecules as receptors [62].

13.4 Bacteriophage Islands in the *Y. pestis* Chromosome

The evolution of *Y. pestis* from *Y. pseudotuberculosis* was accompanied by the loss of many genes and the horizontal acquisition of several genetic elements [61, 63]. In addition to the acquisition of three plasmids (pYV, pFra, and pPla), another horizontally acquired element was the filamentous phage YpfΦ, which was initially almost exclusively present as an extrachromosomal and highly unstable replicon, and subsequently as a stable, integrated prophage, in the 1-ORI branch of the *Y. pestis* phylogenetic tree [41, 61, 62]. YpfΦ plays a role in the ability of *Y. pestis* to multiply and disseminate in mice, conferring on *Y. pestis* an increased fitness during the infectious process and selective advantages under natural conditions [41], which may explain why it was conserved during *Y. pestis* microevolution, despite its instability as an extrachromosomal element in most branches of the phylogenetic tree [62]. Phage YpfΦ produces a filamentous virion that is 1200 nm in length and 8 nm in diameter that contains a circular, positive single-stranded DNA molecule [62]. The 8.7-kb genome of phage YpfΦ shares high sequence identity with the filamentous phage CUS-1 [62]. Like other *Inovirus* phages, the YpfΦ genome comprises three functional modules: the replication module, the structural module, and the assembly/secretion module [61]. Phage YpfΦ has the capacity to infect all three pathogenic *Yersinia* species with variable efficiencies [61]. An endogenous YpfΦ phage confers some level of protection against a superinfection by the same phage [61].

13.5 Application of Bacteriophages in the Detection and Treatment of *Y. pestis*

Bacteriophages have been used in *Y. pestis* identification and plague diagnostics for approximately 80 years [2, 6, 17, 64, 65]. Standard phage lysis assays generally require 24–36 h to complete and are laboratory based [66]. As the rapid detection and diagnosis of plague are essential for a positive prognosis, researchers have made various attempts to reduce the time required to complete these steps [66]. For example, a phiA1122-derived *luxAB* reporter phage transduced a bioluminescent signal response to *Y. pestis* cultures within 10–15 min after phage addition, demonstrating the ability of the reporter phage to rapidly detect *Y. pestis* [19, 66]. The reporter phage could also be used to simultaneously determine the susceptibility of *Y. pestis* to antibiotics within 5 h, directly from blood samples, thereby showing its potential to improve patient prognosis during plague outbreaks [67]. Sergueev developed a rapid and highly sensitive method for the indirect detection of live *Y. pestis* cells based on quantitative real-time polymerase chain reaction (qPCR) monitoring of the amplification of phiA1122 and L-413C sequences [68].

The use of bacteriophages for the treatment of bacterial infections was first proposed by d'Herelle [69]. Regarding the treatment of bubonic plague by phage therapy, the experiments performed by d'Herelle are remarkable; he treated four cases of bubonic plague with lytic phages in 1925 [3, 49]. All four patients recovered, which was considered to be a remarkable achievement. However, because of insufficient knowledge of phage biology and poor experimental design, as well as the discovery of chemical antibiotics, phage therapy was largely abandoned in the West; however, the practice continued in the former Soviet Union and its satellite states. In fact, it is still used routinely today in these regions as an adjunct to antibiotics in human medicine [69]. In 1963, Martinevskii described experiments in which he followed plague-specific bacteriophages in healthy and plague-infected *R. opimus* tissues and studied the possible pathways of transmission under experimental conditions [70]. Filippov has contributed much to the study of phage therapy [71], and the results of his study on bacteriophage therapy of experimental bubonic plague in mice showed that a single injection of phage phiA1122 protected 20–40 % mice against 1000 median lethal doses of *Y. pestis* [72]. In addition, phages phiA1122 and L-413C demonstrated high stability and safety in animal studies [51].

Phages vary in their ability to lyse host bacteria, and they might use different receptors for adsorption; thus, a mixture containing multiple phages may have a broader host range than a single phage. At the same time, the risk of developing a resistant phage will be dramatically reduced, thus improving the long-term efficacy of phage therapy [1]. Rajanna demonstrated that the "phage cocktail" YPP-100, which included phages YpP-G, YpP-Y, YpP-R, and YpsP-PST, had the ability to decontaminate various hard surfaces that were contaminated with *Y. pestis* [1]. Such applications are important because of the emergence of multidrug-resistant strains of *Y. pestis* [73, 74].

Phage-encoded peptidoglycan-degrading activity is responsible for the cell wall-hydrolyzing action of bacteriophages [4]; therefore, purified phage lysins have been produced to specifically target Gram-positive pathogens [75–77]. Gram-negative bacteria are resistant to externally applied phage lysins because the outer membrane protects the peptidoglycan from degradation [75].

Y. pestis produces a bacteriocin, pesticin, which requires the TonB-dependent transporter FyuA for its transport across the outer membrane to its target to kill pesticin-sensitive *Yersinia* strains [78–80]. Pesticin comprises an amino-terminal translocation domain, which binds to FyuA, and a carboxyl-terminal active domain that is structurally similar to lysozyme homologs [81]. To exploit the properties of pesticin, a chimeric protein was engineered that consisted of the amino-terminal domain of pesticin fused to phage T4 lysozyme, which replaced the carboxyl-terminal domain of pesticin. The lysozyme-pesticin product was transported across the outer membrane and attacked the peptidoglycan layer, subsequently killing *Y. pestis* strains that expressed FyuA on their cell surface [75]. It has been suggested that the therapeutic use of endolysins in a controlled fashion appears to be advantageous compared with their uncontrolled release [4].

Another phage-derived product has been developed recently as a potential plague vaccine. The capsular protein (capsule antigen fraction 1 (Caf1) or F1, 16 kDa) and the low calcium response V antigen (LcrV, 37 kDa) have been used as vaccines against bubonic and pneumonic plague [82, 83]. However, the polymeric nature of F1 affects vaccine efficacy and generates varied immune responses in humans [82,

84–88]. Recombinant F1 and V proteins induce strong antibody responses, but poor cellular immune responses [89]. For these reasons, a mutant F1-V gene that encodes a fully soluble, monomeric F1-V fusion protein was constructed, and this gene was fused to the 5′ end of the gene encoding the small outer capsid protein (Soc) of T4, which results in the production of an F1-V-Soc fusion protein. As a result, the F1-V protein was displayed on the phage T4 capsid [90, 91]. Both soluble (purified F1-V mixed with Alhydrogel® from EM Sergeant Pulp & Chemical Co., Inc., USA) and T4-decorated F1-V (no adjuvant) provided mice and rats with 100 % protection against pneumonic plague [90]. Furthermore, phage T4-mediated vaccine delivery into dendritic cells induced the abundant expression of F1-V, as well as its presentation to the immune system for a long period, which led to the continuous stimulation of T cells [91]. The above results demonstrate that phage T4 appears to be a good platform for the development of effective "prime-boost" vaccines [91].

13.6 CRISPRs of *Y. pestis* from Phages or Prophages

Clustered regularly interspaced short palindromic repeats (CRISPRs) are present in bacterial and archaeal genomes, and they consist of arrays of conserved 21–47-bp repeats, which are interspaced with similarly sized variable spacer sequences that are derived from foreign replicons such as phages or plasmids [92]. Fragments of invading DNA are maintained as spacers, and the spacer composition in any bacterium reflects its evolutionary history [93]. In the *Y. pestis* genome, three CRISPR elements (YP1, YP2, and YP3) are found at three distinct loci [94]. The same 28-bp repeat sequence is separated by 32- or 33-bp spacers in all three CRISPR elements [94]. Pourcel et al. sequenced 109 alleles of the three *Y. pestis* CRISPRs and found new spacers at a single locus, which is located inside a 46-kb region that corresponds to a defective lambdoid prophage [94]. CRISPR-CRISPER-associated (Cas) loci have recently been shown to be an immunity system that targets foreign nucleic acids, including phage genomes and plasmids [95]. One possible explanation for the finding that *Y. pestis* CRISPR loci have acquired new spacers from a prophage DNA could be that CRISPRs take up foreign DNA as part of a defense mechanism [94]. In a recent study, Koskela et al. analyzed the CRISPR locus sequences of 335 *Y. pseudotuberculosis* strains, and the results demonstrated that *Y. pestis* has a relatively low number of spacers compared with *Y. pseudotuberculosis*, which suggests that *Y. pestis* has lived a protected or secluded life after diverging from *Y. pseudotuberculosis* and that it has not encountered many foreign DNAs [93].

References

1. Rashid MH, Revazishvili T, Dean T, Butani A, Verratti K, Bishop-Lilly KA, Sozhamannan S, Sulakvelidze A, Rajanna C. A Yersinia pestis-specific, lytic phage preparation significantly reduces viable *Y. pestis* on various hard surfaces experimentally contaminated with the bacterium. Bacteriophage. 2012;2(3):168–77.
2. Garcia E, Elliott JM, Ramanculov E, Chain PS, Chu MC, Molineux IJ. The genome sequence of Yersinia pestis bacteriophage phiA1122 reveals an intimate history with the coliphage T3 and T7 genomes. J Bacteriol. 2003;185(17):5248–62.
3. Summers WC. Bacteriophage therapy. Annu Rev Microbiol. 2001;55:437–51.
4. Anisimov AP, Amoako KK. Treatment of plague: promising alternatives to antibiotics. J Med Microbiol. 2006;55(Pt 11):1461–75.
5. Flu PC. Sur la nature du bacte'riophage. C R Soc Biol. 1927;96:1148–9.
6. Garcia E, Chain P, Elliott JM, Bobrov AG, Motin VL, Kirillina O, Lao V, Calendar R, Filippov AA. Molecular characterization of L-413C, a P2-related plague diagnostic bacteriophage. Virology. 2008;372(1):85–96.
7. Zhao X, Wu W, Qi Z, Cui Y, Yan Y, Guo Z, Wang Z, Wang H, Deng H, Xue Y, et al. The complete genome sequence and proteomics of Yersinia pestis phage Yep-phi. J Gen Virol. 2011;92(Pt 1):216–21.
8. Zhao X, Cui Y, Yan Y, Du Z, Tan Y, Yang H, Bi Y, Zhang P, Zhou L, Zhou D, et al. Outer membrane proteins ail and OmpF of Yersinia pestis are involved in the adsorption of T7-related bacteriophage Yep-phi. J Virol. 2013;87(22):12260–9.
9. Molnar DM, Lawton WD. Pasteurella bacteriophage sex specific in Escherichia coli. J Virol. 1969;4(6):896–900.
10. Hertman I. Bacteriophage common to Pasteurella pestis and Escherichia coli. J Bacteriol. 1964;88:1002–5.
11. Smith DA, Burrows TW. Phage and bacteriocin investigations with Pasteurella pestis and other bacteria. Nature. 1962;193:397–8.
12. Larina VS, Anisimov PI, Adamov AK. A novel strain of plague bacteriophage for identification of Pasteurella pestis. Probl Particularly Danger Infect. 1970;11:132–6.
13. Larina VS. Lysogenic clones of wild-type plague bacteria and characteristics of the phages produced by them. Zh Mikrobiol Epidemiol Immunobiol. 1976;1976(1):119–22.
14. Novosel'tsev NN, Marchenkov VI. *Y. pestis* phage of a new serovar. Zh Mikrobiol Epidemiol Immunobiol. 1990;1990(11):9–12.
15. Novosel'tsev NN, Marchenkov VI, Arutiunov Iu l. **Phages of the IV serovar of Yersinia pestis**. Zh Mikrobiol Epidemiol Immunobiol. 1994;6:9–10.
16. Pokrovskaya MP. Bacteriophage and its practical applications for treatment and prophylaxis of summer children's diarrhea. Dysentery Surg Infect. 1940;1940:55–6.
17. Gunnison JB, Larson A, Lazarus AS. Rapid differentiation between Pasteurella pestis and Pasteurella pseudotuberculosis by action of bacteriophage. J Infect Dis. 1951;88(3):254–5.
18. Plague manual–epidemiology, distribution, surveillance and control. Releve epidemiologique hebdomadaire/Section d'hygiene du Secretariat de la Societe des Nations = Weekly epidemiological record/Health Section of the Secretariat of the League of Nations. 1999;74(51–52):447.
19. Schofield DA, Molineux IJ, Westwater C. 'Bioluminescent' reporter phage for the detection of category a bacterial pathogens. J Visual Exp. 2011;53:e2740.
20. Girard G. General characteristics of pseudolysogenic mutants of Pasteurella pestis selected by phages; epidemiological significance. Comptes rendus des seances de la Societe de biologie et de ses filiales. 1957;151(6):1068–71.
21. Gunnison JB, Lazarus AS. Alteration of Pasteurella pestis bacteriophage following successive transfer on Pasteurella pseudotuberculosis and on shigellae. Proc Soc Exp Biol Med. 1948;69(2):294–6.

22. Cavanaugh DC, Quan SF. Rapid identification of Pasteurella pestis using specific bacteriophage lyophilized on strips of filter paper; a preliminary report. Am J Clin Pathol. 1953;23(6):619–20.
23. Harrison DN, Cavanaugh DC, Rust Jr JH, Marshall Jr JD. Characteristics of a bacteriophage-infected strain of Pasteurella pestis isolated from a human case of plague. Infect Immun. 1971;4(1):85–7.
24. Duberstein R, Oeschger MP. Growth of bacteriophage H on male and female strains of Escherichia coli. J Virol. 1973;11(3):460–3.
25. Brunovskis I, Hyman RW, Summers WC. Pasteurella pestis bacteriophage H and Escherichia coli bacteriophage phi II are nearly identical. J Virol. 1973;11(2):306–13.
26. Shashaev MA. Biological properties of plague bacteriophages. Zh Mikrobiol Epidemiol Immunobiol. 1964;41:32–5.
27. Shashaev MA, Shapiro IL, Shatalova AL. Duration of the presence of plague and pseudotuberculosis bacteriophages from the organism of Rhombomys opimus. Zh Mikrobiol Epidemiol Immunobiol. 1965;42(3):97–101.
28. Arutiunov Iu I. New modification of the increase of bacteriophage titer reaction (in plague). Laboratornoe delo. 1969;3.
29. Novoseltsev NN. Moderate phage of the agent of plague. Zh Mikrobiol Epidemiol Immunobiol. 1974;1974(8):45–9.
30. Novosel'tsev NN, Marchenkov VI, Sorokin VM, Kravtsov AN, Degtiarev BM. Relation between the Yersinia phage and bacteriophages isolated from the environment. Molekuliarnaia genetika, mikrobiologiia i virusologiia. 1990;1990(8):18–21.
31. Novosel'tsev NN, Arutiunov Iu I, Tokarev SA, Kirdeev VK. Electron microscopic study of Pasteurella phages. Zh Mikrobiol Epidemiol Immunobiol. 1971;48(3):16–9.
32. Novosel'tsev NN, Marchenkov VI, Kravchenko AN, Valentsev VE, Tinker LA. The lytic activity of Yersinia pestis phage P 3d serovar. Zh Mikrobiol Epidemiol Immunobiol. 1990;1990(12):15–8.
33. Dorozhko OV, Zhuravlev I, Rotshil'd EV, Kondrashin Iu I. Lysogeny by phages specific to Yersinia pestis in bacteria of the microbiocenoses of wild plague-carrying rodents. Zh Mikrobiol Epidemiol Immunobiol. 1980;1980(8):38–46.
34. Kunikowska D, Glosnicka R. The action of the Yersinia pestis phage on the plague envelope antigen and the human erythrocytes. Bull Inst Marit Trop Med Gdynia. 1985;36(1-4):103–15.
35. Pak G, Satybaldiev NA, Bakanurskaia TL, Aniskina GA, Soorbekov OS. Properties of strains of the causative agent of plague resistant to the diagnostic phage L-413 "c". Zh Mikrobiol Epidemiol Immunobiol. 1985;1985(8):33–6.
36. Wommack KE, Williamson KE, Helton RR, Bench SR, Winget DM. Methods for the isolation of viruses from environmental samples. Methods Mol Biol. 2009;501:3–14.
37. Castro-Mejia JL, Muhammed MK, Kot W, Neve H, Franz CM, Hansen LH, Vogensen FK, Nielsen DS. Optimizing protocols for extraction of bacteriophages prior to metagenomic analyses of phage communities in the human gut. Microbiome. 2015;3:64.
38. Boulanger P. Purification of bacteriophages and SDS-PAGE analysis of phage structural proteins from ghost particles. Methods Mol Biol. 2009;502:227–38.
39. Gerba CP. Applied and theoretical aspects of virus adsorption to surfaces. Adv Appl Microbiol. 1984;30:133–68.
40. Raya RR, H'Bert EM. Isolation of phage via induction of lysogens. Methods Mol Biol. 2009;501:23–32.
41. Derbise A, Chenal-Francisque V, Pouillot F, Fayolle C, Prevost MC, Medigue C, Hinnebusch BJ, Carniel E. A horizontally acquired filamentous phage contributes to the pathogenicity of the plague bacillus. Mol Microbiol. 2007;63(4):1145–57.
42. Kropinski AM, Mazzocco A, Waddell TE, Lingohr E, Johnson RP. Enumeration of bacteriophages by double agar overlay plaque assay. Methods Mol Biol. 2009;501:69–76.
43. Mazzocco A, Waddell TE, Lingohr E, Johnson RP. Enumeration of bacteriophages by the direct plating plaque assay. Methods Mol Biol. 2009;501:77–80.

44. Mazzocco A, Waddell TE, Lingohr E, Johnson RP. Enumeration of bacteriophages using the small drop plaque assay system. Methods Mol Biol. 2009;501:81–5.
45. Brussaard CP. Enumeration of bacteriophages using flow cytometry. Methods Mol Biol. 2009;501:97–111.
46. Ackermann HW. Basic phage electron microscopy. Methods Mol Biol. 2009;501:113–26.
47. Ackermann HW. Phage classification and characterization. Methods Mol Biol. 2009;501:127–40.
48. Kutter E. Phage host range and efficiency of plating. Methods Mol Biol. 2009;501:141–9.
49. Schwudke D, Ergin A, Michael K, Volkmar S, Appel B, Knabner D, Konietzny A, Strauch E. Broad-host-range Yersinia phage PY100: genome sequence, proteome analysis of virions, and DNA packaging strategy. J Bacteriol. 2008;190(1):332–42.
50. Comeau AM, Arbiol C, Krisch HM. Composite conserved promoter-terminator motifs (PeSLs) that mediate modular shuffling in the diverse T4-like myoviruses. Genome Biol Evol. 2014;6(7):1611–9.
51. Filippov AA, Sergueev KV, Nikolich MP. Can phage effectively treat multidrug-resistant plague? Bacteriophage. 2012;2(3):186–9.
52. Demerec M, Fano U. Bacteriophage-resistant mutants in Escherichia coli. Genetics. 1945;30(2):119–36.
53. Filippov AA, Sergueev KV, He Y, Huang XZ, Gnade BT, Mueller AJ, Fernandez-Prada CM, Nikolich MP. Bacteriophage-resistant mutants in Yersinia pestis: identification of phage receptors and attenuation for mice. PLoS One. 2011;6(9):e25486.
54. Kruger DH, Schroeder C. Bacteriophage T3 and bacteriophage T7 virus-host cell interactions. Microbiol Rev. 1981;45(1):9–51.
55. Sao-Jose C, Baptista C, Santos MA. Bacillus subtilis operon encoding a membrane receptor for bacteriophage SPP1. J Bacteriol. 2004;186(24):8337–46.
56. Sorensen MC, van Alphen LB, Fodor C, Crowley SM, Christensen BB, Szymanski CM, Brondsted L. Phase variable expression of capsular polysaccharide modifications allows campylobacter jejuni to avoid bacteriophage infection in chickens. Front Cell Infect Microbiol. 2012;2:11.
57. Shin H, Lee JH, Kim H, Choi Y, Heu S, Ryu S. Receptor diversity and host interaction of bacteriophages infecting Salmonella enterica serovar Typhimurium. PLoS One. 2012;7(8):e43392.
58. Skurnik M. Yersinia surface structures and bacteriophages. Adv Exp Med Biol. 2012;954:293–301.
59. Dentovskaya SV, Anisimov AP, Kondakova AN, Lindner B, Bystrova OV, Svetoch TE, Shaikhutdinova RZ, Ivanov SA, Bakhteeva IV, Titareva GM, et al. Functional characterization and biological significance of Yersinia pestis lipopolysaccharide biosynthesis genes. Biochemistry (Mosc). 2011;76(7):808–22.
60. Kiljunen S, Datta N, Dentovskaya SV, Anisimov AP, Knirel YA, Bengoechea JA, Holst O, Skurnik M. Identification of the lipopolysaccharide core of Yersinia pestis and Yersinia pseudotuberculosis as the receptor for bacteriophage phiA1122. J Bacteriol. 2011;193(18):4963–72.
61. Chouikha I, Charrier L, Filali S, Derbise A, Carniel E. Insights into the infective properties of YpfPhi, the Yersinia pestis filamentous phage. Virology. 2010;407(1):43–52.
62. Derbise A, Carniel E. YpfPhi: a filamentous phage acquired by Yersinia pestis. Front Microbiol. 2014;5:701.
63. Chain PS, Carniel E, Larimer FW, Lamerdin J, Stoutland P, Regala W, Georgescu A, Vergez L, Land M, Motin V. Insights into the evolution of Yersinia pestis through whole-genome comparison with Yersinia pseudotuberculosis. Proc Natl Acad Sci U S A. 2004;101(38):13826–31.
64. Craigie J. The significance and applications of bacteriophage in bacteriological and virus research. Bacteriol Rev. 1946;10(3-4):73–88.

65. Nunes MP, Suassuna I. Bacteriophage specificity in the identification of Yersinia pestis as compared with other enterobacteria. Revista brasileira de pesquisas medicas e biologicas. 1978;11(6):359–63.
66. Schofield DA, Molineux IJ, Westwater C. Diagnostic bioluminescent phage for detection of Yersinia pestis. J Clin Microbiol. 2009;47(12):3887–94.
67. Schofield DA, Molineux IJ, Westwater C. Rapid identification and antibiotic susceptibility testing of Yersinia pestis using bioluminescent reporter phage. J Microbiol Methods. 2012;90(2):80–2.
68. Sergueev KV, He Y, Borschel RH, Nikolich MP, Filippov AA. Rapid and sensitive detection of Yersinia pestis using amplification of plague diagnostic bacteriophages monitored by real-time PCR. PLoS One. 2010;5(6):e11337.
69. Gill JJ, Summer EJ, Russell WK, Cologna SM, Carlile TM, Fuller AC, Kitsopoulos K, Mebane LM, Parkinson BN, Sullivan D, et al. Genomes and characterization of phages Bcep22 and BcepIL02, founders of a novel phage type in Burkholderia cenocepacia. J Bacteriol. 2011;193(19):5300–13.
70. Martinevskii IL, Shashaev MA, Tarakanov NF, Shapovalov AT. on the fate of plague bacteriophage in the body of healthy and plague-infected rhombomys opimus and possible pathways of transmission under experimental conditions. Zh Mikrobiol Epidemiol Immunobiol. 1963;40:31–4.
71. Sergueev KV, Nikolich MP, Filippov AA. Field and clinical applications of advanced bacteriophage-based detection of Yersinia pestis. Adv Exp Med Biol. 2012;954:135–41.
72. Filippov AA, Sergueev KV, He Y, Huang XZ, Gnade BT, Mueller AJ, Fernandez-Prada CM, Nikolich MP. Bacteriophage therapy of experimental bubonic plague in mice. Adv Exp Med Biol. 2012;954:337–48.
73. Galimand M, Carniel E, Courvalin P. Resistance of Yersinia pestis to antimicrobial agents. Antimicrob Agents Chemother. 2006;50(10):3233–6.
74. Welch TJ, Fricke WF, McDermott PF, White DG, Rosso ML, Rasko DA, Mammel MK, Eppinger M, Rosovitz MJ, Wagner D, et al. Multiple antimicrobial resistance in plague: an emerging public health risk. PLoS ONE. 2007;2(3):e309.
75. Lukacik P, Barnard TJ, Buchanan SK. Using a bacteriocin structure to engineer a phage lysin that targets Yersinia pestis. Biochem Soc Trans. 2012;40(6):1503–6.
76. Schuch R, Nelson D, Fischetti VA. A bacteriolytic agent that detects and kills Bacillus anthracis. Nature. 2002;418(6900):884–9.
77. Loeffler JM, Nelson D, Fischetti VA. Rapid killing of Streptococcus pneumoniae with a bacteriophage cell wall hydrolase. Science. 2001;294(5549):2170–2.
78. Bearden SW, Fetherston JD, Perry RD. Genetic organization of the yersiniabactin biosynthetic region and construction of avirulent mutants in Yersinia pestis. Infect Immun. 1997;65(5):1659–68.
79. Rakin A, Saken E, Harmsen D, Heesemann J. The pesticin receptor of Yersinia enterocolitica: a novel virulence factor with dual function. Mol Microbiol. 1994;13(2):253–63.
80. Pilsl H, Killmann H, Hantke K, Braun V. Periplasmic location of the pesticin immunity protein suggests inactivation of pesticin in the periplasm. J Bacteriol. 1996;178(8):2431–5.
81. Patzer SI, Albrecht R, Braun V, Zeth K. Structural and mechanistic studies of pesticin, a bacterial homolog of phage lysozymes. J Biol Chem. 2012;287(28):23381–96.
82. Anderson Jr GW, Heath DG, Bolt CR, Welkos SL, Friedlander AM. Short- and long-term efficacy of single-dose subunit vaccines against Yersinia pestis in mice. AmJTrop Med Hyg. 1998;58(6):793–9.
83. Heath DG, Anderson Jr GW, Mauro JM, Welkos SL, Andrews GP, Adamovicz J, Friedlander AM. Protection against experimental bubonic and pneumonic plague by a recombinant capsular F1-V antigen fusion protein vaccine. Vaccine. 1998;16(11-12):1131–7.
84. Mizel SB, Graff AH, Sriranganathan N, Ervin S, Lees CJ, Lively MO, Hantgan RR, Thomas MJ, Wood J, Bell B. Flagellin-F1-V fusion protein is an effective plague vaccine in mice and two species of nonhuman primates. Clin Vaccine Immunol. 2009;16(1):21–8.

85. Goodin JL, Nellis DF, Powell BS, Vyas VV, Enama JT, Wang LC, Clark PK, Giardina SL, Adamovicz JJ, Michiel DF. Purification and protective efficacy of monomeric and modified Yersinia pestis capsular F1-V antigen fusion proteins for vaccination against plague. Protein Expr Purif. 2007;53(1):63–79.
86. Goodin JL, Powell BS, Enama JT, Raab RW, McKown RL, Coffman GL, Andrews GP. Purification and characterization of a recombinant Yersinia pestis V-F1 "Reversed" fusion protein for use as a new subunit vaccine against plague. Protein Expr Purif. 2011;76(1):136–44.
87. Powell BS, Andrews GP, Enama JT, Jendrek S, Bolt C, Worsham P, Pullen JK, Ribot W, Hines H, Smith L, et al. Design and testing for a nontagged F1-V fusion protein as vaccine antigen against bubonic and pneumonic plague. Biotechnol Prog. 2005;21(5):1490–510.
88. Parent MA, Berggren KN, Kummer LW, Wilhelm LB, Szaba FM, Mullarky IK, Smiley ST. Cell-mediated protection against pulmonary Yersinia pestis infection. Infect Immun. 2005;73(11):7304–10.
89. Smiley ST. Current challenges in the development of vaccines for pneumonic plague. Expert Rev Vaccines. 2008;7(2):209–21.
90. Tao P, Mahalingam M, Kirtley ML, van Lier CJ, Sha J, Yeager LA, Chopra AK, Rao VB. Mutated and bacteriophage T4 nanoparticle arrayed F1-V immunogens from Yersinia pestis as next generation plague vaccines. PLoS Pathog. 2013;9(7):e1003495.
91. Tao P, Mahalingam M, Marasa BS, Zhang Z, Chopra AK, Rao VB. In vitro and in vivo delivery of genes and proteins using the bacteriophage T4 DNA packaging machine. Proc Natl Acad Sci U S A. 2013;110(15):5846–51.
92. Vale PF, Little TJ. CRISPR-mediated phage resistance and the ghost of coevolution past. Proc Biol Sci/Roy Soc. 2010;277(1691):2097–103.
93. Koskela KA, Mattinen L, Kalin-Manttari L, Vergnaud G, Gorge O, Nikkari S, Skurnik M. Generation of a CRISPR database for Yersinia pseudotuberculosis complex and role of CRISPR-based immunity in conjugation. Environ Microbiol. 2015.
94. Pourcel C, Salvignol G, Vergnaud G. CRISPR elements in Yersinia pestis acquire new repeats by preferential uptake of bacteriophage DNA, and provide additional tools for evolutionary studies. Microbiology. 2005;151(Pt 3):653–63.
95. Labrie SJ, Samson JE, Moineau S. Bacteriophage resistance mechanisms. Nat Rev Microbiol. 2010;8(5):317–27.
96. Filippov AA, Sergueev KV, He Y, Nikolich MP. Bacteriophages capable of lysing Yersinia pestis and Yersinia pseudotuberculosis: efficiency of plating tests and identification of receptors in escherichia coli K-12. Adv Exp Med Biol. 2012;954:123–34.

Chapter 14
Perspectives on *Yersinia pestis*: A Model for Studying Zoonotic Pathogens

Ruifu Yang, Yujun Cui, and Yujing Bi

Abstract *Yersinia pestis* is a typical zoonotic bacterial pathogen. The following reasons make this pathogen a model for studying zoonotic pathogens: (1) Its unique lifestyle makes *Y. pestis* an ideal model for studying host–vector–environment–pathogen interactions; (2) population diversity characters in *Y. pestis* render it a model species for studying monomorphic bacterial evolution; (3) the pathogenic features of bacteria provide us with good opportunities to study human immune responses; (4) typical animal and vector models of *Y. pestis* infection create opportunities for experimental studies on pathogenesis and evolution; and (5) repeated pandemics and local outbreaks provide us with clues about the infectious disease outbreaks that have occurred in human history.

Keywords *Y. pestis*, host–vector–environment–pathogen interactions • Pathogenesis, evolution

Yersinia pestis is a typical zoonotic bacterial pathogen. Since its isolation and identification in 1894 in Hong Kong by Dr. Alexandre Yersin, its ecology, arthropod vectors, rodent reservoirs, biochemical features, and pathogenicity have been studied extensively. As mentioned in previous chapters, genome sequencing has revealed the microevolution and population structure of *Y. pestis*. In addition, omics-based research has enhanced our understanding of plague pathogenesis and is helping us to develop novel diagnostic and therapeutic targets against *Y. pestis*, while ecological studies have strengthened our understanding of its life cycle. Genome sequence-based paleomicrobiology has yielded solid evidence that the three historical pandemic plagues were indeed caused by *Y. pestis* and has helped us to learn more about the evolutionary history of this pathogen. Based on our scientific understanding of *Y. pestis*, this typical zoonotic pathogen can be used as a model bacterium for exploring the biology of other similar pathogens.

R. Yang (✉) • Y. Cui • Y. Bi
Beijing Institute of Microbiology and Epidemiology,
No. 20, Dongdajie, Fengtai, Beijing 100071, China
e-mail: ruifuyang@gmail.com

R. Yang, A. Anisimov (eds.), *Yersinia pestis: Retrospective and Perspective*,
Advances in Experimental Medicine and Biology 918,
DOI 10.1007/978-94-024-0890-4_14

14.1 Its Unique Lifestyle Makes *Y. pestis* an Ideal Model for Studying Host–Vector–Environment–Pathogen Interactions

After evolving from *Y. pseudotuberculosis*, one unique feature obtained by *Y. pestis* is its ability to be transmitted by fleas. In natural plague foci, *Y. pestis* usually circulates between rodents and fleas in addition to interacting with other features of the environment such as soil, plants, soil organisms, and rodents. *Y. pestis* obtained the ability to form biofilms in the flea gut by acquiring some important genes, and *ymt*, a murine toxin (phospholipase D) gene encoded on the *Y. pestis* pMT1 plasmid, provides *Y. pestis* with the ability to transmit itself by fleas [1, 2]. The *Y. pestis* genome has experienced a series of changes that enabled it to adapt to a new lifestyle. During its evolution from *Y. pseudotuberculosis*, which is an intestinal pathogen causing mild diarrhea, the enteropathogenic-related genes in *Y. pestis* were inactivated or lost, the unique pMT1 and pPCP1 plasmids were acquired, and gene mutations occurred in its chromosomal DNA [3]. Biologists who are interested in *Y. pestis* evolution are able to use its acknowledged ancestor, *Y. pseudotuberculosis*, to study its evolutionary past because the genome similarity between these closely related species is >90 %, making it easier to conduct comparative genomics studies or use gene mutation techniques to determine the precise role of a specific *Y. pestis* gene [3–8].

A distinct part of the *Y. pestis* life stages occurs in fleas, but transmission to humans can occur when *Y. pestis*-infected fleas bite them. Despite *Y. pestis* being a relatively young pathogen that only emerged several to tens of thousands of years ago, its wide distribution in Asia, Eurasia, Africa, and America has given it ample opportunities to adapt to different animals and environmental conditions. The complex interactions *Y. pestis* has with its vectors, hosts, and environments have provided many opportunities to understand different aspects of its life cycle. The role of the climate in the evolution and epidemiology of *Y. pestis* has been investigated [9–19]. Climate-driven plague outbreaks in Asian rodent plague reservoirs have promoted the introduction of plague into Europe through the maritime trade network via persistent and repeated import of *Y. pestis* from Asian rodent reservoirs, and this provides an explanation as to why the Black Death persisted in Europe for so long [15]. Climate temperature has also been shown to influence the transmissibility of *Y. pestis* by fleas [20–23]. At low temperatures (<5 °C), flea transmission of plague is most likely inhibited, indicating that the epizootic spread of *Y. pestis* by fleas via early-phase transmission seems unlikely to happen during colder periods of the year. From an environmental flea and reservoir perspective of plague emergence, there should be a complex balance between these factors and *Y. pestis*. Unraveling these complex interactions will help us to understand how a milder pathogen evolves into a highly pathogenic one. Better understanding of the complex evolutionary history of *Y. pestis* from *Y. pseudotuberculosis* will provide more evidence to support the famous epidemiological tenet of Theobald Smith [24], which is able to explain the evolution of most infectious organisms. This tenet states that the best strategy a pathogen can use is to minimize damage to its host, thereby preserving the only environment capable of maintaining that pathogen in nature. However, *Y. pestis*, which causes lethal diseases in both humans and animals, seems to be an exception to this theory. When considering the complex interactions between *Y. pestis* and the environment, vectors, and reservoirs, we still need to take into account the balance between these factors to provide more evidence to confirm or refute that the evolution of *Y. pestis* adheres to the evolutionary tenet of Theobald Smith.

In China, the 15 natural plague foci are associated with the following reservoir types [25, 26]: *Marmota caudate* in the Pamirs Plateau, *M. baibacina–Spermophilus undulatus* in the Tian Shan Mountains, *M. himalayana* in the Qinghai–Gansu–Tibet Grassland, *M. himalayana* in Qilian Mountain, *Apodemus chevrieri–Eothenomys miletus* in the highlands of the Northwestern Yunnan province, *M. himalayana* in the Gangdisi Mountains, *Rattus flavipectus* in the Yunnan–Guangdong–Fujian provinces, *S. dauricus* in the Song-Liao Plain, *Meriones unguiculatus* in the Inner Mongolian Plateau, *S. dauricus alaschanicus* in the Loess Plateau in Gansu and Ningxia provinces, *M. himalayana* in the Kunlun Mountains, *Microtus brandti* in the Xilin Gol Grassland, *Microtus fuscus* in the Qinghai–Tibet Plateau, *M. sibirica* in the Hulun Buir Plateau of Inner Mongolia, and *Rhombomys opimus* in the Junggar Basin of Xinjiang. The names of each plague focus tell us that the landscapes are diverse, ranging from plateaus, mountains, and grasslands to desert basin. The vector fleas in the different plague foci also differ depending on the rodent reservoirs (e.g., *Neopsylla hongyangensis*, *N. abagaitui*, *Rhadinopsylla tenella*, *Leptopsylla segnis*, and *Frontopsylla pawlovskii*, among others). The *Y. pestis* strains isolated from *M. brandti* in the plague focus of the Xilin Gol Grassland and *M. fuscus* in the plague focus of the Qinghai–Tibet Plateau are non-virulent to guinea pigs, rabbits, and humans [27]. Therefore, studying these strains can help us to understand how *Y. pestis* interacts with its reservoirs, flea vectors, and environments. This in turn helps us to understand more about the evolution of its unique lifestyle that has allowed it to adapt to diverse environments, reservoirs, and vectors.

During its evolution from *Y. pseudotuberculosis*, *Y. pestis* acquired two unique plasmids, pMT1 and pPCP1 [28], which confer its unique virulence. Therefore, typical *Y. pestis* generally contains three plasmids: the inherent pCD1 plasmid, pMT1, and pPCP1. However, how plasmid interactions determine the unique virulence of *Y. pestis* is not completely understood. Gaining better understanding on how these plasmids coexist in harmony in the bacterium should enable us to obtain insight into the molecular basis of each plasmid's role in bacterial evolution.

Thus, studying *Y. pestis* should not only shed light on the complex interactions this pathogen has with external parameters, such as its reservoirs, vectors, and the outside environment but should also provide us with opportunities to study the internal interactions between plasmids and the bacterial chromosome. Hence, there is much to be gained from use of *Y. pestis* as a model organism for studying this zoonotic bacterium's evolution and pathogenesis.

14.2 Population Diversity Characters in *Y. pestis* Render It a Model Species for Studying Monomorphic Bacterial Evolution

The *Y. pestis* genome shares more than 90 % homology with its ancestor *Y. pseudotuberculosis* [29, 30]. Although *Y. pestis* appeared at least 5000 years ago [31], it still exhibits limited population diversity, possibly as a result of a relatively slow molecular clock with an estimated value of 2.9×10^{-9}–2.3×10^{-8} single nucleotide polymorphisms (SNPs) per site per year [8]. Only 2326 SNPs were identified from 133 of the global *Y. pestis* isolate genomes sequenced (about one SNP per 2000 bp) [7], thereby confirming the genome conservation in this bacterium. Homologous recombination should potentially increase allele diversity. However, the limited number of SNPs within *Y. pestis* exhibit homoplasy, suggesting that very few recombination events have occurred over time. Therefore, *Y. pestis* has evolved clonally, and this type of evolution can be attributed to genetically monomorphic bacterial pathogens [32]. Because many of the clinically dangerous bacterial pathogens that have evolved recently have similar genetic diversity patterns, evolutionary research on *Y. pestis* should help to provide better understanding of the spread and transmission pattern of infectious diseases caused by such pathogens.

One intriguing evolutionary puzzle in *Y. pestis* is its historically variable molecular clock [7, 33]. There is almost a 40-fold difference between the slowest and fastest branches of the *Y. pestis* genealogy at the whole genome-based level. Variation in the mutation rate is supposed to be caused by alternate cycles of enzootic and epizootic periods of sylvatic plagues [7]. During enzootic periods, *Y. pestis* can survive for long periods in a dormant state, but this situation is rare in both fleas and hosts. In contrast, during the typical epizootic state, *Y. pestis* quickly spreads among its hosts and vectors, and the infected rodents can suffer intense bacteremia [34]. Therefore, even if the mutation rate per bacterial replication cycle is constant and lacks natural selection, epizootics should be associated with much higher rates of bacterial replication per unit time that are typical during enzootic disease outbreaks, and such periods of rapid replication are expected to result in observably faster clock rates per unit time. Similar demographic effects are expected to be true also for other epidemic pathogens, especially for those with similar lifestyles that cause zoonotic diseases, such as *Bacillus anthracis* and *Francisella tularensis*.

After its evolution from *Y. pseudotuberculosis*, *Y. pestis* spread to different suitable geographical regions and established different types of plague foci, where it interacted with different reservoirs, vectors, and environments. These different niches have provided considerable driving forces for the evolution of *Y. pestis* and have shaped its geography-associated genome characters, that is, predominant genotypes occur in most of the plague foci [7, 35–37]. Although in the core genome of *Y. pestis* a region-specific selection signal has not been observed as yet [7], evolutionary selection pressures in the different plague foci might leave traces of the accessory genome. One such example is the region-specific spacers in CRISPR loci, which are often associated with a specific plague focus or clade and are supposed to reveal the interactions between *Y. pestis* and the specific phages that live in different regions of plague foci [35]. Therefore, we can postulate that the evolutionary forces from reservoirs, vectors, and environments could sculpt the genome of *Y. pestis* and drive its evolution. To discover the driving forces shaping the nature of plague foci, we should combine the omics data for *Y. pestis* with environmental data, such as changes in climate, plant landscape (measured by normalized difference vegetation index), precipitation levels, and flood information, and the data on the rodent reservoirs, insect vectors, and disease epidemics. Interdisciplinary studies that combine genomic, ecological, and epidemiological approaches should provide better understanding of the evolutionary history of *Y. pestis* and the driving forces that have shaped it from the ecological–evolutionary perspective.

As well as collecting data from multidisciplinary and in silico analysis studies, positive correlations between gene mutations and the other ecological parameters mentioned above should be validated by laboratory experimentation. If a law for an evolutionary force that drives a specific gene mutation could be proven, it should be possible to build a model to predict the generation of genetic variations in the *Y. pestis* population and the corresponding influence on plague epidemics according to the parameters mentioned above. This type of approach might facilitate proactive plague control instead of the current passive plague surveillance.

14.3 The Pathogenic Features of Bacteria Provide Us with Good Opportunities to Study Human Immune Responses

Interactions between bacterial pathogens and the human immune system are complex. Bacterial pathogens avoid the innate immune defenses of the host through the use of the following strategies: antiphagocytosis mechanisms, surface adherence to host cells prior to invasion, post-invasion intracellular growth, and survival outside of host cells in the blood or tissues using defenses that resist immune cell attack. These processes involve secretion of toxic proteins during the different stages of pathogenesis. *Y. pestis* is a typical bacterial pathogen with obvious pathogenesis features [38]. It is usually transmitted to its host species by the bite of an infected flea. After injection into the tissues under the skin, tissue macrophages phagocytize *Y. pestis*, and it can grow while sheltering in these immune cells. In contrast, when engulfed by neutrophils, *Y. pestis* is conventionally considered to be killed. However, it was found that *Y. pestis* could survive in neutrophils and send signals to macrophages [39]. Therefore, *Y. pestis* can spread to nearby lymph nodes causing bubonic plague. The detail mechanism of movement of *Y. pestis* into local lymph node needs further investigation [40]. After macrophage release, *Y. pestis* is armed with a protein capsule and an active type III secretion system, whose properties are antiphagocytic, promote immune inhibition, and enhance toxic bacterial protein expression. *Y. pestis* can grow in the liver, lungs, and spleen in large quantities, resulting in a systemic plague infection.

Y. pestis has evolved several effective strategies to evade the attack of an activated immune system and promote its replication in the host, the means by which *Y. pestis* establishes a systemic plague infection. The strategies used to establish such infections include those that involve manipulation of innate immune signaling and inflammatory-related cell death, as well as immune evasion by promoting the differentiation of inhibitory immune cells that disrupt the immune balance. These strategies are of interest to immunologists who study human pathogens. First, one of the mechanisms by which *Y. pestis* interacts with innate immune host cells (macrophages, neutrophils), and how it exploits innate immunity to propagate within the host, is by manipulating Toll-like receptor (TLR), Nod-like receptor (NLR), and RIG-I signaling [41]. Innate immune cells can detect pathogens through recognition of pathogen-associated molecular patterns via the pattern recognition receptors (PRRs) located in the cellular and vacuolar membranes or in the cytosol. Membrane-associated PRRs include TLRs, whereas cytosolic PRRs include NLRs. Both TLRs and NLRs mediate recognition of *Y. pestis*, and aspects of their activation are exploited to further establish an infection [42]. Second, *Y. pestis* benefits from host immune cell death. A consequence of bacterial infection is the death of host immune cells (neutrophils and/or macrophages) [43]. Although this process is an effective host strategy for inhibiting intracellular replication of bacteria, cell death can also be exploited by *Y. pestis* to further its dissemination. Different mechanisms of cell death occur during infection with *Y. pestis*, including apoptosis and pyroptosis. Host immune cells die by apoptosis (noninflammatory), while later in the infection, activated neutrophils and macrophages die by pyroptosis (inflammatory). Pyroptosis or inflammatory cell death is characterized by the activation of caspase 1 or caspase 11 and by the secretion of IL-1β and IL-18. One of the mechanisms by which *Y. pestis* triggers pyroptosis is through activation of the inflammasome of host immune cells. Third, *Y. pestis* also benefits from the generation and differentiation of inhibitory immune cells. Induction and differentiation of type 2 macrophages [44–46] contribute to *Y. pestis* survival and enable evasion of host immune defenses.

Thus, when a pathogen invades a human or animal host, it first needs to survive the host's innate immunity and then interact with the immunological cells responsible for acquired immunity. To survive, it should not be eliminated by the host completely, but should develop the ability to proliferate without causing severe damage to its host. *Y. pestis* balances survival and pathogenicity by living in the wild reservoirs of natural plague foci, but it usually causes lethal disease in humans. Therefore, interactions between *Y. pestis* and the host's innate immune cells, such as macrophages, dendritic cells, and natural killer cells, are always an interesting topic that attracts researchers' attention (please refer to Chap. 10, Immunology of *Y. pestis*). The cell-mediated and antibody-mediated immunities to *Y. pestis* infection are also a priority for immunologists. Research in these areas should enrich our knowledge about the pathogenesis of *Y. pestis* and hopefully lead to development of a vaccine.

14.4 Typical Animal and Vector Models of *Y. pestis* Infection Create Opportunities for Experimental Studies on Pathogenesis and Evolution

Different animal models exist for studying bubonic and pneumonic plague, and these include mice, rats, guinea pigs, rabbits, and monkeys. The experimental models are infected with *Y. pestis* using subcutaneous, intradermal, intranasal, intravenous, intraperitoneal, and aerosol routes. An ideal animal for modeling a disease should develop similar signs of infection as those manifested in humans, and animal models of plague develop almost the same clinical manifestations as those seen in human infections. Animal models have been used for pathogenesis research [47–49], evaluation of vaccines [50] and drugs [51–53], and evaluation of virulence in different *Y. pestis* isolates [51, 54].

Y. pestis is transmitted by the bite of an infected flea. How the plague bacterium develops and survives in these arthropods is a fascinating research area in the understanding of *Y. pestis* pathogenesis and lifestyle. The flea model is used widely for studying the role of biofilm formation in *Y. pestis* in fleas [55], as well as studying its microevolution [3], the role of virulence factors that determine the survival of *Y. pestis* in the flea gut [56], evaluation of vaccine efficacy [57], plague pathogenesis [58–62], and plague transmission [63–65].

Both animal and flea infection models have successfully revealed aspects of pathogenesis and evolution in *Y. pestis*. How to combine these biological models with omics strategies to develop a high-throughput technology for holistic representation of *Y. pestis* pathogenesis and evolution needs to be addressed. We already know that two specific plasmids were acquired during the evolution of *Y. pestis*, and we now understand some aspects of each plasmid's role in evolution of this pathogen and in pathogenesis, including their role in fleas. In our laboratory, we have created *Y. pestis* mutants with differing plasmid profiles such as a plasmid-null strain, one-plasmid strains, and two-plasmid strains [66]. These mutants have been evaluated for virulence using the mouse model and subcutaneous infections. We found that *Y. pestis* with plasmids pCD1 and pPCP1 had higher virulence than the wild-type *Y. pestis* with the typical three-plasmid genotype and that *Y. pestis* containing pCD1 and pMT1 had lower virulence than the strain with only pCD1. These results indicate that interactions between plasmids are balanced to maintain virulence in *Y. pestis*. If on the base of several strains representing different branches of

Y. pestis phylogenetic tree we would create mutants for each bacterial gene, and combine the animal and flea infection models, it would be possible to obtain more precisely the roles played by genes in the flea life of *Y. pestis* or pathogenesis in animals. Other omics strategies should also be applied to obtain better understanding of the role of each gene in *Y. pestis* pathogenesis and evolution.

Animal and flea models can also be used for real-time monitoring of the infectious process of *Y. pestis* [67–74]. We developed a plasmid-based luxCDABE bioreporter and applied it to *Y. pestis* plasmid mutants to investigate their dissemination by bioluminescence imaging during primary septicemic plague in a mouse model [74]. The results suggested that the pMT1 plasmid was a critical determinant for colonization of the liver, spleen, and blood, whereas the pPCP1 plasmid appeared to be associated with the colonization of *Y. pestis* in the lungs.

Plague animal and flea models provide us with great opportunities to explore the pathogenesis and transmission of *Y. pestis*. Compared with other bacterial pathogens, like *Salmonella typhi*, *Vibrio parahaemolyticus*, *Helicobacter pylori*, and *Shigella* spp., which lack typical animal models, *Y. pestis* researchers are fortunate because they have not only typical animal models but also typical flea models. Animal models provide opportunities to study the bacterium's pathogenesis characteristics, evaluate prevention and treatment countermeasures against it, as well as study its gene functions in pathogenicity and evolution [52]. Also, flea models allow us to study how *Y. pestis* is transmitted and the roles played by its genes during plague transmission [3, 5].

14.5 Repeated Pandemics and Local Outbreaks Provide Us with Clues About the Infectious Disease Outbreaks That Have Occurred in Human History

There were three major plague pandemics in human history [16, 75, 76]. The Justinian plague spread around the Mediterranean Sea in the sixth century AD, while the Black Death started in Europe in the middle of the fourteenth century and swept across Western Asia, the Middle East, North Africa, and Europe, causing catastrophic population losses in all the regions affected. The third pandemic started in South China during the middle of the nineteenth century and spread from Hong Kong throughout the world by merchandise ships. In modern times, several major plague outbreaks have still occurred; these have included the primary pneumonic plague in North China in 1910 [77], which caused more than 60 thousand deaths, and the 1994 plague epidemic in India [78], which caused panic and partial international isolation of India, with resultant immense economic losses to this country; several outbreaks in Africa have also occurred [79, 80]. Plague outbreaks in wild rodents occur almost every year in different natural plague foci around the world [81–87]. Development of high-throughput sequencing technologies has gifted us unprecedented chances to study infectious diseases in human history. Ancient DNA samples from the remains of human teeth can be sequenced to determine whether the remnants of microbial DNA are present in them. Sequencing data have provided evidence that the three plague pandemics were in fact caused by *Y. pestis* [31, 33, 88–93]. Paleomicrobiology has also been used widely in different bacterial source-tracing and microbiome investigations [94–98]. Prior to the advent of genome sequencing-based paleomicrobiology, it was difficult to study the pathogens associated with infectious diseases in history, making record-based postulation the only option.

Some PCR-based evidence has reported on *Y. pestis* being responsible for plague during the latter part of the last century [88, 89, 99–101]; however, because of the relatively few results obtained, the conclusions might have been biased [102, 103]. In a later study, more target genes were analyzed using PCR sequencing for SNP-based genotyping of *Y. pestis* DNA in ancient relics, and this provided a more solid conclusion that the etiological agent of the Black Death is *Y. pestis*, thereby providing key information about the evolution of this pathogen and the plague spread by it during the Middle Ages [104]. The use of whole genome sequencing in paleomicrobiological research is a sophisticated way of uncovering the secrets of infectious diseases during the process of human civilization. Applying whole genome sequencing to ancient samples not only helps us to understand the etiology of historical plague pandemics but also the evolution of virulence in *Y. pestis* in ancient times [6, 31, 33, 90, 92, 93, 105].

The strategies developed for extracting DNA from ancient samples that avoid DNA contamination when sequencing ancient *Y. pestis* DNA could be applied to the treatment of all ancient DNA samples [106]. The bioinformatics developed for genome analysis of ancient samples of *Y. pestis* and its downstream comparison with more recent samples is a useful reference for tracing the evolutionary history of other pathogens. We believe that sequencing ancient microbial DNA provides us with an unprecedented opportunity to study the evolution of microorganisms from ancient samples such as fossils, and for studying the coevolution of microorganisms with humans, animals, and plants.

In summary, comprehensive studies on *Y. pestis* and the different aspects of the diseases caused by it have increased our knowledge about this pathogen and enhanced our understanding of relationships this pathogen has with nature. The results and conclusions drawn from research on *Y. pestis* further our understanding of other similar pathogens, providing the impetus for more interdisciplinary research activity in this area.

Acknowledgments We thank Dr. Guangwei Liu from Beijing Normal University for his helpful discussion on this chapter.

References

1. Zhou D, Yang R. Formation and regulation of Yersinia biofilms. Protein Cell. 2011;2(3):173–9.
2. Hinnebusch BJ, Perry RD, Schwan TG. Role of the *Yersinia pestis* hemin storage (*hms*) locus in the transmission of plague by fleas. Science. 1996;273(5273):367–70.
3. Sun YC, Jarrett CO, Bosio CF, Hinnebusch BJ. Retracing the evolutionary path that led to flea-borne transmission of Yersinia pestis. Cell Host Microbe. 2014;15(5):578–86.
4. Sun YC, Hinnebusch BJ, Darby C. Experimental evidence for negative selection in the evolution of a *Yersinia pestis* pseudogene. Proc Natl Acad Sci U S A. 2008;105(23):8097–101.
5. Sun YC, Guo XP, Hinnebusch BJ, Darby C. The Yersinia pestis Rcs phosphorelay inhibits biofilm formation by repressing transcription of the diguanylate cyclase gene hmsT. J Bacteriol. 2012;194(8):2020–6.
6. Zimbler DL, Schroeder JA, Eddy JL, Lathem WW. Early emergence of Yersinia pestis as a severe respiratory pathogen. Nat Commun. 2015;6:7487.
7. Cui Y, Yu C, Yan Y, Li D, Li Y, Jombart T, Weinert LA, Wang Z, Guo Z, Xu L, et al. Historical variations in mutation rate in an epidemic pathogen, Yersinia pestis. Proc Natl Acad Sci U S A. 2013;110(2):577–82.
8. Morelli G, Song Y, Mazzoni CJ, Eppinger M, Roumagnac P, Wagner DM, Feldkamp M, Kusecek B, Vogler AJ, Li Y, et al. Yersinia pestis genome sequencing identifies patterns of global phylogenetic diversity. Nat Genet. 2010;42(12):1140–3.
9. Ari TB, Gershunov A, Tristan R, Cazelles B, Gage K, Stenseth NC. Interannual variability of human plague occurrence in the Western United States explained by tropical and North Pacific Ocean climate variability. Am J Trop Med Hyg. 2010;83(3):624–32.
10. Ben-Ari T, Neerinckx S, Gage KL, Kreppel K, Laudisoit A, Leirs H, Stenseth NC. Plague and climate: scales matter. PLoS Pathog. 2011;7(9):e1002160.
11. Davis S, Begon M, De Bruyn L, Ageyev VS, Klassovskiy NL, Pole SB, Viljugrein H, Stenseth NC, Leirs H. Predictive thresholds for plague in Kazakhstan. Science. 2004;304(5671):736–8.
12. Easterday WR, Kausrud KL, Star B, Heier L, Haley BJ, Ageyev V, Colwell RR, Stenseth NC. An additional step in the transmission of Yersinia pestis? ISME J. 2012;6(2):231–6.
13. Kausrud KL, Viljugrein H, Frigessi A, Begon M, Davis S, Leirs H, Dubyanskiy V, Stenseth NC. Climatically driven synchrony of gerbil populations allows large-scale plague outbreaks. Proc Biol Sci. 2007;274(1621):1963–9.
14. Samia NI, Kausrud KL, Heesterbeek H, Ageyev V, Begon M, Chan KS, Stenseth NC. Dynamics of the plague-wildlife-human system in Central Asia are controlled by two epidemiological thresholds. Proc Natl Acad Sci U S A. 2011;108(35):14527–32.
15. Schmid BV, Buntgen U, Easterday WR, Ginzler C, Walloe L, Bramanti B, Stenseth NC. Climate-driven introduction of the Black Death and successive plague reintroductions into Europe. Proc Nat Acad Sci USA. 2015;112:3020–5.
16. Stenseth NC, Atshabar BB, Begon M, Belmain SR, Bertherat E, Carniel E, Gage KL, Leirs H, Rahalison L. Plague: past, present, and future. PLoS Med. 2008;5(1):e3.
17. Stenseth NC, Samia NI, Viljugrein H, Kausrud KL, Begon M, Davis S, Leirs H, Dubyanskiy VM, Esper J, Ageyev VS, et al. Plague dynamics are driven by climate variation. Proc Natl Acad Sci U S A. 2006;103(35):13110–5.
18. Xu L, Liu Q, Stige LC, Ben Ari T, Fang X, Chan KS, Wang S, Stenseth NC, Zhang Z. Nonlinear effect of climate on plague during the third pandemic in China. Proc Natl Acad Sci U S A. 2011;108(25):10214–9.
19. Xu L, Schmid BV, Liu J, Si X, Stenseth NC, Zhang Z: The trophic responses of two different rodent-vector-plague systems to climate change. Proc Biol Sci. 2015;282(1800):20141846
20. Tripp DW, Gage KL, Montenieri JA, Antolin MF. Flea abundance on black-tailed prairie dogs (Cynomys ludovicianus) increases during plague epizootics. Vector Borne Zoonotic Dis. 2009;9(3):313–21.

21. Schotthoefer AM, Bearden SW, Holmes JL, Vetter SM, Montenieri JA, Williams SK, Graham CB, Woods ME, Eisen RJ, Gage KL. Effects of temperature on the transmission of *Yersinia pestis* by the flea, *Xenopsylla Cheopis*, in the late phase period. Parasit Vector. 2011;4:191.
22. Schotthoefer AM, Bearden SW, Vetter SM, Holmes J, Montenieri JA, Graham CB, Woods ME, Eisen RJ, Gage KL. Effects of temperature on early-phase transmission of Yersina pestis by the flea, Xenopsylla cheopis. J Med Entomol. 2011;48(2):411–7.
23. Williams SK, Schotthoefer AM, Montenieri JA, Holmes JL, Vetter SM, Gage KL, Bearden SW. Effects of low-temperature flea maintenance on the transmission of yersinia pestis by Oropsylla montana. Vector Borne Zoonotic Dis. 2013;13(7):468–78.
24. Smith T. Parasitism and disease. Princeton: Princeton University Press; 1934.
25. Zhou D, Han Y, Song Y, Huang P, Yang R. Comparative and evolutionary genomics of *Yersinia pestis*. Microb Infect/Inst Pasteur. 2004;6(13):1226–34.
26. Zhang Y, Dai X, Wang X, Maituohuti A, Cui Y, Rehemu A, Wang Q, Meng W, Luo T, Guo R, et al. Dynamics of *Yersinia pestis* and its antibody response in great gerbils (*Rhombomys opimus*) by subcutaneous infection. PLoS ONE. 2012;7(10):e46820.
27. Zhou D, Yang R. Molecular Darwinian evolution of virulence in Yersinia pestis. Infect Immun. 2009;77(6):2242–50.
28. Hu P, Elliott J, McCready P, Skowronski E, Garnes J, Kobayashi A, Brubaker RR, Garcia E. Structural organization of virulence-associated plasmids of *Yersinia pestis*. J Bacteriol. 1998;180(19):5192–202.
29. Achtman M, Morelli G, Zhu P, Wirth T, Diehl I, Kusecek B, Vogler AJ, Wagner DM, Allender CJ, Easterday WR, et al. Microevolution and history of the plague bacillus, *Yersinia pestis*. Proc Natl Acad Sci U S A. 2004;101(51):17837–42.
30. Achtman M, Zurth K, Morelli G, Torrea G, Guiyoule A, Carniel E. *Yersinia pestis*, the cause of plague, is a recently emerged clone of *Yersinia pseudotuberculosis*. Proc Natl Acad Sci U S A. 1999;96(24):14043–8.
31. Rasmussen S, Allentoft Morten E, Nielsen K, Orlando L, Sikora M, Sjögren K-G, Pedersen Anders G, Schubert M, Van Dam A, Kapel Christian Moliin O, et al. Early divergent strains of Yersinia pestis in Eurasia 5,000 years ago. Cell. 2015;163(3):571–82.
32. Achtman M. Evolution, population structure, and phylogeography of genetically monomorphic bacterial pathogens. Annu Rev Microbiol. 2008;62:53–70.
33. Wagner DM, Klunk J, Harbeck M, Devault A, Waglechner N, Sahl JW, Enk J, Birdsell DN, Kuch M, Lumibao C, et al. Yersinia pestis and the plague of Justinian 541–543 AD: a genomic analysis. Lancet Infect Dis. 2014;14(4):319–26.
34. Gage KL, Kosoy MY. Natural history of plague: perspectives from more than a century of research. Annu Rev Entomol. 2005;50:505–28.
35. Cui Y, Li Y, Gorge O, Platonov ME, Yan Y, Guo Z, Pourcel C, Dentovskaya SV, Balakhonov SV, Wang X, et al. Insight into microevolution of *Yersinia pestis* by clustered regularly interspaced short palindromic repeats. PLoS ONE. 2008;3(7):e2652.
36. Li Y, Dai E, Cui Y, Li M, Zhang Y, Wu M, Zhou D, Guo Z, Dai X, Cui B, et al. Different region analysis for genotyping *Yersinia pestis* isolates from China. PLoS ONE. 2008;3(5):e2166.
37. Li Y, Cui Y, Hauck Y, Platonov ME, Dai E, Song Y, Guo Z, Pourcel C, Dentovskaya SV, Anisimov AP, et al. Genotyping and phylogenetic analysis of *Yersinia pestis* by MLVA: insights into the worldwide expansion of Central Asia plague foci. PLoS ONE. 2009;4(6):e6000.
38. Zhou D, Han Y, Yang R. Molecular and physiological insights into plague transmission, virulence and etiology. Microb Infect/Inst Pasteur. 2006;8(1):273–84.
39. Fukuto HS, Bliska JB. Editorial: Yersinia pestis survives in neutrophils and sends a PS to macrophages: bon appetit! J Leukoc Biol. 2014;95(3):383–5.
40. Gonzalez RJ, Miller VL. A deadly path: bacterial spread during bubonic plague. Trends Microbiol. 2016;24(4):239–41.

41. Du Z, Yang H, Tan Y, Tian G, Zhang Q, Cui Y, Yanfeng Y, Wu X, Chen Z, Cao S, et al. Transcriptomic response to Yersinia pestis: RIG-I like receptor signaling response is detrimental to the host against plague. J Genet Genom = Yi chuan xue bao. 2014;41(7):379–96.
42. Viboud GI, Bliska JB. *Yersinia* outer proteins: role in modulation of host cell signaling responses and pathogenesis. Annu Rev Microbiol. 2005;59:69–89.
43. Bergsbaken T, Cookson BT. Macrophage activation redirects yersinia-infected host cell death from apoptosis to caspase-1-dependent pyroptosis. PLoS Pathog. 2007;3(11):e161.
44. Bi Y, Wang X, Han Y, Guo Z, Yang R. Yersinia pestis versus Yersinia pseudotuberculosis: effects on host macrophages. Scand J Immunol. 2012;76(6):541–51.
45. Bi Y, Du Z, Han Y, Guo Z, Tan Y, Zhu Z, Yang R. Yersinia pestis and host macrophages: immunodeficiency of mouse macrophages induced by YscW. Immunology. 2009;128(1 Suppl):e406–17.
46. Bi Y, Zhou J, Yang H, Wang X, Zhang X, Wang Q, Wu X, Han Y, Song Y, Tan Y, et al. IL-17A produced by neutrophils protects against pneumonic plague through orchestrating IFN-gamma-activated macrophage programming. J Immunol. 2014;192(2):704–13.
47. Pechous RD, Sivaraman V, Stasulli NM, Goldman WE. Pneumonic plague: the darker side of Yersinia pestis. Trends Microbiol. 2015;24(3):190–7.
48. Pradel E, Lemaitre N, Merchez M, Ricard I, Reboul A, Dewitte A, Sebbane F. New insights into how Yersinia pestis adapts to its mammalian host during bubonic plague. PLoS Pathog. 2014;10(3):e1004029.
49. Anderson DM, Ciletti NA, Lee-Lewis H, Elli D, Segal J, DeBord KL, Overheim KA, Tretiakova M, Brubaker RR, Schneewind O. Pneumonic plague pathogenesis and immunity in Brown Norway rats. Am J Pathol. 2009;174(3):910–21.
50. Wang X, Zhang X, Zhou D, Yang R. Live-attenuated Yersinia pestis vaccines. Expert Rev Vaccine. 2013;12(6):677–86.
51. Lawrenz MB. Model systems to study plague pathogenesis and develop new therapeutics. Front Microbiol. 2010;1:119.
52. Layton RC, Brasel T, Gigliotti A, Barr E, Storch S, Myers L, Hobbs C, Koster F. Primary pneumonic plague in the African Green monkey as a model for treatment efficacy evaluation. J Med Primatol. 2011;40(1):6–17.
53. Layton RC, Mega W, McDonald JD, Brasel TL, Barr EB, Gigliotti AP, Koster F. Levofloxacin cures experimental pneumonic plague in African green monkeys. PLoS Negl Trop Dis. 2011;5(2):e959.
54. Tian G, Qiu Y, Qi Z, Wu X, Zhang Q, Bi Y, Yang Y, Li Y, Yang X, Xin Y, et al. Histopathological observation of immunized rhesus macaques with plague vaccines after subcutaneous infection of Yersinia pestis. PLoS ONE. 2011;6(4):e19260.
55. Hinnebusch BJ, Erickson DL. *Yersinia pestis* biofilm in the flea vector and its role in the transmission of plague. Curr Top Microbiol Immunol. 2008;322:229–48.
56. Johnson TL, Hinnebusch BJ, Boegler KA, Graham CB, MacMillan K, Montenieri JA, Bearden SW, Gage KL, Eisen RJ. Yersinia murine toxin is not required for early-phase transmission of Yersinia pestis by Oropsylla montana (Siphonaptera: Ceratophyllidae) or Xenopsylla cheopis (Siphonaptera: Pulicidae). Microbiology. 2014;160(Pt 11):2517–25.
57. Jarrett CO, Sebbane F, Adamovicz JJ, Andrews GP, Hinnebusch BJ. Flea-borne transmission model to evaluate vaccine efficacy against naturally acquired bubonic plague. Infect Immun. 2004;72(4):2052–6.
58. Sebbane F, Jarrett C, Gardner D, Long D, Hinnebusch BJ. Role of the *Yersinia pestis* yersiniabactin iron acquisition system in the incidence of flea-borne plague. PLoS ONE. 2010;5(12):e14379.
59. Comer JE, Sturdevant DE, Carmody AB, Virtaneva K, Gardner D, Long D, Rosenke R, Porcella SF, Hinnebusch BJ. Transcriptomic and innate immune responses to Yersinia pestis in the lymph node during bubonic plague. Infect Immun. 2010;78(12):5086–98.

60. Zhang SS, Park CG, Zhang P, Bartra SS, Plano GV, Klena JD, Skurnik M, Hinnebusch BJ, Chen T. Plasminogen activator Pla of Yersinia pestis utilizes murine DEC-205 (CD205) as a receptor to promote dissemination. J Biol Chem. 2008;283(46):31511–21.
61. Comer JE, Lorange EA, Hinnebusch BJ. Examining the vector-host-pathogen interface with quantitative molecular tools. Methods Mol Biol. 2008;431:123–31.
62. Vadyvaloo V, Jarrett C, Sturdevant D, Sebbane F, Hinnebusch BJ. Analysis of Yersinia pestis gene expression in the flea vector. Adv Exp Med Biol. 2007;603:192–200.
63. Jarrett CO, Deak E, Isherwood KE, Oyston PC, Fischer ER, Whitney AR, Kobayashi SD, DeLeo FR, Hinnebusch BJ. Transmission of *Yersinia pestis* from an infectious biofilm in the flea vector. J Infect Dis. 2004;190(4):783–92.
64. Hinnebusch BJ. Transmission factors: Yersinia pestis genes required to infect the flea vector of plague. In: Skurnik M, Bengoechea JA, Granfors K, editors. The genus Yersinia. New York: Kluwer Academic/Plenum Publishers; 2003. p. 55–62.
65. Chouikha I, Hinnebusch BJ. Yersinia-flea interactions and the evolution of the arthropod-borne transmission route of plague. Curr Opin Microbiol. 2012;15(3):239–46.
66. Ni B, Du Z, Guo Z, Zhang Y, Yang R. Curing of four different plasmids in Yersinia pestis using plasmid incompatibility. Lett Appl Microbiol. 2008;47(4):235–40.
67. Gonzalez RJ, Weening EH, Frothingham R, Sempowski GD, Miller VL. Bioluminescence imaging to track bacterial dissemination of Yersinia pestis using different routes of infection in mice. BMC Microbiol. 2012;12:147.
68. Nham T, Filali S, Danne C, Derbise A, Carniel E. Imaging of bubonic plague dynamics by in vivo tracking of bioluminescent Yersinia pestis. PLoS ONE. 2012;7(4):e34714.
69. Schofield DA, Molineux IJ, Westwater C. 'Bioluminescent' reporter phage for the detection of category a bacterial pathogens. J Vis Exp. 2011;53:e2740.
70. Sha J, Rosenzweig JA, Kirtley ML, van Lier CJ, Fitts EC, Kozlova EV, Erova TE, Tiner BL, Chopra AK. A non-invasive in vivo imaging system to study dissemination of bioluminescent Yersinia pestis CO92 in a mouse model of pneumonic plague. Microb Pathog. 2013;55:39–50.
71. Sun Y, Connor MG, Pennington JM, Lawrenz MB. Development of bioluminescent bioreporters for in vitro and in vivo tracking of Yersinia pestis. PLoS ONE. 2012;7(10):e47123.
72. Uliczka F, Pisano F, Kochut A, Opitz W, Herbst K, Stolz T, Dersch P. Monitoring of gene expression in bacteria during infections using an adaptable set of bioluminescent, fluorescent and colorigenic fusion vectors. PLoS ONE. 2011;6(6):e20425.
73. Warawa JM, Lawrenz MB. Bioluminescent imaging of bacteria during mouse infection. Methods Mol Biol. 2014;1098:169–81.
74. Zhou J, Bi Y, Xu X, Qiu Y, Wang Q, Feng N, Cui Y, Yan Y, Zhou L, Tan Y, et al. Bioluminescent tracking of colonization and clearance dynamics of plasmid-deficient Yersinia pestis strains in a mouse model of septicemic plague. Microb Infect/Inst Pasteur. 2014;16(3):214–24.
75. Anisimov AP, Lindler LE, Pier GB. Intraspecific diversity of *Yersinia pestis*. Clin Microbiol Rev. 2004;17(2):434–64.
76. Perry RD, Fetherston JD. *Yersinia pestis*–etiologic agent of plague. Clin Microbiol Rev. 1997;10(1):35–66.
77. Nishiura H. Epidemiology of a primary pneumonic plague in Kantoshu, Manchuria, from 1910 to 1911: statistical analysis of individual records collected by the Japanese Empire. Int J Epidemiol. 2006;35(4):1059–65.
78. Panda SK, Nanda SK, Ghosh A, Sharma C, Shivaji S, Kumar GS, Kannan K, Batra HV, Tuteja U, Ganguly NK, et al. The 1994 plague epidemic of India: molecular diagnosis and characterization of Yersinia pestis isolates from Surat and Beed. Curr Sci. 1994;71(10):794–9.
79. Isaacson M, Levy D, Pienaar BJ, Bubb HD, Louw JA, Genis GK. Unusual cases of human plague in Southern Africa. S Afr Med J. 1973;47(44):2109–13.

80. Shepherd AJ, Hummitzsch DE, Leman PA, Hartwig EK. Studies on plague in the eastern Cape Province of South Africa. Trans R Soc Trop Med Hyg. 1983;77(6):800–8.
81. Davis DH. Plague in Africa from 1935 to 1949; a survey of wild rodents in African territories. Bull World Health Organ. 1953;9(5):665–700.
82. Gordon DH, Isaacson M, Taylor P. Plague antibody in large African mammals. Infect Immun. 1979;26(2):767–9.
83. Shepherd AJ, Leman PA. Plague in South African rodents 1972–1981. Trans R Soc Trop Med Hyg. 1983;77(2):208–11.
84. Quan SF, Von Fintel H, Mc MA. Ecological studies of wild rodent plague in the San Francisco Bay area of California. II. Efficiency of bacterial culture compared to animal inoculation as methods for detecting Pasteurella pestis in wild rodent fleas. Am J Trop Med Hyg. 1958;7(4):411–5.
85. Quan SF, Kartman L, Prince FM, Miles VI. Ecological studies of wild rodent plague in the San Francisco Bay area of California. IV. The fluctuation and intensity of natural infection with Pasteurella pestis in fleas during an epizootic. Am J Trop Med Hyg. 1960;9:91–5.
86. Quan SF, Miles VI, Kartman L. Ecological studies of wild rodent plague in the San Francisco Bay area of California. III. The natural infection rates with Pasteurella pestis in five flea species during an epizootic. Am J Trop Med Hyg. 1960;9:85–90.
87. Hudson BW, Goldenberg MI, Quan TJ. Serologic and bacteriologic studies on the distribution of plague infection in a wild rodent plague pocket in the San Francisco Bay area of California. J Wildl Dis. 1972;8(3):278–86.
88. Drancourt M, Aboudharam G, Signoli M, Dutour O, Raoult D. Detection of 400-year-old *Yersinia pestis* DNA in human dental pulp: an approach to the diagnosis of ancient septicemia. Proc Natl Acad Sci U S A. 1998;95(21):12637–40.
89. Drancourt M, Raoult D. Molecular insights into the history of plague. Microb Infect/Inst Pasteur. 2002;4(1):105–9.
90. Bos KI, Schuenemann VJ, Golding GB, Burbano HA, Waglechner N, Coombes BK, McPhee JB, DeWitte SN, Meyer M, Schmedes S, et al. A draft genome of *Yersinia pestis* from victims of the Black Death. Nature. 2011;478(7370):506–10.
91. Tran TN, Raoult D, Drancourt M. Yersinia pestis DNA sequences in late medieval skeletal finds, Bavaria. Emerg Infect Dis. 2011;17(5):955–7; author reply 957.
92. Bos KI, Stevens P, Nieselt K, Poinar HN, Dewitte SN, Krause J. Yersinia pestis: new evidence for an old infection. PLoS ONE. 2012;7(11):e49803.
93. Harbeck M, Seifert L, Hansch S, Wagner DM, Birdsell D, Parise KL, Wiechmann I, Grupe G, Thomas A, Keim P, et al. *Yersinia pestis* DNA from skeletal remains from the 6(th) century AD reveals insights into Justinianic Plague. PLoS Pathog. 2013;9(5):e1003349.
94. Cano RJ, Rivera-Perez J, Toranzos GA, Santiago-Rodriguez TM, Narganes-Storde YM, Chanlatte-Baik L, Garcia-Roldan E, Bunkley-Williams L, Massey SE. Paleomicrobiology: revealing fecal microbiomes of ancient indigenous cultures. PLoS ONE. 2014;9(9):e106833.
95. Darling MI, Donoghue HD. Insights from paleomicrobiology into the indigenous peoples of pre-colonial America – a review. Mem Inst Oswaldo Cruz. 2014;109(2):131–9.
96. Dagli N, Dagli R, Baroudi K, Tarakji B. Oral paleomicrobiology: study of ancient oral microbiome. J Contemp Dent Pract. 2015;16(7):588–94.
97. Fournier PE, Drancourt M, Aboudharam G, Raoult D. Paleomicrobiology of Bartonella infections. Microb Infect/Inst Pasteur. 2015;17(11–12):879–83.
98. Huynh HT, Verneau J, Levasseur A, Drancourt M, Aboudharam G. Bacteria and archaea paleomicrobiology of the dental calculus: a review. Mol Oral Microbiol. 2015;31:234–42.
99. Drancourt M, Roux V, Dang LV, Tran-Hung L, Castex D, Chenal-Francisque V, Ogata H, Fournier PE, Crubezy E, Raoult D. Genotyping, orientalis-like Yersinia pestis, and plague pandemics. Emerg Infect Dis. 2004;10(9):1585–92.

100. Drancourt M, Raoult D. Molecular detection of Yersinia pestis in dental pulp. Microbiology. 2004;150(Pt 2):263–4; discussion 264–5.
101. Raoult D, Aboudharam G, Crubezy E, Larrouy G, Ludes B, Drancourt M. Molecular identification by "suicide PCR" of Yersinia pestis as the agent of medieval black death. Proc Natl Acad Sci U S A. 2000;97(23):12800–3.
102. Drancourt M, Signoli M, Dang LV, Bizot B, Roux V, Tzortzis S, Raoult D. Yersinia pestis Orientalis in remains of ancient plague patients. Emerg Infect Dis. 2007;13(2):332–3.
103. Vergnaud G. *Yersinia pestis* genotyping. Emerg Infect Dis. 2005;11(8):1317–8; author reply 1318–19.
104. Haensch S, Bianucci R, Signoli M, Rajerison M, Schultz M, Kacki S, Vermunt M, Weston DA, Hurst D, Achtman M, et al. Distinct clones of *Yersinia* pestis caused the Black Death. PLoS Pathog. 2010;6(10).
105. Schuenemann VJ, Bos K, DeWitte S, Schmedes S, Jamieson J, Mittnik A, Forrest S, Coombes BK, Wood JW, Earn DJ, et al. Targeted enrichment of ancient pathogens yielding the pPCP1 plasmid of Yersinia pestis from victims of the Black Death. Proc Natl Acad Sci U S A. 2011;108(38):E746–52.
106. Tran TN, Aboudharam G, Raoult D, Drancourt M. Beyond ancient microbial DNA: nonnucleotidic biomolecules for paleomicrobiology. BioTech. 2011;50(6):370–80.